International Conference on Screw Machines (ICSM 2018)

IOP Conference Series: Materials Science
and Engineering Volume 425

Dortmund, Germany
18 – 19 September 2018

ISBN: 978-1-5108-7507-4
ISSN: 1757-8981

Printed from e-media with permission by:

Curran Associates, Inc.
57 Morehouse Lane
Red Hook, NY 12571

Some format issues inherent in the e-media version may also appear in this print version.

Printed by Curran Associates, Inc. (2019)

For permission requests, please contact the Institute of Physics
at the address below.

Institute of Physics
Dirac House, Temple Back
Bristol BS1 6BE UK

Phone: 44 1 17 929 7481
Fax: 44 1 17 920 0979

techtracking@iop.org

Additional copies of this publication are available from:

Curran Associates, Inc.
57 Morehouse Lane
Red Hook, NY 12571 USA
Phone: 845-758-0400
Fax: 845-758-2633
Email: curran@proceedings.com
Web: www.proceedings.com

TABLE OF CONTENTS

MULTIPHASE FLOW I

DESIGN AND MANUFACTURING I

MULTIPHASE FLOW II

DESIGN AND MANUFACTURING II

SIMULATION (EU MOTOR PROJECT)

IOP Conf. Series: Materials Science and Engineering **425** (2018) 011001 doi:10.1088/1757-899X/425/1/011001

technische universität dortmund

ICSM2018

Dear friends of the screw machines,

Welcome to the International Conference on Screw Machines 2018 (ICSM 2018). This worldwide unique conference with the exclusive focus on research and development of screw machines is now taking place for the 10th time at the TU Dortmund University. I am particularly pleased that this internationally orientated conference brings together authors and participants from 13 countries around the globe. This is a unique opportunity for the community to grow together, from young engineers to the "old hands".

My highly esteemed colleague Prof. Knut Kauder, who had been doing research in the field of screw machines since 1975, had the idea for the first conference ("Schraubenmaschinen '84"), which took place in 1984. I am particularly pleased that one of the speakers at the first conference and certainly a very well-known scientist in the field of screw machines, Prof. L. Rinder, is faithful to this conference and is present today. Another prominent representative of the developers of screw machines is Jack Sauls, who gave his first presentation at this conference in 1990 and is also present today.

Despite these years of research and development, many scientific questions regarding screw machines are still unanswered today. These include, for example, the modelling or simulation of the multiphase flow, the potentials offered by new manufacturing possibilities of the rotors or the special challenges of designing screw spindle vacuum pumps. The 33 presentations and papers of the 10th ICSM deal with these questions and many other topics. The contributions of all nine previous conferences and the contributions of the 2018 conference are freely available and can be downloaded on the conference website (www. icsm.tu-dortmund.de). Additionally, the papers of the ICSM 2018 are published in the open access IOP (Institute of Physics) conference series.

Finally, I would like to thank everyone involved for their pleasant and constructive cooperation. Firstly, the sponsors of the conference, PTG Holroyd and Kapp Niles, without whom the small conference fee would not have been possible. Furthermore I would like to thank the authors, who have put a lot of time and effort into preparing the papers. The program committee and the reviewers, who ensured the high scientific standard of the conference, are to be mentioned immediately afterwards. My very special thanks to the entire chair of fluidics team, who prepared and organized the conference with a lot of time and effort. From this group I would like to thank Mrs. I. Kokott and Mr. C. Huck personally for their above-average commitment and excellent organizational talent.

Prof. Andreas Brümmer

Conference Chair

Peer review statement

All papers published in this volume of *IOP Conference Series: Materials Science and Engineering* have been peer reviewed through processes administered by the proceedings Editors. Reviews were conducted by expert referees to the professional and scientific standards expected of a proceedings journal published by IOP Publishing.

IOP Conf. Series: Materials Science and Engineering **425** (2018) 012001 doi:10.1088/1757-899X/425/1/012001

Numerical analysis of oil injection effects in a single screw expander

S Randi[1], A Suman[1], N Casari[1], M Pinelli[1] and D Ziviani[2]

[1]Dipartimento di Ingegneria, Università degli Studi di Ferrara, ITALY
[2]Ray W. Herrick Laboratories, Purdue University, West Lafayette, IN, USA

E-mail: saverio.randi@unife.it

Abstract. The pursuit of higher efficiency for compression and micro-power generation systems has pushed the researchers to an in-depth analysis of positive displacement machines. Single-screw machines, among the others, are gaining attention in the Organic Rankine Cycle (ORC) systems as expanders, thanks to their extended maintenance intervals and compactness. The performances of such devices are strongly affected by the working conditions, and especially the presence of oil has major effects on the operability. The main advantage of adopting an oil-injected device consists in the lube sealing effect, which permits better performance (greater shaft power for assigned boundary conditions) as well as higher reliability. The choice of whether using an oil-free configuration or not is related to the working fluid cleanness, system complexity (oil separator, filters, recovery pump), flow rate and pressure ratio. In this paper, the full 3D numerical simulation of an oil-injected single-screw expander operating with R245fa refrigerant is presented. Oil is injected together with the working fluid at the inlet of the machine. Oily droplets are tracked over the admission duct to show how the oil droplets reach the inlet ports of the screw machine. Different behaviors related to different oil droplet diameters in the range of (0.5 – 50) μm are studied, for the same operating point. The proper distribution of the oil droplets on the screw inlet ports are directly related to the single screw expander performance. In addition, a particular screw position is analyzed for studying the effects of leakages on the oil injection and oil film evolution over the time.

Nomenclature			
C	specific heat [kJ/kg K]	ν	kinematic viscosity [m²/s]
CFC	chlorofluorocarbon	ODP	ozone depletion potential
CFD	computational fluid dynamics	ORC	organic Rankine cycle
FF	fluid film	σ	surface tension [N/m]
HCFC	hydrochlorofluorocarbon	ρ	density [kg/m³]

1. Introduction

The increasing energy demand, together with the need of a cleaner way of production, has raised the attention of industry and researches towards the recovery of energy from low-quality heat sources.

Among the others, the use of Organic Rankine Cycles (ORCs) is constantly growing, ranking this technology as the most employed one for generating power from low-grade heat [1]. Evidence of the increasing interest into this topic is the number of research papers: whereas almost 100 research papers were published from 2005 to 2014, in the sole 2015 the same number of publications has been reached [2]. Within the low-grade heat recovery framework, a great part of the applications regards source power that are lower than 100 kWe. To this range, the class of so called micro and mini ORCs applies. This work deals with this class of power plants. The research about this topic regards mainly the improvement of the overall efficiency.

Mini and micro-ORCs overall efficiency is strongly affected by the efficiency of two of their components: the expander and the feeding pump. In light of this remark, it is obvious that such components have been thoroughly studied. For small scale cycles, usually volumetric pump and expanders are used, and this last machine is the component that is receiving attention the most, see for example [3-5]. Due to the smaller volume flows inside these systems, non-conventional expansion technologies such as screw expanders become more interesting. Particularly, due to their cost effectiveness, screw machine has been widely employed for this kind of application [4,6], mostly deriving the expander from compressors used the other way around [7].

Screw-type machine are divided in two main families: twin and single screw devices. The vast majority of the applications see the employment of the first one. The use of single screw devices is still marginal, even if they have some pros with respect to the twin-ones: for example, the force-balanced operation thanks to their mechanical layout [8] is a highly appreciated feature of this machine. Contrarily, a weakness of the single screw configuration is related to the relatively high flow leakages between regions of the machine at different pressures. One way to overcome such an issue is to inject a certain amount of oil within the working fluid. The oil acts as a sealant agent between the various parts of the device, closing leakages paths, lubricating and, moreover, cooling the machine. It is then necessary to study how device working conditions are affected by the presence of such a "sealant", in terms of efficiency thickness of the oil film and zones in which fluid can stagnate or affect the normal behaviour of the expander.

Several studies regarding the development of dynamic models of oil injected and oil flooded ORCs are reported in literature, like [4] and [5], mainly with the objective of increasing the thermodynamic efficiency of the system. Stosic et al. [9] developed a model to account for the injection of oil during the working cycle of a twin-screw compressor, finding that oil injection ports position strongly affects power and volumetric efficiencies. Stosic et al. [10] ran several multivariable optimizations to design twin-screw compressors capable of obtaining better efficiencies than the ones designed using classical approaches: in their calculations, they calculated optimum conditions for oil flow and temperature, together with the angular position of the injector inside the casing, for a machine working with air and another using R134a. De Paepe et al. [11] showed that the atomization of oil inside a screw compressor permitted to cool down the device, but did not strongly affect its global performance, thus excluding this approach to obtain an isothermal compression. Ziviani et al. [12], when studying a micro-ORC whose expander has been adapted from a single-screw compressor, observed that by halving leakages and friction losses, isentropic efficiency could reach values of above 70 %: to reduce friction and leakages, one way is to increase the amount of oil circulating in the expander.

To the writer knowledge, very few works have been done to analyze oil thickness on the different parts of the machine, and no one involved the analysis of a screw machine. For example, Minami et al. [13] compared oil film thickness for a rolling piston compressor when using R410a in place of R22 refrigerant, while Hotta et al. [14] evaluated lubricant distribution inside a swash plate compressor by means of an experimental campaign, differently from the approach of Minami et al. [13], that was numerical. In both cases, they concluded that for a reliable operation during the whole machine life cycle, the oil film on the meshing parts has to remain constant and it must not break. The advantages of having a constant film height on the moving parts consists in the possibility of removing low friction materials from the contacting components, thus saving money and weight.

Looking at ORC applications, depending on the refrigerant used in the cycle, several types of oil are available. Provided that CFC and HCFC refrigerants are no more employed due to their non-zero Ozone Depletion Potential (ODP), ester-based oils are used when dealing with HFC refrigerants [15]. Since oil influences properties of the working fluid in which it mixes, different works has been done to investigate properties of such mixtures, such as the ones by Zhelezny et al. [16] or Marsh and Kandil [17]. For example, Zhelezny et al. [16] studied concentration dependencies for vapor pressure, density, surface tension and capillary constant for a R245fa-polyolester compressor oil mixture. The usefulness of this work is due to the fact that different oil concentration in refrigerant are used in the works reported in literature, e.g. Ziviani et al. [18] studied the behavior of a single-screw expander using, as working fluid, a mixture where lubricant concentration was 3 % vol., while Hiwata et al. [19] analyzed how the performance of a CO_2-scroll compressor are affected by the injection of up to 25 % wt. of oil in the working fluid.

In light of these considerations, in this work a 3D transient CFD simulation of an oil-injected single screw expander has been carried out. Experimental characterization of such machine has already been done by Ziviani et al. [12], thus data for mass flow rate, temperature and pressure values of the working fluid at the inlet and outlet sections of the device are known. Furthermore, the same expander has already been modelled by means of CFD techniques, when operating without lubricant oil [20]: both in that work and the present one, Simcenter STAR-CCM+® has been the software used to carry on the simulations. The present work reports two different analyses. The first analysis is related to the droplet and oil film evolution inside the inlet distributor (that is characterized by a particular shape due to position of the ports) according to different droplet size for the same gas and oil mass flow rate. The second analysis is related to a particular geometric condition for which the position and the shape of the screw determine the complete closure of both inlet ports. In this condition, the instantaneous mass flow rate operated by the expander is only due to leakages between screw, case and star wheels, and for this reason, the oil droplets are affected by different behavior compared to those highlighted in the first analysis.

2. Single-screw expander geometry and CFD model

The machine considered in this work is an 11 kWe air compressor which is used as an expander for micro organic Rankine cycle (ORC) systems. Ziviani et al. [21] have already depicted in detail the dimensions of the device; however, for the purposes of this approach, the real geometry (Figure 1a) has been simplified in several details. All the bearings and main rotor and starwheels shafts, have been simplified, as well as the sealings; on the other hand, the external housing has not been changed, in order to predict flow behavior from the main inlet port of the machine to the screw, throughout the admission duct characterized by a curved shape. In addition, all the clearances have been modelled. According to Ziviani et al. [20], the clearances between the screw and case are in the range of (0.03 – 0.06) mm, while the clearances between the starwheel and the screw are in the range of (0.02 – 0.06) mm.

Since, within the considered simulations, oil is introduced at the main inlet of the machine, the two injectors already present in the geometry have not been used. Instead, a cone injector has been modelled and placed coaxially to the main inlet (represented by the violet circle in Figure 1b), in order to uniformly spread lubricant particles in the bulk flow of gas. Following the gas path, the mixture (refrigerant gas and oil droplet) is then injected in the expansion chamber of the screw rotor by means of two admission ports, represented in Figure 2. As can be seen from Fig. 2, the two admission ports are not symmetric with respect to the screw rotor and for this reason, the oil droplet behavior and the oil film evolution over the entire admission duct will be different according to the side. According to the geometry, the upper port is closer to the main inlet section while, the gas path of the lower port is characterized by two narrow bends.

2.1. Numerical models

To simulate the behavior of this kind of machine, different models have been used in STAR-CCM+®. While the capability of this software to deal with single-screw expander moving parts has already been proven (Ziviani et al. [20]) this time, while no rotating parts have been modelled by means of overset

IOP Conf. Series: Materials Science and Engineering **425** (2018) 012001 doi:10.1088/1757-899X/425/1/012001

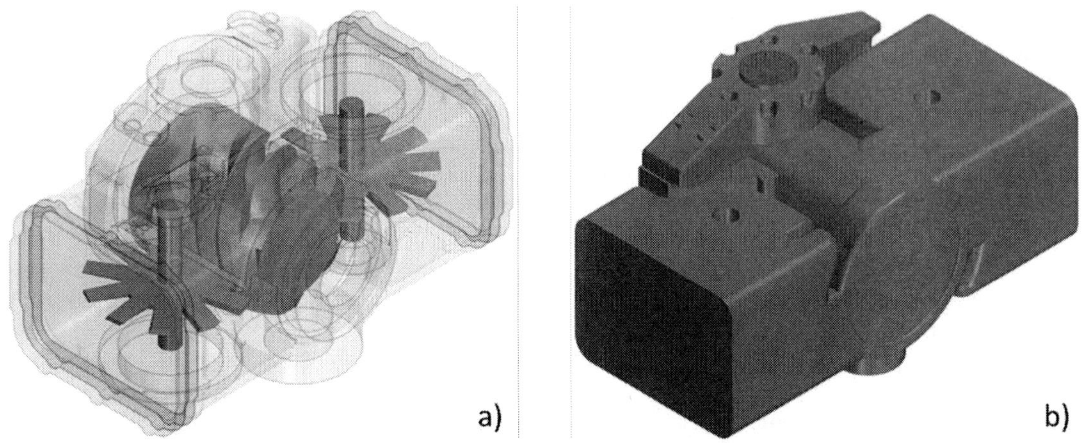

Figure 1. Single screw expander: a) CAD model (external case, screw rotor and two starwheels) and b) fluid model (the main inlet port is highlighted in violet).

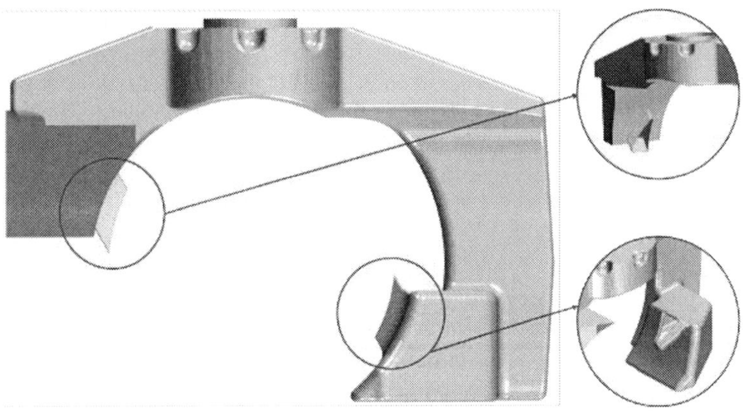

Figure 2. Front view of the inlet duct with the upper and lower admission ports (cyan).

meshes (screw rotor and starwheels are considered fixed), the presence of dispersed liquid lubricant droplets inside the bulk flow represented another challenge. To deal with this peculiarity, the working fluid has been modelled as real gas with Redlich-Kwong equation of state (EoS). The Eulerian approach has been coupled with Lagrangian one, to track oil droplets from the injection point to machine surfaces: since these droplets were predicted to form a film in the expander parts, also the fluid film and impingement models were enabled, together with the edge stripping one to take into account oil stripping from the different parts of the machine. In this paragraph, a short description of each of these models will be presented.

Redlich-Kwong EoS is one of the several equations which have been proposed in the last decades to consider the behavior of gases in those regions of the p-v-T diagram characterized by temperatures near the critical one and pressures far from the critical value. In particular, this one models gas by modifying the ideal gas law with the introduction of two coefficients a and b, which are both functions of the gas temperature and pressure at its critical point. It has to be noticed that physical conditions of R245fa in our case are very close to the ideal gas behavior at the outlet of the machine (Z=1.01), while at the inlet a compressibility factor near 0.83 indicates that an ideal gas approach would not be appropriate. The gas bulk flow receives oil droplets which are modelled with a lagrangian approach, which is the most suited in case of dispersed droplets in a continuous phase. This approach allows the tracking of each single droplets, highlighting the single behavior in terms of velocity, dimension, temperature, etc. Since solving a set of equations for each droplet would be excessively time consuming, the CFD software adopts a statistical approach in which all the droplets are grouped in clusters, which are representative group of droplets which share the same properties. For the sake of the analysis presented in this work, the so-called "two-way coupling" model has been enabled, in order to take into account the exchange of momentum and energy of the droplets with the working fluid, not only the vice versa. Furthermore, since it is expected that these droplets collide with expander surfaces, an impingement model had to be used. In the case of a wet boundary, the liquid particle transfers its mass, momentum and kinetic energy to the fluid film, making it necessary to evaluate the latter's properties by means of the fluid-film (FF) model. Moreover, this is necessary because it is expected that, when reaches a sharp edge, liquid will drip on the underlying surfaces.

The fluid film model solves equations for mass, momentum and energy conservation, together with the computation of the volume of liquid film in each of the cells of the numerical grid. The results are FF pressure, temperature, distribution and thickness. In our case, FF accumulates due to the impinging particles, while its thickness is reduced due to film stripping. This latter phenomenon is modelled thanks to the approaches proposed by Friedrich et al. [22] when considering the liquid film break-up criterion, and the one suggested by Maroteaux et al. [23] for calculating the diameter of droplets which result from stripping. Fluid will separate from an edge when the ratio of inertial force to the surface tension and gravitational force is greater than one; this ratio also affects the number of droplets which are ejected from the bulk flow, where their velocity is the same as the film one. All the simulations performed in the present work are realized by considering the conjugate heat transfer between gas, oil droplets and oil film. The walls are considered adiabatic.

In Table 1, the values of the materials properties and boundary conditions defined in the current work are reported. The transient simulation was performed by imposing a time-step of 5e-7 s that corresponds to a maximum Courant number of 1. To model turbulence, the two-equations, shear-stress transport (SST) k-ω model of Menter [24] has been adopted. Oil is introduced in the numerical domain with an injector modelled as a solid cone, which means that lubricant was injected in the form of uniformly distributed parcels from the base of the cone in seven random directions. This configuration has been considered because premixing oil and refrigerant upstream the expansion chamber is the simplest way of retrofitting a pre-existing machine with a lubrication system, since the alternative is to place an injector near the meshing zone of each starwheel with the screw. The temperature of oil has been chosen to be lower than the refrigerant one in order to take into account the possible lowering of machine components temperature due to the impinging fluid. To initialize lubricant film calculations, a height of 1 μm has been imposed on all the boundaries which were expected to be covered with oil, namely the inlet ducts, the starwheels and the screw. The minimum sharp edge angle at which oil is permitted to strip has been set to 10 degrees. The lubricant properties were defined as functions of temperature, using data presented by Zhelezny et al. [16] and reported in the oil datasheet. Relations for the temperature-varying properties are listed below.

$$\nu = 1.5748\text{e}39\ T\exp(-17.258)$$

$$\rho = 3.5987\text{e}3\ T\exp(-0.23055)$$

$$\sigma = 2.2684\ T\exp(-0.77639)$$

$$C = 6.7533e\text{-}2\ T\exp(0.57540)$$

Table 1. Model properties and boundary conditions.

Properties	Value	Properties	Value
R245fa Inlet Temperature [K]	398.05	Oil Droplets Diameter [μm]	0.5 – 50
R245fa Dynamic Viscosity [Pa s]	1.3723e-5	Oil Inlet Velocity [m/s]	4
R245fa Specific Heat [J/kg K]	1102.3	Injector Cone Angle [rad]	2
R245fa Thermal Conductivity [W/m K]	0.021519	Injector Streams [-]	7
Oil Inlet Temperature [K]	340	Engaging Ratio [-]	11/6

2.2. Mesh

Two numerical domains have been meshed according to the analysis type. As mentioned, the first analysis is devoted to the comprehension of the oil droplet and film behavior inside the inlet distributor. The numerical domain (reported in Fig. 2) was discretized by means of half million hexahedral elements with dimensions spanning from 0.5 to 2.0 mm. The numerical grid realized for the inlet distributor is reported in Fig. 3a. In relation to the second analysis, the numerical domain (reported in Fig. 1b) was discretized by means of hexahedral elements, with dimensions spanning from 0.01 mm to 2.0 mm, as reported in a previous work [20] in order to account for the gaps between screw rotor, case and starwheels. This led to the generation of a near two-million-elements mesh, which has guaranteed a good compromise between computational and solution accuracy. The present meshes are chosen after a grid sensitivity analysis. In Figs. 3a-b, both a section of the domain, and a representation of the numerical grid on the moving parts of the machine are shown.

The combination between elements dimensions and flow characteristics led to y^+ values in the range 0.5-100; this fact suggested to choose an all-y^+ approach, leading to the solution of the viscous sub-layer for those value of $y^+ < 1$ and the use of wall functions for values outside the transition zone.

3. Results

3.1. Inlet distributor analysis

This analysis was carried out in order to highlight the different behaviors which affect the oil droplets characterized by different diameter. In particular, three diameters have been considered in the range of (0.5 – 50) μm. This range is selected according to the literature values [25, 26], and in addition, this wide droplet size range is useful for performing a detailed sensitivity analysis. In the present analysis, the mass flow rate at the main inlet is imposed equal to 0.3 kg/s according to the experimental results reported by Ziviani et al. [12], while, an outlet static pressure equal to 1.3 MPa is imposed on both inlet

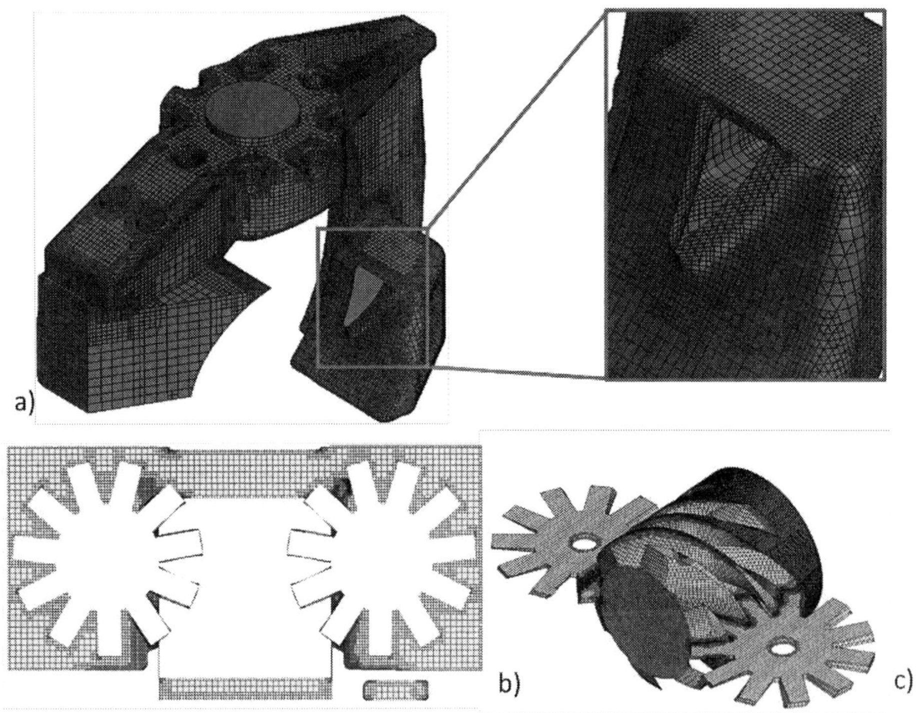

Figure 3. Numerical grids: a) on distributor duct, b) section plane perpendicular to wheels axes and c) on rotating parts.

ports. At the same time, the oil mass flow rate was imposed equal to 0.003 kg/s that corresponds to the 10 % in weight. This weight rate is imposed according to the analysis reported in [4]. The oil droplets were injected after the resolution of the flow field in order to realize a transient analysis of the oil film evolution over the time.

According to these conditions, Table 2 reports the results obtained for the three analyses according to the three particle diameters. As shown in the table, the number of droplets at the inlet ports of the screw changes according to the droplet diameter. The particular shape of the inlet distributor determines several differences of flow field over the flow path. Considering the streamlines reported in Figure 4, it is possible to see that the lower inlet port is affected by eddy structures due to the bent shape of the duct, while the upper port is located immediately downstream of the of the high-radius 90° curve (see Fig. 2 for completeness). The deviation upstream the lower port, influences the droplets trajectories and, for bigger particles, the droplet inertia determines huge deviations from the gas streamlines, as reported in Figure 5. For this reason, the biggest droplets are not able to reach the lower port. In the case of smaller droplets, the trajectories (see Fig. 5a-b) appears more similar to the gas streamlines (see Fig. 4) and in turn, the oil droplets are able to reach the upper and lower port.

In an actual application, the oil injector is responsible of a certain droplet diameter distribution (see for example [25] and [26]) and, in the light of the present results, the proper analysis of the droplet distribution, the location of the injector and the inlet distributor shape is needed. In a general way, and in the presence of a general droplet diameter distribution, smaller droplets are able to follow the gas streamlines, but they are responsible only for a minimum part of the oil mass flow rate. This fact could generate several issues related to the friction losses and wear in a specific region of the machine.

At the inlet ports of the screw, not only the oil droplets are responsible of the lubricant action of the

Table 2. Oil droplet analysis for the inlet distributor.

Oil droplet diameter [μm]	Oil droplets @ upper port [%]	Oil droplets @ lower port [%]
0.5	0.99	0.03
5.0	3.39	0.05
50	5.69	0.00

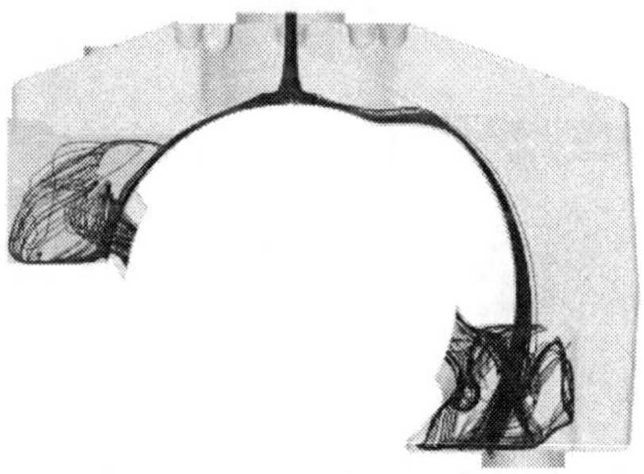

Figure 4. Gas streamline in the inlet distributor.

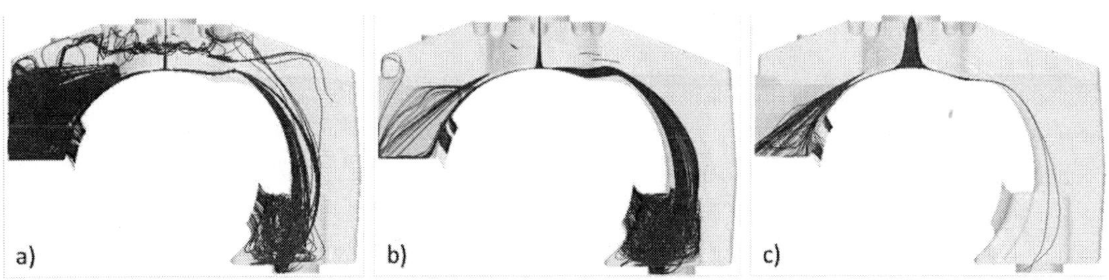

Figure 5. Oil droplet trajectories: a) 0.5 μm b) 5.0 μm and c) 50.0 μm.

oil on the screw walls but also the oil film. This is generated in the inlet distributor by the oil droplets, and moves towards the ports, thanks to the gravity action and to the drag action provided by the gas. Considering the snapshots reported in Figure 5, the trajectories of the oil droplets released by the injector at the main inlet are clearly visible. In Figure 6, the oil film distribution in the inlet distributor are reported. The oil film is representative of 0.4 s of oil injection. It is clearly visible how the oil film grows in the front of the injector and moves down, according to the distributor walls. At the screw inlet ports, the oil film strips and it is dragged towards the screw. In the next section, an analysis related to the interaction between oil film and the rotor screw will be reported in detail.

3.2. Screw rotor interaction analysis

This second analysis is carried out in order to highlight a particular geometric condition for which the position and the shape of the screw determine the complete closure of both inlet ports. As mentioned, in this configuration, the instantaneous mass flow rate that flows over the single screw expander is only due to leakages between screw, case and star wheels. Therefore, the oil droplets are affected by different behavior compared to those highlighted in the first analysis. In the present analysis, the static inlet pressure equal to 1.3 MPa is imposed at the main inlet, while a static pressure equal to 0.3 MPa is imposed at the single screw outlet in agreement with Ziviani et al. [12]. In addition, the rotational velocity is applied to the screw and starwheel walls equal to 3000 rpm and 1636 rpm respectively. In the same way of the first analysis, the oil droplets were injected after the resolution of the flow field in order to realize a transient analysis of the oil film evolution over the time. The relative position of the screw rotor and the upper port is reported in Figure 7 (the upper port trace is depicted in red). In this

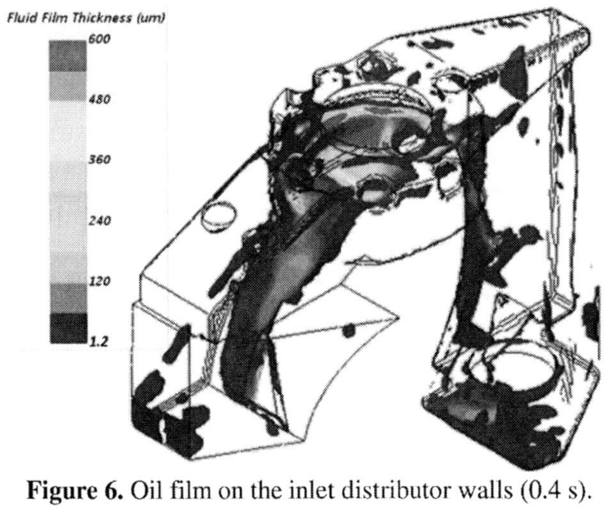

Figure 6. Oil film on the inlet distributor walls (0.4 s).

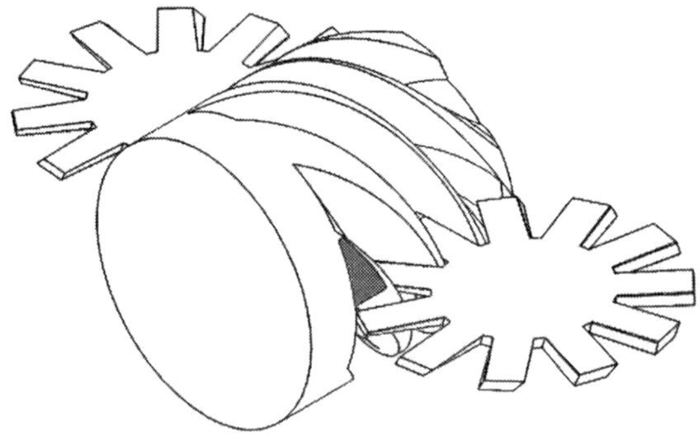

Figure 7. Relative position between inlet port (in red) and screw: the screw thread totally occludes the inlet port.

configuration, the inlet port is affected by a not negligible stagnation and the consequent back pressure. In the following, the oil film thickness variation over the time, pressure and velocity fields are proposed with the aim of highlight the fluid dynamic phenomena given by the contemporary presence of real gas expansion and fluid film behavior.

The results reported here refer to a transient analysis of 0.4 physical seconds. This limited period was sufficient to obtain information about oil distribution and its effect on the flow field on the inlet distributor and on the screw rotor walls. In Figure 8, the oil film thickness on the screw walls, obtained with oil droplet of 50.0 μm, is reported. In addition, the position of the upper port is highlighted in pink. As can be seen from Figure 8, on the lower part of the screw walls the oil film is thicker and involves the screw groves in the proximity of the starwheels surfaces, even if, the maximum thickness seems to be located downstream the engagement zone. This fact is due to the oil droplet trajectories. As reported in Figure 9, the oil droplets escape towards the lower part of screw rotor according to the pressure gradient. As can be seen in Figure 10, the inlet pressure characterizes only the screw region immediately downstream of the upper port (the red region with the triangular shape), while lower pressure values (the discharge pressure of the single screw expander) characterizes the other screw regions.

Taking into consideration the oil film distribution, reported in Figure 8, and the pressure field reported in Figure 10, another important consideration can be done. Part of the oil droplets is dragged towards the rear part of the screw due to the difference of pressure between the incoming working fluid and the lower pressure existing in that part of the domain. Looking at Figure 8, which depicts oil thickness, it is apparent that the two phenomena are related. In that zone, the height of the lubricant film is comparable with respect to the entire domain. As depicted in Figure 10, the inner screw region (in the

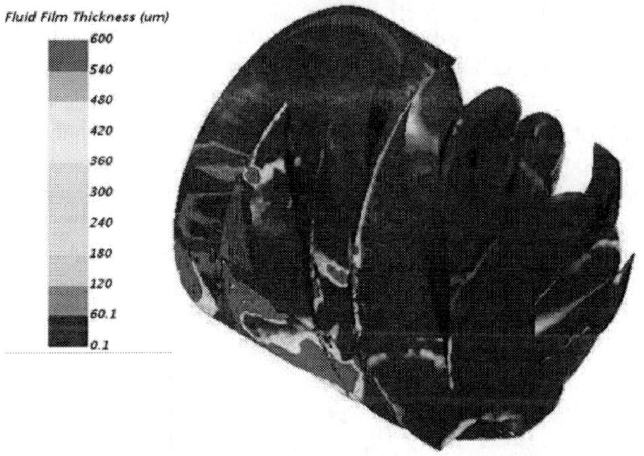

Figure 8. Oil film thickness on the screw walls (the upper port is highlighted in pink).

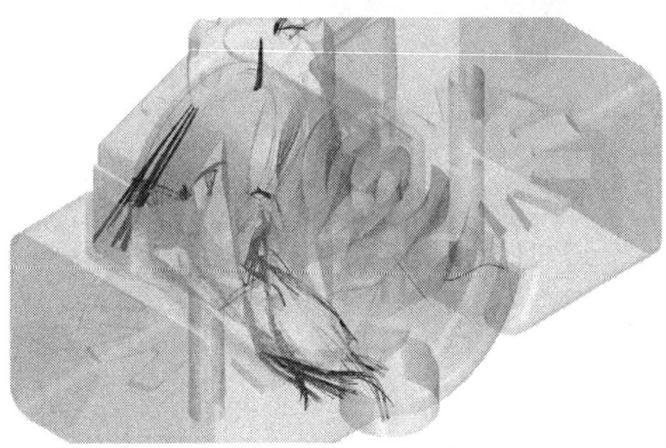

Figure 9. Oil droplet trajectories on the single screw expander.

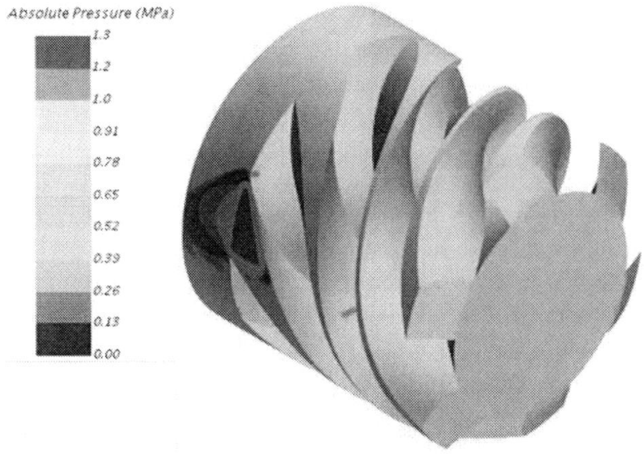

Figure 10. Pressure field on the single screw expander.

proximity of the inlet port) is characterized by huge pressure gradient (equal to about 1.3 MPa) generated by the effect of the gaps between screw and casing and the stagnation provided by the screw thread. This is confirmed by looking at Figure 11, in which the gas streamline colored by velocity are tracked over the single screw expander. Looking at the upper port, the leakage flow of refrigerant is particularly strong immediately downstream the admission port, giving an opposite direction with respect to the usual one. When the mixture oil-refrigerant exits the admission port, it is split in two directions, upstream and downstream the position of the port.

For the present machine setup, in this localized region, the gap between the casing and the screw determines a not negligible leakage. The combination of this radial gap and the pressure field (see Figure 10) determines a relevant oil stagnation. On the other hand, this stagnation causes a reduction of nearly 60 % of the gap between the screw and the casing (calculated considering the height of the oil film with respect to the total clearance of roughly 1 mm), thus reducing leakages in the rear part of the machine. Unfortunately, by injecting the oil at the inlet port, this beneficial effect is not obtained in other zones of the screw, or in the meshing area between this component and the starwheels, since oil drips from the inlet duct and it is scavenged only in one direction. This effect is accented by the distribution of the lubricant on the inlet distributor (see for example Figure 6).

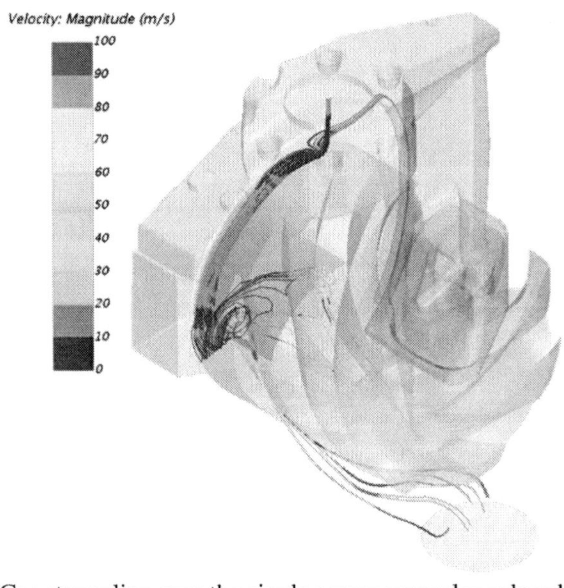

Figure 11. Gas streamline over the single screw expander colored by velocity.

By means of transient simulation, it is possible to discover the oil film progression over the time. Figure 12 depicts different frames, according to the physical time reported for each instant, of the oil film thickness, starting from the initial formation in front of the injector, to the final stripping on the screw rotor walls. At the beginning, the oil film assumes a circular shape according to the injector shape (cone injector) and it appears not equally distributed with respect to screw mid-plane. This result is probably due to the effect of the leakages that affects this particular geometric condition of the single screw expander.

Soon after, the oil film flows towards the upper port (see Figure 2) whose feeding duct is characterized by higher slope than the other one. Therefore, the action of gravity force is predominant over the drag force provided by the incoming gas. For the same reason, the oil droplets are not able of being distributed over the entire duct surfaces and they accumulate mostly in the lower surfaces. When the oil film reaches the admission port, it is split in two ways: one in the lower part of the screw rotor and the other one downstream the inlet port.

Taking into consideration the oil film thickness (see Figures 8 and 12), in the presence of narrow passages, and/or in the presence of high gas velocity, the reduction of the passage area due to the oil film, could generate a sort of blockage with consequent increment of pressure losses. This means that the position of the injector, the oil flow rate and the machine configuration play a double role: on one hand, an increase in the machine efficiency by reducing the friction losses and flow leakages can be expected, but on the other they can have a detrimental effect on machine performance by reducing the flow passage area and causing windage losses.

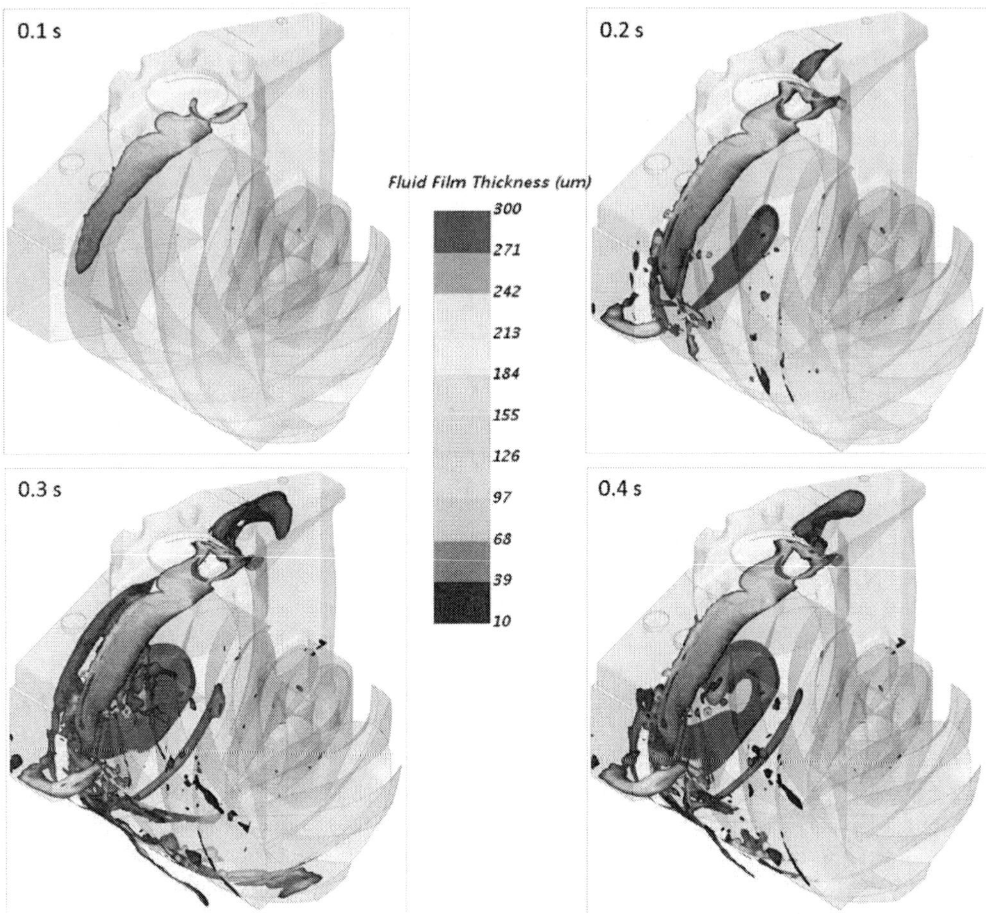

Figure 12. Oil film thickness time evolution from the inlet to the screw rotor.

4. Conclusions

In the present paper, the numerical simulation to study the effects of oil injection on the operating conditions of an actual single-screw expander has been carried out. The lubricant injector has been placed at the inlet port of the machine, with the oil driven by the R245fa to the expansion chamber.
A two-step analysis has been carried out. Firstly, the oil droplets trajectories in the inlet distributor are investigated, as well as the oil film that grows on the wall as oil keeps on impinging. Secondly, the same analysis is carried out on the entire machine, freezing the rotors in the instant where both the inlet ports are completely *shut*. In the first analysis, the effect of three different sizes of droplets has been considered (0.5 μm, 5.0 μm, and 50.0 μm), while for the entire machine, only the 50.0 μm droplets have been simulated.

From the first analysis, it is clear how the particle size affects the overall presence of oil inside the screw machine. Particularly, the high inertia of bigger particles tends to make them impinge on the inner walls of the distributor. Only a small amount of the droplets is able to make its way to the inlet upper port without being entrained in the thin fluid film that flows over the walls.
In light of this is mainly the fluid film that enters the working chamber, being stripped immediately downstream the inlet port. Such film is mainly transported towards the rear part of the screw, making the lubrication of the screw ineffective in the configuration under analysis.

From this work, it is clear that the smallest droplets have the higher probability to reach the actual working chamber, without impinging on the walls. Therefore, an efficient lubrication is obtained by decreasing droplets diameter (choosing an oil with lower viscosity and surface tension or by mounting a nozzle with a different geometry) or adopting a different configuration for the injectors (e.g. by placing them in a position where lubricant drops on the meshing pair, without being dragged by refrigerant flow). The simulation of a fully transient, dynamic mesh simulation is to be considered the next step of this analysis, showing the path the fluid film follows when the dynamic of the machine is considered.

References

[1] Quoilin S and Lemort V 2009 *Proc. 5th European Conference Economics and Management of Energy in Industry* (*Vilamoura*) pp. 459–64.
[2] Landelle A, Tauveron N, Haberschill P, Revellin R and Colasson S 2017 *Appl. Energy* **204** 1172-87
[3] Bao J and Li Z 2013 Renewable and Sustainable Energy Reviews **24** 325-42
[4] Ziviani D, Gusev S, Lecompte S, Groll E A, Braun J E, Horton W T, van den Broek M and De Paepe M 2016 *Appl. Energy* **181** 155-70
[5] Ziviani D, Gusev S, Schuessler S, Achaichia A, Braun J E, Groll E A, De Paepe M and van der Broek M 2017 *Energy Procedia* **129** 379-86
[6] Tang H, Wu H, Wang X and Xing Z 2015 *Energy* **90** 631-42
[7] Papes I, Degroote J and Vierendeels J 2013 *Int. Mechanical Engineering Congress and Exposition* November, 15-21 2013 San Diego
[8] Zimmern B and Patel G C 1972 *Int. Compressor Engineering Conference* West Lafayette
[9] Stosic N, Kovacevic A, Hanjalic K and Milutinovic Lj 1988 *Int. Compressor Engineering Conference* West Lafayette
[10] Stosic N, Smith I K and Kovacevic A 2003 *Appl. Thermal Eng.* **23** 1177-95
[11] De Paepe M, Bogaert W and Mertens D 2005 *Appl. Thermal Eng.* **25** 2764-79
[12] Ziviani D, Gusev D, Lecompte S, Groll E A, Braun J E, Horton W T, van den Broek M and De Paepe M 2017 *Appl. Energy* **189** 416-32
[13] Minami K, Hattori H and Hayano M 1998 *Int. Compressor Engineering Conference* West Lafayette
[14] Hotta T, Inoue T, Matsuda M and Ueda M 2004 *Int. Compressor Engineering Conference* West Lafayette
[15] Whitman B, Tomczyk J, Johnson B and Silberstein E 2013 *Refrigeration & Air Conditioning Technology* (New York: Delmar Cengage Learning)

[16] Zhelezny V P, Semenyuk Yu V, Ancherbak S N, Grebenkov A J and Beliayeva O V 2007 *Journal of Fluorine Chemistry* **128** 1029–38
[17] Marsh K N and Kandil M E 2002 *Fluid Phase Equilibria* **199** 319-34
[18] Ziviani D, Desideri A, Lemort V, De Paepe M and van den Broek M 2015 *IOP Conf. Ser.: Mater. Sci. Eng.* **90** 012061
[19] Hiwata A, Iida N, Futagami Y, Sawai K and Ishii N 2002 *Int. Compressor Engineering Conference* West Lafayette
[20] Ziviani D, Suman A, Gabrielloni J, Pinelli M and De Paepe M 2016 *Int. Compressor Engineering Conference* West Lafayette
[21] Ziviani D, Bell I, De Paepe M and van den Broek M 2016 *Int. Compressor Engineering Conference* West Lafayette Paper n.1486, pp. 1-10.
[22] Friedrich M A, Lan H, Drallmeier J A and Armaly B F 2008 *J. Fluids Eng* **130** (5)
[23] Maroteaux F, Llory D, Le Coz J-F and Habchi C 2002 *J. Fluids Eng.* **124** 565-75
[24] Menter F R 1994 *AIAA Journal* **32** 1598-605
[25] Stosic N, Milutinovic Lj, Hanjalic K and Kovacevic A 1992 *Int. J. Refrig.* **15** 206-20
[26] De Paepe M, Bogaert W, Mertens D 2005 *Appl. Therm. Eng.* **25** 2764-779

Acknowledgments
The research was partially supported by the Italian Ministry of Economic Development within the framework of the Program Agreement MSE-CNR "Micro co/tri generazione di Bioenergia Efficiente e Stabile (Mi-Best)".

IOP Conf. Series: Materials Science and Engineering **425** (2018) 012002 doi:10.1088/1757-899X/425/1/012002

Multiphase-flow simulation of a rotating rectangular profile within a cylinder in terms of hydraulic loss mechanisms

H Vasuthevan[1], A Brümmer[1]

[1]TU Dortmund University, Chair of Fluidics
Leonhard-Euler-Str. 5, 44227 Dortmund, Germany

E-mail: hanushan.vasuthevan@tu-dortmund.de, andreas.bruemmer@tu-dortmund.de

Abstract. In wet running screw compressors, liquid (oil or water) is injected into the machine. The advantage of liquid injection is cooling of the compressed fluid, sealing of clearances, lubrication of rotors and bearings, and reduced noise and vibration. The disadvantage of wet running machines is the loss in internal mechanical efficiency from the hydraulic loss caused by the liquid. In this paper, multiphase flow simulations are performed using the Volume of Fluid method to investigate hydraulic loss mechanisms. The 2D geometric model consists of a rectangular profile rotating in a cylinder with a clearance height that represents the screw machine housing clearance. In this investigation, the distribution and flow pattern of the liquid are examined with regard to hydraulic losses. Furthermore, the influence of the amount of liquid and the circumferential tip speed is determined through calculation of the torque on the rectangular contour. The hydraulic losses increase with increasing liquid mass in a working chamber and with increasing circumferential tip speed. For the simulations points examined, the tip speed has a greater impact on the hydraulic loss than does the amount of liquid.

1. Introduction

Twin-screw machines are widely used as compressors in industry and are the most commonly applied compressor type. Conventional applications are for compression of air, process gases or refrigerants. Furthermore, screw machines can be operated as expanders, which are typically used for waste heat recovery systems. Therefore, an increase in efficiency of screw machines leads to a positive contribution to the reduction of CO2 emissions in electrical power generation.

A crucial factor for the efficiency of screw machines is leakage flow through geometric clearances in the machine. These flows exist due to pressure differences between chambers and relative motion between rotors and housings. The efficiency of a screw machine can be improved by means of wet-running or liquid flooded machines. Here, an auxiliary liquid is brought or injected into the machine to partially seal the clearances, thus reducing gas leakage mass flow. Further beneficial effects due to the presence of liquid in the machine in terms of efficiency are cooling (compressor) or heating (expander) of gas by the liquid, lubrication of rotors, and reduction of vibration and noise. The disadvantage is the increase in internal mechanical losses due to friction and interactions between liquid and rotors or housings, - referred to as hydraulic losses. Important factors affecting hydraulic losses that have been identified are: injected liquid mass, clearance heights, circumferential tip speed, and liquid viscosity [1, 2]. The influence of injected liquid for a screw compressor with oil injection [1] and for a screw expander with water injection [2] are analysed by means of indicator diagrams. It is shown in both cases that injection of liquid improves the performance of the machine due to clearance sealing, while hydraulic losses increase with increasing circumferential tip speed, liquid mass and viscosity. Furthermore, it is suggested in [1] that a liquid surge is pushed in rotational direction at the rotor tip causing additional hydraulic losses. This hypothesis of a liquid surge is examined further in [3], where an analytical model for the distribution of injected oil in the working chamber and in the clearances is developed. The model is compared to experimental results, where the distribution of oil in a screw compressor is examined by means of optical access. An essential result was detection of an oil surge at the rotor tip. Nevertheless, the author described the restrictions of the optical analysis and proposed further investigation with improved optical resources. In a further work, distribution of water and oil in a screw expander is investigated [4].

An analytical model to determine hydraulic losses is developed in [5]; three loss mechanisms are identified: frictional loss, momentum loss, and acceleration loss. Frictional loss describes the viscous shear loss of the liquid. For the momentum loss, it is assumed that an attached liquid film on the housing is periodically accelerated by the passing rotor surface. The third loss mechanism is the acceleration from the injected liquid velocity to the circumferential speed of the working chamber. Further, in a fundamental experiment, one rotor is rotated in a liquid flooded cylinder to measure hydraulic losses, which consist of the frictional and the momentum loss. Comparisons of simulation results with the models are made for a screw compressor with regard to the integral hydraulic loss, but not for the individual loss mechanisms; thus, these analytical models are not validated explicitly. Empirical models for the prediction of hydraulic losses are given in [6]. Further works where the integral hydraulic losses of experimental wet – running screw machines are represented by a mechanical efficiency can be found in [7, 8, 9]. In [10, 11, 12] frictional losses in clearances due to liquid are examined with simplified numerical models, which consist of the superposition of turbulent incompressible Couette and Poiseuille flows. With these analytical models, chamber model simulations are performed to compare the behaviours of dry– and wet – running machines with the auxiliary liquid oil and water.

From the evaluation of the available literature it can be said that there are no validated, generally applicable, detailed methods or models for analysis of hydraulic loss mechanisms in a liquid flooded screw machine. A first step in this direction is made with two-dimensional, multiphase flow simulations to provide a better understanding of the hydraulic loss mechanisms. For a generic examination, a model of a rotating rectangular profile in a cylinder is used in this paper, whereby the cylinder is filled with air and a prescribed amount of water. The rotating rectangular profile can be developed further toward geometries characteristic of actual rotor profiles; it is intended that this be realised in future works. Most of the liquid will be accelerated out to the region of the rotor tip, where a liquid surge is formed [3]. In this paper, the hydraulic losses will be analysed in the steady state condition where the liquid surge is already formed at the contour tip.

First, common fundamentals and influencing parameters for hydraulic losses are described. Then the model and the boundary conditions of the two-dimensional multiphase flow simulation with the Volume of Fluid method (VOF) is presented. Subsequently, simulation results are presented and discussed.

2. Hydraulic losses

In this chapter essential fundamentals and influence factors for the examination of hydraulic losses in screw machines are described. The indicated power from a pressure-volume diagram can be calculated by equation (1); the result is the conversion of energy between the working fluid and the rotors due to the change of volume of a closed working chamber.

$$P_i = n \cdot \oint p dV \tag{1}$$

The effective power P_e is the power which is measured on a rotor shaft by means of rotational speed and torque. Equation (2) shows the effective power, the sum of the indicated power and the mechanical power losses. The mechanical power losses can be divided further into two parts: internal $\dot{\Phi}_{m,i}$ and outer $\dot{\Phi}_{m,o}$ mechanical power losses.

$$P_e = P_i + \dot{\Phi}_m = P_i + \dot{\Phi}_{m,i} + \dot{\Phi}_{m,o} \quad (2)$$

While the internal mechanical power losses $\dot{\Phi}_{m,i}$ describe the losses caused due to friction and interactions between fluid and rotors in the machine, the outer mechanical losses consider the losses of external parts such as bearings and seals.
For dry running screw machines, the internal mechanical power losses $\dot{\Phi}_{m,i}$ are relatively small and therefore not considered, while for liquid-flooded machines the internal mechanical power losses cannot be neglected. Internal mechanical losses in liquid – flooded machines are called hydraulic losses. In the simulations, a rectangular profile rotating within a cylinder is considered; pressure-volume work is not performed and there are no outer mechanical losses. Therefore, the internal mechanical power loss can be analysed by means of the torque acting on the contour and the angular frequency, as shown in equation (3).

$$\dot{\Phi}_{m,i} = M \cdot \omega \quad (3)$$

In the left side of figure 1, a rectangular profile with the surface S is shown to illustrate the calculation of the torque M. Forces acting on the surface in combination with the lever arms of the forces lead to the torque. The force acting on the surface can be divided into two parts: force in the normal direction $\vec{n}$ and force in tangential direction $\vec{t}$. The force normal to the surface is caused by fluid pressure while the force tangential to the surface is caused by the wall shear stress. The calculation of the torque M for the entire contour is given in equation (4).

$$M = \oint_S dM = \oint_S \left\{ -p\left(\vec{r} \times \vec{n}\right) + \tau_W \left(\vec{r} \times \vec{t}\right) \right\} dS \quad (4)$$

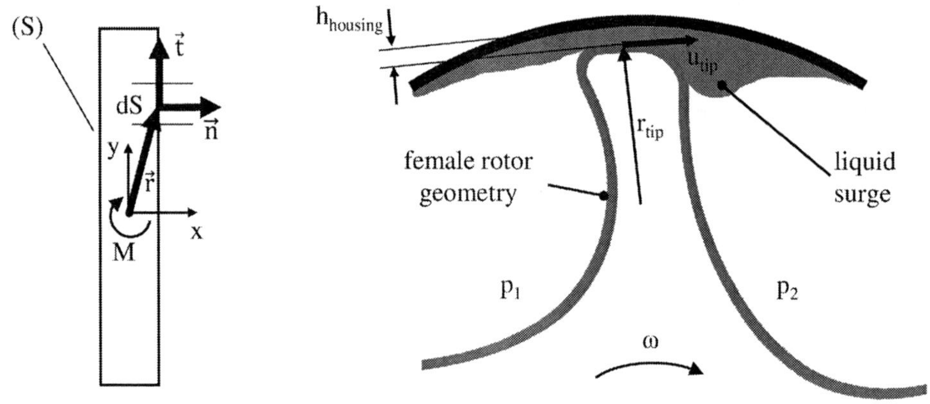

Figure 1. Left: Illustration for the calculation of torque on a rectangular surface S Right: Female rotor tooth with liquid film attached on the housing and a liquid surge at the rotor tip.

The results from the simulation shown in this paper are all time – averaged within a quasi steady state range (from t to t + Δt) and referenced to one tooth $\overline{M_z}$ with a length of 1 m, as shown in equation (5), where z is the number of teeth on the rotor. It is assumed that the time step Δt is great enough that fluctuation of the time dependent torque can be neglected. Furthermore, the torque on one tooth, $\overline{M_z}$, can only be evaluated if the same force conditions exist at every tooth.

$$\overline{M_z} = \frac{1}{z} \cdot \overline{M} = \frac{1}{z} \cdot \frac{1}{\Delta t} \int_t^{t+\Delta t} M(t)dt \quad (5)$$

On the right side of figure 1, a possible condition of liquid surge at the tip of a female rotor in a typical screw machine is shown to identify the influencing parameters. Essential influence factors to describe the torque of a predetermined rotor tooth geometry are presented in table 1. It should be noted that further characteristic geometric parameters can be added as influencing parameters. Here, for a common consideration only the housing radius and the housing clearance height are mentioned as essential geometry parameters. The parameter liquid volume is added for analyses of three dimensional cases. For a two dimensional case, the liquid phase is represented by its surface area.

Table 1. Influencing parameters for the rotor torque.

circumferential tip speed	u	$m\ s^{-1}$
viscosity liquid	ν_{liq}	$m^2\ s^{-1}$
density liquid	ρ_{liq}	$kg\ m^{-3}$
liquid volume	V_{liq}	m^3
housing radius	$r_{housing}$	m
housing clearance height	h_{hc}	m
chamber pressure	p_1; p_2	$N\ m^{-2}$

Using these influencing parameters in application of the Buckingham Pi theorem, a dimensional analysis can be performed and dimensionless characteristic numbers can be derived. Five independent dimensionless numbers are needed to describe the torque on a rotor tooth if liquid is present in the chamber. For every additional geometry parameter in table 1, an additional characteristic number can be developed to describe the geometry. The characteristic numbers so derived are given in table 2.

Table 2. Characteristic numbers for the torque of a rotor tooth in case of liquid presence.

circumferential Reynolds number	$\mathrm{Re} = \dfrac{u h_{hc}}{\nu_{liq}}$
moment coefficient	$C_M = \dfrac{\overline{M_Z}}{V_{liq}\rho_{liq}u_{tip}^2}$
pressure ratio	$\Pi = \dfrac{p_1}{p_2}$
liquid volume ratio	$\varepsilon = \dfrac{V_{liq}}{V_{chamber}}$
geometry number	$G = \dfrac{h_{hc}}{r_{housing}}$

3. Modelling and boundary conditions

For the generic examination of hydraulic losses, a rotating rectangular profile in a closed cylinder is considered; this represents a coarse estimation of a female rotor tooth. For this case, two dimensional multiphase simulations are carried out with the Volume of Fluid (VOF) method using Ansys Fluent 17.1. Details of the mesh, boundary conditions, and solver settings are presented in the following section.

IOP Conf. Series: Materials Science and Engineering **425** (2018) 012002 doi:10.1088/1757-899X/425/1/012002

3.1. Geometrical model

As shown in figure 2, the fluid zone of the geometric model was meshed using a structured grid with hexa elements. The geometric model of a complete closed cylinder is reduced to the two-dimensional model shown based on assumptions of point symmetry of the flow field. In this case the flow entering through one periodic interface is equal to the flow exiting the other periodic interface.

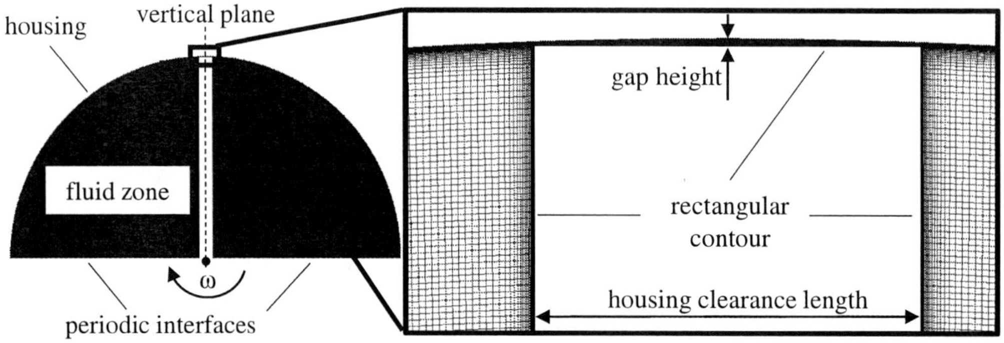

Figure 2. Meshed geometric model of rotating rectangular profile in cylinder.

The mesh was created with the ICEM 17.1 software. In the region of the clearance flow and the solid walls, the mesh contains finer elements, due to higher rates of change of the fluid states in this region. Geometric and mesh information are given in table 3. Skew and aspect ratio are used in assessing mesh quality. Higher aspect ratios in a smaller number of elements (1.7% of the total number of elements) occur away from the clearance region at the housing. Thus, high aspect ratios in these regions are of minor importance. Overall, the mesh quality is in an acceptable range.

Table 3. Geometrical and mesh information.

housing radius	$r_{housing}$	0.07	m
housing clearance height	h_{hc}	0.0001	m
housing clearance length	l_{hc}	0.005	m
element number	$n_{element}$	935961	-
mesh - determinant	-	range: 0,95 -1,0 (100 %)	-
mesh - skew	-	min range: 0,5-0,8 (4.5 %)	-
mesh - aspect ratio (Fluent)	-	max range: 70-270 (1.7 %)	-

3.2. Boundary conditions and solver settings

Transient simulations are performed with the explicit VOF method, where a closed cylinder is filled with air and an amount of liquid water. The air and liquid water are assumed to be incompressible fluids with constant density (air: 1.225 kg m^{-3} and water: 998.2 kg m^{-3}). Furthermore, the surface tension between air and water is asset to the constant value of $\gamma = 0.072$ N m^{-1}. The dynamic viscosities are: air $\eta_{air} = 1.789 \cdot 10^{-5}$ kg m^{-1} s^{-1} and water $\eta_{water} = 1.003 \cdot 10^{-3}$ kg m^{-1} s^{-1}. As a boundary condition, the temperature of air and water are set at the same constant value of 300 K, thus the energy conservation equation of is not used.

The fluid zone and the rectangular contour are calculated in a rotating coordinate frame with a constant angular velocity in clockwise direction, while the housing is rotating in the counter clockwise direction. The simulations are performed with a formulation of the conservation equations in the relative system, where centrifugal and Coriolis forces are considered.

Furthermore, for this generic examination, laminar flow is assumed. It is supposed that laminar flow is sufficient to examine the hydraulic loss mechanisms although a turbulent flow probably is existent. For the consideration of water in the closed cylinder, the amount of liquid is initialised. Due to the examination of a stationary or quasi-stationary solution the location of the water initialisation is not essential for the solution. Nevertheless, the water is assumed to be near the rotor tip and housing. Water is initialised in this region to achieve the stationary solution faster. The absolute pressure in the cylinder in both chambers is initialised with $p_{init} = 0.1$ MPa; thus only a pressure ratio of $\Pi = 1$ over the chambers is examined in the simulations. In table 4 essential solver settings for the simulations with Fluent 17.1 are shown.

Table 4. Solver settings with Fluent 17.1.

Pressure-Velocity-Coupling	SIMPLE
Gradient	Green-Gauss Node Based
Discretization - Pressure	Body Force Weighted
Discretization - Momentum	Second Order Upwind
Discretization – Volume Fraction	Geo-Reconstruct
Transient Formulation	First Order Implicit

A variable time stepping method was chosen for the calculations with a minimum time step of $t_{min} = 10^{-9}$ s, a maximum time step of $t_{max} = 10^{-6}$ s, and a global Courant number of $CFL_G = 0.8$. The maximum number of iterations per time step was set to 30. When all residuals fall below 0.001, the next time step is calculated. Furthermore, the torque on the contour is analysed to observe if a steady or quasi-steady solution is reached. An additional condition for every simulation point is at least one revolution of the rectangular contour.

4. Results

Simulation results are presented in this section. First, a detailed view of the flow pattern for the case of a simulation without liquid is given. Subsequently, the influence of the liquid volume ratio and the circumferential tip speed will be analysed with regard to their effects on hydraulic losses. Finally, results are presented depending on the characteristic numbers in a characteristic diagram. It should be noted that the results of the two-dimensional simulations are all referenced on one tooth with a tooth width of $l = 1$ m.

4.1. Flow pattern

In this section, the flow pattern will be analysed for the case without liquid $\varepsilon = 0$ and for the case with a liquid volume ratio of $\varepsilon = 0.005$ at a constant circumferential tip speed of $u_{tip} = 30$ m s^{-1}.

4.1.1. Case $\varepsilon = 0$: Only air. In figure 3, the flow field for the case without liquid ($\varepsilon = 0$) in the cylinder is shown in the relative system. It should be noted that the full cylinder is only shown here for reference and visualisation. In the relative system, the air flows towards the housing clearance and is then separated into two domains, where the upper part flows through the clearance. The other part flows tangentially at the contour towards the centre and further towards the opposite clearance, where the flow is disturbed slightly by a flow separation dead space before the flow is brought together with the stream of the lower clearance flow. This stream flows towards the upper clearance again and the process is repeated. Due to this flow behaviour, a counter clockwise rotating vortex is formed at the centre of each chamber. The relative tangential velocity at the rectangular contour, which is given in figure 3, can be used in heat transfer modelling.

For the transient simulations with only air, the torque converged to a constant value: $\overline{M_z} = 0.027$ Nm. The torque is mainly influenced by the pressure difference $\Delta p_{max,tip}$ which is 1250 Pa at the contour tip and which decreases with decreasing radius.

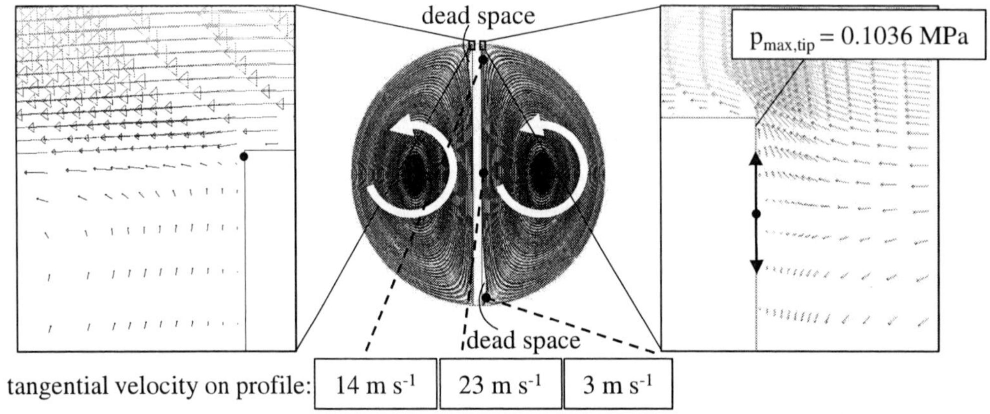

Figure 3. Relative flow pattern without liquid $\varepsilon = 0$ for a circumferential speed of $u_{tip} = 30$ m s^{-1}.

4.1.2. Case: $\varepsilon = 0.005$. The progression of the torque for the case with a liquid volume ratio of $\varepsilon = 0.005$ is shown in figure 4. While in the case without liquid the torque tends to a constant value, the torque with liquid is fluctuating with time. In this case, the mean value of the fluctuation can be used to quantify effects. Furthermore, a liquid surge, which varies with time, is formed at the rotor tip. In figure 4, black areas represent water and grey identifies areas of air. It can be seen that the fluctuation of the torque on the contour is influenced essentially by the shape of the liquid surge at the contour tip. The time averaged torque is $\overline{M_z}$=2.2 Nm and the pressure at the contour tip on the liquid surge side at a point in time t = 0.023 s is about $p_{max,tip}$ = 0.23 Pa, while the pressure on the opposite side of the liquid surge at the contour tip is about $p_{tip,op}$ = 0.102 MPa.

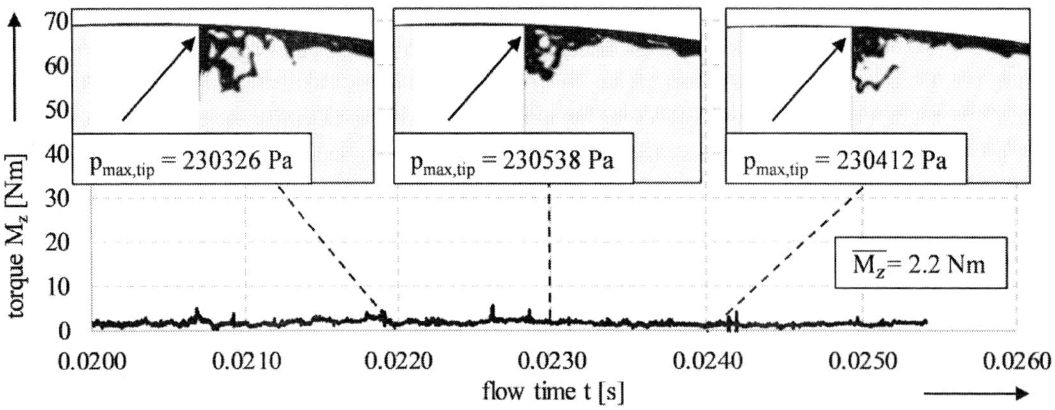

Figure 4. Torque over time for a liquid volume ratio of $\varepsilon = 0.005$ and a circumferential speed of $u_{tip} = 30$ m s^{-1} with flow pattern of liquid surge at different moments (black – water ; grey – air).

IOP Conf. Series: Materials Science and Engineering **425** (2018) 012002 doi:10.1088/1757-899X/425/1/012002

The complete relative flow pattern is presented at time t = 0.023 s in figure 5. The flow field can be divided into two regions, where the one region is similar to the flow field without liquid. In this region, the vortex at the centre of the chamber can be identified. The other region is located in the vicinity of the rotor tip and is disturbed due to the liquid. This region is more complex and consists of smaller regions. The main field of the liquid surge at the contour tip is shown in figure 5b. It should be noted that the rest of the stream lines is not shown here, which mainly consist of air and is attached to the liquid surge. The liquid distribution can be divided into two parts: a liquid surge, which is located directly at the contour tip, and a liquid film, which is attached to the housing.

Figure 5c shows the flow field of the liquid at the contour tip; this will be discussed in the following. The black streamlines show the flow of water and the grey streamlines the flow of air. Similar to the flow of air, liquid flows towards the housing clearance and is separated into two streams. One part of the liquid flows through the upper clearance and flows further as a thin liquid layer on the circumference of the housing towards the lower clearance. The other part flows tangentially on the rectangular contour in the direction of the centre of rotation. This flow is then deflected due to the centrifugal forces into circumferential and further in radial direction, where the liquid is entrained by the liquid stream which then flows towards the upper clearance. Accordingly, a rotating liquid surge is formed at the rotor tip. This rotating liquid surge is surrounded by an outer stream, which flows back away from the clearance.

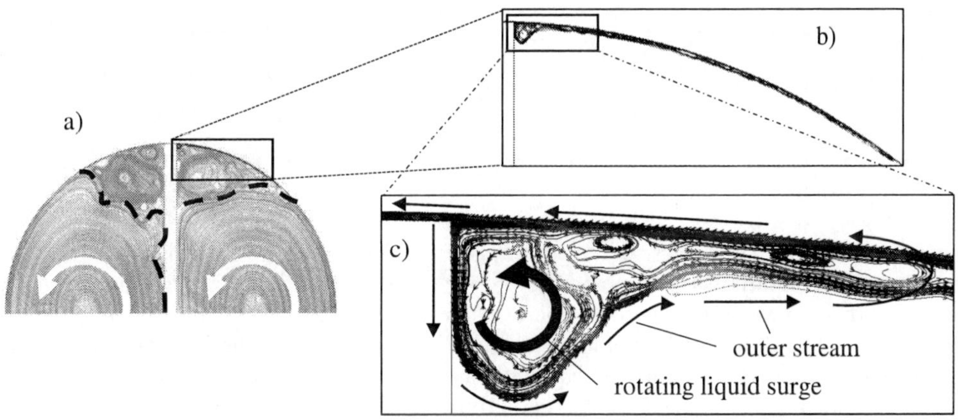

Figure 5. a) Relative flow pattern for a liquid volume ratio ε = 0.005 and a circumferential speed of u_{tip} = 30 m s^{-1} at a time t = 0.023 s b) Extract of the main liquid flow field at the contour tip without the attached flow field c) Detailed view of the liquid surge at the contour tip (black – water; grey – air).

Thus, two liquid layers exist at the housing moving in different circumferential directions: a liquid stream towards the clearance and an outer stream away from the clearance. During the backflow of the outer liquid stream, part of the liquid is entrained by the liquid stream, which flows towards the clearance. Furthermore, dead spaces can be identified between the two liquid streams. The liquid surge and the liquid streams cause essential shear stresses and pressure variation along the contour, which results in resistance to rotation.

A backflow against the relative rotation of the housing seems unlikely. It is assumed that the size of the rotating liquid surge is restricted due to the centrifugal forces acting in the relative system. For the steady state situation, no more liquid can be added to the rotating liquid surge; thus, liquid is forced to flow back around the surge and away from the contour.

4.2. Variation of liquid volume ratio

In this section, the liquid volume ratio ε is varied from 0.25 % to 1 % for a constant circumferential tip speed of u_{tip} = 30 m s^{-1}. The time-averaged torque and the power loss over the liquid volume ratio are shown in figure 6. It can be seen that the total torque on the contour increases almost linearly with increasing liquid volume ratio. Therefore, the hydraulic power loss also increases linearly for a constant circumferential tip speed. Additionally, the time averaged torque and power loss, caused by the flow in the clearance gap, are shown in the diagrams. The torque on the contour caused by shear stresses in the clearance flow is almost constant and independent of the liquid volume ratio. Furthermore, it can be said that the influence of the liquid surge on the hydraulic power loss is greater than the effect of friction in the clearance area. The time-averaged torque and the power loss due to the liquid surge can be determined from the diagrams as the differences between the total and clearance values.

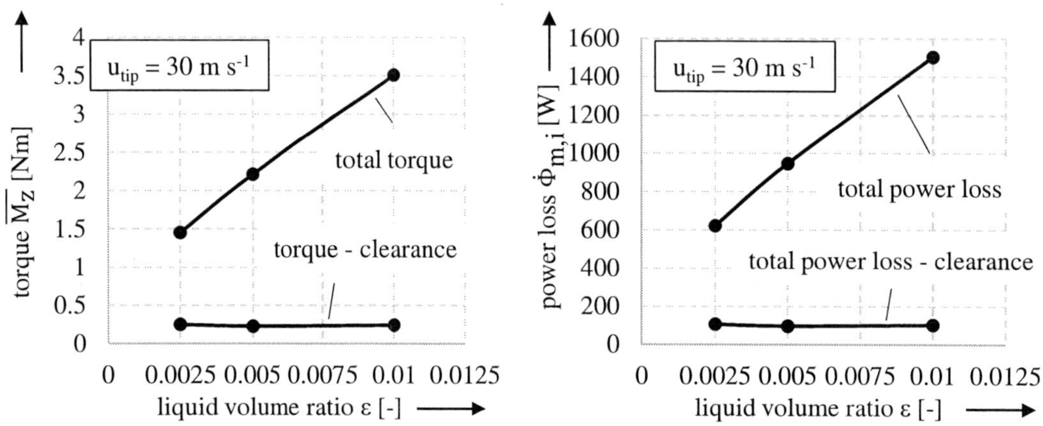

Figure 6. Time averaged torque and power loss over the liquid volume ratio at a fixed circumferential tip speed u_{tip} = 30 m s^{-1}.

The essential reason for the increase of the power loss is the increasing pressure difference at the contour tip. In figure 7 the liquid distribution and the pressure at the contour tip on the liquid surge side for different liquid volume ratios are shown. The black coloured areas consist of water and the grey areas consist of air. It should be noted that the pressure at the contour tip on the opposite side of the liquid surge is about $p_{tip,op}$ = 0.102 MPa for the liquid volume ratios examined in the study.

With increasing liquid volume ratio, the liquid film thickness near the liquid surge increases. Further, it can be said that for low liquid volume ratios, the liquid surge contains air inclusions. The air inclusions decrease with increasing liquid volume ratio, thus the density of the liquid surge increases. This can be a possible reason for the increase of the pressure at the contour tip. Further, is seems that the height of the liquid surge at the contour remains constant for across the range of liquid volume ratio studied.

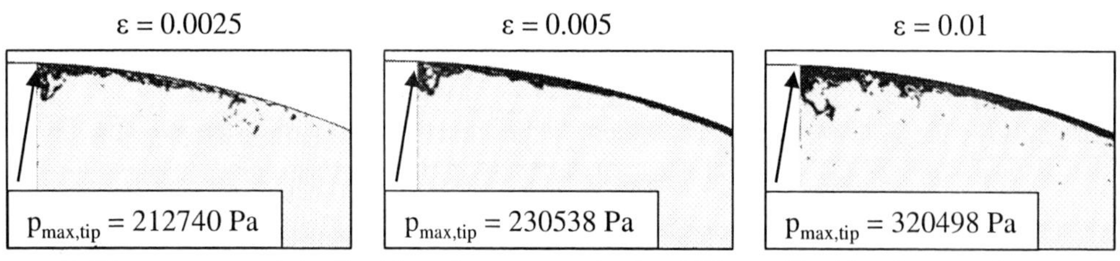

Figure 7. Flow pattern of the liquid at the contour tip for different liquid volume ratios at a fixed circumferential tip speed of u_{tip} = 30 m s^{-1} (black – water; grey – air).

4.3. Variation of circumferential tip speed

Figure 8 shows the time-averaged torque and total power loss over the circumferential tip speed for a constant liquid volume ratio of ε = 0.0025. The torque increases almost quadratically and the hydraulic power loss increases almost cubically with increasing circumferential tip speed. Furthermore, the time averaged torque and power loss caused due to the clearance flow are shown. For increasing circumferential tip speed, the hydraulic power loss due to the clearance flow increases, but the major influence on the total hydraulic power loss is still the liquid surge at the contour tip.

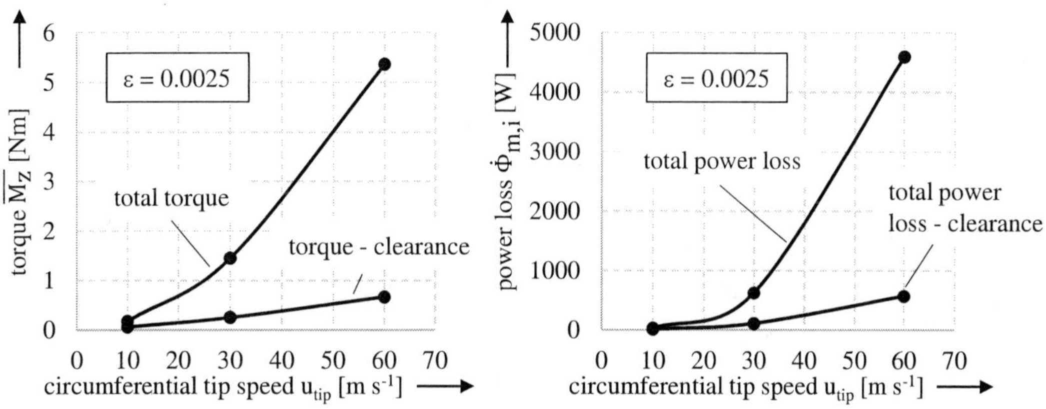

Figure 8. Time averaged torque and power loss over the circumferential tip speed at a fixed liquid volume ratio of ε = 0.0025.

The essential reason for the increase of the power loss is the pressure difference at the contour tip, which depends on the circumferential tip speed. In figure 9, the liquid surge and the pressure at the contour tip on the liquid surge side for different circumferential tip speeds are shown. It should be noted that the pressure at the contour tip on the opposite side of the liquid surge is about $p_{tip,op}$ = 0.102 MPa. The increase in pressure is caused by the increasing circumferential tip speed and the resultant increase in centrifugal forces. A higher circumferential tip speed leads to a characteristic of increasing pressure in radial direction. Equation 6, a balance of forces in the radial direction, is used to examine the qualitative progression of the pressure. This equation is only valid for simplified cases, where, for example, the radial velocity is zero. Nevertheless, the equation shows that the change in pressure in radial direction is influenced quadratically by the circumferential tip speed; this can be seen in the progression of the torque in figure 8. The pressure is influenced additionally by the high density of the liquid, which is mainly located at the contour tip.

$$\frac{dp(r)}{dr} = \rho \frac{u^2(r)}{r} \tag{6}$$

Furthermore, it can be seen that with increasing circumferential tip speed, the height of the rotating liquid surge at the contour tip decreases. This behaviour supports the aforementioned statement that the size of the rotating liquid surge is restricted by the centrifugal forces.

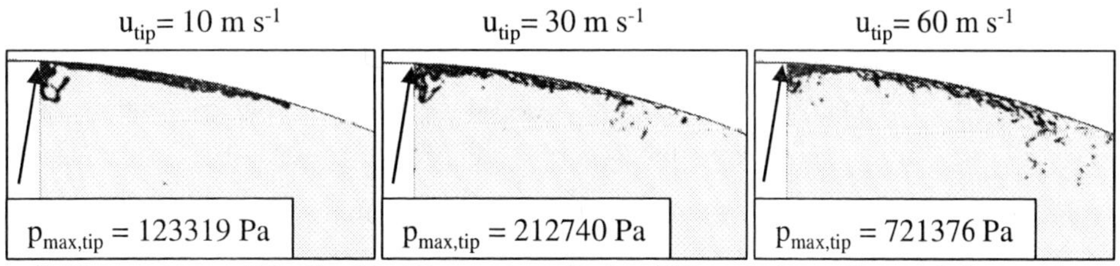

Figure 9. Flow pattern of the liquid at the contour tip for different circumferential tip speed at a fixed liquid volume ratio of ε = 0.0025 (black – water; grey – air).

IOP Conf. Series: Materials Science and Engineering **425** (2018) 012002 doi:10.1088/1757-899X/425/1/012002

4.4. Characteristic diagram

In figure 10, the time averaged torque for the liquid volume ratios studied are shown as functions of the circumferential tip speed. In case of a liquid volume ratio $\varepsilon = 0$, where no liquid is present, the torque due to internal mechanical losses is very small in contrast to the case with liquid.

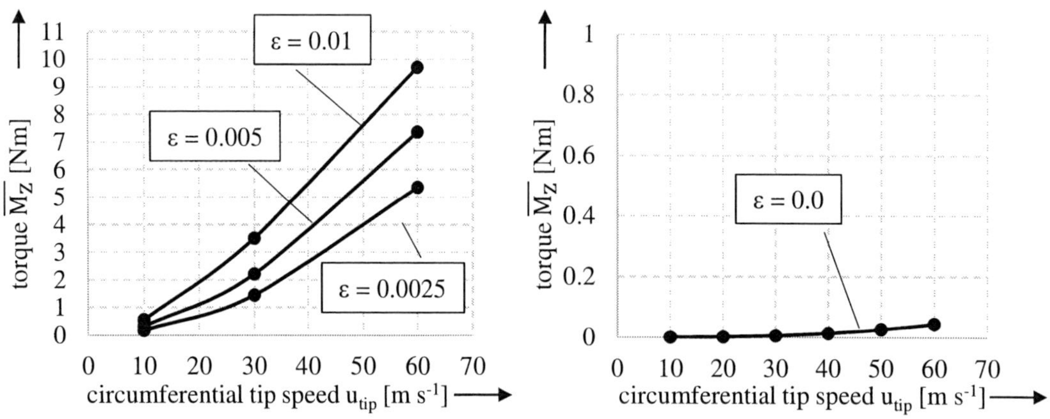

Figure 10. Time averaged torque of the contour for different liquid volume ratios and circumferential tip speeds.

Dimensionless characteristics based on the simulation results are presented in figure 11, where the moment coefficient over the circumferential Reynolds number is shown for different liquid volume ratios. This diagram is only valid for the rectangular profile used here and for a chamber pressure ratio of $\Pi = 1$. The moment coefficient decreases with increasing circumferential Reynolds number. It is important to realise that a smaller moment coefficient does not lead automatically to lower torque or reduced hydraulic losses.

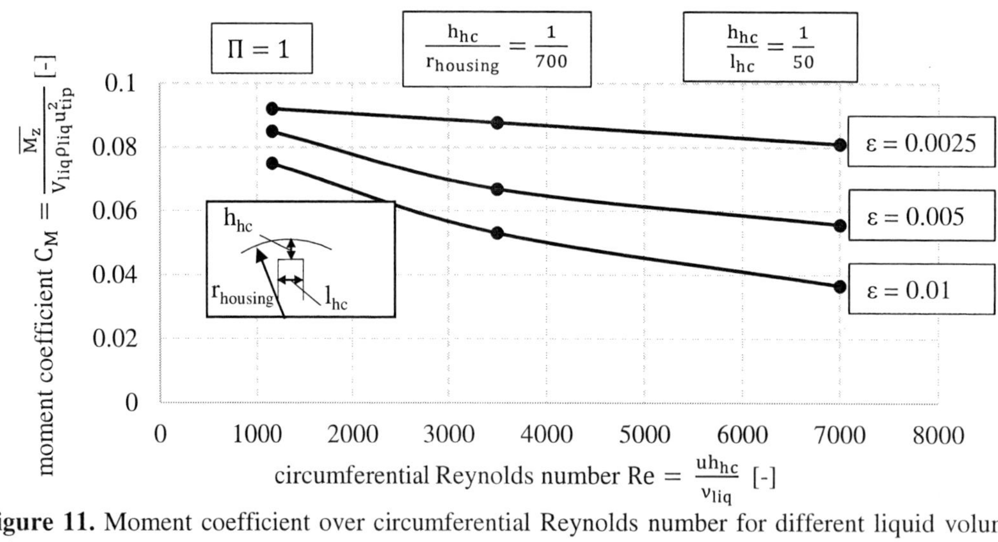

Figure 11. Moment coefficient over circumferential Reynolds number for different liquid volume ratios.

The torque depends mainly on the liquid mass and the circumferential tip speed. For similar cases, the moment coefficient can be determined based on the circumferential Reynolds number to determine torque on the contour. However, the characteristic diagram should be validated first with further similar simulation points, where, for example, geometry and liquid content are varied.

5. Conclusion

In this paper hydraulic loss mechanism as might occur in liquid-flooded screw machines were examined. For the examination, two dimensional transient simulations were performed with the Volume of Fluid method. The geometric model is one of a rotating rectangular profile in a cylinder filled with air and containing various amounts of water. The model is developed in a relative system, where the coordinate system of the contour rotates clockwise with a prescribed angular velocity; the relative motion of the housing is then in the opposite direction. The circumferential tip speed and liquid volume ratio were varied. First the relative flow pattern is analysed to identify the important hydraulic loss mechanisms. Subsequently, the influence of liquid volume ratio and circumferential tip speed are analysed. Finally, simulation results are presented in a dimensionless characteristic diagram, with the recommendation to have this validated in further investigations. The simulation results are analysed in a stationary state.

When there is liquid in the chamber, it is distributed to the rotor tip and casing, forming a rotating liquid surge at the contour tip and resulting in a liquid film becoming attached to the housing surface. The liquid film is transported by the housing towards the rotor-housing clearance, where part of the liquid flows through the clearance and the other part creates the liquid surge at the contour tip. Further, it can be observed that the rotating liquid surge is surrounded by a moving liquid stream which flows away from the clearance. Therefore, the liquid film consists of two liquid streams which flow in contrary circumferential directions. This motion creates additional shear stresses along the liquid film between the two streams. It is proposed that the size of the rotating liquid surge is restricted by the centrifugal forces. Therefore, the separated liquid mass at the contour tip is forced to flow back away from the clearance. During the backflow, the liquid mass is entrained by the liquid stream attached to the housing and flows towards the housing clearance.

The analysis of the torque shows a fluctuating characteristic, which is influenced by the change of the liquid shape at the rotor tip. The time-averaged value of the fluctuation can be used to characterize results. The torque increases almost linearly with increasing liquid volume ratio and quadratically with circumferential tip speed. Furthermore, the hydraulic loss due to the clearance flow between rotor and housing is analysed. It can be observed that the liquid surge at the contour tip is the major component of the hydraulic loss rather than the hydraulic loss due to the clearance flow itself.

Simulation results are presented in a dimensionless characteristic diagram showing the moment coefficient over the circumferential Reynolds number for different liquid volume ratios. For similar cases, the moment coefficient can be determined using the circumferential Reynolds number to determine the hydraulic power loss.

For future works, simulation of screw machine rotor profiles can be examined in a cylinder with different pressure ratios to create several characteristic diagrams. Furthermore, the characteristic diagrams should be validated with several simulation points while the characteristic numbers stay constant. These characteristic diagrams can be implemented, for example, in a chamber model simulation as a database to determine internal mechanical losses. As a matter of course, the simulation model should be validated in an experiment.

List of symbols and abbreviations

Symbols

b	width	(m)
CFL	Courant number	(-)
h	height	(m)
M	torque	(N m)
n	rotational speed	(s^{-1})
P	power	(N m s^{-1})
p	pressure	(N m^{-2})
r	radius	(m)
t	time	(s)
u	circumferential tip speed	(m s^{-1})
V	volume	(m^3)
γ	surface tension	(N m^{-1})
ν	kinematic viscosity	(m^2 s^{-1})
ρ	density	(kg m^{-3})
$\dot{\Phi}$	loss power	(N m s^{-1})
ω	angular frequency	(rad s^{-1})

Abbreviations

i	indicated, internal
e	effective
hc	housing clearance
m	mechanical
o	outer
liq	liquid
init	initialisation
max	maximum
min	minimum
G	global

References

[1] Kauder K 1987 Das Öl im Schraubenkompressor - ein Faktor für optimale Betriebsverhältnisse *Pumpen Vakuumpumpen Kompressoren '87* key no. 71075 pp 38-44 (Düsseldorf)

[2] Nikolov A and Brümmer A 2016 Analysis of Indicator Diagrams of a Water Injected Twin-shaft Screw-type Expander *International Compressor Engineering Conference* (Purdue University Libraries)

[3] Harling H 1993 *Untersuchungen zur Ölverteilung in Schraubenkompressoren mit Schmiermitteleinspritzung* Dissertation University of Dortmund

[4] Zellermann R 1996 *Optimierung von Schraubenmotoren mit Flüssigkeitseinspritzung* Dissertation University of Dortmund

[5] Deipenwisch R 2000 *Ein Beitrag zum Einsatz von Öl als Konstruktionselement in Schraubenmaschinen* Dissertation University of Dortmund

[6] Singh P.J. and Patel G.C. 1984 A Generalized Performance Computer Program for Oil Flooded Twin-Screw Compressors *International Compressor Engineering Conference* (Purdue University Libraries)

[7] Dämgen U 1994 Wassereinspritzung in Schraubenkompressoren *VDI-Berichte – Schraubenmaschinen '94* pp 175-194

[8] Kliem B 2005 *Grundlagen des Zweiphasen-Schraubenmotors* Dissertation University of Dortmund

[9] Nikolov A and Brümmer A 2018 Impact of different clearance heights on the operation of a water-flooded twin-screw expander – experimental investigations based on indicator diagrams *International Conference on Screw Machines* (Dortmund)

[10] Gräßer M and Brümmer A 2014 An analytic model of the incompressible one-phase clearance flow in liquid injected screw expanders *International Conference on Screw Machines – VDI Bericht 2228* pp. 71-89

[11] Gräßer M and Brümmer A 2015 Influence of liquid in clearances on the operational behaviour of twin screw expanders *International Conference on Compressors and their Systems – IOP Conf. Series: Materials Science and Engineering 90*

[12] Gräßer M and Brümmer A 2016 Influence of water and oil clearance flow on the operational behavior of screw expanders *Journal of process mechanical engineering* Part E

Mathematical modeling of working processes of variable frequency screw compressor with differentiated oil supply into the working chamber

V L Yusha[1, a)], G I Chernov[1] and M A Fedorova[1]

[1]Omsk State Technical University, 11, Mira ave., Omsk, 644050, Russia

[a)]Corresponding author: yusha@omgtu.ru

Abstract. Currently, a method of screw compressors regulation by smooth change of rotors spinning frequency, or so-called frequency regulation, is widely used. The operation process of oil-injected screw compressors at frequency regulation modes were examined in this paper in terms of the relationship of a compressor operation modes and its oil system. The calculations based on the operation process mathematical model of oil-injected screw compressor were aimed at defining its integral characteristics when changing the rotor spinning frequency, the quantity of the oil supplied and the discharge pressure.

1. Introduction

It is known that for improving the efficiency of screw compressors working process, the liquid is supplied into their working chamber. This reduces the intensity of gas leakage through the clearances, which is one of the drawbacks of the screw compressor design. In addition, the injected liquid provides intensive cooling of the compressed gas in the working chamber itself and, accordingly, the permissible value of the discharge gas temperature even if the ratio of the discharge pressure to the suction pressure is in the range of 5...15. Usually, oil or water is used as working liquid. Water provides the most efficient gas cooling, but due to a number of indicators (e.g. poor sealing of clearances due to low viscosity; corrosion activity, etc.) is has limited application [1, 2, etc.]. Screw oil-injected air, gas and refrigerating compressors have been widely used; moreover, in the process of operation there is often a change of both specific mode parameters (performance, suction and discharge pressure, boiling and condensation temperatures, capacity) and their various combinations [3 – 5]. The method of regulation of the screw compressor, applied in one or another case, determines not only the efficiency of its operation, but also the reliability of the design, its maintainability, weight characteristics, the cost of the compressor and the compressor unit as a whole. Nowadays, a method of screw compressors regulation by smooth change of rotors spinning frequency, or so-called frequency regulation, is widely used.

In this connection, it is necessary to note, that presence of oil in a working chamber has not only a positive, but also a negative effect, as it leads to increase in indicator losses of power in the discharge process and losses on hydro-mechanical friction in a compressor flowing part, as well as to additional "pumping" power expenditure. Both the positive and the negative effect of oil injection into the working chamber depends on the amount of oil and its viscosity (i.e. the oil temperature). Since the frequency control of the performance is accompanied in most cases by forced injection of oil under pressure, the amount of oil supplied to the compressor is determined by the fixed geometric dimensions of the elements of the oil system flow part and the difference between the values of suction and discharge pressures, which do not depend on the rotation frequency of the compressor rotors. It can be assumed that the change in the specific amount of oil (relative to the number of working gas) will affect the efficiency of the oil-injected compressor. In order to determine the possibility and feasibility of reducing energy consumption in compressor plants, this article investigates the working process of screw oil-injected compressor on frequency control modes concerning the relationship of the compressor operating modes and its oil system.

2. Method of calculation

Computational and theoretical studies were carried out using a mathematical model of oil-injected screw compressor working process and were aimed to define the capacity, efficiency coefficient, performance, flow coefficient and other characteristics at various modes (when changing the rotor speed, the quantity and the temperature of the supplied oil). The analysis of the current methods of calculating the working processes of screw oil-injected compressors [6-14] allowed to develop a methodology that takes into account the interconnection of the compressor stage and its oil system.

Methodology for calculation of the screw compressor under consideration was chronologically developed in two stages. At the first stage, the working process was analyzed with the widest possible list of physical phenomena accompanying it. Based on the results of its implementation, not very important physical phenomena were identified, which allowed, at the second stage, to simplify the analytical methodology for more effective application.

According to the results of the first stage of developing a mathematical model [11, 12] a physical model of the working process of a screw compressor with a heterogeneous gas-liquid working medium can be presented as follows. During the rotor (rotors) spinning, the working chamber volume changes. Gas (the first stage – air) rotates together with the rotor about its axis. During the compression, dropping liquid (at the first stage, water or oil) is supplied into the working chamber. The gas flow acts upon the dropping liquid that causes the change in its velocity, deformation and fragmentation. When interacting with the surfaces of the body, the droplets precipitate on them and spread out in the form of a film. When interacting with the surfaces of the rotating rotor the liquid may either break up into smaller particles, or precipitate on them, depending on the impact velocity. The working process is accompanied by external mass transfer through the clearances, partially sealed by liquid, and through the gas distribution elements; by the external heat gas exchange with the surfaces of the rotor and the body, and by heat and mass transfer between gas and liquid.

A generalized working process scheme in the considered control volume is shown in figure 1. In the control volume mass and heat flows are conventionally presented by generalized values, which reflect the direction and physical nature of these flows. To simplify the mathematical description of the considered working process and their implementation let us make the following assumptions: gas medium is continuous; pressure of the working liquid in the suction and discharge chambers is constant; dropping liquid is a system of spherical shape droplets of the same diameter suspended in the flow; working gas flow through a gas-distribution elements and constructive clearances is taken as adiabatic and quasi-static (where real gas properties and its compressibility are taken into account by the empirical coefficient); the processes of inter-phase mass transfer in the working chambers are quasi-stationary; the heating of the liquid film takes place with a constant temperature on the surface and in the liquid layer; the absolute motion of the gas phase in the working chamber is defined by translational, caused by the rotor spinning and by relative movement caused by a change of the working length of the screw chamber during the rotors interaction; the motion of droplets is not affected by the hydrodynamic interaction between them; impact of the Magnus force, hydro-mechanical, diffusive and reactive forces on a droplet is negligible; during the interaction with the body surfaces the liquid is precipitated in the form of a film, uniformly distributing over the surface; there are no failure of the film from the surface of the working chamber and the coagulation of the droplets in the gas stream; during the crushing of liquid particles by the rotating surfaces of the rotors

IOP Conf. Series: Materials Science and Engineering **425** (2018) 012003 doi:10.1088/1757-899X/425/1/012003

the size of the secondary particles is completely determined by the velocity of collision of the liquid and the rotor.

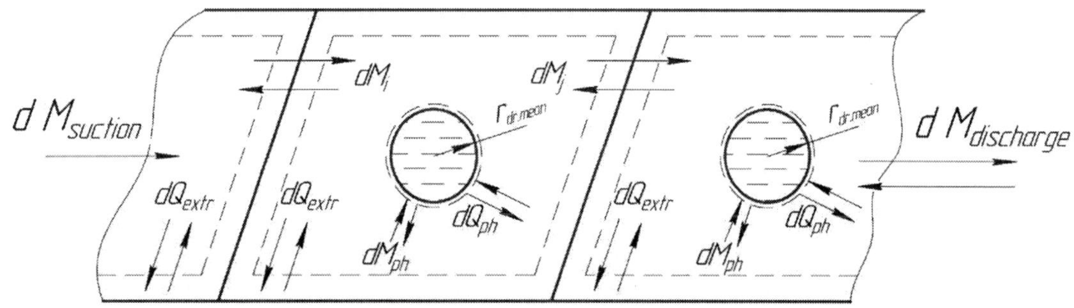

Figure 1. A calculation scheme of a screw compressor with a heterogeneous gas-liquid working medium

Single-valuedness conditions include:

- geometrical conditions (listed below);
- physical conditions (physical characteristics of components of the working medium at the first stage the working medium was a heterogeneous mixture of air and liquid (water or oil), at the second stage - air or R134 and oil;
- initial conditions characterizing the parameters of the working medium in initial time of the working process; for gas these are the settings in the beginning of the process of absorption, which are assumed to be parameters of the gas in the intake chamber; for the liquid–droplet parameters at the moment of entering the coupled chamber from the injection device;
- boundary conditions, characterizing the interaction of working medium with the medium; these are parameters of the gas in the suction and the discharge chambers, surface temperature of compressor parts [11, 13], interacting with the working liquid and liquid parameters before the injection device (injector).

The basic calculation equations used in the mathematical model are the first law of thermodynamics for a body of variable mass, including external and inter-phase mass transfer, law of conservation of mass for gas and liquid component of the working medium; the Newton-Richmann law (external (gas – compressor parts) and inter-phase (gas – liquid) heat exchange); the dependencies for determination of thermo-physical properties of gas and liquids; and the real gas equation [15]. The heat transfer coefficient for a droplet moving relative to the gas phase, in accordance with the assumptions with respect to the averaged parameters of liquid droplets and gas flow, is determined as follows [16]:

$$\alpha = \frac{Nu\lambda}{2r}, \qquad (1)$$

where the criterion Nusselt number according to [17], is defined by the following expression: $Nu = 2\left(1 + \beta \mathrm{Re}^{1/2} \mathrm{Pr}^{1/3}\right)$ (β is the constant coefficient [17]).

The calculation of transient heating of the dropping liquid is carried out by the heat balance method [16]. The change in mass of the gas phase due to the inter-phase mass transfer, taking into account quasi-stationary processes of evaporation and condensation for the liquid droplet can be measured using the formula of Fressling [18]. The flow of the gas-liquid mixture through the working clearances and the gas distribution valve is described by the equation of Saint-Venant and Wantsel with the introduction of the correction factor that takes into account the compressibility as the compressibility of gas and the presence of liquids [11, 14]. The motion of evaporating droplets in the working chamber is described by the differential equation of a variable mass point motion (the equation of Meshchersky). While droplets breaking by the gas flow the size of the secondary droplets is determined by the recommendations given in [19]; while its breaking by the surfaces of the spinning rotor – by ones supposed in [20].

In addition to the current settings during the calculation of the working process integral characteristics, efficiency factor of the supply, power input to the compressor shaft, taking into account the expenditures of liquid injection are determined [11, 21, 22]:

$$W = W_{\mathrm{ind}} + W_T + W_{\mathrm{add}} \tag{2}$$

where, in relation to the object of study, W_{ind} is the indicated power of the compressor stage; W_{sdd} is the power of the pump that feeds the liquid into the working chamber; $W_{\mathrm{T}} = W_{\mathrm{mech}} + W_{\mathrm{mix}}$ is the friction power of rotors on the gas-liquid mixture (W_{mix}), as well as the friction power in the bearings (W_{mech}).

The explanation is required for the calculation of power losses of the rotors friction on the gas-liquid mixture, which is approximately determined on the basis of the available experimental and theoretical data on the friction of simple configuration bodies of rotation.

Friction power of the disc surfaces of the rotor ends is described by the following expression [22]:

$$W_{\mathrm{mix}1} = 0.5 C_{\mathrm{m}} \rho_{mix} \omega^3 r_{\mathrm{r}}^5 \tag{3}$$

where ρ_{mix} is the density of the gas-liquid mixture, ω is the angular velocity of the rotor; r_{r} is the radius of the rotor; $C_{\mathrm{m}} = f(\mathrm{Re}, r_{\mathrm{r}}, \delta)$ is the coefficient of resistance (δ is the rotor-body clearance).

The power expenditures for the friction of the rotor outer surface on the near-wall liquid layer can be determined as follows [23]:

$$W_{\mathrm{mix}2} = \mu_{\mathrm{mix}} \omega^2 r_{\mathrm{r}}^2 F_{cyl} / \delta \tag{4}$$

where μ_{mix} is the viscosity of gas-liquid mixture in the side-wall layer; F_{cyl} is the area of the outer cylindrical surface of the rotors.

The power required for moving the gas-liquid mixture by the side surfaces of the rotor teeth is determined by analogy with the power expenditure to the sparging of the gas-liquid mixture by toothed cylindrical wheels [24]:

$$W_{\mathrm{mix}3} = 0.015 \omega^3 r_{\mathrm{r}}^5 \left(1 + 2.5 L_{\mathrm{r}} / r_{\mathrm{r}}\right) \cdot 10^{-3} \tag{5}$$

where L_{r} is the rotor length.

For solving the systems of differential equations, described above, the iterated Euler numerical method is used [25].

The convergence of the initial and final values of the calculating cycle is provided by the following conditions [21, 26]:

$$P_{n+1(\varphi_1=0)} - P_{n(\varphi_1=\pi/2)} \le \xi_1 \tag{6}$$

$$T_{n+1(\varphi_1=0)} - T_{n(\varphi_1=\pi/2)} \le \xi_2 \tag{7}$$

where ξ_1 and ξ_2 is the pre-set accuracy of calculation; P_n, T_n are the parameters of the working medium at the current phase or at the time layer of the working process; P_{n+1}, T_{n+1} are the parameters of the working medium at the following phase or the time layer of the working process.

If the condition of convergence is not met, the adjustment of the initial conditions for each area is made, for example, $P_{n+1(\varphi 1 = 0)} = P_{n(\varphi 1 = \pi/2)}$. At the end of every calculation cycle the value of the gas temperature in the discharge chamber is specified as the bulk temperature of the discharge gas.
The results of the verification of the mathematical model developed at the first stage, by comparing the calculated and experimental indicator diagrams of the working process of air screw compressor with oil or water injection, as well as its integral characteristics, are presented in [11, 12]. They show that the tested methods correctly describe the quality of physical processes, accompanying the investigated screw compressor operation; however, they accurately ensure the calculation of their quantitative characteristics.

At the first phase of the working process modeling analysis it was proved that during oil injection the influence of evaporation and condensation processes is negligible; it was also shown that convective heat transfer between gas and liquid film, between gas and compressor parts surfaces is also negligible in comparison with convective heat exchange between gas and liquid droplets. When analyzing the integral characteristics it was revealed that the compressor efficiency ratio (i.e. input-output ratio) depends on the frequency of the rotors spinning and the relative amount of liquid supplied into the working chamber. The optimal amount of liquid varies with the ratio of discharge and suction pressure and the frequency of the rotors.

Therefore, at the second stage of the modeling the main aim was to analyze the influence of the oil supply system into the screw compressor on its efficiency at frequency regulation modes. Based on the results of the calculation and parametric analysis at the first stage, the mathematical model of the working process of air and refrigeration screw compressor with oil injection into the working chamber was simplified. This provided the efficiency of its practical application. The simplified system of the main design equations on the second stage of mathematical modeling has obtained the following form.

The system of equations for calculating the gas suction into the working chamber considers mainly oil-free gas and has the following form:

$$\begin{cases} \dfrac{dT}{d\varphi} = (k-1)\cdot\dfrac{T}{m}\cdot\dfrac{\dot{m}}{\omega} + \dfrac{\sum\limits_{i\neq 1}(k\cdot T_i - T)\cdot\dot{m}_i}{m\cdot\omega} - (k-1)\dfrac{T}{V}\cdot\dfrac{dV}{d\varphi} \\ \dfrac{dp}{d\varphi} = z\cdot k\cdot\dfrac{p}{m}\cdot\dfrac{\dot{m}}{\omega} + z\cdot\dfrac{p}{m}\cdot\dfrac{k\cdot\sum\limits_{i\neq 1}T_i\cdot\dot{m}_i}{T\cdot\omega} - k\cdot\dfrac{p}{V}\cdot\dfrac{dV}{d\varphi} \\ \dfrac{dV}{d\varphi} = f(\varphi) \\ \dfrac{dm}{d\varphi} = \dfrac{\sum\dot{m}}{\omega} \\ \dfrac{dL}{d\varphi} = p\cdot\dfrac{dV}{d\varphi} \end{cases} \tag{8}$$

The system of equations for calculating the processes of compression and discharge is written as follows:

$$\begin{cases} \dfrac{dT}{d\varphi} = \dfrac{3 \cdot \alpha \cdot d}{c_V \cdot r \cdot \rho_{oil}} \cdot \dfrac{(T_{oil} - T)}{\omega} + (k-1) \cdot \dfrac{T}{m} \cdot \dfrac{\dot{m}}{\omega} + \dfrac{\sum\limits_{i \neq 1} (k \cdot T_i - T) \cdot \dot{m}_i}{m \cdot \omega} - (k-1) \dfrac{T}{V} \cdot \dfrac{dV}{d\varphi} \\ \dfrac{dp}{d\varphi} = z \cdot (k-1) \cdot \dfrac{3 \cdot \alpha \cdot d}{r \cdot \rho_{oil}} \cdot \dfrac{m}{V} \cdot \dfrac{(T_{oil} - T)}{\omega} + z \cdot k \cdot \dfrac{p}{m} \cdot \dfrac{\dot{m}}{\omega} + \dfrac{p}{m} \cdot \dfrac{k \cdot \sum\limits_{i \neq 1} T_i \cdot \dot{m}_i}{T \cdot \omega} - k \cdot \dfrac{p}{V} \cdot \dfrac{dV}{d\varphi} \\ \dfrac{dT_{oil}}{d\varphi} = \dfrac{3 \cdot \alpha \cdot d}{c_{oil} \cdot r \cdot \rho_{oil}} \cdot \dfrac{(T - T_{oil})}{\omega} \\ \dfrac{dV}{d\varphi} = f(\varphi) \\ \dfrac{dm}{d\varphi} = \dfrac{\dot{m}}{\omega} \\ \dfrac{dL}{d\varphi} = p \cdot f(\varphi) \end{cases} \tag{9}$$

In this system, T is the absolute temperature of the gas in the working chamber, K; P is the gas pressure in the working chamber, Pa; V is the volume of the working chamber, m^3; L is the expansion work, J; ω is the angular speed of the driving rotor rotation, deg/s; T_i is the temperature in the i-th chamber of the compressor; m is the mass of gas in the working chamber, kg/s; k is the coefficient of adiabatic compressed gas; $\dot{m}_i$ is the mass flow of gas leakage from the i-chamber to the working one, kg/s; z is the compressibility coefficient (which takes into account gas reality [15]); c_v is the specific mass isochoric heat capacity, J/(kg · k); L is the expansion work, J; α is the coefficient of heat transfer between a droplet of liquid and gas; T_{oil} is the liquid temperature; P_{oil} is the liquid density; m_l is the mass of liquid in the working chamber; r is the droplet radius; c_{oil} is the specific mass of liquid heat capacity, J/(kg · K).

Like the first stage, the structure of the calculated costs of power included the indicator power, determined by the numerical integration and taking into account the indicator losses on pushing the gas-oil mixture through the discharge port; pumping power costs for oil supply to the working chamber; hydro-mechanical losses on the movement of the gas-oil mixture in the working chamber. Such integral characteristics as the productivity, the average temperature of the discharge gas, and the indicator power were determined by numerical integration according to the known scheme, applied at calculation of volumetric action compressors [27]. According to the results of these calculations, the efficiency coefficient of the screw compressor is determined.

3. Results and discussion

The calculation of the working processes of the screw oil-injected compressor and its integral characteristics was carried out for the following initial data: the diameter of the driving rotor is 142 mm; the diameter of the secondary rotor is 110 mm; the length of the rotors is 236 mm; the main (male) rotor has five teeth, the gate (female) one has six spans (the spans profile was designed by "NIIturbokompressor" company, Kazan, Russia), the average height of the profile clearances is 80 µm, of the end clearances – 50 µm, on the tops of teeth – 100 µm; working gas is air or freon R134a.

The main calculation results are presented in figures 2 – 10. In the form of three-dimensional surfaces, the data on calculations of the efficiency coefficient dependence on the injected oil temperature T_{oil} and the mass share of oil d which was supplied into the working chamber of the compressor at the beginning of the compression process by an independent pump (or that the same, by displacement) at different speeds of the driving rotor.

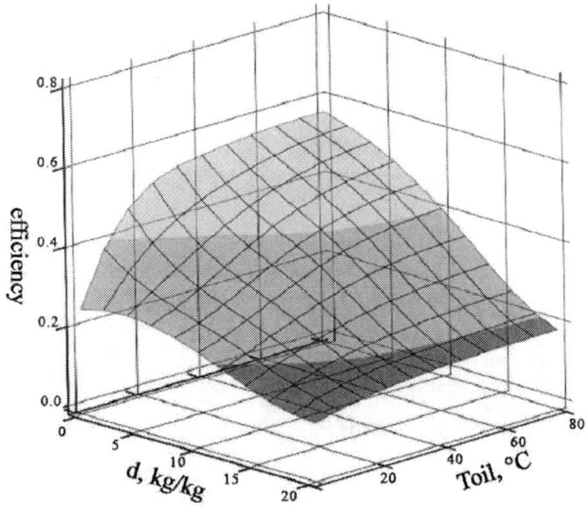

Figure 2. Dependence of the air screw compressor efficiency coefficient on the specific amount of oil in the working chamber d and the temperature of the injected oil Toil at the speed of the screw compressor driving rotor 3000 rpm; the suction pressure 0.1 MPa, and the discharge pressure 0.5 MPa.

As can be seen from the results presented in figure 1, it is not always necessary to strive for efficient cooling of oil in oil coolers. At low temperature of the injected oil its viscosity increases and therefore the pumping costs of power, hydro-mechanical losses and indicator losses in the injection process increase sharply. When there is a small amount of oil (at *d* from 1 to 5 kg/kg), the efficiency coefficient of the working process is drastically reduced by cooling the oil below 40^0 C, at *d* from 5 to 10 kg/kg this effect is also considerable. The large enough clearances in the working chamber, accepted for the calculation, allow to show the negative influence of excessive quantity of oil on the efficiency coefficient. As it is clearly seen in figure 1, at high rotor speeds, the increase in *d* results in a reduction of the compressor performance due to the rise of pumping power expenditure, hydro-mechanical losses in the working chamber and indicator losses during the discharge process. Only in a small range of low oil temperatures a slight increase in *d* provides a slight increase in efficiency by improving the sealing of the clearances. In the remaining range of *d* and T_{oil} at high rotor speeds, the positive effect of the oil on the clearances sealing is worse than the influence of the above losses.

Figures 2 and 3 present the results of the dependence of the screw oil-injected compressor efficiency coefficient on the temperature of the injected oil T_{oil} and the mass fraction of the oil *d* at the rotational speed of the driving rotor 2000 rpm and 1000 rpm. As it can be observed, at decrease in frequency of rotors speed at the expense of frequency regulation the character of oil quantity and temperature influence on the working processes efficiency changes noticeably. It is well seen that the reduction of the rotor speed results in the displacement of the optimal amount of oil towards the increase of *d*. It appears that the increase in *d* and the decrease in the speed make a non-linear reverse ratio: and non-linearity can be provided by automatic regulation of the amount of the oil supplied, if necessary. Note that during performance spool and throttle regulation at constant speed of rotors and constant oil consumption, the value *d* increases, which leads to deviation of the operation mode from a nominal one. The effect of T_{oil} when reducing the speed of rotation is the same: the recommended temperature of the injected oil is in the range of 40 – 80^0 C.

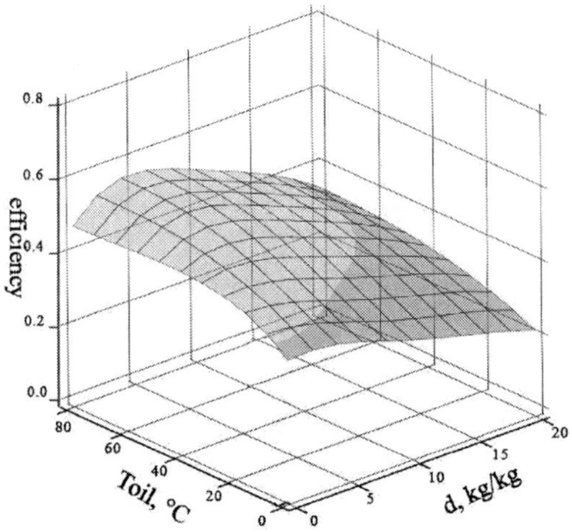

Figure 3. Dependence of the air screw compressor efficiency coefficient on the specific amount of oil in the working chamber d and the temperature of the injected oil Toil at the speed of the screw compressor driving rotor 2000 rpm; at the suction pressure 0.1 MPa, and the discharge pressure 0.5 MPa.

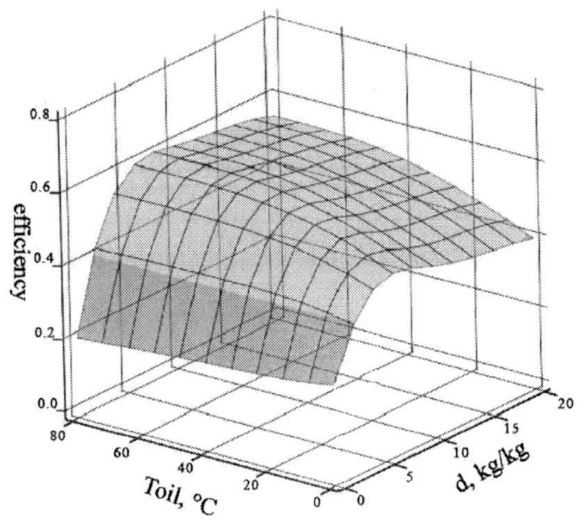

Figure 4. Dependence of the air screw compressor efficiency coefficient on the specific amount of oil in the working chamber d and the temperature of the injected oil Toil at the speed of the screw compressor driving rotor 1000 rpm; the suction pressure 0.1 MPa, and the discharge pressure 0.5 MPa.

Let us assess the influence of the specific amount of oil in the working chamber d and the temperature of the injected oil T_{oil} on the efficiency coefficient of the screw oil-injected compressor when compressing various working gases: air and freon R134a. Figures 4 and 5 present the results of calculations at the same suction pressure 0.196 MPa (which for freon R134a corresponds to the boiling point-10^0 C; at the same time, the gas temperature at the inlet of the compressor was taken as overheated by 5^0 C in relation to the boiling temperature); and at the discharge pressure of 1.0 MPa.

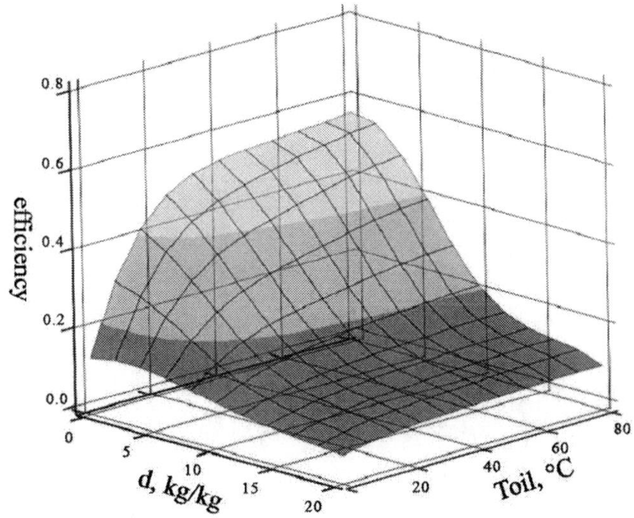

Figure 5. Dependence of the refrigeration screw compressor efficiency coefficient on the specific amount of oil in the working chamber d and the temperature of the injected oil Toil at the speed of the screw compressor driving rotor 3000 rpm; the suction pressure 0.196 MPa, and the discharge pressure 1.0 MPa.

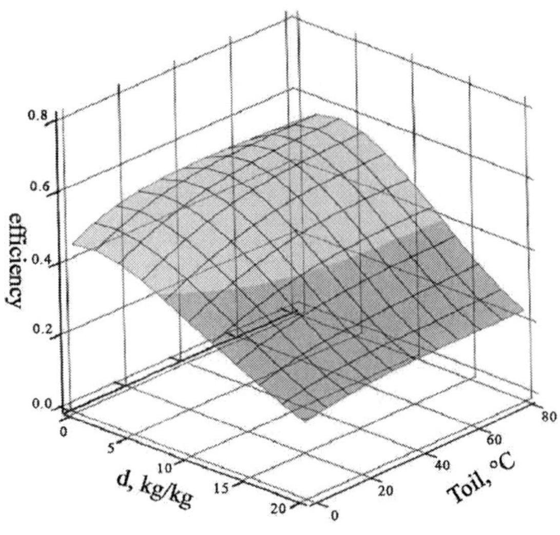

Figure 6. Dependence of the air screw compressor efficiency coefficient on the specific amount of oil in the working chamber d and the temperature of the injected oil Toil at the speed of the screw compressor driving rotor 3000 rpm; the suction pressure 0.196 MPa, and the discharge pressure 1.0 MPa.

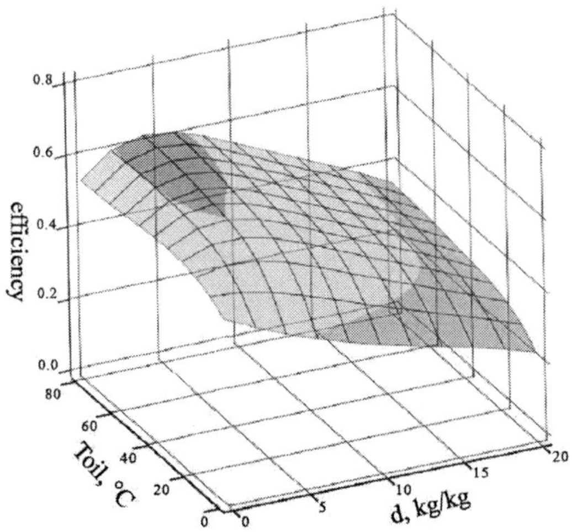

Figure 7. Dependence of the refrigeration screw compressor efficiency coefficient on the specific amount of oil in the working chamber d and the temperature of the injected oil Toil at the speed of the screw compressor driving rotor 1000 rpm; the suction pressure 0.196 MPa, and the discharge pressure 1.0 MPa.

As can be seen from the presented calculation results, in the refrigeration compressor, apparently, because of the working gas greater density, at smaller *d*, higher work efficiency is reached, comparing with air. Still, the qualitative influence of *d* and T_{oil} change on the efficiency coefficient of a screw oil-injected compressor stays the same for both working gases both at a nominal speed, and at its reduction. Note that the determination of the optimal *d* and T_{oil} for the nominal rotation speed is a priority, as the deviation of *d* and T_{oil} from the optimum can be even more critical when reducing the speed of rotation [5].

The influence of *d* and T_{oil} on the temperature of discharge gas is not less important. Let us consider the results presented in figures 8 and 9.

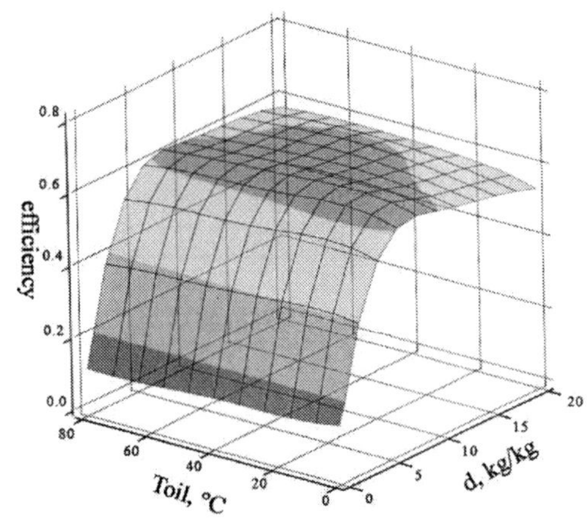

Figure 8. Dependence of the air screw compressor efficiency coefficient on the specific amount of oil in the working chamber d and the temperature of the injected oil Toil at the speed of the screw compressor driving rotor 1000 rpm; the suction pressure 0.196 MPa, and the discharge pressure 1.0 MPa.

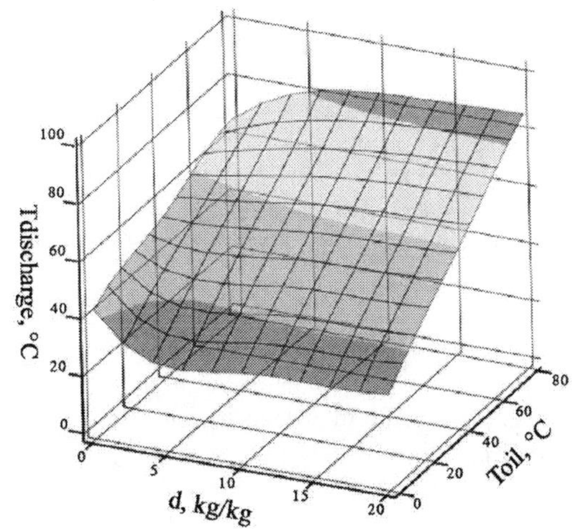

Figure 9. Dependence of the average discharge temperature of the refrigeration screw compressor on the specific amount of oil in the working chamber d and the temperature of the injected oil Toil at the speed of the screw compressor driving rotor 3000 rpm; the suction pressure 0.196 MPa, and the discharge pressure 1.0 MPa.

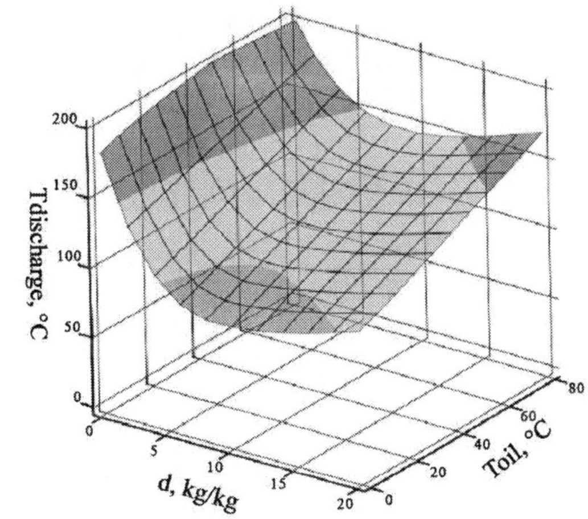

Figure 10. Dependence of the average discharge temperature of the air screw compressor on the specific amount of oil in the working chamber d and the temperature of the injected oil Toil at the speed of the screw compressor driving rotor 3000 rpm; the suction pressure 0.196 MPa, and the discharge pressure 1.0 MPa.

Analyzing the temperature mode, it can be noted that the temperature of the oil, of course, affects the temperature of the discharge gas. Therefore, the choice of the optimal value T_{oil} may in each case be determined depending on the prevailing criterion, in this case, either on the efficiency of work or on the temperature mode. The effect of the oil amount on the discharge temperature for the air compressor is more obvious. At low d the insufficient quantity of heat from compressed gas is transferred; at high *d* the pressure of the gas in the working chamber increases sharply, which leads to the growth of its temperature.

4. Conclusion

Thus, even taking into account the known errors of theoretical calculations, connected with accepted assumptions and uncertainty on certain conditions of ambiguity, qualitative analysis of a screw oil-injected compressor efficiency at frequency control modes has shown that with independent oil feed to the flow part (supply system or system with independent oil pump) has an advantage in comparison with spool regulation or regulation by throttling gas at suction. It means that with proper selection of the oil supply system flow characteristics, the rotors speed change does not result in any significant change in the screw compressor efficiency coefficient, because the value *d* is automatically changed and is slightly different from the optimal values over the entire range of speed variation. However, it sets strict requirements for the determination of the optimal value *d* at nominal rotors speed. The failure to comply with these requirements may lead to reduced efficiency by tens of percent both at nominal rotors speed and its decreasing in the process of compressor frequency control. Essential is the cooling system of oil, which temperature affects both the efficiency of the screw compressor and the temperature of the discharge gas; besides, the effects of injected oil temperature change for these characteristics may not be the same.

References

[1] Westpfall C R, Bell H S 1978 *Des. and Oper. Ind. Compressors Conf.* (Glasgow – London) pp pp 53–63
[2] Ushiota H and Mitsuhashi H 1987 *Hitachi Review* **36** p 147–154
[3] Mosemann D, Krienke S, and Nowotny S 1988 *Proc. of the International Compressor Engineering Conf.* (Purdue, USA) pp 250–255
[4] Wu H, Peng X, Xing Z and Shu P 2004 *Applied Thermal Engineering* **24** pp 1491–1500
[5] Yusha V L, Vasil'ev V K, Chernov G I and Panyutich A A 2017 *Chemical and Petroleum Engineering.* **52** (9-10) pp 590 – 596
[6] Stosic N, Smith I K and Kovacevic F 2005 *Screw Compressors: Mathematical Modelling and Performance Calculation (*Springer)
[7] Sangfors B 1998 *International Compressor Engineering Conference*, (Purdue, USA) **I** pp 595-600
[8] Kovacevic A, Stosic N, and Smith I 2007 *Screw Compressors: Three Dimensional Computational Liquid Dynamics and Solid Liquid Interaction* (Springer-Verlag Berlin Heidelberg)
[9] Nagam Seshaiah 2006 *Experimental and Computational Studies on Oil Injected Twin-Screw Compressor* (Rourkela, National Institute of Technology)
[10] Tian Y, Shen J, Wang C, Xing Z, and Wang X 2017 *International Journal of Refrigeration* **83** pp 75–87
[11] Yusha V L 1987 *Improving the efficiency and safety of screw compressor with gas-liquid working medium*: Diss. ... cand. tech. sci. (Leningrad)
[12] Yusha V L 2006 *Cooling and gas distribution systems of volumetric compressors* (Novosibirsk: Science)
[13] Mustafin T N, Yakupov R R, Burmistrov A V, Khamidullin M S and Khisameev M S 2016 *Proc. Eng. Int. Conf. on Oil and Gas Engineering* **152** (2016) pp 264-269
[14] Vernyi A L and Kupriyanov A 1982 *Design and study of compressor machinery: Proc* (Kazan) pp 9-17
[15] Handbook of physical and technical fundamentals of cryogenics 1985 ed. Malkova M P (Moscow, Energoatomizdat)
[16] Isachenko V P, Osipova V A and Sukomel A S 1981 *Heat Transfer* (Moscow: Energoizdat)
[17] Kostin A K, Larionov V V and Mikhailov L I 1979 *Heat density of internal combustion*

engines. Ref. manual (Leningrad: Mech. engineering)
[18] Fuks N A 1955 *Mechanics of aerosols* (Moscow AS USSR)
[19] Kogarko S M 1971 *Dynamics of droplet of liquid breakup in gas flow* (DAN USSR) T 198 11 pp 71-73
[20] Nazarov O I, Povarov OA, Jatcheni I A 1975 *Power engineering* **4** pp 47-49
[21] Amossov P E, Bobrikov N I , Schwartz A I et al. 1977 *Screw compressor machines.* Guide (Leningrad: Mechanical engineering)
[22] Dorfman L A *Hydrodynamic resistance and heat transfer of rotating bodies* 1960 (Moscow State Publ. house of phys.-math. lit)
[23] Yemtsev B T *Technical liquid mechanics* 1978 (Moscow: Mechanical engineering)
[24] Balashov B A, Galper R R, Garkavy A M et al. *Gearboxes of energy machines* 1985 Guide (Leningrad: Mechanical engineering)
[25] Demidovich V P, Maron A I, Shuvalov E Z *Numerical methods of analysis* 1967 (Moscow: Science)
[26] Tarasov A M, Egorov V G 1970 Power plant engineering **6** p 43-45
[27] Yusha V L, Busarov S S and Gromov A Yu 2017 *Chemical and Petroleum Engineering* **53 (7-8)** available at: https://doi.org/10.1007/s10556-017-0362-2

IOP Conf. Series: Materials Science and Engineering **425** (2018) 012004 doi:10.1088/1757-899X/425/1/012004

Effects of surface condensation in an idealised steam-driven screw expander

Manuel Grieb and Andreas Brümmer
TU Dortmund University, Chair of Fluidics
Leonhard-Euler-Str. 5, 44227 Dortmund, Germany

E-mail: manuel.grieb@tu-dortmund.de, andreas.bruemmer@tu-dortmund.de

Abstract. Condensation and its effects on turbo machinery operation are well understood and have been widely investigated. However, little scientific work on condensation in rotary positive displacement machines has been published. Since the robustness of screw expanders allows for expansion of slightly superheated vapours or even two-phase fluids, condensation on machine parts needs to be considered during design and simulation of these machines. In this paper the general effects of surface condensation of water on the machine parts of an idealised screw expander are discussed. Diabatic chamber model simulation is used for the thermodynamic simulation of operational behaviour. The effect of surface condensation on energy conversion and the delivered mass flow rate is analysed. Furthermore, a comparison of adiabatic and diabatic simulation of steam expansion in screw expanders is given in order to quantify condensation losses. Typical operating parameters are widely varied in simulation so as to identify influential factors on the condensation process. It is found that surface condensation, which is driven by heat exchange from the working fluid to adjacent machine parts, slightly raises the mass flow rate of the machine. For low expander speeds a reduction in isentropic efficiency can be expected due to a condensation induced pressure drop during the expansion phase.

1. Introduction

Rotary twin screw machines are robust and are therefore commonly operated with two-phase working fluids such as gas-oil mixtures in compressor applications. Due to the design similarities of screw compressors and expanders, the presence of condensate in the working chambers of expanders is not an issue for reliable machine operation; this allows expansion of vaporous fluids with minor superheating or even wet vapours. Thus, utilising screw expanders for power generation in Rankine cycles with low temperature heat sources is a promising concept. However, for performance prediction and correct design, it is necessary to take condensation into account in simulations. It has been found that screw expanders can be affected by the effects of spontaneous condensation. However, theoretical considerations indicate the impact on energy conversion is relatively small [1]. In this study, a simulation based investigation into the general influence of surface condensation in a steam-driven screw expander is reviewed.

2. Surface condensation

Condensation generally describes the phase change of a fluid from gas to liquid, whereby a significant release of latent heat occurs. This process is further differentiated into surface condensation and spontaneous condensation, where in the latter, a dispersed homogeneous phase emerges within the fluid, e. g. formation of clouds. Condensation phenomena on surfaces, which are the main subject of discussion in this paper, are characterised by the accumulation of droplets or a condensate film, which wet the cooler surface, and the transfer of latent heat from the fluid to the adjacent solid.

Surface condensation occurs when a vaporous fluid is in contact with a cooler wall whose temperature is below the saturation temperature of the fluid in thermodynamic equilibrium. In contrast to spontaneous condensation, where the latent heat of the fluid is released to the surrounding gas phase, the latent heat is transferred through the wall. If a condensate film is formed, the condensate layer acts as an additional thermal resistance and phase transition of the vapour takes place on the vapour-liquid interface. If the surface conditions prohibit complete wetting, the solid is only partially covered with condensate droplets, which grow in time. Due to external forces or shear stresses, droplets will eventually move along the wall while absorbing other droplets along their way. This mechanism again creates a partially dry surface with a subsequent re-initialisation of the dropwise condensation process. In general, maximum heat transfer is reached in dropwise condensation.

Fundamental work on film condensation was carried out by Nußelt, who presented an analytical model for laminar filmwise condensation on a vertical wall [2]. His model was later improved by consideration of the temperature-dependency of film fluid states [3], formation of waves on the film surface [4], and film turbulence [5]. An approach to modelling the improvement of heat transfer in cocurrent flows of condensate and vapour was presented by Numrich et al. [6]. Condensation heat transfers coefficients are typically determined from dimensionless heat transfer equations based on empirical studies. Since the physical mechanisms, and thus choice of the correct empirical formulation, highly depend on the mode of the flow, Tandon et al. proposed a flow regimes map for condensing two-phase flow [7].

It has been found that film condensation is the predominant effect in condenser applications. However, in the case of low surface energy of the solid or when non-wetting layers or hydrophobic materials are added to surfaces, dropwise condensation, characterised by heat transfer coefficients up to ten times higher than for film condensation, can be achieved [8]. One model proposed postulates that vapour condenses on a microscopic film between droplets, which acts as condensate collector. Due to its small length scale, the effect of the thermal boundary layer is negligible; this explains the extremely high heat transfer [9]. Other models are based on nucleation processes in surface cavities. However, the question of whether or not a thin film forms between droplets is as yet unresolved. Furthermore, an integral heat transfer equation explaining all tendencies of published experimental data has yet to be developed [10].

3. Diabatic simulation of rotary positive displacement machinery

Chamber model simulation has been found to be an appropriate method for the thermodynamic simulation of rotary positive displacement machines. In such simulations, geometry (e. g. chamber volume history, machine port cross-sectional area, leakage paths) as a function of the angular rotor position is provided as input to a non-dimensional model. Chamber states for each time step are determined on the basis of the conversation of mass and energy, where transport processes of energy and mass are calculated with a quasi-stationary approach.

Keller investigated thermal deformation of the rotors and casing of screw expanders, and proposed a model to calculate heat exchange between the working chambers and the surrounding solids [11]. A representative velocity was derived from the machine speed and rotor geometry; this approach allowed use of dimensionless heat transfer correlations for turbulent gas flow. Von Unwerth extended the model by considering convective heat transfer for the casing of screw expanders [12]. In addition, Rohe used plane wall correlations to calculate the heat exchange between the working fluid and machine parts of screw vacuum pumps [13]. On the basis of these investigations, Hütker determined related losses and their effect on efficiency ratings of

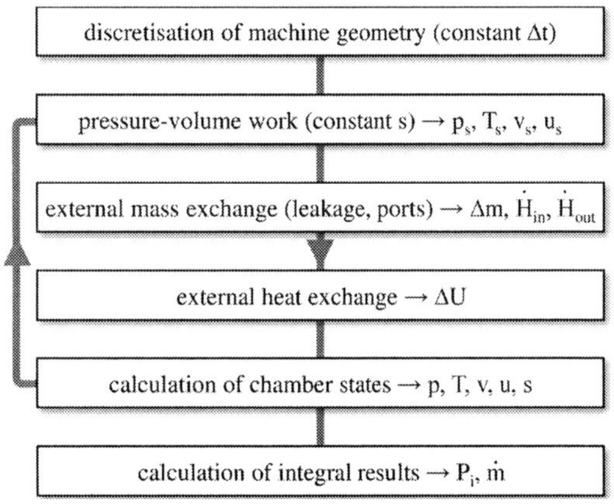

Figure 1. Process for diabatic chamber model simulation.

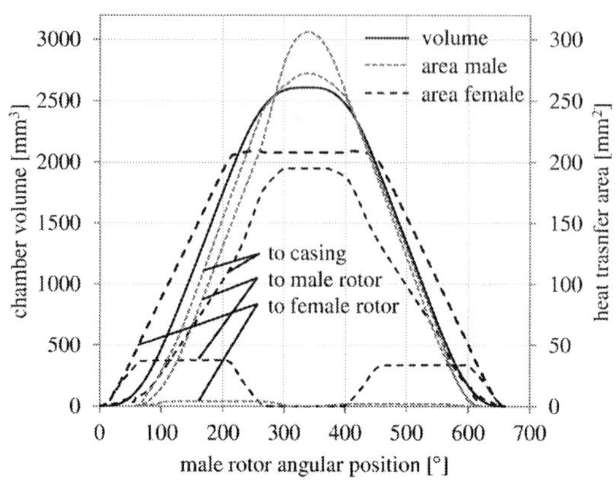

Figure 2. Chamber volume and heat transfer areas on male and female rotor chamber.

screw expanders. Losses investigated included external heat loss and increasing leakage caused by thermally induced changes of clearance heights [14]. However, these investigations were performed for dry-running screw machines, and, consequently, the proposed heat exchange models are only valid for single-phase gaseous fluids.

4. Simulation model

In this investigation, chamber model simulation is used to analyse a screw expander process with external heat exchange. An idealised expander cycle is investigated in order to show the full impact on operation and to isolate surface condensation from typical loss effects such as leakage and throttling. The following changes of state are postulated for the idealised cycle: an isobaric inlet phase at inlet pressure, an expansion phase with constant chamber mass, and an isobaric discharge phase at a given back pressure. In the case where heat transfer is suppressed, the idealised process becomes isentropic. Thermodynamic properties are provided from a fundamental equation of state where the fluid states are calculated from an empirical formulation of Helmholtz free energy.

In this theoretical study, the exchanged heat Q between the working fluid and surrounding machine surfaces A within one simulation time step Δt is considered during the inlet and expansion phases by applying heat fluxes $\dot{q}$ corresponding to typical surface condensation modes. α is the heat transfer coefficient, and ΔT is the temperature difference between fluid and solid.

$$\dot{q} = \frac{Q}{A\,\Delta t} = \alpha\,\Delta T \tag{1}$$

In dimensionless notation heat transfer is expressed with the Nußelt number Nu, where $\mathcal{L}$ is a characteristic length and λ is the thermal conductivity of the fluid.

$$Nu = \frac{\alpha\,\mathcal{L}}{\lambda} \tag{2}$$

The calculation routine for the diabatic chamber model simulation is depicted in **figure 1**. Changes of state in the working chamber are calculated in two successive steps. Initially, an isentropic expansion with constant trapped chamber mass m is performed. Based on the isentropic change of states, mass inflow and outflow as well as heat transfer are calculated; these lead to changes in specific internal energy u and specific volume v. The final chamber states

Table 1. Expander data

characteristic	value
rotor profile	SRM A
number of lobes	4 (male) / 6 (female)
maximum chamber volume	$26254\,\mathrm{mm}^3$
built-in volume ratio	4.0
male rotor tip diameter	$58.9\,\mathrm{mm}$
axis to axis distance	$46.2\,\mathrm{mm}$

are ultimately calculated from the equation of state with those inputs. For the idealised filling period at constant inlet pressure, the inlet mass increment is iteratively calculated. Indicated work $W_i = -\oint p\,\mathrm{d}V$ and indicated power $P_i = n\,z_{male}\,W_i$ are later determined by the integration of pressure history with respect to the chamber volume. The basic principle of chamber model simulation and formulations for the conservation of energy and mass are well documented, for instance, in [15].

4.1. Simulation parameters and machine geometry

The calculation model presented in this study requires both numerical and machine parameters. The expansion phase is discretised with a defined time step. A sensitivity analysis with regard to indicated power P_i implies $\Delta t = 1.0 \cdot 10^{-6}\,\mathrm{s}$ is sufficiently small for the parameter range under consideration. The fluid model selected is Pollak's fundamental equation of state; fluid states are calculated in thermodynamic equilibrium [16].

Relevant information on the machine geometry, which determines the volume curve and links tip speed with rotational speed, is shown in **table 1**. The corresponding plots of chamber volume and heat transfer areas to the rotors and casing are depicted in **figure 2**. It can be noted that heat transfer areas on the male rotor side are, in general, greater than on the female rotor side, but areas connecting the female chamber to the female rotor emerge at an earlier time. In addition, with the male rotor operating at higher speeds, greater heat transfer coefficients can be expected. This imbalance of thermal load is further amplified since the surface connection from the female rotor chamber to the male rotor is greater than vice versa.

5. Expander cycle with surface condensation

Initially, the general influence of surface condensation on an idealised expander cycle is discussed. There are two fundamental differences in comparison of heat exchange between working fluid and machine parts in dry-running screw machines with single-phase fluids. On the one hand, the release of latent heat in close proximity to the heat transfer surface induces significantly higher heat transfer coefficients. On the other hand, condensation includes mass transfer from the gaseous to the liquid phase, where, due to the difference in density, a substantial part of the trapped mass might be accumulated in a small liquid volume within the chamber. This becomes obvious when plotting the vapour volume fraction ϵ of wet steam over the steam quality, **figure 3**. Even with a substantial reduction of steam quality, by far the largest part of the chamber volume is still filled with vaporous fluid.

The conservation of energy for a working chamber of the screw expander, neglecting gravitation effects, can be expressed as follows:

$$\mathit{\Delta} U = Q + W + H_{t,in} - H_{t,out} \tag{3}$$

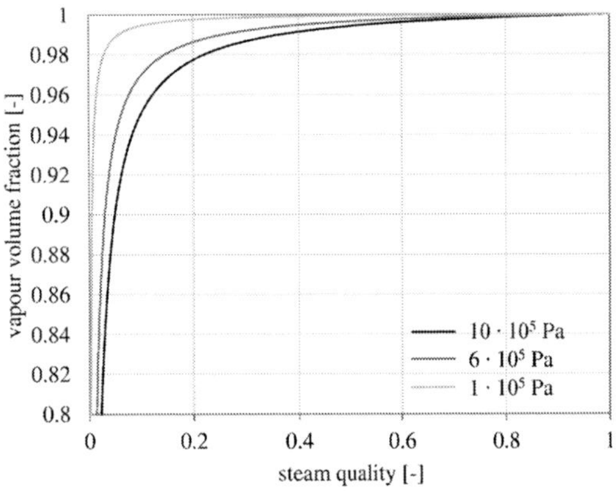

Figure 3. Vapour volume fraction plotted as function of steam quality for various pressures.

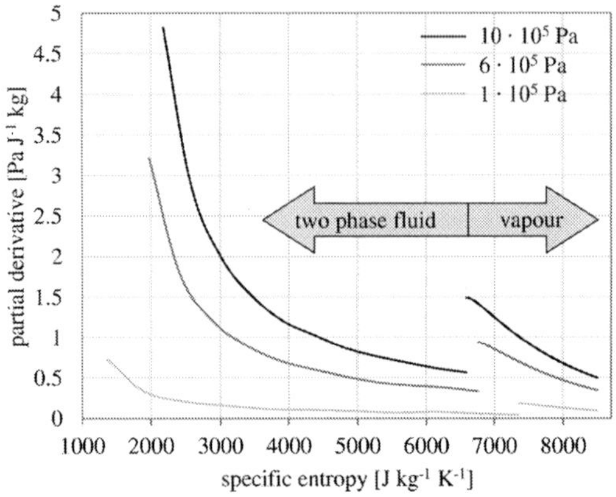

Figure 4. Partial derivative of pressure with respect to specific internal energy at constant specific volume of superheated and wet steam for various pressures.

Surface condensation is always combined with heat transfer to the surrounding surface ($Q < 0$), thus reducing internal energy of the fluid within the working chamber. For an idealised process, the trapped chamber mass remains constant during the expansion phase. Therefore, condensation induced heat transfer results in an additional pressure drop. In order to quantify this effect, the partial derivative of pressure with respect to specific internal energy at constant specific volume (constant trapped mass and constant chamber volume) is evaluated:

$$\left(\frac{\partial p}{\partial u}\right)_v \tag{4}$$

The result is depicted in **figure 4**. The impact of surface condensation on pressure history and, in turn indicated work, is potentially higher at greater pressure and for lower steam quality. When comparing the effect of heat output from the working chamber on the pressure drop in the two phase region and for superheated fluid states, greater sensitivity can be expected for moderately superheated steam. The general influence of heat outflow on chamber pressure decreases again for well superheated states.

During the inlet phase, when the working chamber is, by definition, not a closed thermodynamic system, surface condensation decreases the specific volume of the working fluid and hence raises the expander mass flow. However, due to the high condensation enthalpy, this effect requires extremely high heat output for a considerable impact. The heat loss induced pressure drop during the inlet phase is eventually compensated for by a higher inflow.

A comparison between the idealised indicator plot of an adiabatic working cycle and that of a diabatic working cycle for the selected twin screw expander is given in **figure 5**. Here, the heat flux is set to an exemplary value. Obviously, only minor differences in pressure history are observed. Despite a substantial amount of heat being transferred from the working fluid to the casing and rotors, a notable drop in chamber pressure is only seen late in the expansion phase. Consequently, the (ideal) indicated work of the machine is only slightly reduced. Since the expander is fed with saturated steam, heat losses during the inlet phase reduce steam quality in the working chamber. The resulting reduction of the specific volume increases the amount of fluid being trapped at the start of expansion, which in turn increases the mass flow rate. However, due to the high specific latent heat of water, high heat outflow is required for a

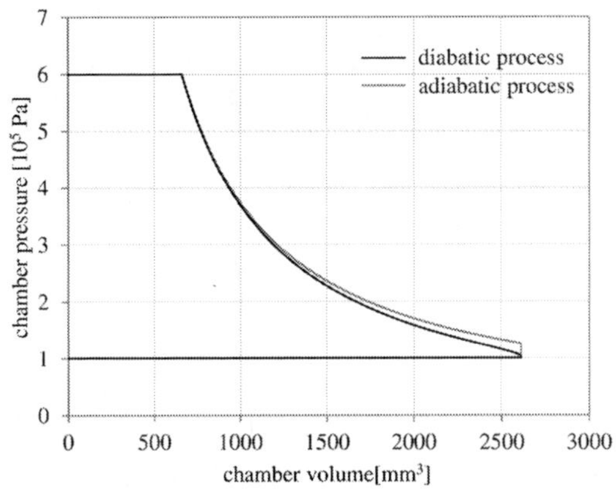

Figure 5. Comparison of indicator plots for adiabatic and diabatic idealised expander cycle, $p_{in} = 6 \cdot 10^5\,\mathrm{Pa}$, $p_{out} = 1 \cdot 10^5\,\mathrm{Pa}$, $T_{in} = 431.98\,\mathrm{K}$ (sat. steam), $u = 60\,\mathrm{m\,s^{-1}}$, $\dot{q} = 6 \cdot 10^5\,\mathrm{W\,m^{-2}}$.

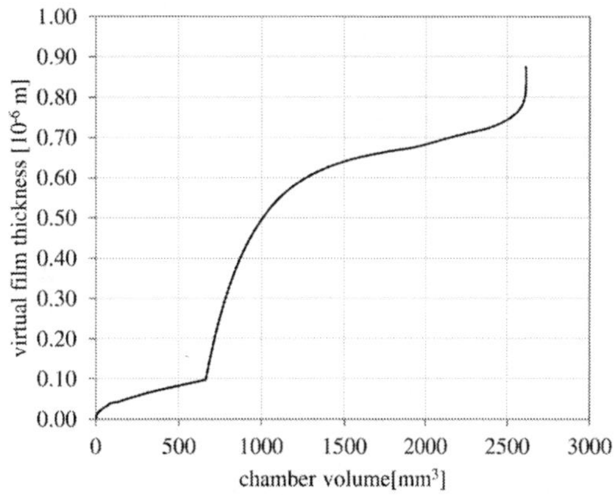

Figure 6. Virtual thickness of condensate film during inlet and expansion phase, $p_{in} = 6 \cdot 10^5\,\mathrm{Pa}$, $p_{out} = 1 \cdot 10^5\,\mathrm{Pa}$, $T_{in} = 431.98\,\mathrm{K}$ (sat. steam), $u = 60\,\mathrm{m\,s^{-1}}$, $\dot{q} = 6 \cdot 10^5\,\mathrm{W\,m^{-2}}$.

substantial increase in mass flow rate. At this exemplary operating point, the simulated mass flow rate is only increased by 1.2 % compared to the adiabatic idealised expander cycle.

In order to quantify emerging liquid during expansion, a virtual film thickness δ_v is introduced. This is the thickness of the film that would be formed if all of the trapped condensate volume V_c were uniformly distributed over the heat transfer surfaces.

$$\delta_v = \frac{V_c}{A} = \frac{(1-\epsilon)\,V}{A} \tag{5}$$

Here, ϵ is the vapour volume fraction. The characteristic of this virtual film thickness vs. chamber volume for the inlet and expansion phases is plotted in **figure 6**. The absolute value of heat flow in this exemplary simulation is $-11.4 \cdot 10^3\,\mathrm{W}$, which is equivalent to 68.6 % of the machine's isentropic power. Nevertheless, the vapour volume fraction being close to unity results in small values of virtual film thickness, and only little interaction between the emerging condensate and the machine parts is expected (e. g. seal of clearances).

From this theoretical consideration, two effects of surface condensation on the working cycle of a screw expander are identified: the reduction of the specific volume of the fluid during the inlet phase, and the heat transfer induced pressure drop during expansion. In addition to the influence of fluid properties, both mechanisms are driven by the amount of heat being transferred from the working fluid to the surrounding surfaces. Here, measurements of surface areas, which depend on the rotor dimension and geometry, temperature differences between working fluid and machine parts, which, to some extent, can be affected by insulation or cooling, the duration of the working cycle, which results from machine speed, and the heat transfer coefficient, which is mainly affected by the flow mode within the expander, are the main factors impacting heat outflow. In the subsequent theoretical study, the magnitude of the heat flow is systematically varied by altering heat fluxes $\dot{q} = \alpha\,\Delta T$. Time dependence is considered by altering machine speed.

6. Variation of expander parameters

The influence of decreasing specific volume during the inlet phase due to heat outflow can be quantified by the delivery rate λ_L, which is the ratio of actual mass flow rate $\dot{m}$ to the theoretical

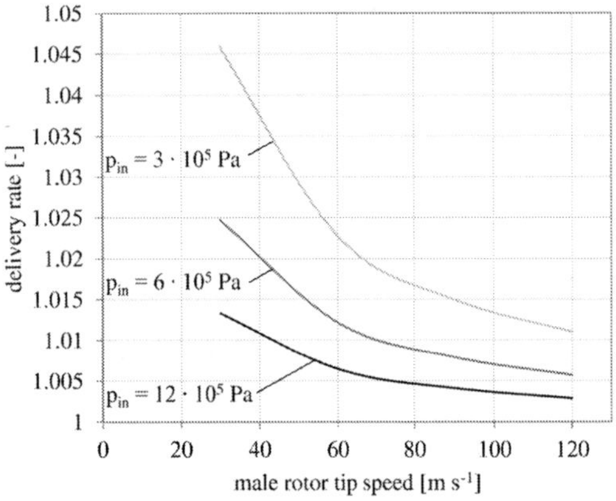

Figure 7. Impact of surface condensation on delivery rate for various expander speeds and inlet pressures, $\Pi = \frac{p_{in}}{p_{out}} = 6$, $\dot{q} = 6{\cdot}10^5\,\mathrm{W\,m^{-2}}$, $T_{in} = 431.98\,\mathrm{K}$ (sat. steam).

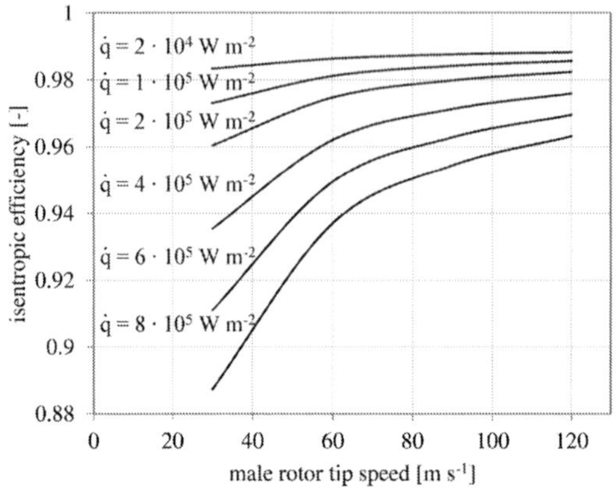

Figure 8. Influence of heat transfer on isentropic efficiency of an idealised expander cycle, $p_{in} = 6 \cdot 10^5\,\mathrm{Pa}$, $p_{out} = 1 \cdot 10^5\,\mathrm{Pa}$, $T_{in} = 431.98\,\mathrm{K}$ (sat. steam).

mass flow rate $\dot{m}_{th}$. The theoretical flow rate is calculated from expander speed n, number of lobes on the male rotor z_{male}, the chamber volume at the beginning of expansion phase V_{ex} and the specific volume at inlet conditions v_{in}.

$$\lambda_L = \frac{\dot{m}}{\dot{m}_{th}} = \frac{\dot{m}\, v_{in}}{n\, z_{male}\, V_{ex}} \tag{6}$$

The impact of surface condensation on the delivery rate of an idealised diabatic expander cycle is plotted in **figure 7** for various inlet pressures and expander speeds. Despite choosing high absolute values of heat flux for the simulation, the specific volume at the start of expansion is only slightly affected. Consequently, the impact on delivery rate is moderate for all operating points displayed. In addition to the influence of expander speed, which correlates with the absolute value of heat being transferred from the working fluid to the adjacent solids, the delivery rate increases with decreasing inlet pressures. Although the specific latent heat of steam is lower for higher pressures, the mass trapped in the chamber is dependent on the steam density, which decreases with pressure. Consequently, with the boundary condition of a constant heat flux (heat outflow is proportional to expander speed), the relative change in specific internal energy is greater for lower pressures.

The increasing delivery rate in combination with the surface condensation induced pressure drop (reduced indicated work) during expansion result in a reduced isentropic efficiency η_s of the machine under consideration. The isentropic power P_s is calculated from the actual mass flow rate and the specific isentropic enthalpy difference Δh_s. The isentropic efficiency can furthermore be expressed with W_i being the indicated work and $W_{s,th}$ being the theoretical isentropic work of the idealised expander cycle.

$$\eta_s = \frac{P_i}{P_s} = \frac{P_i}{\dot{m}\,\Delta h_s} = \frac{n\, z_{male}\, W_i}{\lambda_L\, \dot{m}_{th}\, \Delta h_s} = \frac{W_i}{\lambda_L\, W_{s,th}} \tag{7}$$

Figure 8 shows the isentropic efficiency characteristics for a process with surface condensation in thermodynamic equilibrium. The inlet pressure and back pressure are set to fixed values. Expander tip speed is varied from $u = 30$ to $120\,\mathrm{m\,s^{-1}}$. Heat flux is varied from $\dot{q} = 2 \cdot 10^4$ to $8 \cdot 10^5\,\mathrm{W\,m^{-2}}$. Isentropic efficiency of the adiabatic process is $\eta_s = 0.99$,

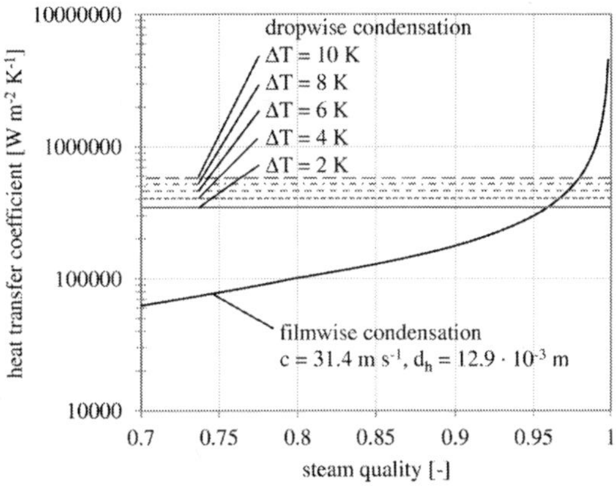

Figure 9. Calculated values of heat transfer coefficient for annular film condensation [8] and dropwise condensation [17] for steam, $p = 6 \cdot 10^5\,\mathrm{Pa}$.

which is the small loss being caused by under-expansion (see figure 5). The influence of surface condensation on indicated power is moderate for expander operation with saturated steam and efficiency decreases both with increasing heat flux and with reduced expander speed. However, for the operating points displayed, the influence on pressure history is predominant compared to the changing mass flow rates due to heat outflow during the inlet phase. Despite choosing high absolute values of heat flux, the mass flow rate is maximally increased by 3.2 % for the operating points depicted in figure 8 ($u = 30\,\mathrm{m\,s^{-1}}$, $\dot{q} = 8 \cdot 10^5\,\mathrm{W\,m^{-2}}$).

7. Validity of condensation models for screw expanders

For a precise simulation of surface condensation in screw expanders, the application of a realistic heat transfer model is required. Initially, the mode of condensation (separated or dispersed flow, filmwise or dropwise condensation) needs to be identified. Flow regime maps, as provided by Tandon et. al [7], allow for a first estimate of the condensation process. However, information on the steam flow within the working chamber needs to be provided. The principle of chamber model simulation cannot provide such information, since in this model, thermophysical properties are calculated for homogeneous chamber states at rest. A reference velocity along the helical working chambers, which is in the range of the rotor tip speed, can be calculated from the expander speed and its geometry [11]. These relatively high velocities dominate over gravitational force and indicate the mode being an annular condensation flow or, for high expander speeds, even a mist flow.

Heat transfer models for annular condensation proposed in the literature focus on steady state flow and validation is most often provided for complete phase transition in condenser pipes. However, only partial condensation is expected during the expansion of steam in screw expanders and empirical correlations might provide extremely high values of heat transfer coefficient for the onset of annular condensation due to small film thickness. **Figure 9** shows calculated values for annular film condensation in a male rotor chamber and dropwise condensation for saturated steam at $p = 6 \cdot 10^5\,\mathrm{Pa}$. The heat transfer correlation in [17] for dropwise condensation is obtained from experimental studies. For the annular film condensation, the working chamber is treated as a helically wound condenser pipe. Heat transfer coefficients are calculated from a Nußelt correlation documented in [8]. Due to the small film heights, transport properties of the condensate film are calculated for saturation points. The reference vapour velocity along the helix is calculated with Keller's proposal according to the expander geometry [11] for a male rotor tip speed of $u = 30\,\mathrm{m\,s^{-1}}$. The hydraulic diameter d_h of the pipe is established from a plane orthogonal to the velocity vector within the rotor cavity. The boundary conditions for the film condensation model lead to heat transfer coefficients being greater than for dropwise condensation at high steam quality, which is a questionable result.

Furthermore, it is a debatable point whether or not the assumption of a steady state condensation process is applicable, since at least the surfaces of the housing bore are only in contact with the expander chamber for a finite time frame. Moreover, condensate entrainment is expected in the high velocity flow of the housing clearances, which removes liquid from the housing bore as the rotor tips pass by. This re-exposure of dry surfaces is comparable to the effects occurring in dropwise condensation and potentially increases heat transfer. In addition, centrifugal forces might affect the condensation process, especially on rotor surfaces. In summary, it can be stated that adopting Nußelt correlations for surface condensation documented in literature to the thermodynamic simulation of rotary displacement machines can only lead to a rough approximation of the physical effects expected during the working cycle.

8. Conclusion

The general influence of steam condensation on the surfaces of an idealised screw expander is discussed. Surface condensation in rotary positive displacement machines can be expected when external heat transfer occurs in the two phase region. External heat transfer has two major effects on the energy conversion: The specific volume of the fluid is decreased during the inlet phase, which results in a higher mass being trapped within the working chamber and, consequently, increases the mass flow rate. Furthermore, heat transfer induces an additional pressure drop during the expansion phase. However, the high specific condensation enthalpy of water minimises these two effects. Validation of these findings against experimental data is still pending. Numerous Nußelt correlations for surface condensation are documented in literature. Since those models are developed for steady state flow and present most often a solution for complete phase transition in condenser applications, their validity is limited for surface condensation in rotary displacement machines. Estimation of heat transfer based on these models is, however, possible and indicates high values of heat transfer coefficients for steam-driven screw expanders. Experimental work on condensation processes in rotating systems and consideration of forced condensate entrainment might help gain a deeper insight into the actual surface condensation mechanisms in rotary positive displacement machines.

References

[1] Grieb M and Brümmer A 2017 Effects of spontaneous condensation in steam-driven screw expanders *IOP Conf. Ser.: Mater. Sci. Eng.* **232** 012078
[2] Nußelt W 1916 Die Oberflächenkondensation des Wasserdampes *Zeitschrift des Vereins deutscher Ingenieure* **60** 569–575
[3] Voskresenskij K D 1948 Calculation of heat transfer during film condensation with condensate properties as a function of temperature (in russian) *Izv. Akad. Nauk SSSR* 1023–1028
[4] Kutateladze S S and Gorgonin I I 1979 Heat transfer in film condensation of slowly moving vapor *Int J Heat Mass Transfer* **22** 1593–1599
[5] Müller J 1992 *Wärmeübergang bei der Filmkondensation und seine Einordnung in Wärme- und Stoffübertragungsvorgänge bei Filmströmungen* (Düsseldorf: Fortschritt-Berichte VDI Reihe 3 Nr. 270)
[6] Numrich R 1990 Influence of gas flow on heat transfer in film condensation *Chem Eng Technol* **13** 136–143
[7] Tandon T N, Varma H K and Gupta C P 1990 A new flow regimes map for condensation inside horizontal tubes *Journal of Heat Transfer* **104** 763–768
[8] Numrich R and Müller J 2013 *Filmkondensation reiner Dämpfe* (Berlin, Heidelberg: Springer-Verlag) pp 1035–1052 VDI-Wärmeatlas 11th ed
[9] Koch G 1996 *Untersuchungen zur Tropfenkondensation auf metallischen, hartstoffbeschichteten Oberflächen* (Erlangen: Berichte zur Energie und Verfahrenstechnik H. 96.4 ESYTEC)
[10] Merte H 1973 Condensation heat transfer *Advanced Heat Transfer* **9** 181–272

[11] Keller G 1998 *Simulationsgestützte Entwicklung des Motors einer Heißgasschraubenmaschine* (Düsseldorf: Fortschrittsberichte VDI Reihe 6 Nr. 387)
[12] von Unwerth T 2002 *Experimentelle Verifikation eines Simulationssystems für eine GASSCREW* (Dortmund: Universität Dortmund)
[13] Rohe A 2004 *Wärmehaushalt von Schraubenspindel-Vakuumpumpen* (Dortmund: Universität Dortmund)
[14] Hütker J 2016 *Energiewandlung in trockenlaufenden Schraubenmotoren* (Dortmund: Technische Universität Dortmund)
[15] Kauder K, Janicki M, Rohe A, Kliem B and Temming J 2002 Thermodynamic simulation of rotary displacement machines *Schraubenmaschinen 2002, VDI Berichte* **1715**
[16] Pollak R 1974 *Die thermodynamischen Eigenschaften von Wasser - dargestellt durch eine kanonische Zustandsgleichung für die fluiden homogenen und heterogenen Zustände bis 1200 Kelvin und 3000 bar* (Bochum: Ruhr-Universität Bochum)
[17] Rose J W 1998 Interphase matter transfer, the condensation coefficient and dropwise condensation *Proc 11th Heat Transfer Conference Kyongju, Korea* **13** 89–104

List of symbols

Symbol	Unit	Property
A	$\mathrm{m^2}$	area
c	$\mathrm{m\,s^{-1}}$	velocity
d	m	diameter
h	$\mathrm{J\,kg^{-1}}$	specific enthalpy
H	J	enthalpy
$\dot{m}$	$\mathrm{kg\,s^{-1}}$	mass flow rate
m	kg	mass
n	$\mathrm{s^{-1}}$	expander speed
Nu	–	Nußelt number
p	Pa	pressure
P	W	power
$\dot{q}$	$\mathrm{W\,m^{-2}}$	heat flux
Q	J	heat
s	$\mathrm{J\,kg^{-1}\,K^{-1}}$	specific entropy
t	s	time
T	K	temperature
u	$\mathrm{m\,s^{-1}}$	male rotor tip speed
u	$\mathrm{J\,kg^{-1}}$	specific internal energy
U	J	internal energy
v	$\mathrm{m^3\,kg^{-1}}$	specific volume
V	$\mathrm{m^3}$	volume
W	J	work
z	–	number of lobes
α	$\mathrm{W\,m^{-2}\,K^{-1}}$	heat transfer coefficient
δ	m	film thickness
ϵ	–	volume fraction
λ	$\mathrm{W\,m^{-1}\,K^{-1}}$	thermal conductivity
λ_L	–	delivery rate
η	–	efficiency
$\mathcal{L}$	m	characteristic length

Subscript	Property
c	condensate
ex	expansion
h	hydraulic
i	indicated
s	isentropic
sat	saturated
t	total
th	theoretical
v	virtual
Δ	difference, increment

IOP Conf. Series: Materials Science and Engineering **425** (2018) 012005 doi:10.1088/1757-899X/425/1/012005

The workflow of rotor machine development: design phases, steps and tools of the development process

H Österman, A Edrisi and I Lashgari[1]

Svenska Rotor Maskiner (SRM), Fannys Väg 3, 131 54 Nacka, Stockholm, Sweden

[1]Email: iman.lashgari@rotor.se

Abstract. In this work we present an overview of the workflow of rotor machine[§] development, focusing on twin-screw compressors and expanders. The rotor machine development process consists of an orchestrated activity of specification, design and verification where individual phases, steps and tools contribute significantly in developing cutting-edge technology and delivering sustainable solutions. Our goal is to present those phases, steps and tools with relevant examples, highlight the development challenges and argue the inevitable deviations of real-life condition from the original specification. We believe that this contribution will shed light upon the essential elements of the rotor machine development process, something that benefits the entire community from academia and suppliers to end users.

1. Introduction

This report presents an overview of the workflow of rotor machine development, based on SRM's (Svenska Rotor Maskiner) experience, including more than 500 individual machine designs. Most of these designs have been developed all the way from specification to full-scale prototype tests in SRM laboratories. We will discuss the individual design phases, steps and tool packages which are essential in developing forefront technology and viable solutions. We emphasize that state-of-the-art rotor machine development is guaranteed only by considering all these individual elements in the chain of the development process.

Rotor machines are categorized as positive displacement rotary machines which serve in many industrial applications such as refrigeration, air conditioning, process gas, chemical, automobile, airborne and waste heat recovery. These machines are important tools for engineers worldwide to find industrial solutions with low environmental impact. Even though these machines have relatively simple mechanical structure, they can tolerate high rotational speeds, wide range of operating flowrates and pressures and many different fluids and gases. After more than 80 years of continues development, rotor machine technology still has a competitive market because of the substantial improvement potential which depends on many design and manufacturing parameters from rotor and housing to bearing, lubrication, sealing and injection systems. That is why top developing companies invest significantly on optimization procedures to offer competitive high efficiency - reasonable cost products. The two most common products of rotor machines are twin-screw compressor and twin-screw expander which are the focus of the present paper.

Screw compressors are the most developed type of rotor machines which have extensively replaced other types such as reciprocating and centrifugal compressors due to their compact, simple, efficient and reliable design. Opposite to reciprocating compressors, screw compressors do not need suction or discharge valves and timing of filling, compression and discharge processes are imposed by the geometry. To meet the market needs, there exists a large variation of screw compressor designs such as dry-running or liquid-flooded, synchronized or rotor-derived, one or multiple stages, with or without capacity control and many more which reflects the broad application range and development possibilities of these machines.

Screw expander, even though with similar technology core as the screw compressor, has still a relatively low amount of market penetration in most applications. These machines offer an appropriate solution in power generation applications where opposite to the classical turbines they can operate even with phase change of working fluid within the machine. Screw expanders can be integrated in Organic Rankine Cycles (ORC) which are the baseline of decentralized small and medium-scale energy production units from low-grade waste heat sources. In near future, heat recovery thermodynamic cycles with screw expanders are going to serve a wider range of applications such as biomass CHP, geothermal power, solar power and internal combustion engines [1].

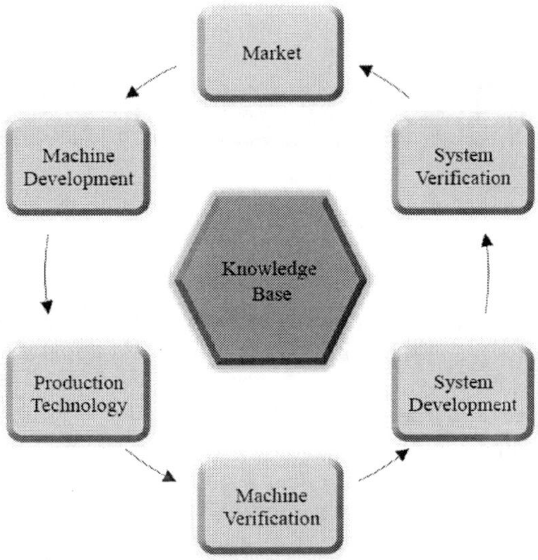

Figure 1. Workflow of rotor machine development.

In this work, following SRM development loop of rotor machines, we will discuss the individual development phases as illustrated in figure 1. These phases are: Market, Machine Development, Production Technology, Machine Verification, System Development and System Verification. The heart of the development process is the knowledge base that contributes decidedly in each phase. The knowledge base is established, over an extended period of developing various machines, facing technical challenges and assessing the final product. Knowledge base is an integrated archive of machine and system database, design and calculation manuals, production and test procedures, internally and externally developed software, patents, technical reports and many more. The arrows indicate the workflow direction of the development process as well as the order of the sections in the present paper. Note however that the connections between individual phases or steps are much more complex in real life, including many crossflow iterations and reevaluations. Despite the intricacy of rotor machine development, the work will be presented in a pedagogical fashion as our goal is to engage the readers of various professional background and interests.

2. Market

The main purpose of rotor machine technology is to meet the market needs, so the development loop should start and end here. Application range, capacity and manufacturing capability of rotor machines have been increased drastically. Commercial machines can now stand operating pressure of more than 60 bar and developing machines well over 100 bar, power consumption of oil injected screw compressors has passed 2 MW and machines can be manufactured conveniently with relatively low costs while meeting the requirement of very small tolerances. The global growth of rotor machine market poses significant opportunities, and at the same time challenges, to the top technology developing companies. The more specific demands on energy- and cost-optimized rotor machines has forced the market to pursue only highly reliable, feasible and tailored technical solutions. Therefore, it is crucial to bind strongly the technology experience, as the main input and output of the development process, with the customer specific requirements.

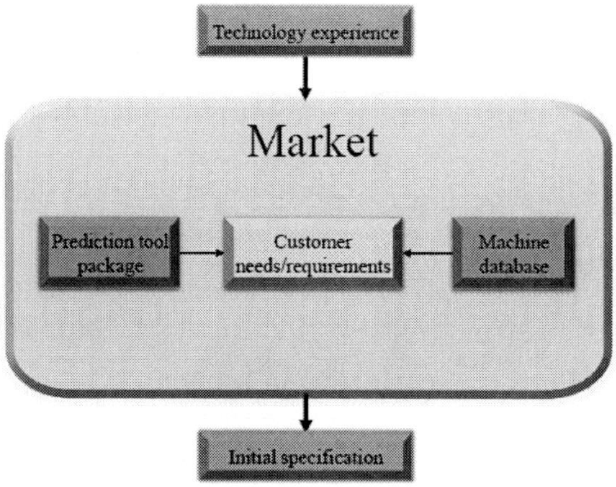

Figure 2. Workflow of market phase

The development process in the market phase is shown in figure 2. In market, technology experience combines with machine database and prediction package to meet customer needs/requirements. Machine database is a documented collection of previously developed machines which helps the market experts in understanding the possibilities and challenges of the upcoming project. The prediction tool package however contains computer programs to estimate the machine performance under different conditions. These programs are based on efficient numerical algorithms with extensive correlation with the laboratory data. We will discuss more the prediction package in connection to the simulation programs in the next section. The outcome of the market phase, initial specification, is not too detailed at the beginning, but it develops gradually through an iterative procedure between market and design experts. The challenge here is to keep the specification alive and relevant while identifying and analyzing the actual customer needs.

3. Machine Development

The machine development phase is the first technical step toward the development of a new machine based on the initial specification, see figure 3. This phase normally consists of two main steps, the initial calculation and layout and detail design. The outcome of the former is the starting point of the latter.

3.1. Initial calculation and layout

In this part the essential machine components such as rotors, ports, bearing, injection and sealing systems are designed and optimized. The heart of the rotor machine, the rotors, are designed based on an often-patented profile. Theses profiles have been constantly developed since 1973 when the first profile with commercial viability, the “A” profile, was introduced by SRM [2]. As shown in figure 3, profile tool package provides designers with necessary routines to compute and optimize the rotor geometrical properties such as lobe shape specifications, sealing line and volume curve. In addition, the desired clearances and manufacturing tolerances are taken into consideration. This task is even more challenging for dry gas compressors and screw expanders because they normally operate at higher temperature range and therefore greater thermal expansion. We note that the choice of rotor profile and other geometrical properties are finalized combining the outcome of profile, simulation and design tool packages.

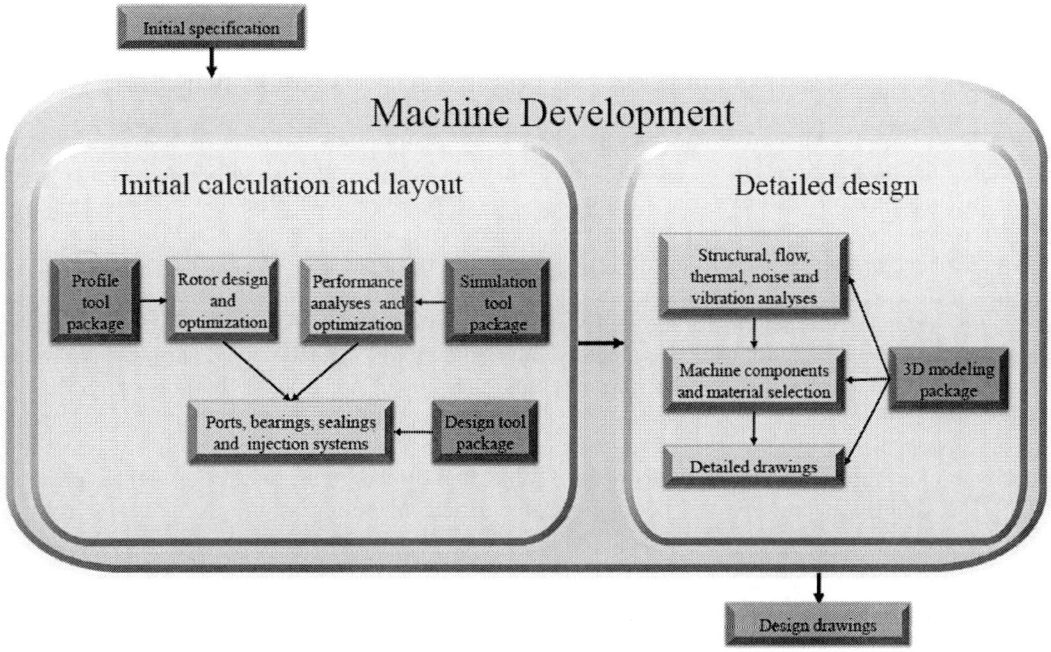

Figure 3. Workflow of machine development

In the next step, but in real life more conducted in parallel, the machine performance is estimated using the simulation tool package. The programs included in the package are developed combining mathematical models, numerical algorithms and experimental correlations. The so-called Thermodynamic Chamber Model (TCM) often is the baseline of the mathematical modeling programs where differential equations of mass and energy conservations together with the equations of states are solved, under quasi-steady assumption, to compute the variation in thermodynamic properties (such as pressure, temperature and phase quality) of working fluid as a function of rotation angle and thereafter machine performance [3]. Mathematical modeling of screw machine has started around 35 years ago where the pioneers are among others Fujiwara et al, 1972 and Sangfors, 1982 [4,5]. Thermodynamic simulation package which covers a wide range of features such as different working fluids and geometries, dry-running or fluid-flooded machine, capacity regulation, economizer and many more, is still among the most important development tool both in industry and academia (see e.g. [6,7,8]).

The initial design calculations are connected directly to the outcome of geometrical and thermodynamic calculations. Designer get assistance from specialized design programs to evaluate various components/systems within the machine (ports, bearing, injections, sealing, gears and so on). This step is fulfilled with the consideration to the experience from former solutions, requirements in specifications and the existing production technology. Calculation of each component/system is usually accompanied with a multi-variable optimization. For example, an efficient oil injection system provides optimal oil flow into rotor and bearing housing under various running conditions to assure proper cooling, sealing, lubrication and noise reduction while avoiding undesired friction losses. Another example is the port design when parameters such as energy losses, mismatch between the internal and external compression/expansion and noise level are optimized simultaneously.

3.2. Detail design

The detailed design step includes a range of 3D modelling and computations which complement the initial calculations and produce the final design drawings. The complexity level of these computations has a direct connection to the machine application. For example, a refrigeration compressor often has much more complex structure in comparison to an air compressor where features like balance piston and slide valves are also included in the design [9].

The rotor machine is a mechanical structure that ought to be analyzed from three main aspects of machine design: structural, fluid flow and vibration. Structural analyses often occur through a series of FEM (Finite Element Method) computations where stress distribution and rotor and casing distortions are among the greatest interest. In addition, FEM analysis is often coupled with a thermal analysis of the entire machine, to estimate the distribution of temperature over machine components under operation (see e.g. [10]). The fluid flow analyses, generally via Computational Fluid Dynamics (CFD), help understanding the details of the fillings, compression or expansion and discharge processes (see e.g. [11]). In addition, force distribution on rotors, friction loses, and injection process are simulated employing reliable flow analyzing tools. This information is also employed to further develop the simulation and prediction tool packages. Finally, noise and vibration analyses are performed to interest both environmental concerns and machine life time. The main noise and vibration sources can be divided into mechanical and fluid sources. The mechanical sources such as rotor rattling, or improper gearing system are the main reason of the machine vibration. The fluid sources such as gas pulsation contribute in noise which is more evident in the filling and discharge processes through the ports (see [12]). The simulation package based on the mathematical modeling of the excitement forces and pressure waves helps the designer with the optimization of, among others, transmission mechanism and port design to reduce the noise level (see [13]).

Next is the selection of machine components depending on the specification, production technology and above-mentioned analyses. Components such as bearings and sealings, also affecting the machine function and performance [14], are often provided by external distributors. These components are verified in the quality control step of the machine verification phase, see section 5. Finally, the choice of material and surface treatment are decided in this step. As regards rotors, improper material selection may lead to strain-hardening and adhesion tendency in the production process while improper surface treatment intensifies wear due to contacts or abrasive particles and therefore results in a reduction of service time especially for the rotor-derived machines (see [15] for more detail). Different rotor coating may be used not only to treat the surface but also to achieve the best possible clearance configuration between the rotors under the operation as the rotor contacts remove partly the coating materials.

The last step in machine development is the production of detailed drawings that often fulfilled employing advanced, but conventional 3D modelling package. The design drawings and technical solutions are the baseline of the machine development process. These solutions will be evaluated in the verification step and compared to the initial specification.

4. Production Technology

This phase reflects the fact that the design should be consistent to the available production technology. It is important to note that this phase may be skipped in the development process if the available production technology is suitable for the technical solution. All the potential steps in this phase are accomplished employing Computer Aided Manufacturing (CAM) systems. Here we focus only on the production technology of rotors, as shown in figure 4, which is both complex and costly. The rotor production technology usually consists of two main steps, special tool development and machining technology. The former deals with the production of grinding wheel and cutters while the latter focuses on rotor production from rounded bars through milling and grinding technology. The main challenge is to generate high quality surfaces within required tolerances in an optimal production time and cost.

In the milling technology, we either deal with special milling machines with disc cutters or relatively modern multi-functional flexible milling machines with bar, torus or finger cutters. The former is more suitable for mass production of small and medium size rotors while the latter is employed in manufacturing large rotors of small numbers or prototyping [16]. The grinding technology, also known as polishing technology, may also be fulfilled using flexible machines when high quality surface is needed. The grinding wheels often show high shape accuracy and wear tolerance which make them an excellent choice for grinding without defects like crack or burn. The outcome of this phase is the production method which is an input to the machine verification phase together with the detailed drawings.

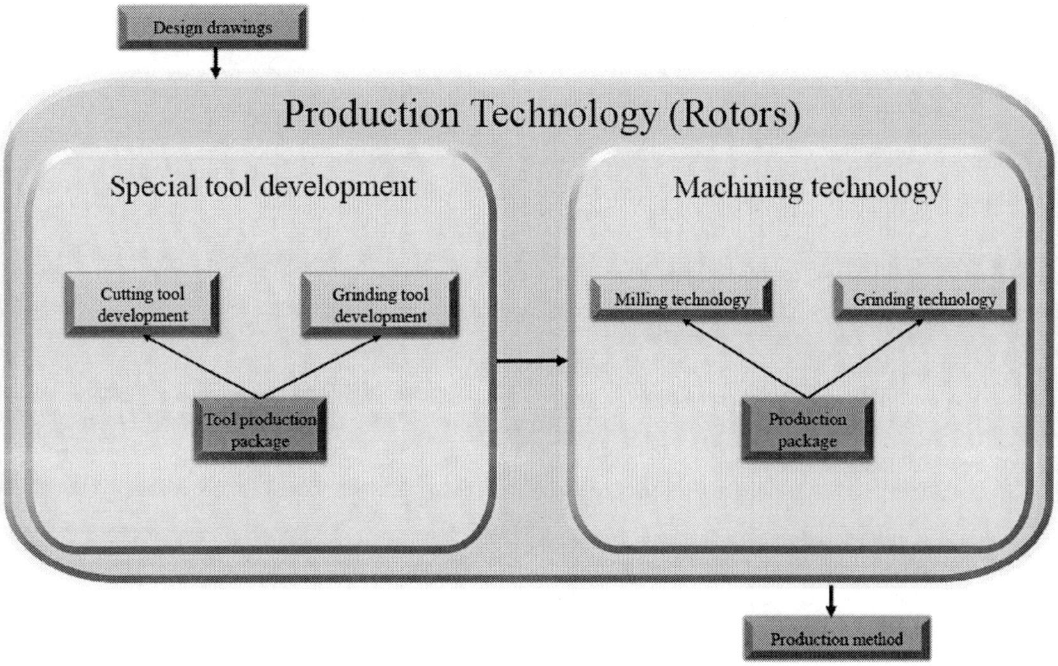

Figure 4. Workflow of production technology

5. Machine verification

The development process of rotor machine cannot be successful without verification. In real life the verification phase is closely connected to the machine development phase. The technical solution suggested in the machine development phase should now be verified and compered to the initial specification. This phase covers quality control of machine components, assembly, and function and performance tests in the laboratories. The workflow of this phase is shown in figure 5.

5.1. Quality control

To be able to verify that the function and performance is according to the specification, one must first verify that all individual machine components are according to the design drawings. The quality control step is more sophisticated for the rotors and housing. The geometrical properties of rotors are usually measured point by point employing Coordinate Measuring Machine (CMM). Rotor inspection is not limited to the individual rotor but also the rotor-pairing when the clearance between the rotors in several points is measured, using rotor-pairing measurement tools, and compared with the design protocol. This step is extremely important as the quality of inter-lobe

IOP Conf. Series: Materials Science and Engineering **425** (2018) 012005 doi:10.1088/1757-899X/425/1/012005

clearance affects significantly the machine performance, noise level and reliability. Next, dynamic balancing of rotors is accomplished using balancing machines, not shown in figure. All these inspections safeguard the high-speed rotations (sometime up to several tens of thousands RPM) of the rotors with minimum friction, rattling, buzz and distortion [17]. CMM machine is also employed for housing control to ensure all relative dimensions are correct, for example the desired clearance between the rotor tips and the housing. As mentioned before, other machine components, provided by external distributors, are quality checked in this step based on the specifications and design drawings.

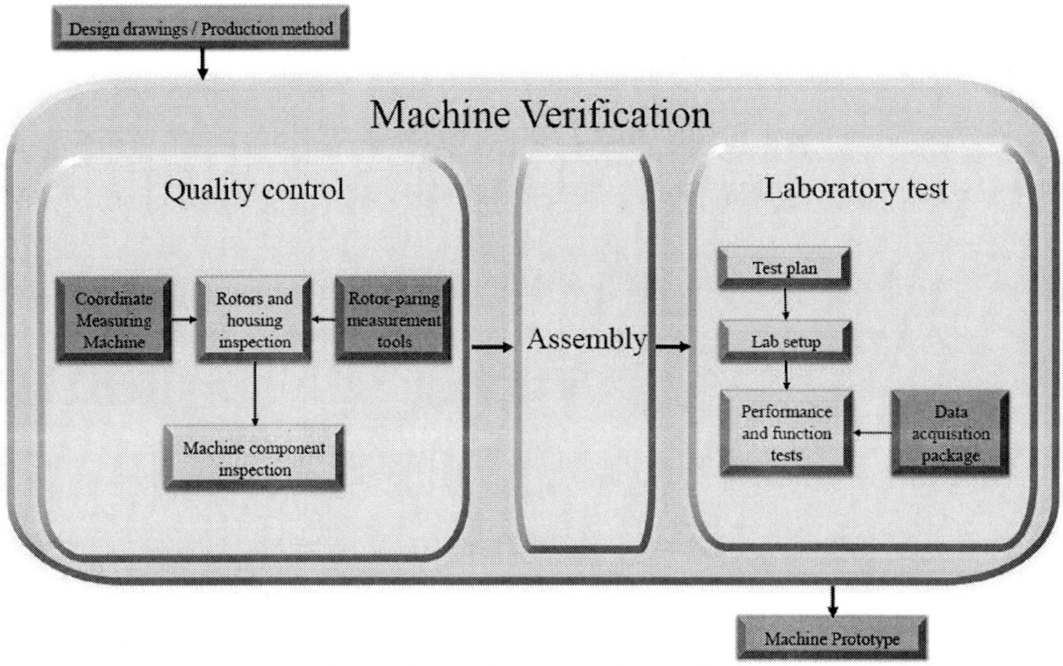

Figure 5. Workflow of machine verification

5.2. Assembly

Before the function of the machine can be verified one must also ensure that it is assembled properly. The assembly of rotor machines is a sensitive task, though is not fully automated. Therefore, the skill of assembly experts directly affects the final product. Assembly of rotor machines is performed step by step following the design drawings and includes many pre-installation procedures such as cleaning, cooling or warming machine components. The actual back-lash and end-plan clearances is reported in this step and compared to the detailed drawings. In real life it is also very common that many deviations are found during the prototype assembly, iterating back to the detail design step of the machine development phase.

5.3. Laboratory tests

The last step in machine verification is laboratory test which generally follows certain ISO standards such as ISO 917, ISO 5167-1 and ISO 1217. Prior to the experimental verification often a test plan is created to indicate a full-scale test range on machine with actual working fluid and over a wide range of running conditions, also taking into account what has previously been tested in the knowledge base. Experimental test rigs are design-specific where the system components, piping, instruments and measurement programs are adjusted to the machine application and working fluid. Data acquisition programs process the signals from the measurement devices, calibrate data and perform thermodynamic calculations such that the operator can monitor the performance of the machine in real time. Accumulated data is then analysed and documented in the final prototype report and also employed to correlate simulation/prediction programs.

6. System development

System development is the next phase in the rotor machine development process. The main concern here is to design a system that exploits the maximum potential of the rotor machine. The integration of the rotor machine in the system is essential to secure the target performance of the final product. In other word, a highly efficient machine may malfunction or poorly function in a system which does not provide the necessary conditions.

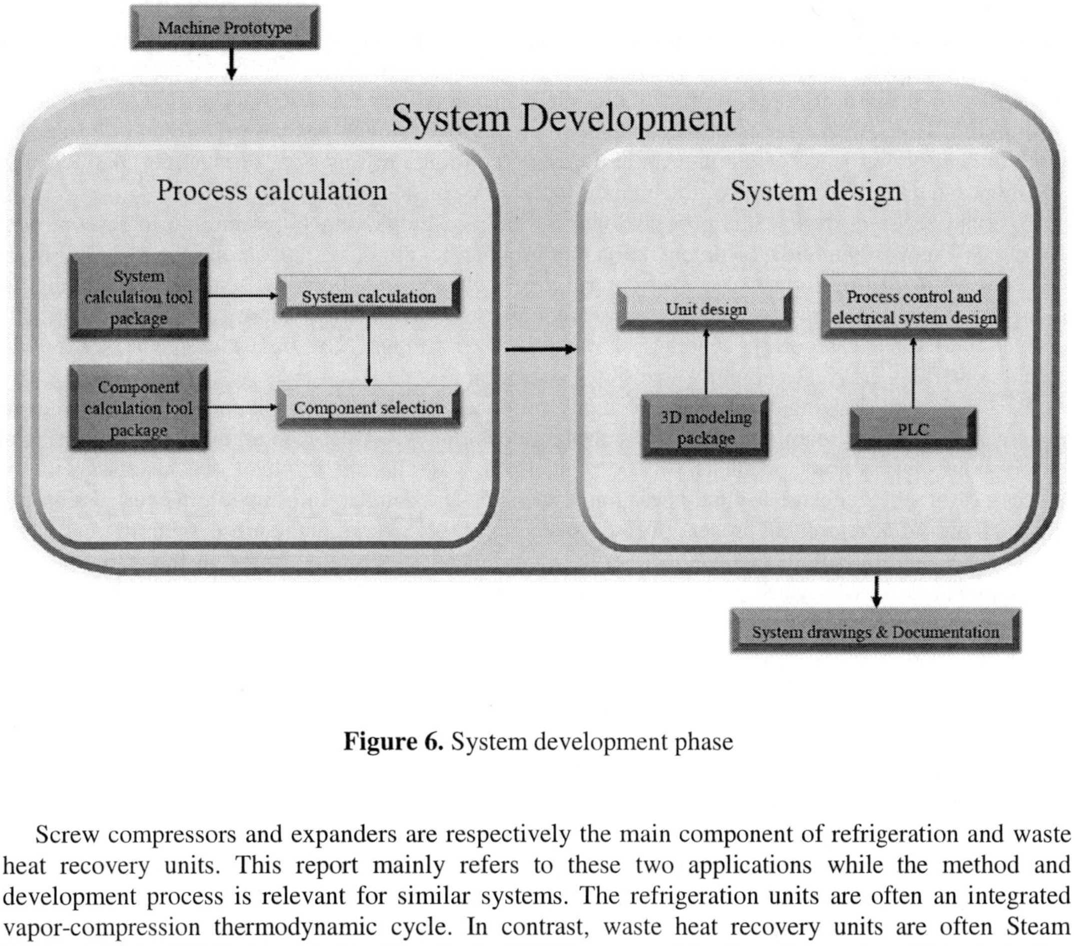

Figure 6. System development phase

Screw compressors and expanders are respectively the main component of refrigeration and waste heat recovery units. This report mainly refers to these two applications while the method and development process is relevant for similar systems. The refrigeration units are often an integrated vapor-compression thermodynamic cycle. In contrast, waste heat recovery units are often Steam Rankine Cycle (SRC), Organic Rankine Cycle (ORC) or a combination of these two. Note that other cycles such as Trilateral Flash Cycle (TFC) and Kalina Cycle (KC) have still negligible global market penetration [18]. System development consists of the process calculation and system design steps as shown in figure 6. The former deals with system calculations and component selection and the latter focuses on unit design and process control and electrical system design.

6.1. Process calculation

Process calculation often starts with conceptual design where simulation tool package is employed to predict the thermodynamic status of specific points on the cycle as well as parameters like power and flowrate under different loads. This information is then employed in basic component design using component calculation tool package. These components can be of the main cycle such as evaporator and condenser or auxiliary cycles such as oil and cooling water pumps. In selecting various components, generally a balance approach is needed to trade-off efficiency, controllability, pressure drop, size, tightness, durability and cost. As an example, heat exchanger should have high heat transfer efficiency, low pressure drops and high resistant against corrosion and leakage while at the same time reasonable price as it affects the total cost considerably. In addition, the process logic is developed in connection with piping calculations, sensor arrangement and controller sequences. Process calculation is usually reported in the form of function specification together with piping and instrumentation diagram (P&ID).

6.2. System design

System design usually occurs after the system calculation. System design can be divided into unit design and process control and electrical design. Unit design starts with the three-dimensional modelling of the entire system with actual component properties such as size, weight, material and positioning. Thereafter, a series of mechanical and structural calculations are performed to ensure the function and performance of the unit under design conditions. In this step, production, maintenance, transport and total cost are some of the main aspects to be considered.

In control system design, the process logic is finalized employing programming logic controller (PLC). CPU controller cards, communication modules and logging units are among the main hardware of PLCs while start-up sequence, power control, trip sequence and shutdown sequence are among the main logics. It is of immense importance to equip PLCs with proper alarm lists which normally divided into three categories, tripping, shutdown and warnings. The control sequences keep the machine working efficient within its window of operation.

Beside system drawings which are normally produced by three-dimensional modelling package, documentation is the main outcome of the system development. Documentation includes a series of descriptions, instructions, manuals, safety regulations and checklists which are necessary in the journey from system design to final operation at the site. The detailed information in every document, although part of development process of rotor machine, is not covered in the present report.

Here we again highlight that the system development is an essential phase in the rotor machine development process because only an appropriate and efficient system can exploit the maximum capacity of an efficient machine. The high-quality machine and system design solutions together result in the best final product.

7. System verification

The last step in the development process is system verification. This is the phase in which we evaluate whether or not the inevitable deviation of the final product from initial specification is within the acceptable range. Ideally, this phase can be done in a laboratory environment, but more often most of the work has to be conducted on the field. Similar to machine verification phase, here we start with the quality control of both internally and externally produced components, see figure 7. In the next step system is produced (assembled) based on the system drawings. The system is now ready for the test based on the protocols which are regarded as Factory Acceptance Test (FAT) and Site Acceptance Test (SAT). The sub steps here are test plan, test setup and performance and function tests which are similar to those of the machine verification. Finally, system is evaluated at the field (site) in two main sub-steps. First, system is quality controlled including regulations and safety checked following documentation. Second, data is logged to monitor system control, stability and performance. Data logging is also important to track flaws if something goes wrong. In this step data mining and analysing package creates a baseline in computing indicator parameters such as machine and system efficiency. The collected data is further employed to develop correlation functions and prediction models and optimize future machines/systems.

We finally note that the most important output of this phase is the technology experience. Rotor machine development from initial specification to system verification has now completed one round where technology and experience are developed and integrated in the knowledge base. At this stage, usually the initial specification has to be updated because the real-life product may have other requirements comparing to what was originally specified.

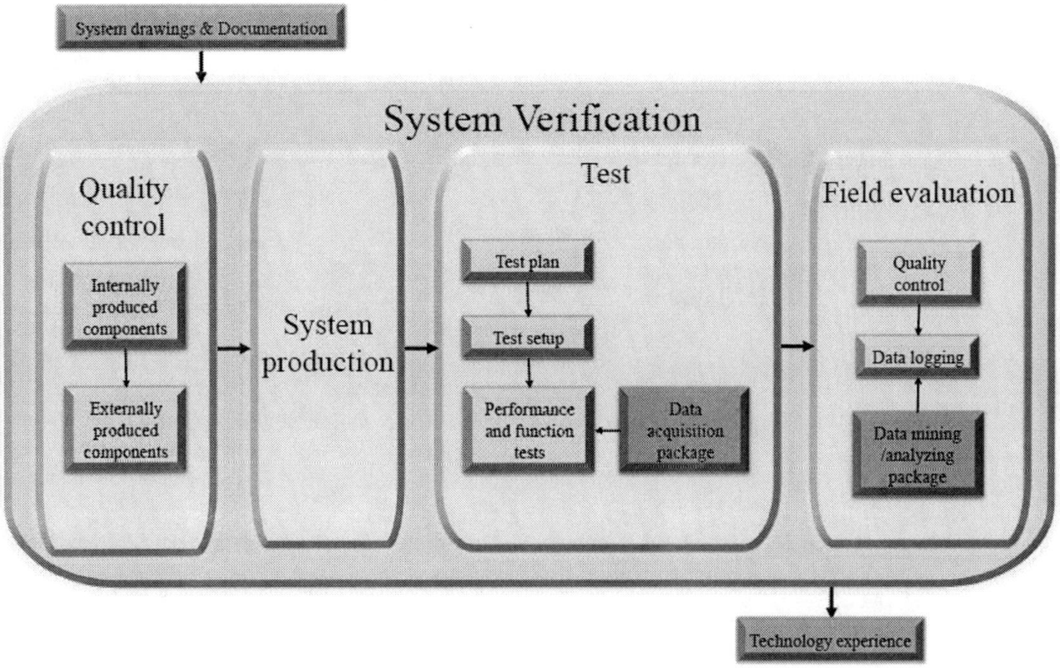

Figure 7. Workflow of system verification

8. Conclusion

The workflow of rotor machine development is an orchestrated activity of many phases and steps where every detail counts; the manipulation of each phase, step or even sub-step may lead to distinctive improvement. Following SRM development loop, the workflow is divided into 6 phases, Market, Machine Development, Production Technology, Machine Verification, System Development and System Verification. In this report, steps and tools under each phase has been discussed thoroughly together with examples and challenges. In figure 8, we show the detailed workflow of rotor machine development by merging all the figures presented in this report. We again note that the production technology may be skipped if the available production method is suitable for the technical solution.

We finally note that even though the connections between individual phases or steps are much more complex in real life, including many crossflow iterations and reevaluations, we pursued a simple method to describe the work for readers of various professional background and interests. We emphasize again that the success in the development process lies on looking up to the individual details, innovations, cross-team collaboration, iterations and optimizations alongside the development loop. The chain of the rotor machine development is of course not stronger than its weakest link.

IOP Conf. Series: Materials Science and Engineering **425** (2018) 012005 doi:10.1088/1757-899X/425/1/012005

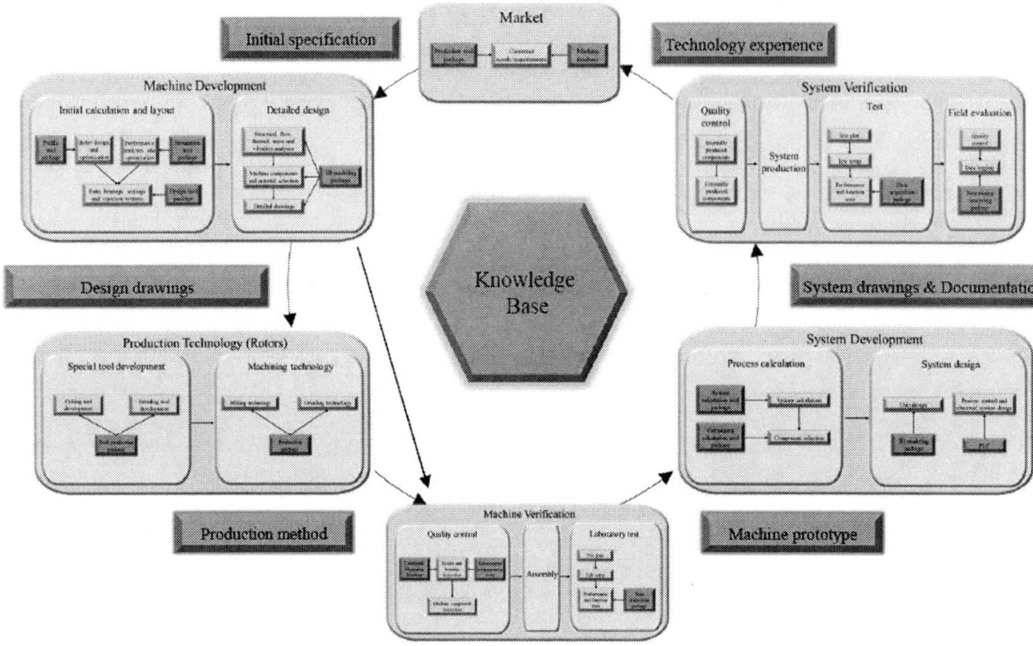

Figure 8. Detailed workflow of rotor machine development

References

[1] Quoilin S, Van Den Broek M, Declaye S, Dewallef P and Lemort V 2013 Techno-economic survey of organic rankine cycle (ORC) systems *Renewable and Sustainable Energy Reviews* **22** 168–186

[2] Shibbie 1979, *US Patent* **4,140,445**.

[3] Stosic N, Smith I K, Kovacevic A and Mujic E 2011 Review of mathematical models in performance calculation of screw compressors *Inter. J. Fluid Machiney and Systems* **4** 2

[4] Fujiwara M, Mori H and Suwama T 1974 Prediction of the oil free screw compressor performance using digital computer *Inter. Compressor Engineering Conf. at Purdue* **186**

[5] Sangfors B 1982 Analytical modeling of helical screw machine for analysis and performance prediction *Inter. Compressor Engineering Conf. at Purdue* **144**

[6] Vasuthevan H and Bruemmer A 2016 Thermodynamic modeling of screw expander in trilateral flash cycle *Inter. Compressor Engineering Conf. at Purdue* **2475**

[7] Bianchi G, Kennedy S, Zaher O, Tassou S A, Miller J and Jouhara H 2017 Two-phase chamber modeling of a twin-screw expander for trilateral flash cycle applications *Energy Procedia* **129** 347–354

[8] Nikolov A, Bruemmer A 2017 Investigating a small oil-flooded twin-screw expander for waste-heat utilisation in organic Rankine cycle systems *Energies* **10** 869

[9] Hou F, Zhao Z, Zhiqiang Y and Xing Z 2017 Experimental study of the axial force on rotors in a twin-screw refrigeration compressor *Inter. J. Refrigeration* **75** 155–163

[10] Powell G, Weathers B and Sauls J 2006 Transient thermal analysis of screw compressors, Part I-III *Inter. Compressor Engineering Conf. at Purdue* **1811-13**

[11] Andres R, Hesse J, Babic H, Salecker U, Spille-Kohoff A, Nikolov A and Brummer A 2016 CFD simulation of a twin screw expander inclusing leakage flows *Inter. Compressor Engineering Conf. at Purdue* **2497**

[12] Mujic E, Kovacevic A, Stosic N and Smith I K 2011 Noise generation and suppression in twin-screw compressors *Proc. of the Inst. of Mech. Eng. Part E* **225** 127–148

[13] Sangfors B 2000 Modeling, measurment and analysis of gas-flow generated noise from twin-screw compressors *Inter. Compressor Engineering Conf. at Purdue* **1487**

[14] Meyers K 1997 Creating the right environment for compressor bearings *Inter. J. Evolution SKF* **4** 21

[15] Muiznieks G, Gerins E, Katasevs A, Ozolins J and Pikurs G 2012 Evaluation of coating material structure of screw type compressor rotor contact surface *Proc. World Congress on Engineering* **III**

[16] Zabel A 2014 Modern machining processes for the manufacturing of screw machine components *Inter. Confrenece on Screw Machines* **2228** 267–278

[17] Holmes C S 2004 Inspection of screw compressor rotors for the prediction of performance, reliability and noise *Inter. Compressor Engineering Conf. at Purdue* **1692**

[18] Öhman H 2015 Low temperature difference power systems and implications of multi-phase screw expanders in organic Rankine cycles *Trita-Pefr Raport 15/02, KTH, Stockholm, Sweden* ISBN 978–91–7595–827–9

---

§ In this work we decide to use the term "rotor machine" instead of "screw machine" as the former is employed historically in SRM development process. In addition, in the context of development process, we believe that "rotor machine" is the better match.

An analysis of manufacturing factors' influences on the actual screw compressor rotors' profile clearances

T N Mustafin[1], R R Yakupov[1], M S Khamidullin[1], I G Khisameev[1], V A Alyaev[1], E R Ibragimov[2]

[1]Kazan National Research Tehnological University, Russian Federation;

[2]HMS Group NIIturbokompressor, Russian Federation

Email: mustimur@rambler.ru

Abstract. One of the most important parameters, which have high influence on screw compressor characteristics, are rotors' profile clearances. They determine in general the smooth of the rotors' gearing and influence on the screw compressor vibration level. Analyse of actual rotor profile clearances depending on the actual rotor temperature fields and on the manufacturing imperfection was done. Specially profiled milling tools were chosen as rotor profile manufacturing tools. Following four types of the manufacturing error, which have the highest influence on the accuracy and quality of rotors' profile surface, were chosen to analyses: the centre distance deviation between the rotor and the milling tool axises (dAi), the departure of the milling tool position from their basic plate (dZf), the milling tool mounting position angle departure (dβ) and the milling tool radii departure (dRf).The actual screw compressor profile coordinates can be obtained by taken into account all factors presented above. They determine the actual rotor profile clearances on "cold" state when screw compressor is on shutdown conditions and can be measured during a compressor assembling or a compressor stripping. However, it should be noted that they are changed depending on the screw compressor working conditions. To correctly determining actual profile clearances depending on working condition also needs to take into account rotors' thermal deformation. The presented multifaceted approach of the screw compressor actual rotor profile clearances calculation improves an accuracy compressor characteristic obtained by their mathematical model. A deep analysis of the obtained results can pointed to the rotors' manufacturing factors, which should be improved firstly during the compressors' mass production.

1. Introduction

At present, the rotary compressors have become more widespread in the area of small and medium-size unit capacities. The profile surfaces of rotors made with high accuracy and determining the main power characteristics of the compressor are the main working element of these compressors. The optimization of both the theoretical and actual rotor profiles is one of the major directions in the development of rotary compressors. At the same time, for this optimization, the actual operating clearances in the compressor should be known and, consequently, the actual position of the rotors' profile surfaces relative to the casing and to each other should also be known.

It has to be noted that the main method of surface preparation is milling and, therefore, it is especially important to assign tolerances to both disk mills and to their settings in the processing machine. At present, qualitative analysis of these errors influence is provided in the studies [1, 2, 3, 4]. They mainly analyze the resulting profile of rotors, but their mutual meshing is actually not analyzed. The purpose of the study is to fill this gap and justify the choice of tolerances.

2. Mathematical Description of meshing clearances

As a rule, the coordinates of the rotor actual profile are obtained by derating the coordinates of the theoretical profile or calculating based on actual milling tool profile coordinates. For the convenience of the coordinates description and subsequent use, it is rational to present their nominal values in a parametric form via the radius vector:

$$\vec{r}_{Mn} = r_{Mn}\left(X_{Mn}\left(t_{Mn};\theta_1\right);Y_{Mn}\left(t_{Mn};\theta_1\right);\, p_{Mn}\cdot\theta_1\right) = r_{Mn}\left(t_{Mn};\theta_1\right), \tag{1}$$

$$\vec{r}_{Gn} = r_{Gn}\left(X_{Gn}\left(t_{Gn};\theta_2\right);Y_{Gn}\left(t_{Gn};\theta_2\right);\, p_{Gn}\cdot\theta_2\right) = r_{Gn}\left(t_{Gn};\theta_2\right) \tag{2}$$

where X, Y are the profile coordinates; t – the parameter of the profile shape; θ – the angle of the rotor rotation; p – the screw parameter; the "Mn" index and index "1" mean that the parameter refers to the male rotor; the "Gn" index and index "2" mean that the parameter refers to the free rotor. The required initial coordinates of the rotor profile for subsequent analysis are calculated on the basis of the equation [4] of the shaping by a disk mill:

$$\begin{cases} X_{Mn(Gn)} = \left(Rf_{Mn(Gn)}\cdot\cos\gamma_{Mn(Gn)} + A_{i_{Mn(Gn)}}\right)\cdot\cos\left(\theta_{1(2)}\right) - \left(Zf_{Mn(Gn)}\cdot\sin\beta_{Mn(Gn)} - Rf_{Mn(Gn)}\cdot\sin\gamma_{Mn(Gn)}\cdot\cos\beta_{Mn(Gn)}\right)\cdot\sin\left(\theta_{1(2)}\right) \\ Y_{Mn(Gn)} = -\left(Rf_{Mn(Gn)}\cdot\cos\gamma_{Mn(Gn)} + A_{i_{Mn(Gn)}}\right)\cdot\sin\left(\theta_{1(2)}\right) + \left(Zf_{Mn(Gn)}\cdot\sin\beta_{Mn(Gn)} - Rf_{Mn(Gn)}\cdot\sin\gamma_{Mn(Gn)}\cdot\cos\beta_{Mn(Gn)}\right)\cdot\cos\left(\theta_{1(2)}\right) \\ \theta_{1(2)} = -\dfrac{Rf_{Mn(Gn)}\cdot\sin\gamma_{Mn(Gn)}\cdot\sin\beta_{Mn(Gn)} + Zf_{Mn(Gn)}\cdot\cos\beta_{Mn(Gn)}}{p_{Mn(Gn)}}, \end{cases} \tag{3}$$

where Rf and Zf are the milling tool profile coordinates (Fig. 1), β – the angle between the milling axis and the rotor axis, A_i – the distance between the milling tool axis and the rotor axis, γ - the milling tool rotation angle around its axis.

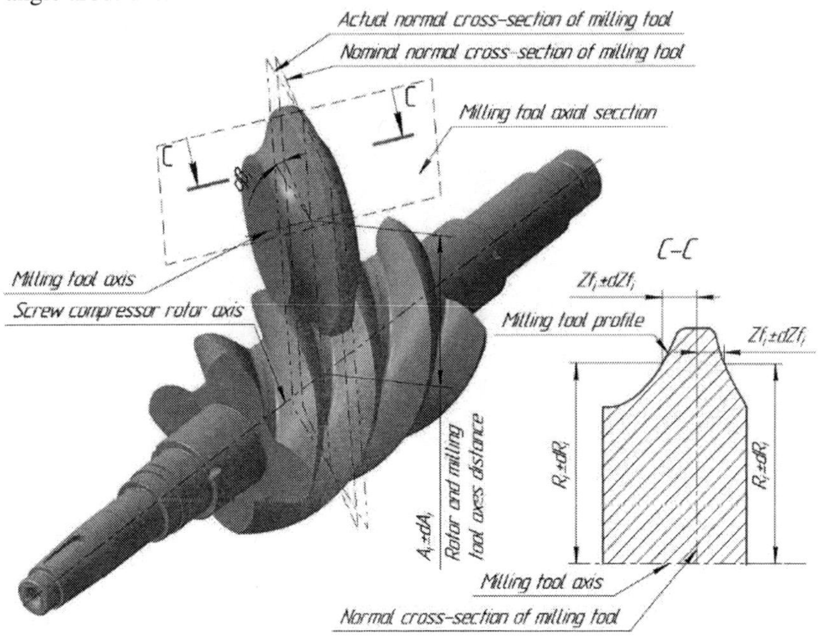

Figure 1. Schematic of determining the impact of major manufacturing deviations.

The free rotor position is determined by the extreme positions when the rotors contact one of the sides, or by some intermediate positions when moving to a particular extreme position. To determine the positions, it is necessary to determine the frameworks within which the free rotor can rotate till contacting the female rotor [5, 6]. Let's consider a similar approach in the example of determining the female rotor correction angle till contacting the free rotor. To do this, let's consider the position of the rotors when the male rotor is turned at an angle θ_1 (Fig. 2), while the female rotor, respectively, must be nominally turned at an angle of:

$$\theta_2 = \theta_1 \frac{z_1}{z_2} \tag{4}$$

where z_1, z_2 – the number of teeth of the male and female rotors. The schematic shown in Figure 1 is given for the nominal positions of the rotor centers and does not require a fundamental change to account for their movement in the presented procedure.

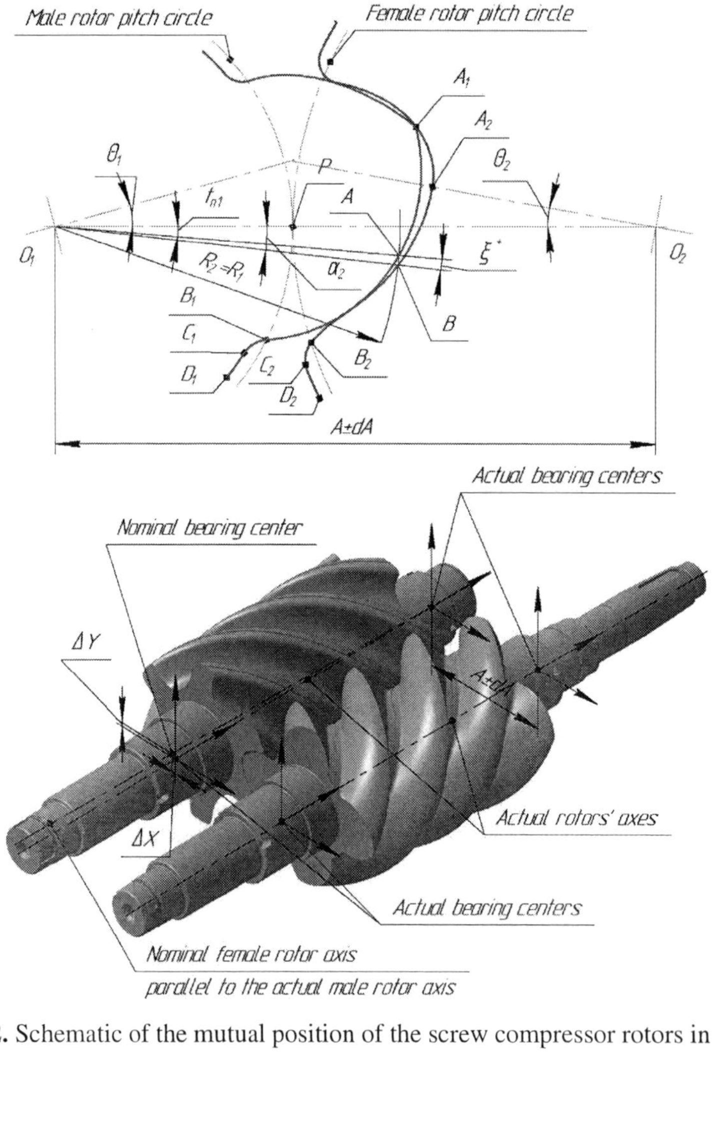

Figure 2. Schematic of the mutual position of the screw compressor rotors in meshing.

Let us choose an arbitrary point A on the section A_1D_1 with angular coordinate α_1. At the front of the female rotor profile (section A_2D_2) rotated by an angle θ_2, we find the point B, the radial coordinate R_2 of which is equal to R_1. The difference of the angular coordinates β of A and B points gives the value of the male rotor correction angle to the touching of the female rotor profile by point A

$$\beta = \alpha_2 - t_{n1} . \tag{5}$$

It is worth noting that:

$$R_2 - R_1 = 0 . \tag{6}$$

Thereby the radial coordinate of point A is determined by the formula:

$$R_1 = S_1(t_{n1}, \theta_1) , \tag{7}$$

where S_1 is the function describing the male rotor profile. And the dependence of point B for determining the coordinates is as follows.

$$R_2 = f\left[S_2(\alpha_2), \theta_2\right], \tag{8}$$

where S_2 is the function describing the female rotor profile. By solving equations (5) – (8) together, we obtain the dependence

$$\xi^+ = f(t_{n1}, \theta_1) , \tag{9}$$

The solution to these equations is found by numerical methods.

Having found the minimum of the function (9) with respect to the variable α_1, we obtain the value of the male rotor correction angle before it touches the profile of the female rotor for the considered position:

$$\xi^+_{\min\alpha} = f(\theta_1)., \tag{10}$$

For the entire range of the correction angle θ_1 from dependence (10), within the limits of the contact line along the front side of the rotor tooth profile, we determine the minimum correction angle of the male rotor ξ^+_{min}.

Thus, when calculating the total profile clearance, the male rotor will be additionally turned at an angle ξ_{min}, emitting the contact of the rotors along the front side of the profile.

The total meshing clearance will be determined on the back side of the profiles. For this it is necessary to find the distance between the intersection points of the profiles with the straight line passing through the contact point and the meshing pole.

So for an arbitrary point A (x_1, y_1) (Figure 2) chosen on the drive profile, the equation of the straight line passing through point A and the meshing pole P (x_0, y_0) is as follows

$$y = f\left[S_1(\alpha_1), \theta_1, \xi_{\min}, x\right]. \tag{11}$$

Let us find point B (x_2, y_2) as an intersection of the straight line described by formula (11) with the profile of the female rotor.

For this, it is necessary to solve jointly the equations (4), (8), and (11) and determine the distance between points A and B

$$\delta = \sqrt{(x_2 - x_1)^2 - (y_2 - y_1)^2} . \tag{12}$$

Solving equation (12), we obtain the dependence

$$\delta = f(\alpha_1, \theta_1) . \tag{13}$$

The initial data for the analysis are the results of calculating the screw compressor profile by taking into account the errors in cutting the profile part of the rotors.

3. Calculation Results

For the calculations, four types of errors in the technological system were chosen, which had a significant impact on the made screw surfaces of the rotor: dA_i – deviation of the distance between the

IOP Conf. Series: Materials Science and Engineering **425** (2018) 012006 doi:10.1088/1757-899X/425/1/012006

rotor axes and tool, dZf_i – deviation of the tool position relative to reference plane, $d\beta_i$ – deviation of the tool setting angle, and dRf_i – deviation of the tool radius. For making the calculations, the profiles of a screw compressor with a 200 mm external diameter and SKBK [3, 5, 6, 7, 8] profile were chosen.

A typical graph of the total profile clearance variation from the male rotor rotation angle is shown in Figure 3a, and the clearance size in relation to the contact line on the male rotor surface is shown in Figure 3b.

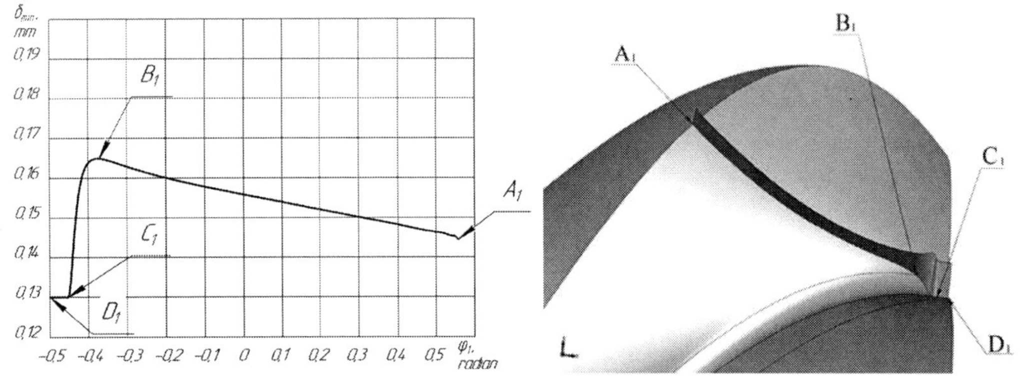

Figure 3. Clearance in the meshing of rotors.

The first section C_1D_1 is determined by the cylindrical sections in the male rotor convex and female rotor tip. The clearance at this section is determined by the radial lowering of the male rotor.

After that, the dependence curve has a sharp spike of the clearance variation at the C_1B_1 section; this is explained by the fact that in the accepted method of determining the clearance, the side clearance is defined as a sum of the clearances on the front and rear faces of the rotors. A change from radial to double side lowering occurs at this section.

The clearance value at point B_1 is determined by the total side lowering. Section B_1A_1 is determined by an algorithm of the lowering of a small section of the female rotor below the initial circle. The results of calculating the impact of the position deviation and the cutting tool profile on the resulting meshing clearance are shown in Figures 4 – 7.

From the above graphs it is apparent that the influence of various technological system errors on the profile clearance change is not the same. In general, a common similarity of the influence of the technological errors of the cutting tool setting on the resulting meshing clearance, when cutting the female and male rotors of the screw compressor, should be noted. At the same time, the influence of the tool profile deviations dRf_i and dZf_i is significantly different for the cutting of the male and female rotor. These data should be taken into account when assigning tolerances in the technology of profile cutting with the help of profile milling cutters. Also it should be noted that due to some rotors' manufacture errors contact between rotors occurs at the point A_1. In this case the profile clearance value dependence on the rotation angle is changed dramatically. It explains some appear steps in the diagrams (Figures 4-7)

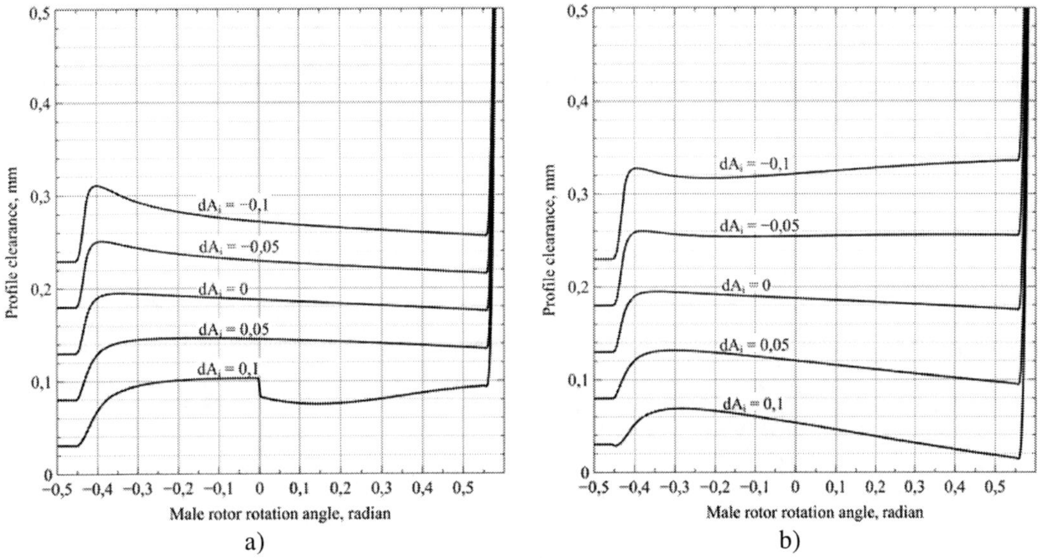

Figure 4. Impact of the distance deviation between the rotor and tool axes dA$_i$ on the profile clearance when cutting the rotors:
a) female rotor cutting;
b) male rotor cutting

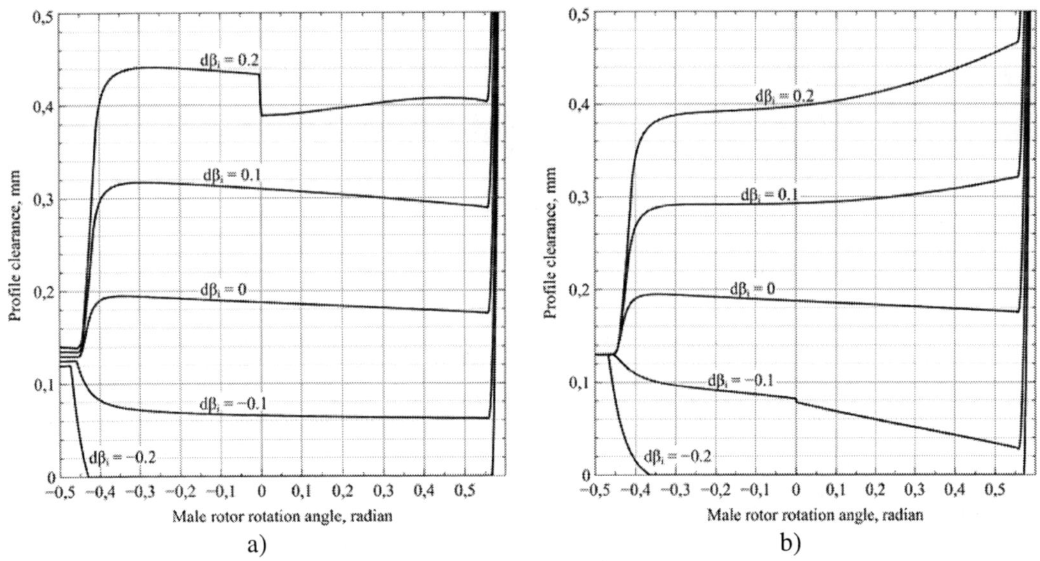

Figure 5. Impact of the tool setting angle deviation dβ$_i$ on the profile clearance when cutting the rotors:
a) female rotor cutting;
b) male rotor cutting

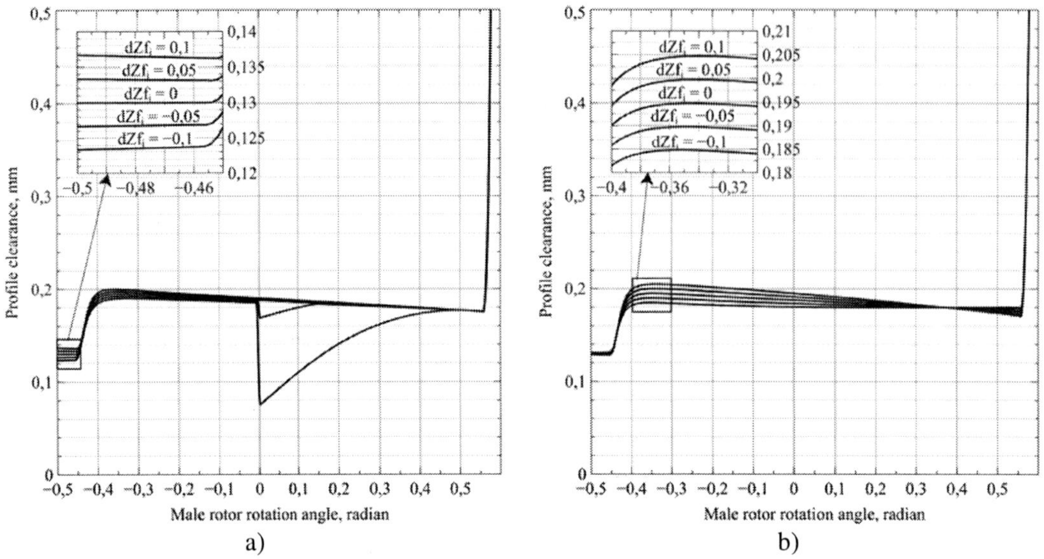

Figure 6. Impact of the tool profile deviation dZf_i on the profile clearance when cutting the rotors:
a) female rotor cutting;
b) male rotor cutting

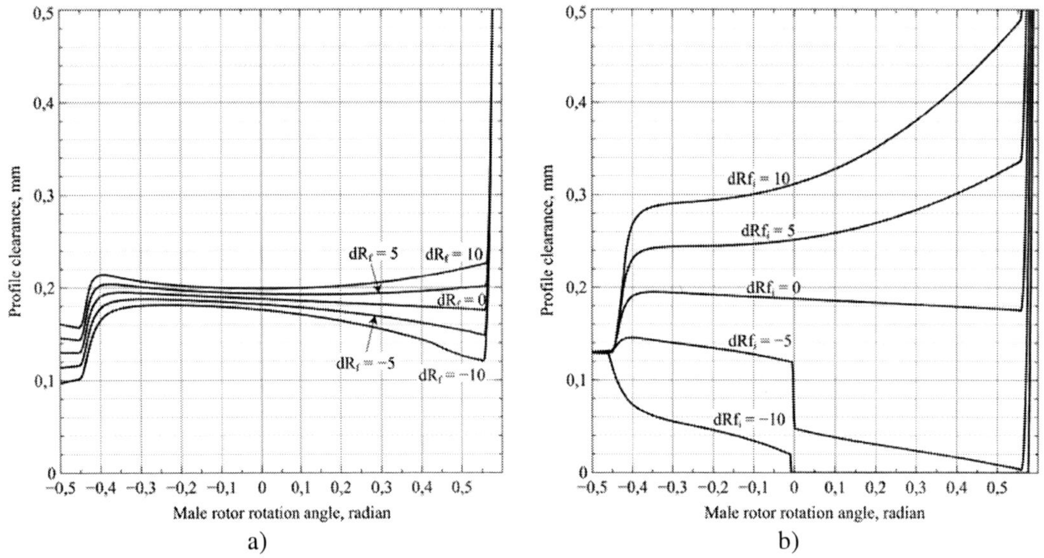

Figure 7. Impact of the tool profile deviation dRf_i on the profile clearance when cutting the rotors:
a) female rotor cutting;
b) male rotor cutting

The profile clearance in the meshing of rotors is also heavily influenced by the error in the screw compressor casing parts manufacturing, namely the error in the position of the bearing supports. For making these calculations, three types of error in the manufacture of casing parts were chosen: deviation (tolerance zone) of the center-to-center distance between the rotors dA, parallel alignment of the rotor axes ΔX, misalignment of the rotor axes ΔY (Figure 2). Their effect on the meshing profile clearance for the same pair of rotors, for which the previous calculations were performed, is shown in Figures 8-9.

IOP Conf. Series: Materials Science and Engineering **425** (2018) 012006 doi:10.1088/1757-899X/425/1/012006

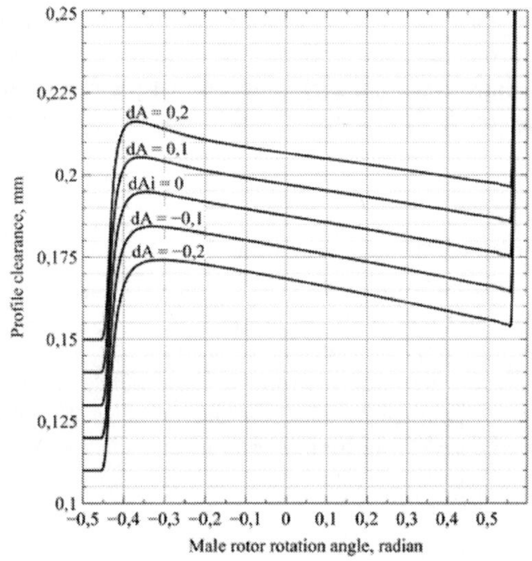

Figure 8. Influence of the center-to-center distance deviation between the rotors *dA* on the profile clearance.

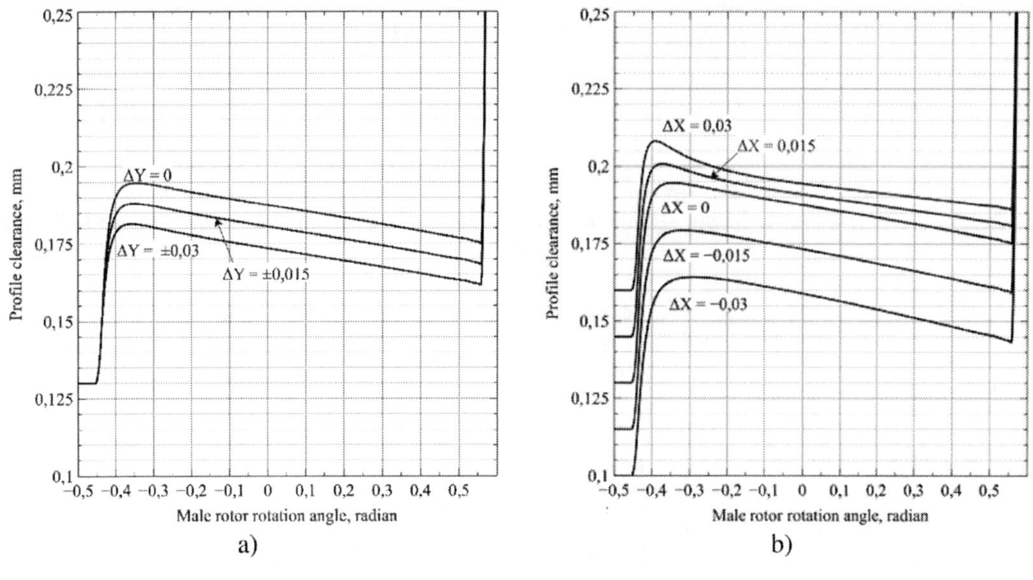

Figure 9. The impact of the rotor axes position deviation on the profile clearance:
a) misalignment of axes ΔY;
b) deviation of axes parallel alignment ΔX

The analysis of the dependences obtained allows us to draw the following conclusions about the influence of errors in the manufacture of casing parts on the meshing profile clearance:

1. The deviation of the center-to-center distance results in a proportional even change in the profile clearance. With the maximum deviation of the center-to-center distance, the change in the profile clearance is 15%.

2. The misalignment of rotors' axes results in a change of the lateral profile clearance and does not cause a change in the radial clearance. With a maximum skew of the rotors' axes, the deviation of the profile clearance is 8%.

3. It can be seen from the graph that at a negative deviation from the parallel alignment of the rotors' axes, the change in the profile clearance is similar to the changes in the deviation of the center-to-center distance. The positive deviations in the parallel alignment of the rotors' axes results in a change of the radial clearance, but the effect on the lateral clearances is much less. The maximum clearances deviation is 23%.

4. Conclusion

The analysis of the results obtained allows to justify and optimize the assigned technological tolerances for both the rotors and the casing parts of screw compressors. The profile clearances obtained as a result of calculation, taking into account their correction with respect to the temperature deformation at the compressor operating conditions [7], can be subsequently used for mathematical simulation of the thermodynamic processes and compressor performance data. This, in turn, will allow us to estimate the design ranges of the compressor characteristics at the design stage, by taking into account not only the nominal dimensions but also the set technological tolerances for the of compressor parts dimensions.

Acknowledgments

This study is a part of scientific investigations supported and funded by V.B. Shnepp NIIturbokompressor (HMS GROUP) CJSC.

References

[1] Stosic N, Smith I K, Kovacevic A and Mujic E, Geometry of screw compressor rotors and their tools 2011 *Journal of Zhejiang University: Science* A, **12 (4)**, pp 310–326

[2] Sauls J, Application of Manufacturing Simulation for Screw Compressor Rotors 2000 *International Compressor Engineering Conference* Paper 1475

[3] Stosic N, Smith I K and Kovacevic A 2005 *Screw Compressors: Mathematical Modeling and Performance Calculation* (Springer Verlag, Berlin, Germany)

[4] Abdreev M M 2005 *Povyishenie effektivnosti izgotovleniya rabochih poverhnostey rotorov vintovyih kompressorov diskovyim instrumentom (The efficiency improvement of a screw compressor rotor working surface manufacturing by cutting with milling tools)* (PhD Thesis), Kazan (in Russian)

[5] Yakupov R R, Mustafin T N, Nalimov V N, Khamidullin M S and Khisameev I G 2013 Discussion of actual profile clearances' calculation method in rotary compressors in the absence of rotor timing units *8th Int. Conf. on Compressors and their Systems (London: City University)* 209

[6] Yakupov R R, Mustafin T N, Nalimov V N, Khamidullin M S and Khisameev I G 2014 Analysis of transmission error depending on compressor working conditions *J of Proc. Mechanical Engineering (Proc. of the Institution of Mechanical Engineers) Part E (London: SAGE Publications)*

[7] Amosov P E, Bobrikov N I, Svarc I I and Vernii A L 1977 *Vintovie Kompresornie Mashinii Spravochnik (Screw Compression Machines-Handbook)* (Mashinstroienie, Leningrad) (in Russian)

[8] Mustafin T N, Yakupov R R, BurmistrovA V, Khamidullin M S, Khisameev I G 2015 Analysis of the screw compressor rotors' non-uniform thermal field effect on transmission error *IOP Conf. Ser.: Mater. Sci. Eng.* 90

IOP Conf. Series: Materials Science and Engineering **425** (2018) 012007 doi:10.1088/1757-899X/425/1/012007

An innovative rotor milling method for flexible multi-functional machines

A Bergström

Svenska Rotor Maskiner (SRM), Fannys Väg 3, 131 54 Nacka, Stockholm, Sweden

Email: andre.bergstrom@rotor.se

Abstract. In this work we present an innovative method of rotor milling for flexible multi-functional machines. As an alternative to the disc cutters which are often employed in traditional milling machines, we propose a new tool, a finger cutter, which integrates easily with flexible multi-functional machines. This allows much smoother operation with lower required torque which is in line with the characteristics of flexible machines. This paper describes basic geometrical condition to generate the finger cutter profile from the rotor profile, the mathematical formulation, the basic tool design and grinding procedure of the cutter blades. It also discusses the machining process, productivity and accuracy of the proposed method. We believe that this method opens a new way of manufacturing large rotors with fewer production steps while maintaining required tolerances.

1. Introduction

Screw machines are categorised as positive displacement machines where their functionality is directly connected to the rotation of helical-shaped rotor-pairs in a casing, i.e. suction, compression/expansion and discharge processes are entirely influenced by the geometrical properties of the rotors. Therefore, high quality rotor manufacturing is essential in producing high performance and reliable screw machines. In this work we will discuss an innovative method in manufacturing process of rotors which gives an opportunity to use standard machines with fewer production steps while maintaining acceptable tolerances.

Screw machine rotors are traditionally manufactured employing specialized milling machines with disc cutters. In figure 1, we display a milling operation with this kind of machine where the cutter is rigidly supported on both sides. A disc cutter rotates around its axis which is located outside the rotor body while the cutter blade removes the material from inside of the rotor body. Principally, the whole profile is cut with the same blade (can also be divided into smaller blades). The specialized machines with disc cutter have the advantage of high productivity milling, however, they require high torque due to the large cutter diameter, usually in the same order of magnitude as the rotor diameter. This often induces high vibration and leads to the selection of a larger/stronger milling machine.

Lately, rotor manufacturers tend to employ more flexible multi-functional machines in their production line where fewer production steps is preferred. Multi-functional machines have less maximum torque and lower stability in comparison to traditional milling machines therefore tools with smaller rotating diameters can be of significant help. In this work, we will introduce such a tool, discuss the principle of tool design and argue productivity and accuracy of the proposed method.

Figure 1. Milling a screw rotor with a traditional milling machine using a rough disc cutter equipped with sectional blades (standard inserts)

2. Methodology

Considering the abovementioned shortcoming of multi-functional machines, in this work we introduce newly developed cutters so-called Finger Cutters (FC). FC rotates around its axis which is located inside the rotor body unlike the traditional disc cutter which has its axis outside the rotor body. This results in a significant reduction in the required torque and consequently lower vibration. The schematic of cutter placement inside the rotor is shown in figure 2.

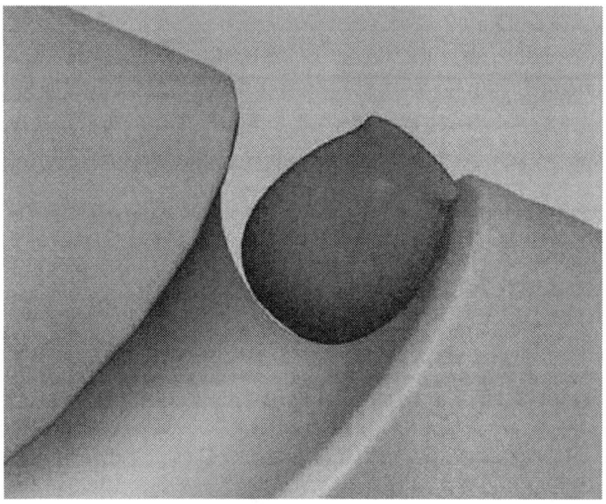

Figure 2. FC rotates around its axis which is located inside the rotor body

Combining a finger cutter with a multi-functional machine, we provide a smooth milling operation where the cutter only needs to be supported from one side due to the lower applied torque and force. The finger cutter in operation is shown in figure 3.

Figure 3. A multi-functional machine using a finger cutter to cut a screw compressor rotor.

Note that an alternative to the presented method is the method of free formed surface machining [1], which has the advantage of higher flexibility and the possibility to use standard tools, but with the disadvantage of much lower productivity.

3. The principal of cutter profile design

The cutter profile is generated when the three-dimensional normal vectors of the rotor profile align with the rotational axis of the cutter as shown in figure 4.

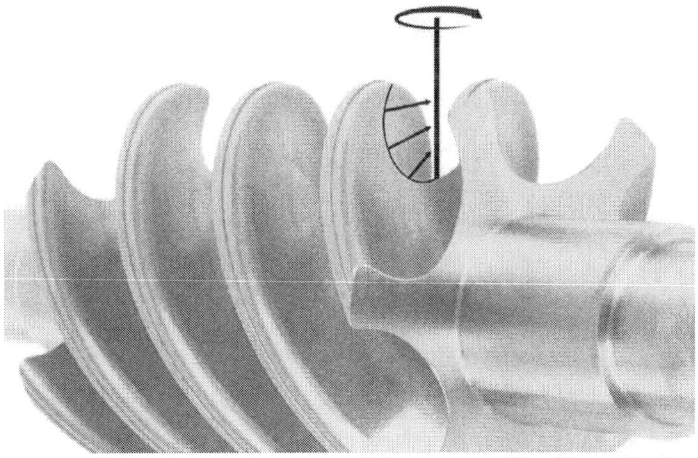

Figure 4. The three-dimensional normal vectors of the rotor profile align with the rotational axis of the cutter

3.1. Theoretical description

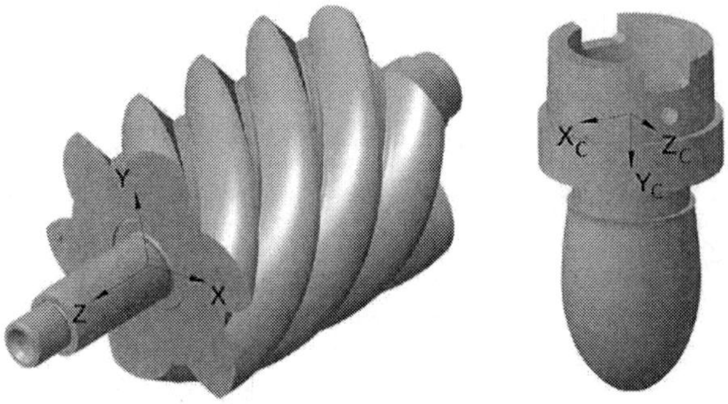

Figure 5. Left: Rotor coordinate system, Right: Finger Cutter coordinate system

The mathematical description of cutter profile is fundamental in the tool design process where the goal is to transform the rotor body coordinates into tool coordinates. The complication of mathematical descriptions is often reflected in the complexity of the tool design (see for example [2]).

Here we describe the mathematics needed to design the cutter profile. It is based on the transformation between a point on the rotor profile into the corresponding point on the cutter. Let's $\boldsymbol{R}_x(\alpha)$, $\boldsymbol{R}_y(\alpha)$ and $\boldsymbol{R}_z(\alpha)$ denote rotation matrixes, 3×3, of angle α around the x-, y- and z-axis respectively (see figure 5). Function $\boldsymbol{O}: \mathbb{R} \times \mathbb{R}^3 \to \mathbb{R}^3$ describes the motion of a point on the profile depending on angle, θ, and position of that point, $\boldsymbol{P}$,

$$\boldsymbol{O}(\theta, \boldsymbol{P}) = \boldsymbol{R}_z(\theta)\boldsymbol{P} + (0,0,l \cdot \theta) \quad (1)$$

where l indicates the lead (length per radian) with which the profile moves along the z-axis. Note that the profile does not have translational movement along the x and y-axis. Changing the coordinate system (x, y, x) to the coordinate system at the cutter position (x_c, y_c, z_c), the new coordinate of the point on the profile becomes,

$$\boldsymbol{C}(\boldsymbol{P}) = \boldsymbol{R}_x(\alpha)\boldsymbol{R}_y(\beta)\boldsymbol{R}_z(\gamma)\boldsymbol{P} - (m_x, m_y, m_z) \quad (2)$$

where the first term on the left-hand side indicates the rotation around x-, y- and z-axis of angle α, β and γ and the second term denotes the displacement vector.

Following the point on the profile along its normal, we define vector $\boldsymbol{F}(\theta, s)$ as a function of angle and displacement along the normal, s,

$$\boldsymbol{F}(\theta, s) = \boldsymbol{C}(\boldsymbol{O}(\theta, \boldsymbol{P}) + \boldsymbol{R}_z(\theta)\mathbf{n} \cdot s) \quad (3)$$

where $\mathbf{n}$ is the normal vector at the point on the profile. To find when this point hits the cutter which is oriented along y_c-axis, we solve the following non-linear system of equations where θ and s are unknown variables for any given point $\boldsymbol{P}$,

$$\boldsymbol{\pi_1}\boldsymbol{F}(\theta, s) = 0, \ \boldsymbol{\pi_3}\boldsymbol{F}(\theta, s) = 0. \quad (4)$$

Here $\boldsymbol{\pi_1}$ and $\boldsymbol{\pi_3}$ are the first and third projection operators. We indicate the solution of these equations by θ^*, s^*. Next, we solve for the angle ω to rotate back the point to the $x_c y_c$-plane of the cutter in order to get the corresponding point on the cutter profile,

$$\boldsymbol{\pi_3}\boldsymbol{R}_y(\omega)\boldsymbol{F}(\theta^*, 0) = 0. \quad (5)$$

The solution of equation (5), ω^*, is then employed to compute the desired point on the cutter profile,

$$\boldsymbol{P_C} = \boldsymbol{R_y}(\omega^*)\boldsymbol{F}(\theta^*, 0). \qquad (6)$$

3.2. Cutter design

The cutter profile is generated from the corresponding rotor profile. Considering that the cutter is rotational symmetric while most of the profiles are asymmetric, a minimum of two cutters is needed for milling an asymmetric profile. The left and right side of the profile generate the corresponding left and right cutter profiles. In practice however, more than two cutter profiles are needed to optimize the milling process of a certain rotor profile. In this case the cutter axis can be tilted in two directions and be moved perpendicular to the center axis of the rotor. The restriction of the setting data (tilting and movement) is governed by the requirement that the mill does not touch the side it will not process.

In particular, milling the root of the male rotor is challenging. Therefore, while the female rotor normally needs three or four cutters, the male rotor needs approximately three cutters more. In figure 6 we display the four female cutters which. The two cutters to the left are used to machine the inner part of the rotor, left and right side. The two cutters to the right are used to machine the outer part of the rotor, left and right side.

Figure 6. Finger cutters for female screw compressor rotors

When designing the cutter there are as usually compromises that need to be considered:

- The profile is normally divided into axial segments, to optimize the position of the blades, to reduce the grinding allowance.
- A higher number of segments might reduce the possible number of effective blades.
- To get as many effective blades as possible the rake angle (see 3.3) will normally become negative. This is not optimal from a cutting perspective especially with rotors of softer steel.

3.3. Grinding the blade

The cutter profile is generated to match the profile of the rotational body and the blades are normally tilted around two axes, shown in figure 7 and 8.

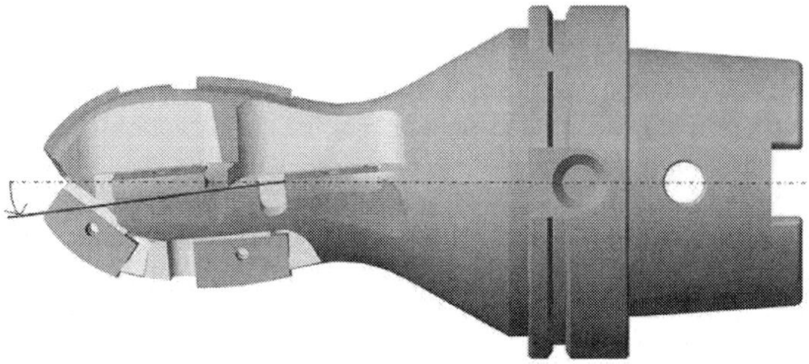

Figure 7. Helix angle

The helix angle is defined as shown in figure 7, a normal value for FC tool is about 20°. The reason for having a helix angle is to get a smoother cut as a result of entering the blades successively.

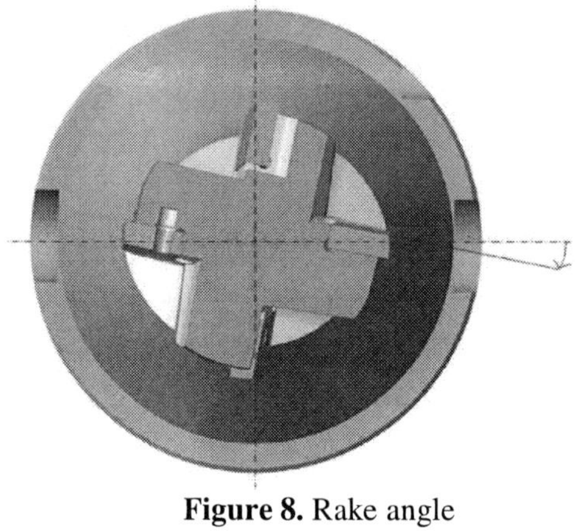

Figure 8. Rake angle

The rake angle is defined as shown in figure 8 and describes the angle of the cutting face relative to the work piece. The figure above shows a negative rake angle, which tends to be used for these FC, due to geometrical limitations. Normally a positive rake angle is preferable, more positive for softer materials.

To grind the blades to the rotational profile a built-in probe in the grinding machine will measure the plane of the blades and the plane will be defined mathematically, see figure 9. During the grinding cycle the cutter will be rotated to ground each point of the profile at the horizontal center plane.

IOP Conf. Series: Materials Science and Engineering **425** (2018) 012007 doi:10.1088/1757-899X/425/1/012007

Figure 9. Measuring the plane of the insert

The cutter blades also need to have clearance angles as shown in figure 10. The preliminary clearance angle (Ω) is normally 5-10° depending of the material. A harder material needs a smaller angle. To avoid back-cut a secondary clearance angle (ω) is needed, a normal value is between 10-20°.

To grind an FC from new rectangular inserts takes about two hours. To regrind or to grind from pre-profiled inserts takes about 30 minutes. Finally, we note that a coated carbide blade has an expected life of 10 cutting meters.

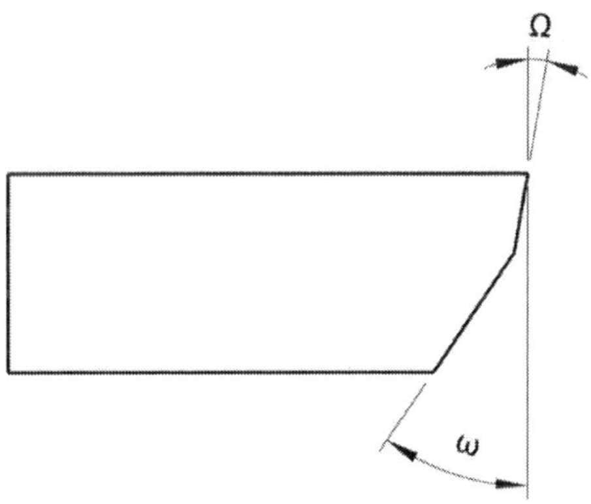

Figure 10. The clearance angles of a cutter blade

4. Productivity and accuracy

In addition to what we discussed so far, the productivity must be evaluated with respect to the complete machining process and available machines. The method is most beneficial for small batches, for larger batches the traditional milling machines are competitive. A significant advantage of using flexible multi-functional machines is that one can manufacture components in single setup, which significantly affects both productivity and accuracy.

The FC is optimized for machining the final profile but can be used for roughing as well. Due to the cost of the FC the rotor should preferably be rough machined with standard cutters of various sizes. It is possible to grind the blades within ±0.002 mm, but the accuracy of the rotor depends, among other factors, on the multi-functional machine and rotor deflection. A reasonable accuracy of the rotor profile is ±0.015 mm regardless of the rotor size. Considering accuracy, size of the cutters and machining time, the presented method is most advantageous for larger rotors (diameter >400 mm).

5. Conclusion

In this work Finger Cutter (FC) is introduced as an innovative tool for rotor milling which can be easily integrated with multi-functional flexible machines. FCs are operating inside the rotor body with the rotational axis almost parallel to the profile surface. The small rotational diameter of FCs, compared to traditional disc cutters, gives the opportunity to employ standard machines with lower required torque. Considering the accuracy and machining time, this method is considered one of the best alternative for manufacturing large rotors (diameter > 400 mm) where smooth operation, lower vibration and acceptable tolerances are desirable simultaneously.

References

[1] Zabel A 2014 Modern machining processes for the manufacturing of screw machine components *Inter. Confrenece on Screw Machines* **2228** 267–278
[2] Wu Y and Hsu W 2014 A general mathematical model for continuous generation machining of screw rotos with worm-shaped tools *Applied Mathematical Modeling* **38** 28–37

IOP Conf. Series: Materials Science and Engineering **425** (2018) 012008 doi:10.1088/1757-899X/425/1/012008

Impact of different clearance heights on the operation of a water-flooded twin-screw expander—experimental investigations based on indicator diagrams

A Nikolov[1] **and A Brümmer**[1]

[1]Chair of Fluidics, TU Dortmund University, Dortmund, Germany

E-mail: alexander.nikolov@tu-dortmund.de

Abstract. Twin-screw expanders offer a high potential for energy conversion in the lower and medium power range, for instance in Rankine cycle systems for exhaust heat recovery. With the aim of minimising internal leakages within the expander and lubricating moving machine parts, an auxiliary liquid can be carried with the main flow or liquid working fluid can be fed to the twin-screw expander. Moreover, the operation of twin-screw expanders in Rankine cycle systems at high liquid mass fractions and in trilateral flash cycles is predicted to be a very promising application for expanders in lower power ranges. Thus, a fundamental understanding of the operation of liquid-flooded twin-screw expanders is mandatory. This paper presents results of an experimental investigation into a water-flooded twin-screw expander prototype SE 51.2 designed at the Chair of Fluidics at TU Dortmund University. On the one hand, the aspect of a two-phase working fluid is discussed considering integral characteristic numbers such as mass flow, delivery rate, and effective isentropic efficiency. On the other hand, in order to explain the influence mechanisms of a two-phase working fluid on the operating behaviour of the twin-screw expander, indicator diagrams are recorded by means of high-resolution pressure transmitters to determine indicated power. Hence, mechanical and hydraulic losses, indicated isentropic efficiency, and mechanical efficiency of the twin-screw expander can be calculated. In order to determine the influence of the narrow clearance on the hydraulic losses and the clearance sealing effect in terms of a two-phase working fluid, a systematic variation of the rotor-tip clearance height is carried out. As a result of the investigations, a water surge at the rotor tip is proved to be a significant mechanism affecting hydraulic losses in a water-flooded twin-screw expander.

1. Introduction

Due to the growing shortage of non-renewable fossil fuel reserves and the consequential increase in primary energy costs, underdeveloped energy potentials are increasingly moving into the focus of economic interest. Available heat sources in the field of decentralised energy systems of small and medium power ranges from 3 kW to 1.5 MW [1, 2]—such as industrial exhaust gases or waste heat in vehicle engines, geothermal or solar thermal energy—can be converted into usable mechanical power by means of expanders or turbines within a Rankine cycle system.

In this context, twin-screw expanders in Rankine cycles possess clear advantages compared with turbomachines or even other volumetric expander concepts, being capable of economically transforming small volume flows at relatively large pressure ratios [3]. In general, twin-screw machines are characterised by relatively high energy density and efficiency, advantageous part load behaviour, and a rather simple design [4]. Furthermore, twin-screw expanders in (organic) Rankine cycles are suitable for wet-vapour operation; thus, an overheating system is not essential [5]. On the one hand, the number of system components and the overall costs can be reduced. On the other hand, liquid in twin-screw expanders could be beneficial with regard to their operation. The auxiliary fluid also reduces noise emissions and thermal stress within the machine parts and seals structurally necessary machine clearances [6].

Investigations into water- and oil-injected twin-screw expanders with timing gears were carried out by Zellermann [7]. Kauder and Zellermann [8] reported a significant increase of hydraulic friction losses in terms of oil-injection at high rotor tip-speeds. Hence, the operating range of wet-running screw machines taking hydraulic losses into account is usually restricted to a maximum rotor tip-speed of 40 $m \cdot s^{-1}$ [4]. Without limitations in the rotor tip-speed, performance reduction due to oil shearing in and at the machine clearances exceeds the positive sealing effect of the injected liquid. In this context, Harling [6] has carried out extensive experimental studies on the distribution of the injected oil in a twin-screw compressor, considering the oil-surge theory of Kauder [9]. Here, losses as a result of momentum exchange between the oil and the rotor surface have been investigated. In order to determine hydraulic friction losses in screw expander clearances using an auxiliary liquid, an analytical approach was presented by Gräßer and Brümmer with regard to water [10] and oil [11, 12]. In [13], Vasuthevan and Brümmer proved by means of numerical simulations that hydraulic losses generated by a water surge at the rotor-tip are more significant compared to water friction within the rotor-tip clearance. In [14], the operational behaviour of a small oil-flooded twin-screw expander in an organic Rankine cycle system has been experimentally and theoretically investigated. Kliem [15] examined a twin-screw expander application in a trilateral flash cycle theoretically and experimentally. Here, overheated water is injected into the working chamber where it expands and evaporates. According to Ohman and Lundqvist [1], this application of twin-screw expanders is predicted to be a promising approach in the medium and low power operating range. In this context, a method for the theoretical investigation into twin-screw expanders in trilateral flash cycles has been presented by Vasuthevan and Brümmer [16]. With regard to thermal deformation, Svigler *et al.* [17] reported that under water-injected conditions, the thermal deformation of the rotor flank in twin-screw compressors corresponds to the manufacturing tolerance magnitude for ground rotors.

Hereinafter, the results obtained from pursuing experimental investigations into the operation of a water-flooded twin-screw expander based on the studies presented in [18] and [19] are discussed. In [18], the effect of water-flooded operation has been presented integrally from an energetic point of view considering effective isentropic efficiency and delivery rate of the test expander. At part load operation and at low rotor tip-speeds in particular, it has been demonstrated that the water-flooded compared to the dry-running twin-screw expander reveals higher effective isentropic efficiency. At high rotor tip-speeds however, expander efficiency declines with increasing amounts of water, since hydraulic losses become more dominant. In [19], indicator diagrams were presented in order to quantify mechanical power losses in the twin-screw expander and to identity the main influence physical mechanisms with respect to water-flooded operation. Generally, mechanical power losses in twin-screw expanders can be attributed to friction losses in the grease-lubricated bearings and in the rotor contact as well as to acceleration of the auxiliary fluid and hydraulic friction in the clearances and in the water surge at the rotor tips. Within the framework of the following study, a methodical variation of the rotor-tip clearance height is carried out in order to detect its impact on hydraulic losses and clearance sealing in terms of water in the working chamber.

2. Experimental setup

A detailed description of the experimental setup within the framework of the investigations has been presented in [18] and [19]. A 3D model of the test twin-screw expander prototype SE 51.2 and the positions of the high-resolution pressure indication transmitters are illustrated in figure 1. SE 51.2 is a twin-screw expander prototype without timing gears. The transmission of torque occurs directly via contact between the rotor flanks. The screw rotors are hardened and have a tough tungsten carbide/carbon (WC/C) wear-protection coating, so seizure can be avoided in dry-running or water-flooded operation. Both fixed and loose bearing sets are grease lubricated, so no oil supply is necessary. Details about the expander geometry parameters are listed in table 1. For the purposes of the experimental investigations, the rotor-tip clearance height has been varied in order to analyse how hydraulic losses change and which are the consequences on the reduction of internal leakages in the expander. The clearance height variation has been achieved by increasing the rotor bore diameter keeping any further expander geometry parameters constant.

Figure 1. Positions of pressure indication transmitters (0…6), inlet and outlet area, as well as volume curve of the test screw expander SE 51.2 including ranges of male rotor rotational angle corresponding to each pressure transmitter.

Table 1. Parameters of the test screw expander SE 51.2.

designation	**unit**	**male rotor (MR)**	**female rotor (FR)**
number of lobes z	[-]	3	5
diameter d	[mm]	71.8	67.5
rotor lead s	[mm]	181.8	303
wrap angle φ	[°]	200	-120
rotor length l	[mm]	101	
rotor profile	[-]	modified asymmetric SRM	
axis-centre distance a	[mm]	51.2	
internal volume ratio v_i	[-]	2.5	
displaced volume per male rotor rotation V	[cm^3]	286	
front clearance height $h_{fc,hp}$ (high pressure)	[mm]	0.1	
front clearance height $h_{fc,lp}$ (low pressure)	[mm]	0.17	
rotor-tip (housing) clearance height h_{hc}	[mm]	0.08, 0.16, 0.24	

The relevant measurement devices installed in the test rig and the twin-screw expander are listed in table 2. To record the indicator diagrams of the test twin-screw expander, two different piezoresistive sensor types are installed flush with the rotor bore surface and on the high-pressure front side. The pressure transducers of the series M5 manufactured by "KELLER AG für Druckmesstechnik" at position 0, 1, and 6 record the absolute static pressure in the inlet port (position 0), in the working chamber in the range of rotational angle between 85° and 130° (position 1), and during the fluid discharge on the low-pressure side of the expander (position 6). With regard to the manufacturer data sheet, a natural frequency of more than 50 kHz can be expected. At positions 2 to 5, XTM-190 pressure transducers manufactured by "Kulite Semiconductor Products, Inc" with natural frequency higher than 425 kHz are installed. Hence, no acoustic or vibration issues at the pressure metering point with regard to eigenfrequencies are expected. Depending on pressure and temperature, the sensitivity of each XTM-190 sensor was calibrated in order to increase its accuracy. Due to a relatively high temperature related zero point offset of the XTM-190 sensors, only the relative change in pressure to a reference pressure level is considered. For this reason, the pressure difference between two adjacent transducers within their overlapping scope (figure 1) is set to zero and the high-resolution static pressure at position 6 is used as a starting reference pressure level.

Table 2. Measuring range and accuracy of installed measurement devices.

parameter	range	accuracy	type (manufacturer)
chamber pressure p	0 to $10 \cdot 10^5$ Pa (absolute)	±0.2 % FS* (linearity)	series M5 (KELLER AG für Druckmesstechnik)
chamber pressure p	0 to $17 \cdot 10^5$ Pa (absolute)	$\pm 10^3$ Pa (linearity calibrated)	XTM-190 (Kulite Semiconductor Products, Inc.)
inlet pressure p_i	0 to $6 \cdot 10^5$ Pa (relative)	±0.1 % FS*	ATM.1ST (STS)
temperature ϑ	-185…300 °C	±1 °C (-40…133 °C)	insulated thermocouples type T
torque M_{MR}	±20 Nm	±0.1 % FS*	ETH DRLF-II-n (ETH-messtechnik)
rotational speed n_{MR}	max. 19,000 min^{-1}	-	
air mass flow $\dot{m}_a$	max. ≈ 3300 kg·h^{-1} (at $p_i = 5.5 \cdot 10^5$ Pa)	±0.35 % o.r.**	Promass 83F (DN40) (Endress+Hauser)
water volume flow	max. 8.2 l·min^{-1}	±(0.4 % o.r.** + 0.02 % FS*)	ProcessMaster300 (ABB)

*FS–full scale; **o.r.–of reading

With regard to the pressure indication, a rotational angle range between 130° and 140° cannot be covered due to lack of space in the casing, so the pressure has to be estimated. Preliminary analysis of the recorded pressure revealed that a linear pressure distribution in the missing range is qualitatively and quantitatively sufficiently accurate. At low rotor speeds, the pressure transmitter at position 0 is assumed to represent the chamber pressure from the beginning of the working cycle at rotational angle of 45° up to rotational angle of 85° as indicated in figure 1. In this way, the assumed chamber pressure is mapped with regard to an ideal chamber filling without pressure drop over the inlet. In theory, this results in an upper limit for the converted work within this working cycle range. At higher rotational speeds, a different approach to map the chamber pressure profile in this section is applied. Here, linear pressure increase starting from the outlet and ending at the pressure level of the transmitter at position 1 is assumed. This method is more appropriate with respect to increasing inlet throttling losses at higher rotor speeds or increasing amount of injected water. Therefore, the pressure level and the indicated work within the missing range of the indicator diagram are expected to be lower than the real one during the chamber growth. This approach represents a lower limit for the indicated work within the first 40° of chamber development the range of rotational angle between 45° and 85°.

Within the framework of this experimental investigation, both the system and the twin-screw expander operating parameters were varied. Water was injected into the high-pressure domain at a superficial water temperature of $\vartheta_{i,w} \approx 60$ °C at constant air inlet pressures of $p_i = 3 \cdot 10^5$ Pa, $p_i = 4 \cdot 10^5$ Pa, $p_i = 5 \cdot 10^5$ Pa as well as superficial air temperature of $\vartheta_{i,a} = 90$ °C. Outlet pressure corresponds to ambient conditions at ca. $p_o \approx 1 \cdot 10^5$ Pa. The resulting mixture temperature at the expander inlet is nearly $\vartheta_i \approx 50$ °C depending on water volume flow and expander rotational speed.

In the style of steam or vapour mass fraction, the amount of injected water into the expander inlet port is represented by the dryness fraction x relating air mass flow to the overall air and water mass flow:

$$x = \frac{\dot{m}_a}{\dot{m}_a + \dot{m}_w}. \tag{1}$$

Here, measured air mass flow $\dot{m}_a$ (Coriolis mass flow meter) can be considered as dry air, since a refrigerant type dryer is used after air compression. Water volume flow is recorded by means of an electromagnetic flow-metering device. Regarding water temperature to ambient conditions, water density and mass flow $\dot{m}_w$ can be determined. Dryness fraction of $x = 1$ corresponds to a dry-running operation and $0 < x < 1$ to the two-phase flow in water-flooded operation. Dryness fraction x is varied in the range between 1.0 and 0.4. For water injection, two nozzle sizes are used at two adjacent water flow ranges with respect to a satisfactorily homogeneous water injection quality in the expander inlet domain. The maximum expander rotational speed is $n_{MR} = 18{,}000$ min^{-1} which equals a male rotor tip-speed of $u_{MR} = 67.7$ m·s^{-1}.

3. Characteristic numbers

Using characteristic numbers, physical processes or the operation of machines can be quantitatively analysed and compared with each other. In the following analysis, different characteristic numbers are used to explain the relevant effects on the operation of the test twin-screw expander.

3.1. Delivery rate

Delivery rate λ_L is a characteristic number that quantifies mass loss mechanisms such as internal or external leakages, throttling losses during the chamber filling, or thermal effects within a displacement machine. Taking the actual measured air mass flow $\dot{m}_a$ and the theoretical air mass flow $\dot{m}_{th}$ of the twin-screw expander into account, delivery rate is defined as:

$$\lambda_L = \frac{\dot{m}_a}{\dot{m}_{th}}. \tag{2}$$

Here, the theoretical mass flow $\dot{m}_{th}$ is referred to dry-air density $\rho_{da,i}$ in the inlet domain taking the chamber volume at the begin of the internal expansion $V_{th,ex}$ (chamber volume at male rotor rotational angle of 201° in figure 1), the male rotor speed n_{MR}, and the number of the male rotor lobes z_{MR} into account:

$$\dot{m}_{th} = \rho_{da,i} \cdot V_{th,ex} \cdot n_{MR} \cdot z_{MR}. \tag{3}$$

The chamber volume displaced by liquid can be ignored due to a relatively high volume fraction of air. As already mentioned in [19], considering a homogeneous two-phase mixture in the working chamber, air volume fraction is more than 98.5 % even at the lowest air mass flow and highest water flow. Thus, $V_{th,ex}$ refers to the working chamber volume after completing the chamber filling. Furthermore, dry air is considered as an ideal gas and its density $\rho_{da,i}$ is determined by means of the calculated subcooled temperature after water was injected. In terms of delivery rate, measured air mass flow $\dot{m}_a$ does not consider evaporated water steam with respect to saturated humid air after water injection, since the Coriolis mass flow meter is installed upstream of the water injection nozzle. In some cases, at high temperature levels and low pressure at the same time, steam fraction has to be taken into account, since it significantly increases. Within the following investigations, steam mass flow increases to a level of up to 3.2 % of metered air mass flow at the lowest inlet pressure and dryness fraction. Since steam mass flow is not measured by the Coriolis flow meter, an additional decrease of delivery rate (section 3.1) of a maximum of 3.2 % may arise at some operating points.

3.2. Indicated and effective isentropic efficiency

Indicated and effective isentropic efficiency ($\eta_{i,s}$ and $\eta_{e,s}$) reveal the energetic potential of the twin-screw expander considering different loss mechanisms. Here, indicated or effective power is related to the isentropic potential of the working fluid considering the inlet and outlet conditions. With respect to water-flooded operation, an isentropic change in state represents the maximum energy available in the two-phase working fluid. Hence, indicated isentropic efficiency is defined as:

$$\eta_{i,s} = \frac{P_i}{\Delta \dot{H}_s} \tag{4}$$

and effective isentropic efficiency respectively as:

$$\eta_{e,s} = \frac{P_e}{\Delta \dot{H}_s}. \tag{5}$$

Indicated power P_i can be calculated by means of the time-dependent working chamber pressure p and volume V, male rotor rotational speed n_{MR}, and number of male rotor lobes z_{MR} as follows:

$$P_i = -n_{MR} \cdot z_{MR} \cdot \oint p \cdot dV. \tag{6}$$

Effective power P_e results from the recorded male rotor rotational speed n_{MR} and torque M_{MR} as:

$$P_e = 2 \cdot \pi \cdot n_{MR} \cdot M_{MR}. \tag{7}$$

Isentropic enthalpy flow $\Delta \dot{H}_s$ for a homogeneous mixture is defined as the product of dry-air mass flow $\dot{m}_{da}$ and the specific isentropic enthalpy difference for humid air $\Delta h_{1+x,s}$ between expander outlet (o) and inlet (i) as follows:

$$\Delta \dot{H}_s = \dot{m}_{da} \cdot \Delta h_{1+x,s} = \dot{m}_{da} \cdot \left(h_{1+x,s,o} - h_{1+x,i}\right). \tag{8}$$

In this context, the specific enthalpy h_{1+x} for a homogeneous humid-air-water mixture in thermodynamic equilibrium can be calculated with reference to water load X [20]. Water load X represents the ratio of water mass flow $\dot{m}_w$ to dry-air mass flow $\dot{m}_{da}$:

$$X = \frac{\dot{m}_w}{\dot{m}_{da}}. \tag{9}$$

Since water load of the pressurised and preheated air at expander inlet upstream of water injection is negligibly low, dry-air mass flow $\dot{m}_{da}$ can be related to measured air mass flow $\dot{m}_a$. The specific enthalpy of unsaturated humid air ($X < X_{sat}$) is defined as:

$$h_{1+x} = c_{p,da} \cdot \vartheta + X \cdot \left(r_w + c_{p,ws} \cdot \vartheta\right). \tag{10}$$

Here, water load X_{sat} corresponds to humid air at a relative humidity of 100 %. The specific enthalpy of saturated humid air ($X \geq X_{sat}$) with reference to water in the mixture can be calculated as:

$$h_{1+x} = c_{p,da} \cdot \vartheta + X_{sat} \cdot \left(r_w + c_{p,ws} \cdot \vartheta\right) + (X - X_{sat}) \cdot c_w \cdot \vartheta. \tag{11}$$

For temperatures $\vartheta < 0$ °C, the calculation of the specific humid air enthalpy is referred to ice condensation as follows:

$$h_{1+x} = c_{p,da} \cdot \vartheta + X_{sat} \cdot \left(r_w + c_{p,ws} \cdot \vartheta\right) - (X - X_{sat}) \cdot (r_{ice} - c_{ice} \cdot \vartheta). \tag{12}$$

The constant fluid properties required to calculate the enthalpy of the humid-air-water mixture are listed in table 3. Specific enthalpy of humid air at the expander inlet can be calculated by means of the recorded inlet temperature ϑ_i and pressure p_i. Isentropic fluid mixture temperature and enthalpy at the expander outlet is calculated with regard to a constant specific entropy at the expander inlet and outlet according to [20].

Table 3. Constant fluid properties for the fluid state calculation of humid air in thermodynamic equilibrium [20].

designation	unit	value
dry-air heat capacity $c_{p,da}$	[J·kg^{-1}·K^{-1}]	1004.6
water-steam heat capacity $c_{p,ws}$	[J·kg^{-1}·K^{-1}]	1863
liquid water heat capacity c_w	[J·kg^{-1}·K^{-1}]	4185
triple temperature ϑ_{tr}	[°C]	0.01
ice heat capacity c_{ice} at ϑ_{tr}	[J·kg^{-1}·K^{-1}]	2070
specific solidification enthalpy r_{ice} at ϑ_{tr}	[kJ·kg^{-1}]	333.4
specific vaporisation enthalpy r_w at ϑ_{tr}	[kJ·kg^{-1}]	2500.9

3.3. Mechanical efficiency

Mechanical efficiency η_m is a characteristic number that quantifies mechanical power losses in twin-screw expanders relating effective and indicated power to each other:

$$\eta_m = \frac{P_e}{P_i}. \tag{13}$$

In the twin-screw expander investigated, mechanical power losses result from the contact between both screw rotors, within the bearings and sealing elements, as well as from hydraulic losses considering acceleration and hydraulic friction of the injected auxiliary fluid in and at the clearances and rotor surfaces. Mechanical power loss $\dot{\Phi}$ can be calculated as follows:

$$\dot{\Phi} = P_i - P_e. \tag{14}$$

Hence, mechanical efficiency of the twin-screw expander can be calculated as:

$$\eta_m = \frac{P_e}{P_i} = \frac{P_i - \dot{\Phi}}{P_i} = 1 - \frac{\dot{\Phi}}{P_i}. \tag{15}$$

By means of equation (4) and equation (5), mechanical efficiency η_m can be also determined from the ratio of effective isentropic to indicated isentropic efficiency as:

$$\eta_m = \frac{\eta_{e,s}}{\eta_{i,s}}. \tag{16}$$

4. Results

In this section, results from the experimental investigations into the water-flooded twin-screw expander considering effects of rotor-tip clearance height variation are presented. First, indicator diagrams are discussed revealing the impact of different clearance situations on the working chamber pressure during a working cycle under dry-running and water-flooded conditions. After that, expander operating maps based on characteristic numbers are used to quantify the arguments discussed initially.

4.1. Indicator diagrams

Recording the chamber pressure during a working cycle and generating indicator diagrams is a common method to evaluate the available fluid energy within the working chamber of a reciprocating machine and respectively of a twin-screw expander. In this way, indicated power takes throttling losses, leakages, under- and overexpansion, as well as the interaction between the working and auxiliary fluid and the working chamber boundaries into account. Moreover, the different influence mechanisms can be quantitatively illustrated by variation of the expander operation parameters.

In order to illustrate the influence of water on the chamber pressure at different rotor-tip clearance heights, indicator diagrams at a high and a low male rotor tip-speed are presented in figure 2 and figure 3. Each operating point corresponds to a constant inlet pressure and superficial air and water temperatures. Generally, at both rotor tip-speed levels, water in the working chamber mostly initiates an increase in chamber pressure during internal expansion and a greater indicated power respectively (except for rotor-tip clearance height of 0.08 mm in figure 3). As explained in [19], this effect can be primarily traced back to the extensive heat exchange between water and the expanding gaseous fluid, so the polytropic expansion exponent is reduced compared to a pure gaseous operation. Additionally, a sealing effect resulting from water in the clearances is expected, while this can be only partially proved by means of the indicator diagrams.

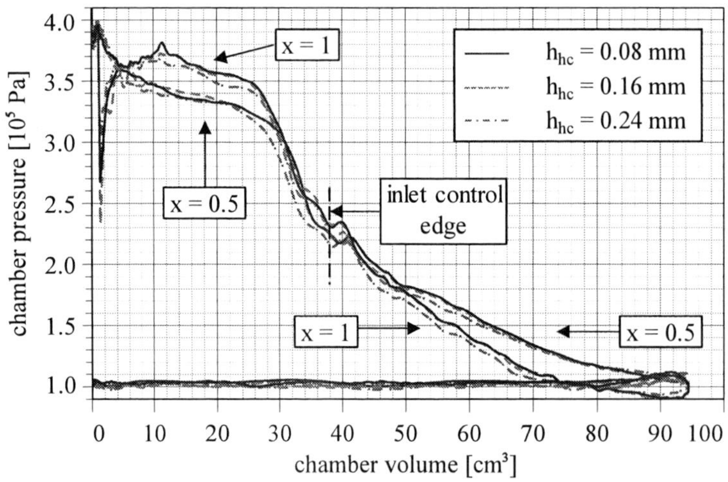

Figure 2. Indicator diagrams at a male rotor tip-speed of 67.7 m·s^{-1}, an inlet pressure of 4·10^5 Pa, a superficial air temperature of 90 °C, and a superficial water temperature of 60 °C; *dryness fraction: 1 and 0.5; rotor-tip clearance height: 0.08 mm, 0.16 mm, and 0.24 mm.*

Generally, at high rotor tip-speeds, the internal leakages related to the expander air mass flow can be neglected, so no significant impact of clearance height on the indicator diagrams is expected. In fact, inlet throttling and hydraulic losses become dominant. In figure 2, a significant drop in pressure during the chamber filling can be observed under water-flooded conditions due to increasing inlet throttling. The main reason for the pressure increase during the internal expansion at high rotor tip-speeds is the heat flow from the liquid phase into the gaseous fluid. Hence, overexpansion can be avoided. In a rotor-related coordinate system considering the clearance sealing effect, it is more likely to have a boundary-driven (Couette) water back-flow from the leading chamber as a consequence of a relative motion between the rotor tip and the housing boundary. The pressure-driven (Poiseuille) water flow plays a subordinate role and merely reduces the Couette water flow from the leading chamber. At lower dryness fractions, a liquid surge is expected to form at the rotor-tip clearance outlet on the leading chamber side. On the one hand, this would reduce the clearance cross-section for the gaseous flow and seal the clearance (which is in fact not a dominant effect at higher rotor tip-speeds). Clearance sealing in terms of water-flooded operation is discussed in section 4.2 using delivery rate as defined in equation (2). On the other hand, hydraulic losses become dominant. In section 4.3, hydraulic friction losses resulting from liquid friction in the clearances and in the water surge at the rotor tip in the leading working chamber are analysed by means of mechanical power loss and mechanical efficiency.

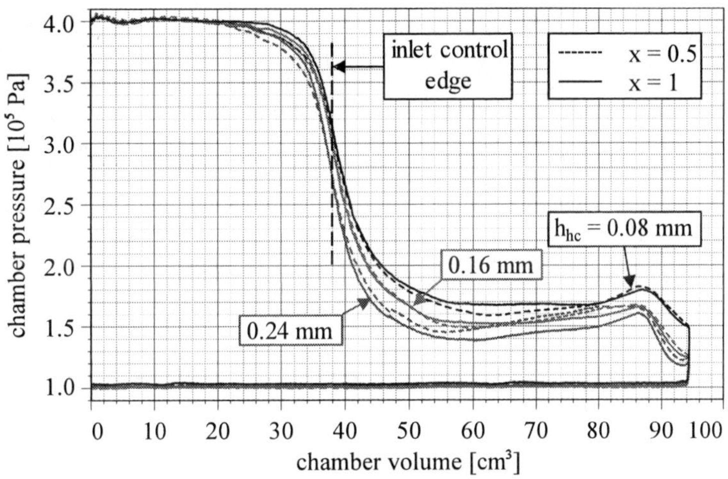

Figure 3. Indicator diagrams at a male rotor tip-speed of 3.8 $m \cdot s^{-1}$, an inlet pressure of $4 \cdot 10^5$ Pa, a superficial air temperature of 90 °C, and a superficial water temperature of 60 °C; *dryness fraction: 1 and 0.5; rotor-tip clearance height: 0.08 mm, 0.16 mm, and 0.24 mm.*

In contrast, at low rotor tip-speeds, internal leakage becomes more relevant, while the influence of inlet throttling declines or disappears as illustrated in figure 3. Due to internal leakage, higher pressure is revealed in the working chamber during the internal expansion unlike at high rotor tip-speed in figure 2. A chamber refilling is the result. In terms of inlet throttling, no significant pressure drop as a function of clearance height and only a minor influence of dryness fraction on chamber filling are observed. Under water-flooded conditions, expander operation leads to less significant boundary-driven (Couette) water flow in the clearances, so now the pressure-driven (Poiseuille) water flow predominates. For different clearance heights and approximately comparable pressure difference at the rotor tip, the pressure-driven water clearance volume flow toward the leading chamber increases for greater clearance cross-sections. Therefore, less water might remain at the clearance inlet to effectively seal the clearance. In this context, at a clearance height of 0.08 mm in figure 3, the clearance sealing in the water-flooded expander is relatively effective and overshadows the heat exchange effect. Thus, the pressure in the chamber declines under the levels seen in dry-running operation. At a clearance height of 0.16 mm, a slight decrease in sealing effect can be observed, since the pressure-driven (Poiseuille) water volume flow in the rotor-tip clearance increases. Moreover, heat exchange in the leading chamber intensifies due to the increased amount of water. Eventually, the lowest sealing effect can be detected at a clearance height of 0.24 mm, resulting in the greatest air and water mass and, thus, enthalpy flow into the leading chamber under water-flooded conditions. Here, a chamber pressure increase can be detected compared to dry-running operation due to a maximum amount of water and heat flow from water to air respectively in the expanding working chamber.

4.2. Mass flow and delivery rate

In figure 4, expander air mass flow and delivery rate are illustrated as functions of the male rotor tip-speed at different dryness fractions. As expected, expander air mass flow increases at higher rotor tip-speeds under dry-running and water-flooded conditions. Reducing dryness fraction, expander air mass flow declines due to reduced internal leakages, increased heat flow from water into air, or greater inlet throttling losses depending on the rotor tip-speed. Since, according to [18], the inlet temperature of the two-phase mixture depends on the amount of water, lower dryness fraction results in higher inlet temperatures and lower air inlet density, since heat flux from water into the gaseous working fluid increases. As a consequence, expander air mass flow may decline depending on dryness fraction, too. To quantify the sealing and inlet throttling effects, delivery rate is considered. At a fixed low rotor tip-speed, a significant drop in delivery rate as a consequence of declining dryness fraction and, hence, increasing clearance sealing effect, can be perceived. In contrast, at a constant high rotor tip-speed, delivery rate drops mainly due to increasing inlet throttling in terms of water-flooded operation as demonstrated in figure 2. As mentioned in section 3.1, an additional decrease in air mass flow and delivery rate due to water steam in the saturated gaseous phase of the two-phase mixture results from water-flooded operation. Nevertheless, at the lowest dryness fraction considered, air mass flow decreases not more than 3.2 % at the lowest and 2.2 % at the highest inlet pressure investigated. Therefore, this effect is neglected within the framework of this study.

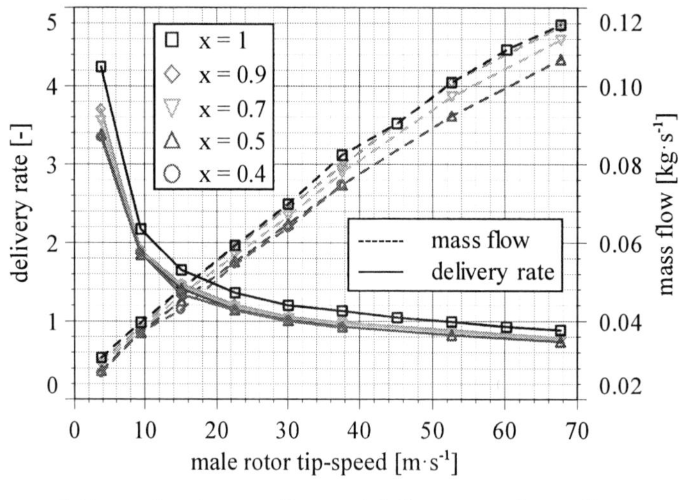

Figure 4. Expander air mass flow and delivery rate depending on male rotor tip-speed at an inlet pressure of $4 \cdot 10^5$ Pa, a superficial air temperature of 90 °C, a superficial water temperature of 60 °C, and a rotor-tip clearance height of 0.08 mm; *dryness fraction: 1, 0.9, 0.7, 0.5, and 0.4.*

In figure 5, the impact of different rotor-tip clearance heights on the expander air mass flow and delivery rate is presented for a constant dryness fraction of 0.5. At low rotor tip-speeds and increasing clearance height, the sealing effect decreases due to increasing pressure-driven (Poiseuille) water volume flow in the high-pressure rotor-tip clearances, so expander air mass flow and delivery rate increase. Up to a rotor tip-speed of about 35 m·s^{-1}, the sealing effect of the middle (0.16 mm) compared to the lowest (0.08 mm) clearance height increases continuously according to expander air mass flow and delivery rate. Then, sealing becomes maximum as expander air mass flow and delivery rate at both clearance heights are identical. As explained in section 4.1, this fact can be traced back to the boundary-driven (Couette) water flow providing more liquid to seal the clearance cross-section at rising rotor tip-speed. As a result of the larger clearance cross-section at a clearance height of 0.24 mm in relation to 0.08 mm and 0.16 mm, the sealing effect appears to be relatively inefficient.

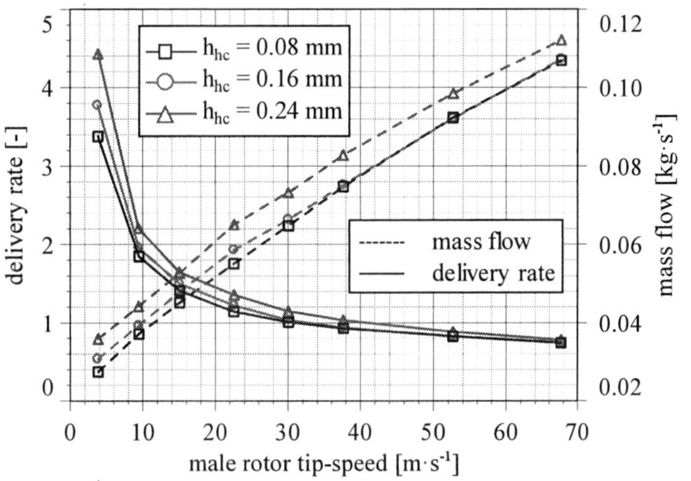

Figure 5. Expander air mass flow and delivery rate at a male rotor tip-speed of 67.7 m·s^{-1}, an inlet pressure of $4 \cdot 10^5$ Pa, a superficial air temperature of 90 °C, a superficial water temperature of 60 °C, and a dryness fraction of 0.5; *rotor-tip clearance height: 0.08 mm, 0.16 mm, and 0.24 mm.*

4.3. Mechanical power loss and expander efficiency

Hereinafter, to analyse the influence mechanisms on the operation of the test twin-screw expander from energetic point of view, mechanical power loss and mechanical efficiency respectively as well as indicated and effective isentropic efficiency are considered. Mechanical power loss and different efficiencies depending on rotor tip-speed and clearance height under dry-running and water-flooded conditions are illustrated in figure 6. For dry-running and water-flooded operation, mechanical power loss increases exponentially as a function of male rotor tip-speed. At a dryness fraction of 1, frictional losses in the bearings and in the rotor contact account for the entire mechanical power loss. Under water-flooded conditions, hydraulic shear stress in the narrow clearances and in the water surge at the rotor tip contribute additionally to the overall mechanical power loss. Within the framework of the experimental investigation, mechanical power losses with regard to the water flow in the clearances and the water surge can only be considered integrally. Here, compared to dry-running operation, mechanical power losses in terms of a water-flooded expander significantly increase at each rotor-tip clearance height.

In this context, hydraulic losses at the rotor tip develop differently depending on the rotor position and the applied pressure difference between two adjacent working chambers. On the one hand, during the chamber filling, a pressure-driven (Poiseuille) water flow from the high-pressure to the expanding working chamber inhibits the forming of a water surge considering low rotor tip-speeds, so hydraulic losses can be avoided. Moreover, the Poiseuille water flow drives the rotor by transferring momentum to the rotor tip and contributes to the expander power, since both pressure-driven water flow and rotor-tip motion work in the same direction. Increasing rotor tip-speed results in higher boundary-driven (Couette) against the pressure-driven (Poiseuille) water flow. As a consequence, the effect of the Poiseuille water flow declines, clearance water flow stagnates, and a water surge starts forming at the rotor tip in the leading working chamber after the Couette water flow rate becomes higher than the pressure-driven flow. At this turning point, the water surge at the high-pressure rotor-tip clearances might reveal a maximum size. A further increase in rotor tip-speed provides higher boundary-driven (Couette) water flow from the leading into the following chamber, so the water surge size declines. On the other hand, at low pressure levels during a working cycle, boundary-driven (Couette) water flow dominates especially at high rotor tip-speeds. In terms of under-expansion at low rotor tip-speeds, a pressure-driven (Poiseuille) flow reduces the boundary-driven (Couette) water flow, which initiates an increase in the water surge size. Nevertheless, the absolute hydraulic loss is relatively low due to lower rotor tip-speed. In this context, hydraulic friction in the clearances and with respect to the water surge have been analysed separately by means of theoretical investigations in [12] and [13]. In [13], hydraulic losses generated by the water surge at the rotor tip are proved to be more significant compared to water friction within the rotor-tip clearances.

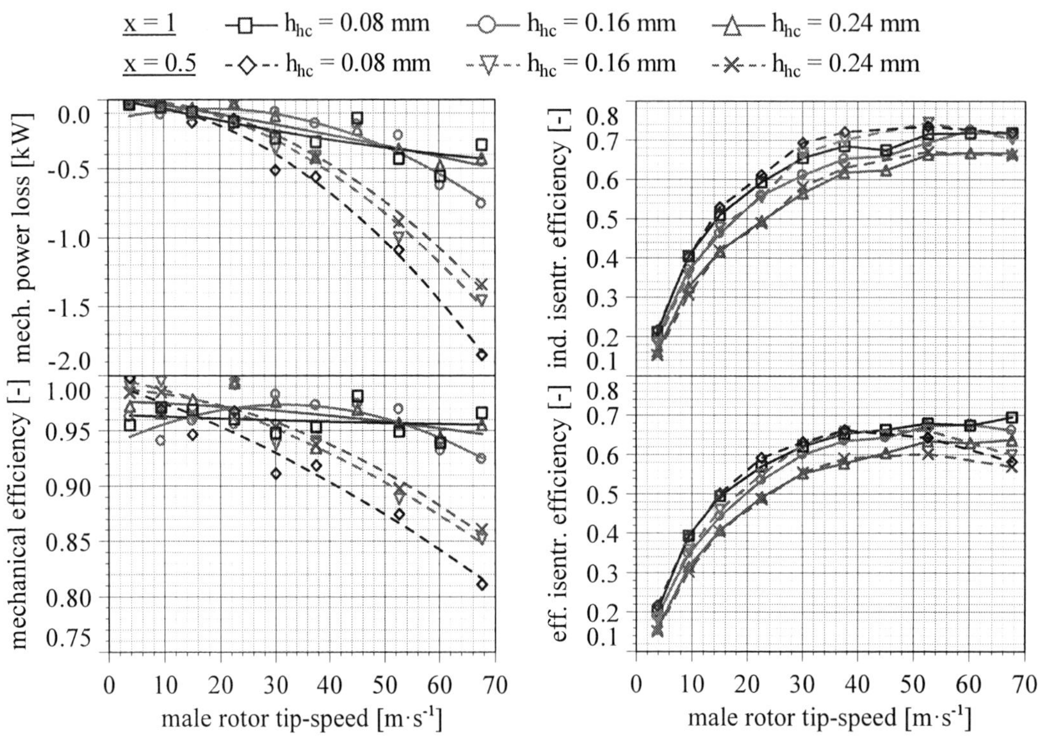

Figure 6. Mechanical power loss, mechanical efficiency, indicated isentropic efficiency, and effective isentropic efficiency depending on male rotor tip-speed at an inlet pressure of $4 \cdot 10^5$ Pa, a superficial air temperature of 90 °C, and a superficial water temperature of 60 °C; *dryness fraction: 1 and 0.5; rotor-tip clearance height: 0.08 mm, 0.16 mm, and 0.24 mm.*

With regard to increasing rotor-tip clearance height in figure 6, mechanical power losses under water-flooded conditions decline significantly, since the boundary-driven Couette water volume flow increases in relation to the clearance cross-section, so less water remains at the rotor-tip in the leading chamber and the water surge declines or disappears completely. At the same time, hydraulic shear stress within the narrow rotor-tip clearances declines due to increased clearance height that additionally contributes to less friction at the low-pressure clearances or less drive momentum at the clearances corresponding to higher pressure levels. Due to the smaller dimensions of the high-pressure compared to the low-pressure rotor-tip clearances, hydraulic losses due to water surge might significantly overshadow the drive effect in terms of a Poiseuille water flow at the high-pressure rotor-tip clearances.

Relating mechanical power losses to the available energy potential of the fluid in the working chamber, no substantial change in mechanical efficiency due to the lack of hydraulic losses at dry-running operation can be observed as a function of rotor tip-speed (figure 6). Considering a two-phase working fluid at low rotor tip-speeds, no significant influence of water on mechanical efficiency compared to dry-running operation can be detected, either. In contrast, at constant dryness fraction of 0.5, mechanical efficiency declines at increasing rotor tip-speed in consequence of rising hydraulic losses as explained before. Greater rotor-tip clearance heights reveal higher mechanical efficiency levels due to declining size of the water surge and reduced liquid shear stress in the rotor-tip clearances.

In order to quantify different loss mechanisms relative to the available two-phase working fluid potential of the twin-screw expander in figure 6, indicated and effective isentropic efficiency are presented. Considering indicated isentropic efficiency at rising male rotor tip-speed, increases can be identified under dry-running and water-flooded conditions due to the reduced influence of clearance flows on expander operation. Moreover, under-expansion is reduced and the expander operation approaches the design operation conditions when chamber pressure at the end of the expansion equals the discharge pressure. At highest rotor tip-speeds in dry-running operation, indicated isentropic efficiency declines as a consequence of overexpansion and increasing inlet throttling losses. An exception is seen for a rotor-tip clearance height of 0.24 mm where the indicated isentropic efficiency remains relatively constant for dry-running and water-flooded operation due to the slightly lower overexpansion, as presented in figure 2, resulting from the highest internal leakages.

Comparing dry-running and water-flooded operation with each other, basically higher indicated isentropic efficiency at a dryness fraction of 0.5 can be detected due to heat transfer from water into the gaseous working fluid as well as reduced internal leakages with regard to different ranges of rotor tip-speed. The lowest clearance height of 0.08 mm reveals higher indicated isentropic efficiency in the rotor tip-speed range between 10 $m \cdot s^{-1}$ and 60 $m \cdot s^{-1}$, whereas at the greatest clearance height efficiency increases for rotor tip-speeds from 30 $m \cdot s^{-1}$ to 50 $m \cdot s^{-1}$ in water-flooded operation. At the maximum rotor tip-speed, similar indicated isentropic efficiencies are disclosed for both operational modes, since relatively high inlet throttling at a dryness fraction of 0.5 eliminates the advantages of water in the working chamber. In contrast, at lower rotor tip-speed, efficiency even declines as result of relatively high energetic losses in terms of internal leakages. Considering increasing rotor-tip clearance height, indicated isentropic efficiency declines at each dryness fraction considered. This can be again traced back to the increase in internal leakage, expander air mass flow, and delivery rate as presented in section 4.2.

Despite increasing mechanical efficiency at greater rotor-tip clearance heights, a conclusive estimation of the energy conversion made by means of effective isentropic efficiency reveals, as expected, a less advantageous operation under both dry-running and water-flooded conditions. Nevertheless, at constant low clearance height, an increase in effective isentropic efficiency up to male rotor tip-speed of ca. 40 $m \cdot s^{-1}$ to 45 $m \cdot s^{-1}$ can be detected considering a two-phase working fluid. In general, this covers the part load operational range of the test twin-screw expander. Under overload operational conditions, lower effective isentropic efficiencies are determined in a water-flooded compared to a dry-running expander.

Since the interaction between the pressure-driven (Poiseuille) and boundary-driven (Couette) water clearance flow appears to have a significant impact on the development of hydraulic losses as a function of rotor tip-speed, mechanical power loss and different expander efficiencies at three inlet pressure levels are presented in order to prove this fact, figure 7. Here, for dry-running operation, no significant dependency of mechanical power loss on inlet pressure can be observed. Considering a constant dryness fraction of 0.5, increasing mechanical power loss is unveiled at higher inlet pressure levels. Hence, the pressure level in the working chamber might have a relatively strong impact on the development of the hydraulic losses and the water surge in particular.

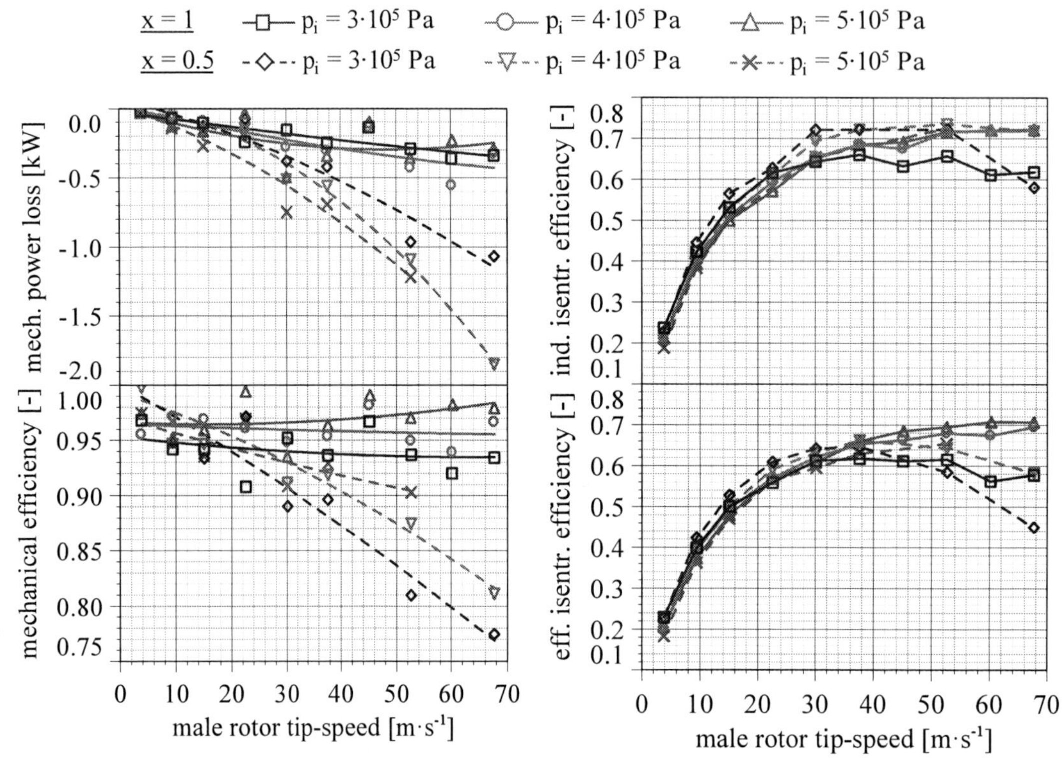

Figure 7. Mechanical power loss, mechanical efficiency, indicated isentropic efficiency, and effective isentropic efficiency depending on male rotor tip-speed at a superficial air temperature of 90 °C, a superficial water temperature of 60 °C, and a rotor-tip clearance height of 0.08 mm; *dryness fraction: 1 and 0.5; inlet pressure: $3\cdot10^5$ Pa, $4\cdot10^5$ Pa, and $5\cdot10^5$ Pa.*

Additional analysis of mechanical power loss depending on injected water volume flow and amount of water in the working chamber respectively at different inlet pressures and rotor-tip clearance heights are presented in figure 8. Here, increasing inlet pressure at rotor-tip clearance height of 0.08 mm unveils a significant rise in mechanical power losses even at lower rotor tip-speeds. In contrast, considering a rotor-tip clearance height of 0.24 mm, this dependency mostly disappears except at the highest rotor tip-speed presented. Since the expander has a fixed internal volume ratio, higher inlet pressure results in increasing pressure difference and Poiseuille water flow at the rotor-tips. The consequence is increasing water surge size at the low-pressure rotor-tip clearances due to a blocking effect of the Poiseuille flow. At the high-pressure rotor-tip clearances, the water surge might decline or disappear at increasing pressure levels, since a greater pressure difference and higher pressure-driven (Poiseuille) water volume flow respectively is present inhibiting the water surge development. Hydraulic shear stress in the narrow rotor-tip clearances at increasing inlet pressure might play a subdominant role compared to the impact of the water surge. Nevertheless, at higher rotor tip-speeds, less shear-stress induced hydraulic losses at the low-pressure rotor-tip clearances due to stagnating water flow could be expected. At the same time at low rotor tip-speeds, drive momentum transferred from the water clearance flow to the high-pressure rotor tips might increase. Bellow the line, as explained before, mechanical power losses under water-flooded operation increase, since the impact of low-pressure clearances with regard to water surge development overshadows the effect of the Poiseuille flow at high-pressure rotor-tip clearances due to greater clearance width and cross-section respectively.

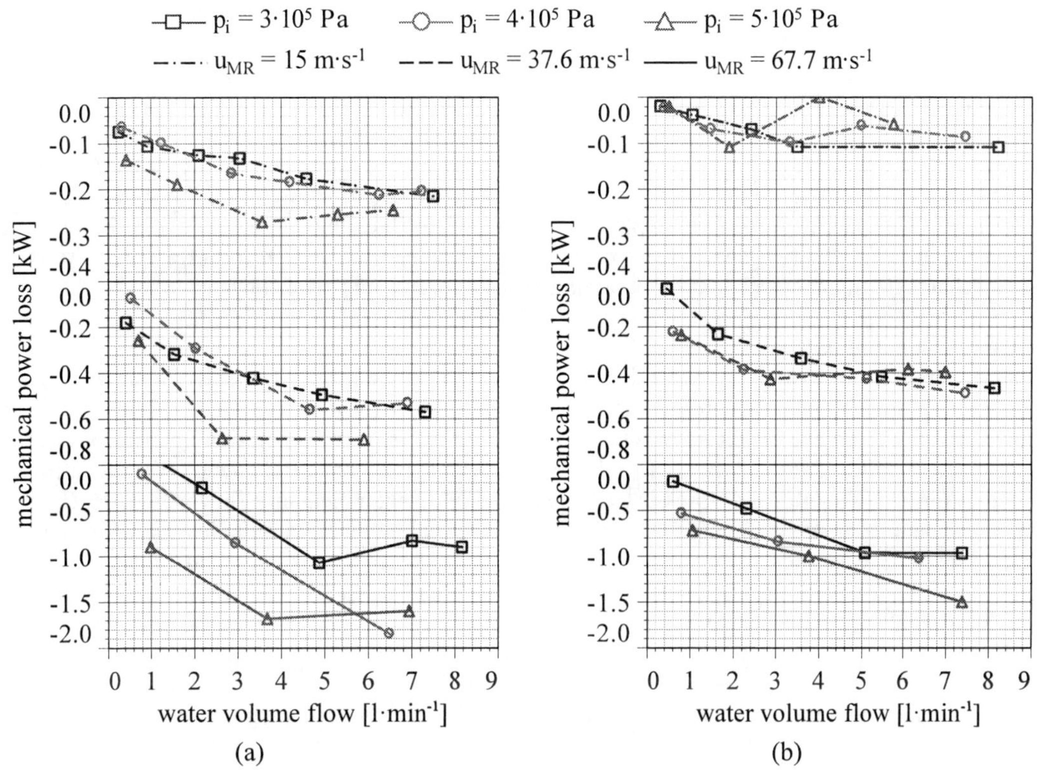

Figure 8. Mechanical power loss as a function of water volume flow at a superficial air temperature of 90 °C and a superficial water temperature of 60 °C; *inlet pressure: $3{\cdot}10^5$ Pa, $4{\cdot}10^5$ Pa, and $5{\cdot}10^5$ Pa; rotor tip-speed: 15 $m{\cdot}s^{-1}$, 37.6 $m{\cdot}s^{-1}$, and 67.7 $m{\cdot}s^{-1}$; rotor-tip clearance height: 0.08 mm (a) and 0.24 mm (b).*

For dry-running operation, no significant dependency of mechanical efficiency on rotor tip-speed can be detected at each inlet pressure presented in figure 7. Under water-flooded conditions, mechanical efficiency decreases at higher rotor tip-speed as discussed before according to figure 6. At constant rotor tip-speed, in contrast to mechanical power loss, mechanical efficiency increases at higher inlet pressure levels under both dry-running and water-flooded conditions. This can generally be traced back to the simultaneously increasing indicated power at higher inlet pressures that even compensates the increase in mechanical power losses under water-flooded conditions. Therefore, at constant rotor tip-speed and water-flooded operation, maximum mechanical efficiency is detected basically at an inlet pressure of $5{\cdot}10^5$ Pa despite maximum mechanical power losses.

In terms of indicated and effective isentropic efficiency at increasing inlet pressure in figure 7, the maximum values generally move to higher rotor tip-speeds for both dryness fractions presented. At constant internal volume ratio, this can be explained, on the one hand, by a greater rotor tip-speed according to transition from under to overexpansion and internal leakages. On the other hand, at high rotor tip-speeds, overexpansion in terms of inlet throttling declines at higher inlet pressure.

Comparing dry-running to water-flooded operation at the maximum inlet pressure of $5{\cdot}10^5$ Pa, a relatively constant indicated isentropic efficiency can be determined. In terms of a two-phase working fluid, the increasing chamber pressure due to heat flow from water into the gaseous working fluid can explain this fact. Thus, a greater under-expansion at constant expander internal volume ratio is the consequence. Moreover, at low rotor tip-speeds, this effect is additionally sustained by a relatively high pressure-driven (Poiseuille) water volume flow and, hence, less effective clearance sealing. Considering effective isentropic efficiency under water-flooded conditions, lower values compared to dry-running operation result from rising hydraulic losses at higher rotor tip-speeds.

In contrast, a significant increase in indicated isentropic efficiency at an inlet pressure of $3 \cdot 10^5$ Pa is achieved under water-flooded conditions related to dry-running operation. As opposed to the operation at an inlet pressure of $5 \cdot 10^5$ Pa, a more effective clearance sealing due to lower clearance pressure difference and lower pressure-driven (Poiseuille) water flow corresponds to higher efficiency values. At maximum rotor tip-speed, increased overexpansion according to greater inlet throttling reduces the efficiency. Considering effective isentropic efficiency, the sealing effect at low rotor tip-speeds dominates in terms of a two-phase working fluid. At the same time at high rotor tip-speeds, relatively low mechanical efficiency account for declining effective isentropic efficiency.

5. Conclusion and outlook

This paper presents the results of experimental investigations into a water-flooded twin-screw expander with varying operational parameters considering three different rotor-tip clearance heights. This work shows, that the clearance height has a relevant impact on the expander operation in terms of clearance sealing effect and hydraulic losses. In this context, two significant mechanisms were identified—a boundary-driven (Couette) and a pressure-driven (Poiseuille) water clearance flow which, depending on the operational parameters, impact the level of clearance sealing and hydraulic loss induction. Hence, greater rotor-tip clearance height results in lower hydraulic friction losses and higher mechanical efficiencies due to declining shear stress in the clearance and water surge size at the rotor tip. At the same time, analysis of mechanical losses at different inlet pressures and rotor-tip clearance heights have evidenced the significant effect of water surge on the operation of the water-flooded twin-screw expander. Even though a greater rotor-tip clearance height provides an increase in mechanical efficiency, indicated and effective isentropic efficiency reveal worse energy conversion considering both dry-running and water-flooded operation. Finally, the advantages of water-flooded twin-screw expander were identified at part load operation, whereas at higher rotor tip-speeds and overload operation, hydraulic losses significantly affect the effective isentropic efficiency.

In future works, the effect of each twin-screw expander clearance type—rotor-tip clearance, front clearance, interlobe clearance, and blow hole in particular—should be clarified. A further goal is to quantify the determined influence mechanisms by means of characteristic numbers in order to generate a characteristic map of liquid-flooded twin-screw expanders.

References

[1] Ohman H and Lundqvist P 2015 SCREW EXPANDERS IN ORC APPLICATIONS, REVIEW AND A NEW PERSPECTIVE *3rd International Seminar on ORC Power Systems (Brussels, Belgium, 12-14 October)* ed ASME ORC 2015

[2] Hütker J 2016 Energiewandlung in trockenlaufenden Schraubenmotoren *PhD thesis* TU Dortmund University

[3] Kauder K (ed) 1993 *Schraubenmaschinen Nr. 1: Betriebssicherheit und Steuerungen von Schraubenmaschinen (Schraubenmaschinen* Nr. 1*)* (Dortmund: Univ., FG Fluidenergiemaschinen)

[4] Rinder L 1979 *Schraubenverdichter* (Wien, New York: Springer-Verlag)

[5] Brümmer A 2012 Energy efficiency – waste heat utilisation with screw expanders *Prozesstechnik und Komponenten* 120–6

[6] Harling H-B 1994 Untersuchungen zur Ölverteilung in Schraubenkompressoren mit Schmiermitteleinspritzung *PhD thesis* TU Dortmund University

[7] Zellermann R 1996 Optimierung von Schraubenmotoren mit Flüssigkeitseinspritzung *PhD thesis* TU Dortmund University

[8] Kauder K and Zellermann R 1994 Einspritzflüssigkeiten im Schraubenmotor *Schraubenmaschinen '94: Schraubenkompressoren, Schraubenlader, Schraubenmotoren; Tagung Dortmund, 5. und 6. Oktober 1994 (VDI-Berichte* vol 1135*)* ed VDI (Düsseldorf: VDI-Verlag) pp 153–74

[9] Kauder K 1987 Das Öl im Schraubenkompressor - ein Faktor für optimale Betriebsverhaeltnisse *Pumpen Vakuumpumpen Kompressoren, Nürnberg: Harnisch* 38–44

[10] Gräßer M and Brümmer A 2014 An analytic model of the incompressible one-phase clearance flow in liquid injected screw expanders *International Conference on Screw Machines 2014 (TU Dortmund University, Germany, 23rd-24th September) (VDI-Berichte* vol 2228*)* (Düsseldorf: VDI-Verl.) pp 71–89

[11] Gräßer M and Brümmer A 2015 Influence of liquid in clearances on the operational behaviour of twin screw expanders *IOP Conference Series: Materials Science and Engineering* **90** 12060

[12] Gräßer M and Brümmer A 2016 Influence of water and oil clearance flow on the operational behavior of screw expanders *Proceedings of the Institution of Mechanical Engineers, Part E: Journal of Process Mechanical Engineering* **231** 38–46

[13] Vasuthevan H and Brümmer A 2018 Multiphase-flow simulation of a rotating rectangular profile within a cylinder in terms of hydraulic loss mechanisms *International Conference on Screw Machines (TU Dortmund University, Germany, 18th-19th September)* ed IoP

[14] Nikolov A and Brümmer A 2017 Investigating a Small Oil-Flooded Twin-Screw Expander for Waste-Heat Utilisation in Organic Rankine Cycle Systems *Energies* **10** 869

[15] Kliem B 2005 Grundlagen des Zweiphasen-Schraubenmotors *PhD thesis* TU Dortmund University

[16] Vasuthevan H and Brümmer A 2016 Thermodynamic Modeling of Screw Expander in a Trilateral Flash Cycle *23rd International Compressor Engineering Conference at Purdue: West Lafayette, Indiana, USA, 11-14 July 2016 (Purdue)* ed Purdue University Libraries

[17] Svigler J, Rinder L, Rehounek L and Kuchler M 2005 Temperaturdeformation der Rotoren bei nass laufenden Schraubenkompressoren *Schraubenmaschinen: KaSim, Simulationen, Temperaturdeformationen, Energiewandler in BHKW's, GASSCREW, SCREW (Schraubenmaschinen)* ed K Kauder pp 17–36

[18] Nikolov A and Brümmer A 2014 Influence of water injection on the operating behaviour of screw expanders: Experimental investigation *International Conference on Screw Machines 2014 (TU Dortmund University, Germany, 23rd-24th September) (VDI-Berichte* vol 2228*)* (Düsseldorf: VDI-Verl.) pp 43–60

[19] Nikolov A and Brümmer A 2016 Analysis of Indicator Diagrams of a Water Injected Twin-shaft Screw-type Expander *23rd International Compressor Engineering Conference at Purdue: West Lafayette, Indiana, USA, 11-14 July 2016 (Purdue)* ed Purdue University Libraries

[20] Baehr H D and Kabelac S 2012 *Thermodynamik: Grundlagen und technische Anwendungen (Springer-Lehrbuch)* 15th edn (Berlin u.a.: Springer Vieweg)

IOP Conf. Series: Materials Science and Engineering **425** (2018) 012009 doi:10.1088/1757-899X/425/1/012009

Effect of oil-injection on twin screw compressor performance

N Basha, A Kovacevic, N Stosic, I Smith

Centre for Compressor Technology, City, University of London, London, UK

E-mail: nausheen.basha@city.ac.uk

Abstract. Oil-injection improves twin-screw compressor performance significantly and its overall effect depends on the working fluid mass flow rate, the shaft speed, the injection position and the injected oil temperature. However, the individual contribution of each of these variables to the machine performance is yet not fully researched. Therefore, tests have been carried out on an air compressor with a male rotor diameter of 98mm, with injection nozzle sizes of 2, 3 and 4mm and with a nozzle of multiple holes. Tests were carried out with discharge pressures of 6.5, 8.5, 10.5 and 12.5 bar and rotor shaft speeds of 3000, 4000 and 5000 RPM. The results from the tests were compared with performance estimates at the same operating conditions as measurements against the standard injection nozzle of 5mm diameter. It was found that there is an optimum quantity of oil injected to obtain the best performance.

1. Introduction

The effect of oil injection on screw compressor performance is dependent on the compressor tip speed, the injected fluid mass flow rate, temperature, viscosity, and the fineness of its atomisation and the position of the oil injection port. It is well known that, the compressor efficiency is improved by this means, but as shown by Deipenwisch and Kauder [1], an excess amount of oil leads to frictional and momentum losses that increase power consumption. Other investigators have carried out analytical studies on how each of these parameters affect performance.

In 1986, Singh and Bowman [1] developed a mathematical model to evaluate the effect of oil droplet size on compressor performance. They reported that a droplet of 1000 μm diameter reduces the specific power by 7% whereas a droplet of 100 μm diameter reduces the specific power by 8.3%. Although the authors state these results agreed with experimental data, the details were not reported. Stosic et al. (1988) [2] used a mathematical model to investigate the effect of parameters such as oil inlet temperature, oil to gas mass ratio, oil injection position and oil viscosity. The effect of outlet temperature was shown for 100 μm, 500 μm and 1000 μm diameter droplets. It was concluded that the oil temperature closely follows that of the gas and atomisation only has a slight effect on performance although they did not consider droplet diameters below 100 μm. The authors concluded that the oil mass flow and viscosity have a surprisingly small effect on specific power compared to the oil injection temperature and the position of the oil port. Thus, modifying the injection position in the test rig improved the specific power by 7% compared to that of the basic design [3][4]. They reported a comparison with experimental data for a single operating pressure when running at 2490 RPM. However, little was published to confirm these predictions.

Peng et. al (2000) [5] studied the effect of oil to gas ratio on the isentropic and volumetric efficiencies at varying rotational speeds. Also, the oil droplet distribution was viewed through a transparent test chamber, but the droplet sizes were neither estimated nor measured. De Pape et al (2005) [6] introduced a factor called the cooling effectiveness and studied it with nozzles of varying diameter in a cylindrical test section with the capability of generating a smallest droplet size of Sauter Mean Diameter (SMD) of 278 μm. The cooling effectiveness was introduced as a factor in a thermodynamic model, and by this means it was shown that raising the oil flow by 10% lowered the specific power by 0.2% and raising the cooling effectiveness by 10 points lowered the specific power by 0.05%. However, these results were not confirmed with experimental data. The cooling effectiveness was expressed as a function of temperatures which were different in the cylindrical test rig and the actual compressor as the heat transfer between the oil and gas is dependent upon the oil residence time within the chamber.

In recent years, the effect of oil injection has been studied in positive displacement rotary vane machines both experimentally and mathematically. The optimum value of liquid to gas mass ratio was estimated mathematically and the droplet diameter was estimated from experimental data. As a result, an advanced orifice was developed from consideration of these parameters [7]. A comparison was made between a conventional and advanced orifice in a rotary vane compressor at design conditions which indicated that the specific power was reduced by 2.4% by means of the advanced orifice [8]. In addition, pressure swirl atomisers were tested in the vane machine resulting in a saving of 0.5kW/(m^3/min) in specific power [9].

Most of the screw compressor experimental studies did not include the variation in oil related parameters over a range of discharge pressures and rotor speeds, although these were considered, to a good extent, in the mathematical models.

In order to gain a better understanding of the effects of oil injection on specific power over a range of operating conditions, the authors carried out tests on an oil-injected air compressor of 98mm male rotor diameter using oil injection nozzles of 2mm, 3mm and 4mm diameter at discharge pressures of, 6.5 bar, 8.5 bar, 10.5 bar and 12.5 bar and male rotor rotational speeds of 3000 RPM, 4000 RPM, and 5000 RPM. In addition to this, a multiple holes nozzle with an equivalent cross sectional area of a 4 mm diameter was used. The test results obtained were normalised with the existing design nozzle of 5 mm diameter and then compared, in order to gain a better understanding of the significance of each variable nozzle.

2. Case Study

2.1. *Compressor Specification*

The reference oil-injected compressor has a radial suction port and an axial discharge port with oil injection port on the female side of the rotor. The lobe combination is 4-5 and blow-hole area is relatively large. The rotor centre distance is 67.5 mm, the outer diameter of the male rotor is 98.8 mm and the outer diameter of the female rotor is 77.8 mm. The length of the rotors is 153.140 mm with the male rotor wrap angle of 306.647° and volume index of 4.6.

2.2. *Experimentation*

The air test rig at City, University of London is shown in Figure 1. The test rig was built to meet CAGI and PNEUROP test standards where the testing procedures are carried out according to ISO 1706. The compressor is driven by a variable speed electric motor through a belt drive. Speed variation is obtained by use of a frequency inverter to allow electric motor of 75 kW. Oil is injected into the compressor from main oil supply manifold. The oil injection process is driven by the difference in the pressure at the after discharge and the pressure in the cavity at the point of injection. This limits injection characteristics unless an external system is used to control the injection pressures. However, the power required to drive the external system should balance the savings in power achieved through improved oil injection.

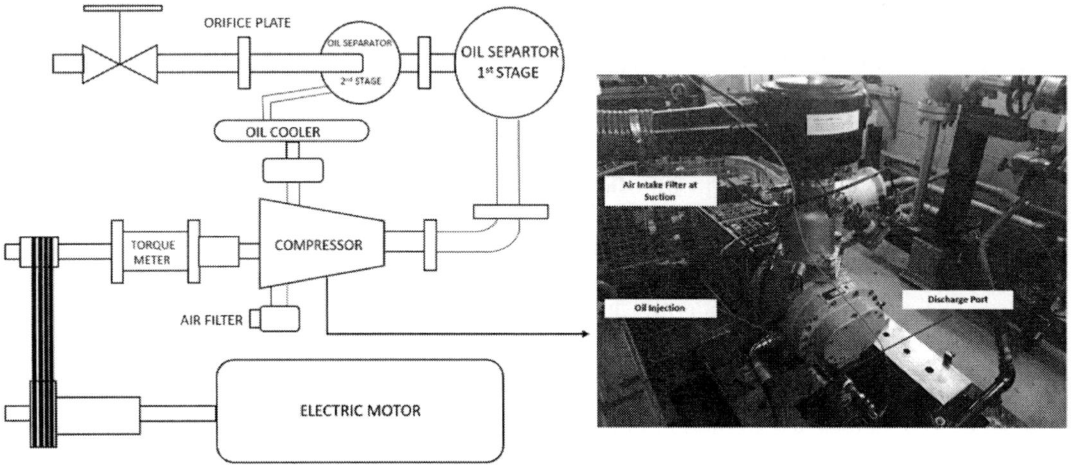

Figure 1. Air test rig.

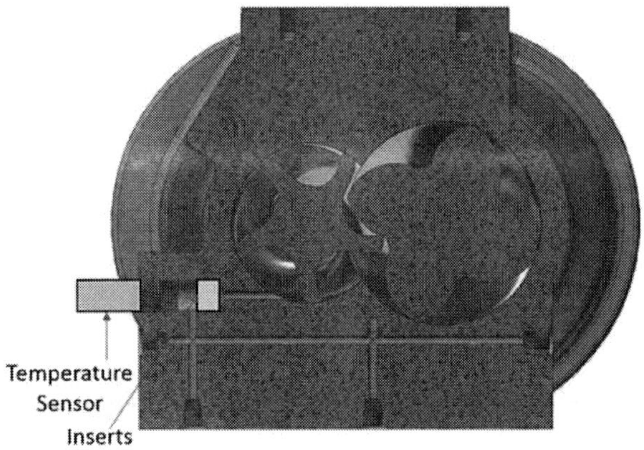

Figure 2. Cross sectional view of compressor with inserts.

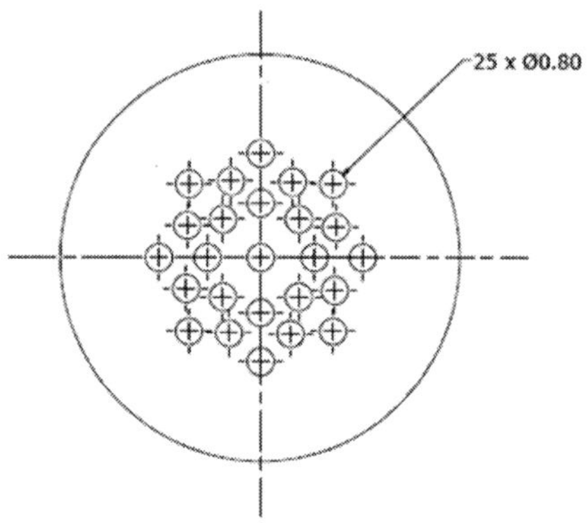

Figure 3. Atomiser or Insert with equivalent cross sectional area of 4 mm (25 x 0.8 mm).

An orifice plate is used to determine the air flow rate and measured according to BS 5600. The plate is, sandwiched between two flanged pipes, located between the globe valve and exit of the second stage oil separator. The details on the installed instruments in the test rig are shown in Table 1. The existing compressor test rig is modified to add the temperature sensor and the insert which can be easily replaced to test the influencing parameters. Inserts of perspex material with diameters of 2 mm, 3mm, 4mm and 0.8mm x 25 holes (Figure 3) will be used for this study. The operating conditions for the test are shown in Table 1.

Table 1. List of installed instruments.

MEASURED	INSTRUMENT	SPECIFICATIONS
Compressor Speed, N	Shaft Encoder (BHG 16.25W.3600-B2-5)	360 pulses per revolution Accuracy= 2.7%
Compressor Torque, Tq	TRP-500 torque meter (strain gauge transducer)	max capacity: 500Nm, Calibration level: 335Nm Range = 0 - 6000 rpm, Supply volt=10v dc, Accuracy= 0.25 % of max capacity
inlet pressure, P_1	PDCR 110/w -pressure Transducer	Operating range = 3.5bar(abs) Excite voltage=10V dc, Accuracy =0.6%,
inlet temperature, T_1	Platinum Resistance Thermometer	Range= -750C to 3500C, Accuracy= ±0.50C
outlet pressure, P_2	PDCR 922-pressure transducer	Operating range =15 bar (abs) Excite voltage=10V dc, Output voltage= 100 mV Accuracy =0.6%
outlet temperature, T_2	Platinum Resistance Thermometer	Range= -750C to 3500C, Accuracy= ±0.50C
orifice plate inlet press, P_{op}	PDCR 922-pressure transducer	Operating range =15 bar (abs) Excite voltage=10V dc, Output voltage= 100 mV
orifice plate inlet Temp, T_{op}	Platinum Resistance Thermometer	Range= -750C to 3500C, Accuracy= ±0.50C
Orifice plate differential pressure, dp	PDCR 2120- pressure Transducer	Pressure diff= 0.35 bar, Excitation Voltage=10V dc, Accuracy= 0.6%,
Oil Temp @ inlet of the cooler– $T_{oil,IN}$	K- type thermocouple (based on Ni/Cr-Ni/Al alloy)	Range= -2000C to 13000C, Accuracy= ± 10C sensitivity = 41 μV/°C
Oil Temp @ outlet of the cooler–$T_{oil,OUT}$	Platinum Resistance Thermometer	Range= -750C to 3500C, Accuracy= ±0.50C
Water Temp @ inlet & outlet of the cooler– $T_{w,IN}$ & $T_{w, OUT}$	K- type thermocouple (based on Ni/Cr-Ni/Al alloy)	Range= -2000C to 13000C, Accuracy= ± 2.20C sensitivity = 41 μV/°C

The performance of the compressor is measured in terms of power (kW), flow rate (m^3/min) and specific power (kW/m^3/min). The flow is measured using an orifice plate while the power is calculated based on the measurements of torque and speed according to BS1042 [10].

$$Power = \frac{Tq \times N}{60} \tag{1}$$

$$AF - \frac{C_d e\pi D_i^2 \sqrt{2dp\rho_0}}{4\sqrt{1-\beta^4}} \tag{2}$$

In the above equation 2, C_d is the discharge coefficient, e is the expansion factor, D_i is the orifice diameter (m) and ρ (kg/m^3) is the density of the fluid. β is the ratio of orifice diameter (D_i) to pipe diameter (D).

$$\beta = \frac{D_i}{D} \tag{3}$$

C_d, and e can be further defined as (equation 4 and equation 6):

$$C_d = 0.5959 + 0.0312\beta^{2.1} - 0.1840\beta^8 + 0.0029\beta^{2.5}\left(\frac{10^6}{Re}\right)^{0.75} + \frac{0.09L_1\beta^4}{(1-\beta^4)} - 0.033L_1(\frac{1}{2})\beta^3 \tag{4}$$

Reynolds number (Re) in the above equation can be written as,

$$Re = \frac{4AF}{\pi D_i \mu} \tag{5}$$

In the above equation, the viscosity of the fluid μ in kg/m-s is obtained through Sutherland viscosity law which is dependent upon the orifice inlet temperature (T_{op}) in K [11].

$$\mu = 1.458 \times 10^{-6}\frac{T_{op}^{1.5}}{T_{op} + 110.4} \tag{6}$$

Note that an iterative calculation is needed to get a value for C_d (equation 4), because the upstream velocity needed for Re isn't known until C_d is known.

$$e = 1 - 0.41 + 0.35\beta^4(\frac{dp}{1.4P_{op}}) \tag{7}$$

Lastly, the specific power (SP) in kW/m3/min can be written as,

$$SP = \frac{Power}{AF} \tag{8}$$

Here, power is in kW and flow rate is in m^3/min. Finally, the mass flow of oil can be estimated as:

$$\dot{m}_{oil} = \frac{(Power - \dot{m}_{air}C_{p\,air}(T_2 - T_1))}{C_{p\,oil}(T_2 - T_{oil,IN})} \tag{9}$$

T_2 is the discharge temperature, T_1 is the suction temperature and $T_{oil,IN}$ is the oil temperature. When the oil flow is expressed in lt/min, the density of oil for conversion is taken as 860 kg/m^3.

The instruments used for the experiments have an error percentage and this error is reflected in the macro properties like air flow, power and specific power. The error propagation for air flow measured through orifice plate, power, specific power and oil flow uncertainties are added in a quadrature [12] considering their dependent variables as shown in equation 10, 11, 12 and 13. Error percentage for β value of 0.198 is 4% and C_d is 0.2% [13]. These give an error propagation in air flow (δAF) measurement of 4.7%. Similarly, error propagation for power is 2.7%, specific power is 5.5% and oil flow (OF) is 12.77%

$$\delta Power = \sqrt{(\delta Tq)^2 + (\delta N)^2} \tag{10}$$

$$\delta AF = \sqrt{2(\delta dp)^2 + 2\left((\delta P_{op})^2 + (\delta T_{op})^2\right) + (\delta\beta)^2 + (\delta C_d)^2} \quad (11)$$

$$\delta Sp = \sqrt{(\delta AF)^2 + (\delta Power)^2} \quad (12)$$

$$\delta OF = \sqrt{(\delta Power)^2 + (\delta AF)^2 + (\delta P_1)^2 + 2((\delta T_1)^2 + (\delta T_2)^2) + (\delta T_{oil,IN})^2} \quad (13)$$

Table 2. Compressor testing conditions for inserts.

Discharge Pressure (bar)/Speed (RPM)	6.5	8.5	10.5	12.5
3000	Except 2mm	All inserts	All inserts	All inserts
4000	Except 2mm	All inserts	All inserts	All inserts
5000	Except 2mm	All inserts	All inserts	All inserts

All the inserts used for the tests are mentioned in Table 2. The inserts of 2mm could not be tested at 6.5 bar because of the reduced oil flow and the compressor discharge temperature exceeding limits of 100°C. Considering safety regulations, the test rig cannot be operated beyond this temperature.

The inlet temperatures for oil and air should be ideally kept constant. However, there were limitations with the test rig due to which the inlet temperatures of air or oil could not be kept constant. Hence, 'Temperature Difference Ratio' (TDR) is introduced. TDR is used to quantify the effect of oil acting as a coolant by looking at the compressor discharge temperature (T_2) in relation with the inlet temperature (T_1). Temperature difference ratio is the ratio of temperature difference due to air with temperature difference due to injection of oil.

$$Temperature\ difference\ ratio = \frac{T_2 - T_1}{T_2 - T_{oil,IN}} \quad (14)$$

3. Results and Discussion

3.1. Performance measurements at various operating conditions

Oil mass flow in the compressor can be divided into the main and auxiliary injection. The main injection comes through the direct injection into the chamber and the auxiliary injection of oil seeps through the bearings. By adding in the different size of nozzles, effectively changes the main injection flow rate but the auxiliary oil flow remains the same. The actual prototype design considers oil nozzle diameter of 5mm. The performance values along with the temperatures and volume flow of oil obtained for the design conditions are tabulated in Table 3, Table 4 and Table 5.

Table 3. Performance values with 5mm oil injection nozzle.

Speed (RPM)	Air Flow m³/min				Power kW				Specific Power (kW/m³/min)			
	Discharge Pressure (bar)				Discharge Pressure (bar)				Discharge Pressure (bar)			
	6.5	8.5	10.5	12.5	6.5	8.5	10.5	12.5	6.5	8.5	10.5	12.5
3000	1.76	1.77	1.79	1.79	13.14	16.66	18.58	20.19	7.46	9.42	10.36	11.27
4000	2.44	2.42	2.41	2.40	18.25	21.62	23.68	25.48	7.49	8.93	9.83	10.60
5000	3.08	3.06	3.05	3.04	22.68	26.70	29.27	31.40	7.36	8.73	9.58	10.33

Table 4. Discharge, inlet and oil injection temperature with 5mm oil injection nozzle.

Speed (RPM)	Discharge Temperature (T2)				Inlet Temperature (T_1)				Oil Injection Temperature ($T_{oil,IN}$)			
	°C				°C				°C			
	Discharge Pressure (bar)				Discharge Pressure (bar)				Discharge Pressure (bar)			
	6.5	8.5	10.5	12.5	6.5	8.5	10.5	12.5	6.5	8.5	10.5	12.5
3000	59.9	55.2	56.1	56.1	30.4	31.3	36.8	42.6	31.0	32.8	37.2	39.0
4000	66.4	63.2	64	64.7	29.3	29.5	32.1	36.9	38.6	35.8	41.5	44.3
5000	73.2	68.4	70.6	73.6	28.9	26.4	32.1	40.1	36.6	37.6	44.3	49.4

Table 5. Oil flow with 5mm oil injection nozzle.

Speed (RPM)	Oil Flow			
	lt/min			
	Discharge Pressure (bar)			
	6.5	8.5	10.5	12.5
3000	13.86	22.83	30.59	37.52
4000	19.35	23.78	32.12	38.31
5000	17.81	25.55	33.36	39.63

The main oil flow is varied with orifices of 2mm, 3mm, 4mm and the resulting performance values of power, air flow, specific power, oil flow are normalised according to the values listed in Table 3 and Table 4. For example, the power for 2mm nozzle at 8.5bar- 3000RPM is divided by power for 5mm nozzle at 8.5bar-3000 RPM. Similarly, the air flow for 3mm nozzle at 10.5bar- 5000RPM is divided by the air flow for the 5mm nozzle at 10.5bar-5000RPM. In short, all the performance values for various nozzles are normalised with the values of 5mm nozzle with relevant discharge pressures and male rotor shaft speeds. These normalised values for various nozzles are plotted in the figures below. '4.1mm' nozzle diameter in the below plots represents the nozzle of type describes in Figure 3, which has an equivalent cross sectional area of 4mm with multiple holes.

From Figure 4, Figure 9 and Figure 14 it can be concluded that the variation of oil flow has a higher effect on power than the flow rate. At 6.5bar and 8.5bar, power increases with the increasing nozzle diameters (or oil flow) from 3mm to 5mm. On other hand, power decreases with the increasing nozzle diameter from 3mm to 5mm for higher pressures of 10.5bar and 12.5bar. However, for the nozzle diameter of 2mm higher values of normalised power are observed at all measured discharge pressures (6.5bar, 8.5bar, 10.5bar and 12.5bar) and male rotor shaft speeds (3000RPM, 4000RPM and 5000RPM). This indicates that if less amount of oil is present for the purpose of cooling in the compression chamber then it leads to over compression within the compressor pockets leading to higher consumption of indicated power. Saying that, higher amount of oil in the compression chamber can lead to frictional and momentum losses which in turn result in power losses [14]. Therefore, in order to achieve a better power consumption or specific power an optimal quantity of oil is required for the tested operating conditions.

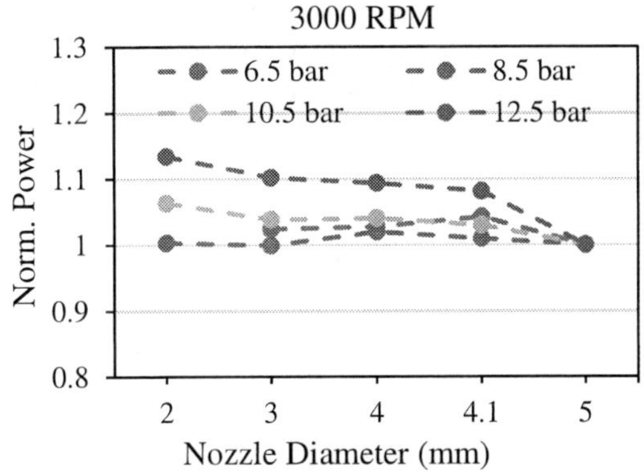

Figure 4. Normalised power with nozzle diameter at 3000 RPM.

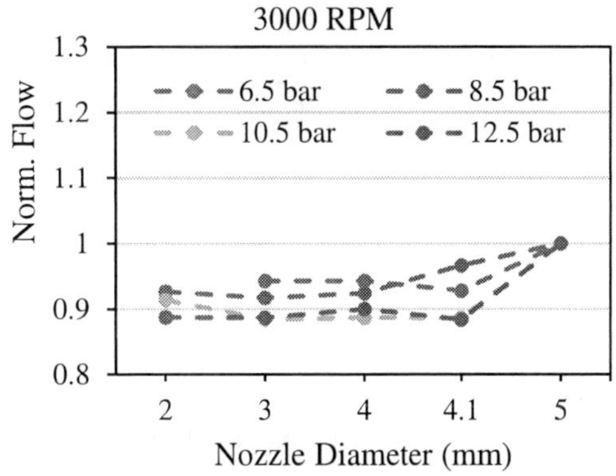

Figure 5. Normalised flow with nozzle diameter at 3000 RPM.

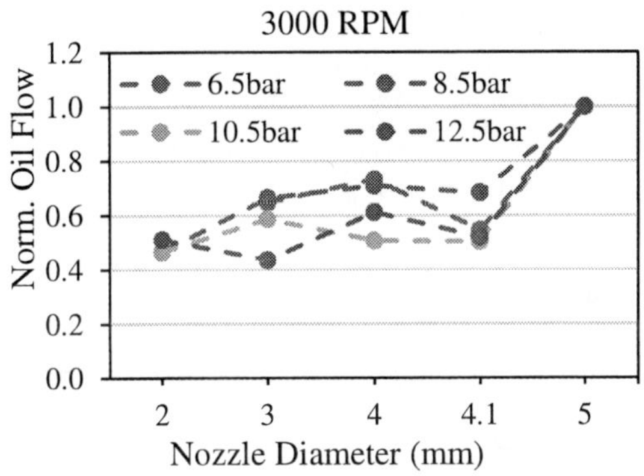

Figure 6. Normalised oil flow with nozzle diameter at 3000 RPM.

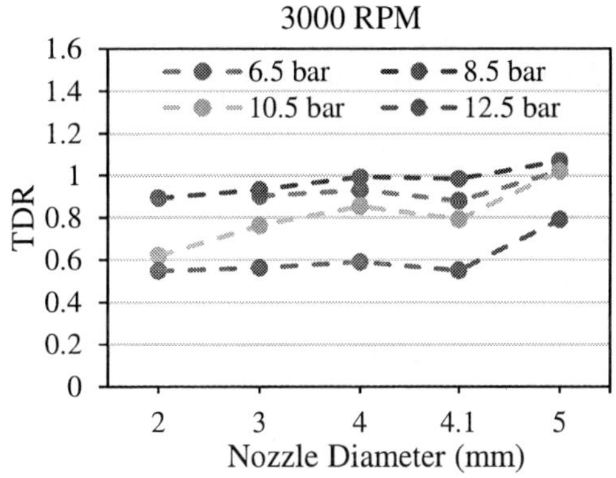

Figure 7. Temperature difference ratio with nozzle diameter at 3000 RPM.

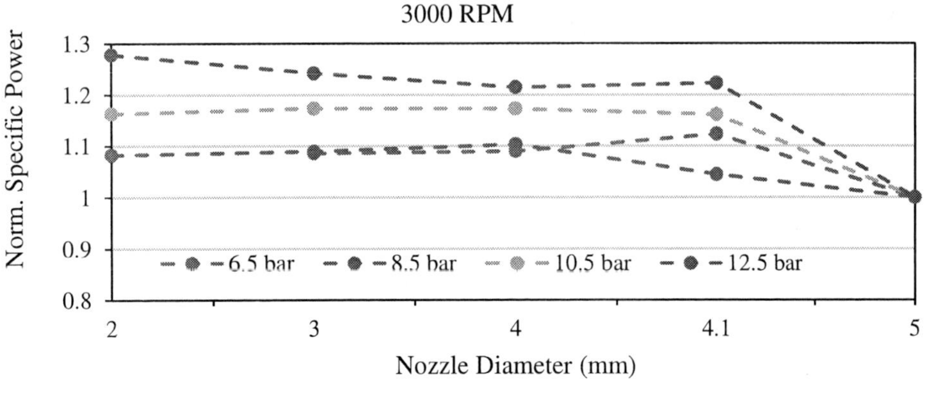

Figure 8. Normalised specific power with nozzle diameter at 3000 RPM.

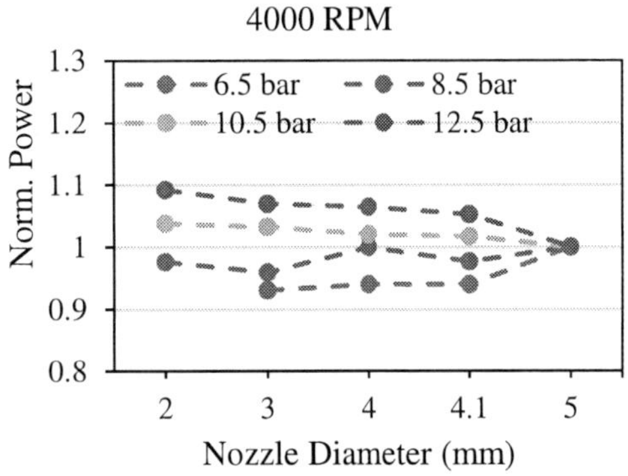

Figure 9. Normalised power with nozzle diameter at 4000 RPM.

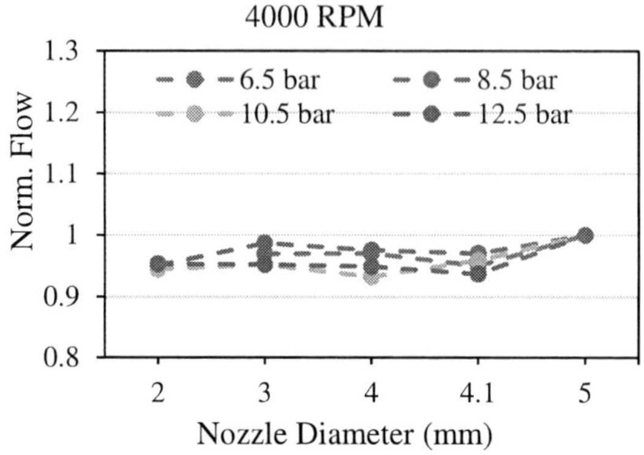

Figure 10. Normalised flow with nozzle diameter at 4000 RPM.

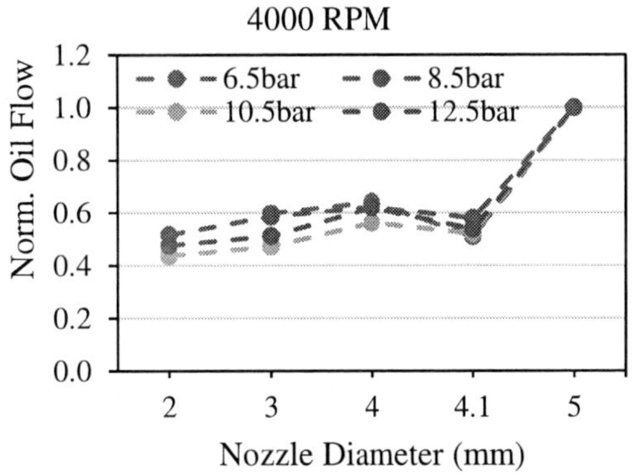

Figure 11. Normalised oil flow with nozzle diameter at 4000 RPM.

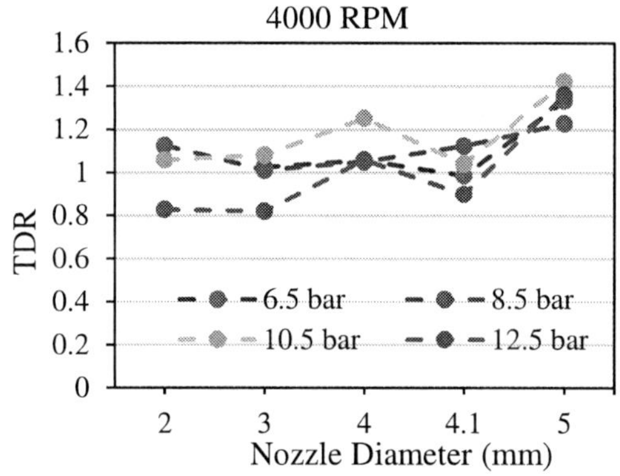

Figure 12. Temperature difference ratio with nozzle diameter at 4000 RPM.

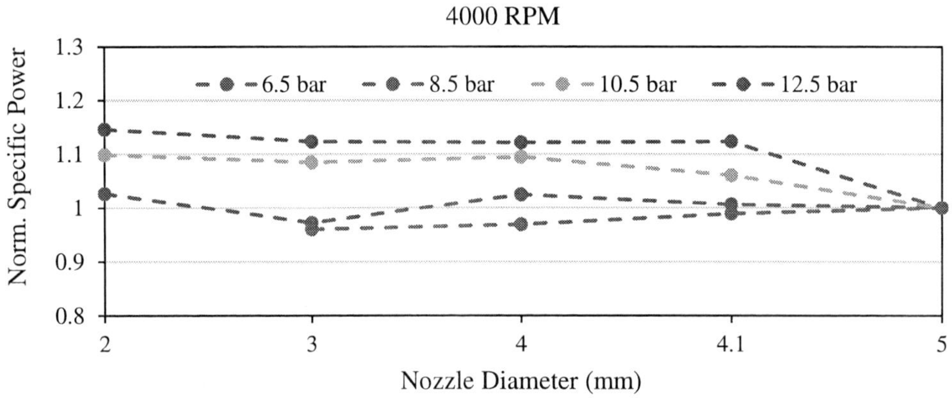

Figure 13. Normalised power with nozzle diameter at 4000 RPM.

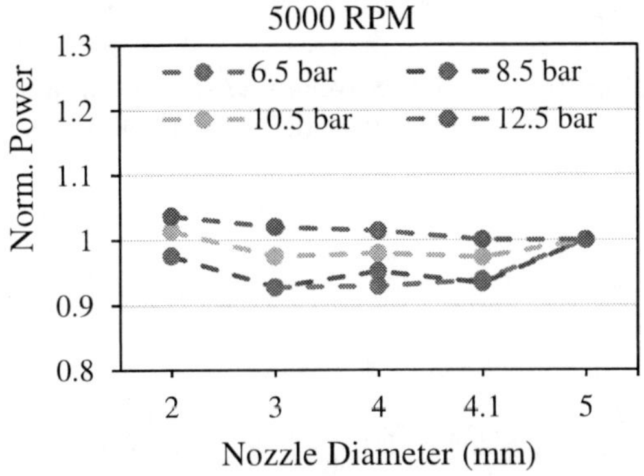

Figure 14. Normalised power with nozzle diameter at 5000 RPM.

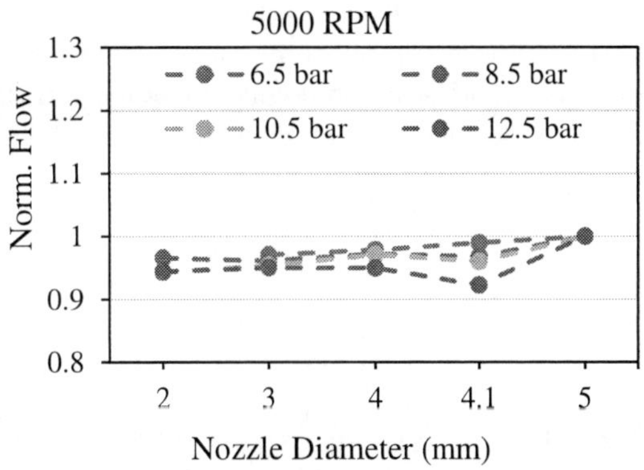

Figure 15. Normalised flow with nozzle diameter at 5000 RPM.

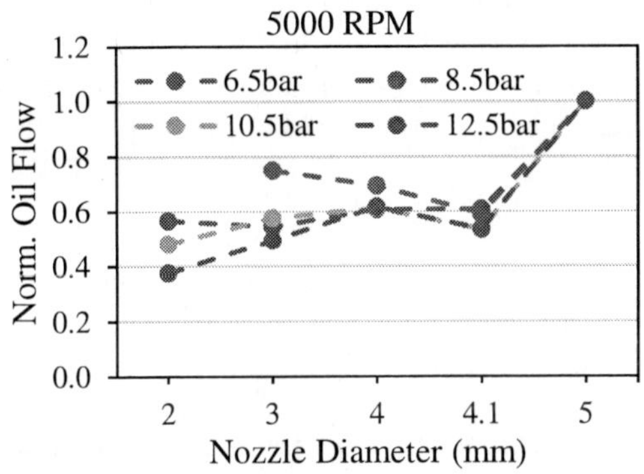

Figure 16. Normalised oil flow with nozzle diameter at 5000 RPM.

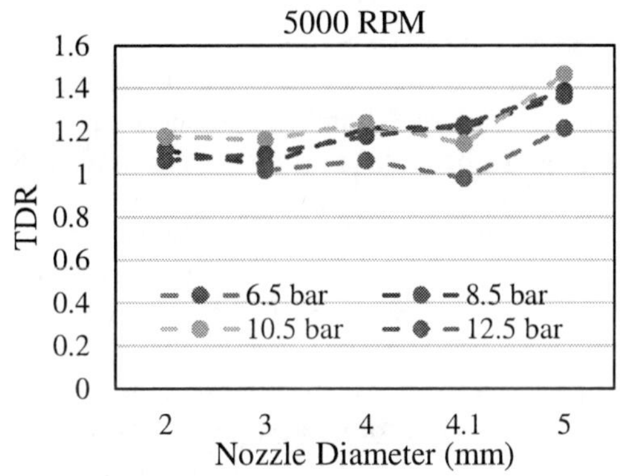

Figure 17. Temperature difference ratio with nozzle diameter at 5000 RPM.

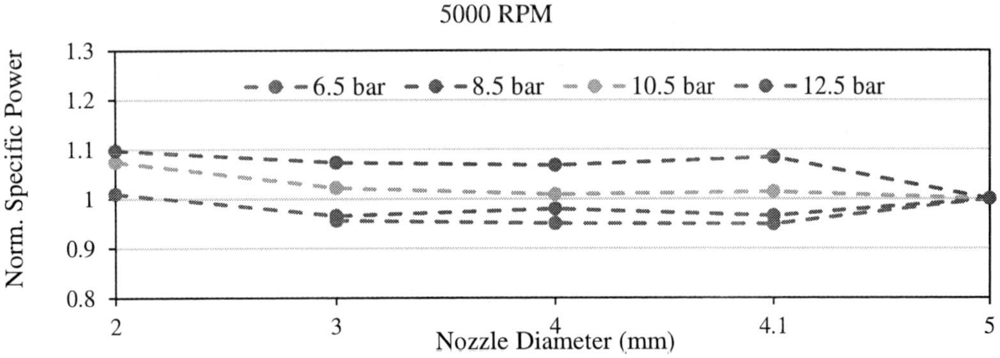

Figure 18. Normalised specific power with nozzle diameter at 5000 RPM.

Additionally, comparing the normalised power values for the male rotor shaft speeds of 3000 RPM, 4000 RPM and 5000 RPM, it can be observed that the overall normalised power consumed at 5000 RPM is lower than at 4000 RPM, while the normalised power at 4000 RPM itself is lower compared to 3000 RPM for the same discharge pressures. The main reason for this is that at higher speeds there are reduced leakages. Another reason might be that the shearing of oil films at higher rotational speeds can lead to formation of annular mist flow within the compression chamber. This annular mist flow has an increased oil droplets of lower mean diameters enhancing heat transfer surface area [15]. Although annular mist flow reduces the droplet diameter sizes which promotes the heat transfer, but it also reduces the droplet residence time within the compression chamber which in turn effects the time required for heat transfer. Nevertheless, in these experiments the effect of annular mist flow related to better heat transfer might weigh over the droplet residence as normalised power consumption decreases with increasing male rotor shaft speed. Altogether, normalised power decreases with the increasing rotational speeds for the same discharge pressures.

Air flow rate at constant speed and pressure in general increases with the increase in oil flow as the increased amount of oil improves sealing (Figure 5, Figure 10 and Figure 15). Also, air flow rate achieved at lower pressures is slightly higher compared to the higher pressures, as at higher pressures the leakage flow rate is higher. Finally, the higher TDR indicates the presence of more oil in the compression chamber and higher cooling effect. Overall, TDR increases with the increase in the oil flow with increasing nozzle diameters (Figure 7, Figure 12 and Figure 17). It can be noticed here that the temperature difference ratio increases with the increase in speed from 3000RPM to 5000RPM (3000RPM- 0.5, 4000RPM- 0.8 and 5000RPM- 1.0). Increasing male rotor shaft speed leads to the higher air flow rate which led to shear of oil films and hence increased surface area of oil particles resulting in a better heat transfer between oil and gas.

3.2. Best nozzle diameter for tested operating condition

For the performed experiments, the atomiser nozzle with equivalent cross-sectional area of 4mm diameter nozzle (Figure 3) achieves better atomisation and cooling within the compression chamber. Figure 19 shows the specific power for 4mm atomiser nozzle compared with the 4mm circular nozzle with the average improvement in specific power of 1.15%. Figure 6, Figure 11 and Figure 16, show that for almost all male rotor shaft speeds and discharge pressures the oil flow decreases for the multi hole '4.1' nozzle compared to 4 mm circular nozzle. As a result TDR slightly decreases with '4.1' nozzle indicating higher temperatures within the compression chamber compared to that of the injection with single diameter nozzle of 4mm.

It is believed that at the lower pressure of 8.5 bar, the slight reduction in power is because of the lower amount of oil flow with 4.1 mm insert. Though lower TDR is observed in that case which indicates the higher temperatures in compression chamber, but the extra amount of oil contributed to the increased the frictional losses. On other hand, at higher pressures (10.5 and 12.5 bar) it is expected that the lower oil flow with 4.1mm insert will reduce the performance. Nevertheless, a counterintuitive trend was observed due to the overpowering oil losses or reduced oil droplet residence time leading to slight improvement in compressor power or specific power.

Lastly, the higher specific power is observed at 6.5 bar and 3000 RPM. This might be due to the 20.1% dip in the oil flow at these conditions. This requires further investigation.

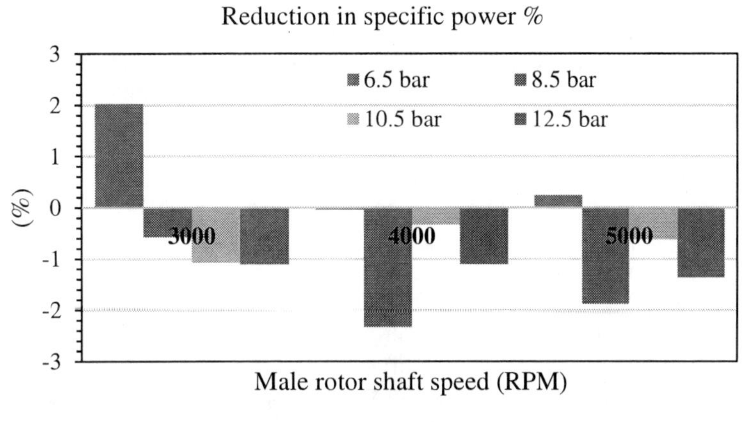

Figure 19. Comparison of specific power % of 4 mm atomiser nozzle ('4.1') against 4 mm single hole nozzle.

Also, all specific powers in Figure 8, Figure 13 and Figure 18 at various discharge pressures and male rotor shaft speeds are summarised to reveal the best nozzle diameter that will achieve least power and specific power. The summarised data is shown in Table 6 and Table 7 along with reduction in oil flow required to achieve improvements in performance.

Table 6. Best nozzle diameter at tested operating conditions resulting in least power consumption.

Shaft Speed (RPM)	Discharge Pressure									
	6.5 bar			8.5 bar			10.5 bar			12.5 bar
	Best Dia (mm)	Reduc-tion in oil flow (%)	Reduc-tion in Power (%)	Best Dia (mm)	Reduc-tion in oil flow (%)	Reduc-tion in Power (%)	Best Dia (mm)	Reduc-tion in oil flow (%)	Reduc-tion in Power (%)	Best Dia (mm)
3000	5	N/A	N/A	5	N/A	N/A	5	N/A	N/A	5
4000	3	40.24	6.95	3	41.42	4.07	5	N/A	N/A	5
5000	3	25.10	7.24	3	45.55	7.23	4.1	39.25	2.63	5

Table 7. Best nozzle diameter at tested operating conditions resulting in least specific power.

Shaft Speed (RPM)	Discharge Pressure							
	6.5 bar			8.5 bar			10.5 bar	12.5 bar
	Best Dia (mm)	Reduction in oil flow (%)	Reduction in Specific Power (%)	Best Dia (mm)	Reduct-ion in oil flow (%)	Reduction in Specific Power (%)	Best Dia (mm)	Best Dia (mm)
3000	5	N/A	N/A	5	N/A	N/A	5	5
4000	3	40.24	4.02	3	41.42	2.81	5	5
5000	3	25.10	4.44	3	45.55	3.50	5	5

4. Conclusion

In this paper, the performance of an oil-injected twin screw compressor is studied by varying the oil flow rate by changing the main nozzle diameters. The experimental study is conducted for discharge pressures from 6.5 bar to 12.5 bar and male rotor shaft speed from 3000 RPM to 5000 RPM. Key conclusions are:

- Variation in oil volume flow rate has a comparatively higher effect on the power consumption than air flow rate for a certain discharge pressure and the male rotor shaft speed.
- For lower discharge pressures (6.5 bar and 8.5 bar), the power consumption increases with the increase in oil flow. Meanwhile, at higher discharge pressures (10.5 bar and 12.5 bar) the power consumption decreases with the increase in the oil flow. Hence, an optimum quantity of oil should be determined for a compressor running at certain design condition.
- It was expected that a lower temperature difference ratio will be achieved by use of the atomiser nozzle with equivalent cross sectional area of 4 mm. However, it was noticed that the atomiser restricted the oil flow and improved performance as a result.
- At 4000 RPM- 6.5 bar and 8.5 bar, by reducing an average oil flow of 40.8% can improve specific power by an average of 3.41%.
- At 5000 RPM- 6.5 bar and 8.5 bar, reducing an average oil flow by 35.3% can give improvement in specific power of 3.97%.
- Better heat transfer characteristics through TDR is observed with increasing the male rotor shaft speed. This indicates that further study towards the relevant multiphase flow regime needs to be conducted to understand oil droplet distribution at various air flow rates while oil flow rate remains the same as well the relevant leakages.

The results demonstrate the effects of an important parameter oil flow which further investigated lead to savings in specific power and therefore conditions to achieve a better performing machine.

References

[1] P. Singh and J. Bowman, "Heat Transfer in Oil-Flooded Screw Compressors," *International Compressor Engineering Conference*, pp. 135–153, 1986.

[2] N. Stosic, A. Kovacevic, K. Hanjalic, and L. Milutinovic, "Mathematical modelling of the oil influence upon the working cycle of screw compressors," in *International Compressor Engineering Conference*, 1988.

[3] N. Stošic, L. Milutinović, K. Hanjalić, and A. Kovačević, "Investigation of the influence of oil injection upon the screw compressor working process," *International Journal of Refrigeration*, vol. 15, no. 4, pp. 206–220, 1992.

[4] N. Stosic, L. Milutinovic, K. Hanjalic, and A. Kovacevic, "Experimental Investigation of the Influence of Oil Injection Upon the Screw Compressor Working Process," in *International Compressor Engineering Conference*, 1990.

[5] X. Peng, Z. Xing, T. Cui, and P. Shu, "Experimental Study of Oil Injection and Its Effect On Performance of Twin Screw Compressors," in *International Compressor Engineering Conference*, 2000.

[6] M. De Paepe, W. Bogaert, and D. Mertens, "Cooling of oil injected screw compressors by oil atomisation," vol. 25, pp. 2764–2779, 2005.

[7] G. Valenti *et al.*, "Thermal effect of lubricating oil in positive-displacement air compressors," *Applied Thermal Engineering*, vol. 51, no. 1–2, pp. 1055–1066, 2013.

[8] G. Valenti, S. Murgia, G. Contaldi, and A. Valenti, "Experimental evidence of the thermal effect of lubricating oil sprayed in sliding-vane air compressors," *Case Studies in Thermal Engineering*, vol. 4, pp. 113–117, 2014.

[9] G. Bianchi and R. Cipollone, "Experimental assessments on a pressure swirl oil atomizer for positive displacement vane compressors," vol. 82, pp. 286–293, 2015.

[10] Chattopadhyay, "Flowmeters & Flow Measurement," *Flowmeter and Flow Measurement*, p. 452, 2006.
[11] J. H. Arnold, "Vapor Viscosities and the Sutherland Equation," *Journal of Chemical Physics*, vol. 1, no. 2, p. 170, 1933.
[12] "A Summary of Error Propagation," *Harvard University*, 2013. [Online]. Available: http://ipl.physics.harvard.edu/wp-uploads/2013/03/PS3_Error_Propagation_sp13.pdf. [Accessed: 09-May-2018].
[13] M. Reader-Harris, *Orifice Plates and Venturi Tubes*. Springer, 2015.
[14] R. Deipenwisch and K. Kauder, "Oil as a design parameter in screw-type compressors: oil distribution and power losses caused by oil in the working chamber of a screw-type compressor," in *International Conference on Compressor and their Systems*, 1999, pp. 49–58.
[15] J. Xu, "Oil Flow From Compressor Valve to Discharge Tube," University of Illinois at Urbana-Champaign, 2016.

IOP Conf. Series: Materials Science and Engineering **425** (2018) 012010 doi:10.1088/1757-899X/425/1/012010

Thermal expansion in liquid-injected screw compressors

Ulrich Daemgen[1], Patrick Hadamitzky[2], Joachim Dohmann[3]
[1] Boge Kompressoren, Germany; correspondence: u.daemgen@boge.de
[2] Technische Universität Braunschweig, Germany;
[3] Ostwestfalen-Lippe-University-of-Applied-Sciences, Germany

Abstract. In screw compressors gaps between rotors and casing have to be small, in the range of 30 to 300 µm. These gaps are influenced by thermal expansion when the screw compressor is in operation. The amount of thermal expansion is about the same size as the gaps, even in oil-injected screw compressors. Especially when starting a cold compressor, there are temperature differences. These temperature differences were measured in a 22 kW oil-injected screw compressor for industrial compressed air as an easily manageable example. Measuring the temperatures in the rotor teeth was done with Pt1000-sensors and transmission with Wireless-LAN. The results showed lower temperature differences than expected. The influence of the oil-injection-temperature is dominant. In bigger screw compressors the temperature differences tend to be larger because of dimensional reasons: A simple numerical model for the temperature differences in a screw compressor with fluid injection allows an extrapolation for larger rotors.

1. Introduction

The subject of this investigation is liquid-injected screw compressors. These include the oil-injected screws for industrial compressed air, the widest spread air compressor in industry. And there are also the process gas screw compressors in the chemical industry, often very large machines. They often have an injection with water, oil or a process liquid.

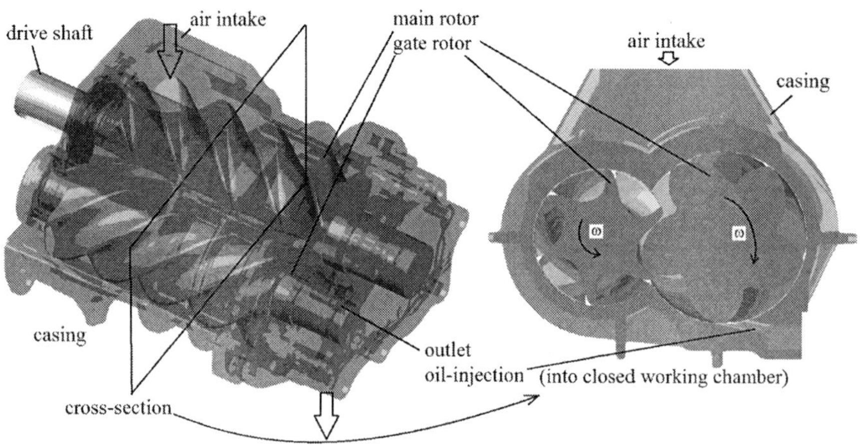

Figure 1. View on an oil-injected screw-compressor block.

IOP Conf. Series: Materials Science and Engineering **425** (2018) 012010 doi:10.1088/1757-899X/425/1/012010

Figure 1 shows such a screw compressor block. There are no special seals for the working chambers, only the littleness of the gaps gives the sealing effect. The gap size is about 30µm in small compressors and up to 300µm in big compressors. Gaps too big cause increased leakage between the working chambers and lower efficiency, gaps too small cause jamming and often total damage of the screw compressor block. The amount of thermal expansion is about the same size as the gaps. As long as rotors and the casing have the same coefficient of thermal expansion and their temperatures are rising and falling simultaneously, the gap widths remain practically unchanged. But that is not secured. In order to clarify that, this investigation was made. As we can see in figure 1, the teeth of the gate rotors are slimmer than those of the main rotor. As thin parts are more prone to fast heating and cooling, the investigation concentrated on these teeth and on the casing

2. Type of screw compressor

The machine for the tests was a typical oil-injected screw compressor for industrial air with the type name "S29", with 29 HP shaft power, fixed-speed (about 4200 r.p.m. for 10 bar (g)), and cooled with ambient air, figure 2. The screw compressor block casing is made of grey-cast iron; the rotors are of spheroidal cast iron (practically the same coefficient of thermal expansion for both).

It was chosen because of being typical and manageable for a test: not too big, what had caused higher costs, not too small, allowing standard temperature probes. Figure 3 shows the scheme of air and oil flow. The oil is separated from the compressed air, cooled and recirculated. The machine starts on compressed-air demand and shuts down, when air demand is satisfied. Starting and stopping can be repeated many times per hour. The heating of the machine comes from the heat of compression and from losses.

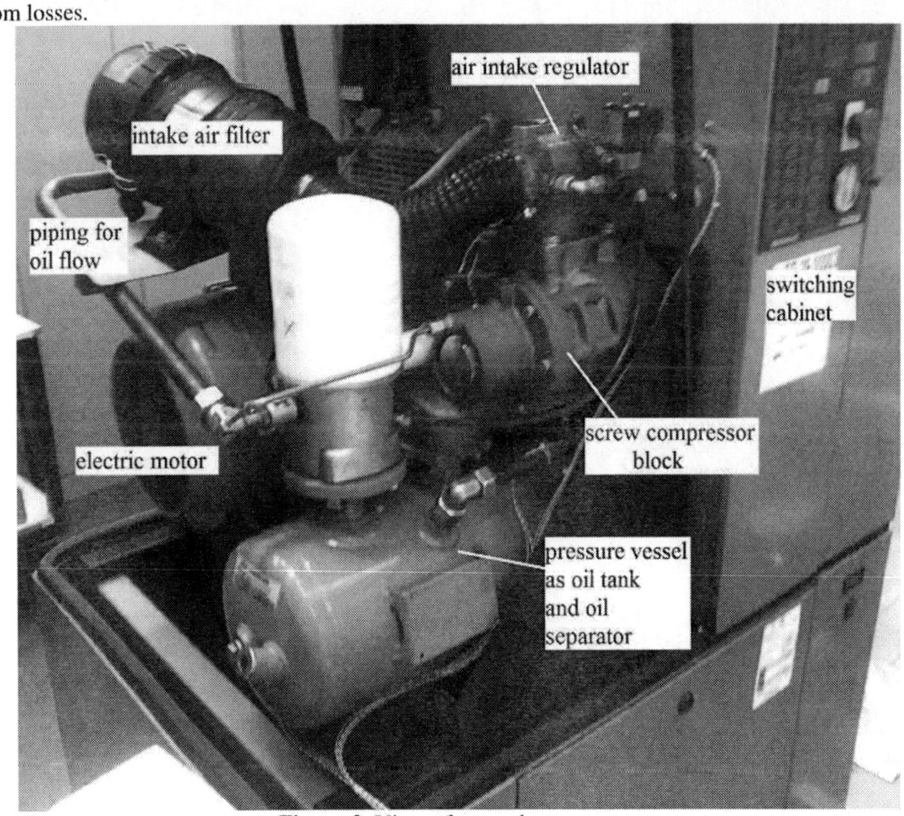

Figure 2. View of opened compressor.

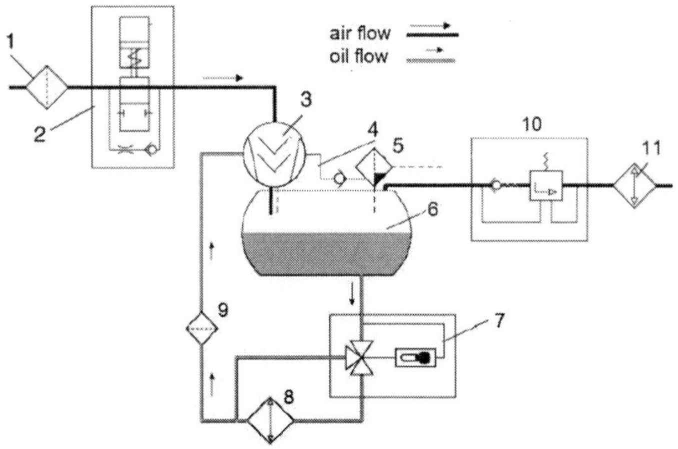

Figure 3: 1 intake air filter for ambient air
2 air intake regulator
3 screw compressor block
4 small pipe for returning oil
5 oil mist separating filter
6 pressure vessel as oil tank and oil separator
7 thermostatic valve of the melting-wax-type, opens the way through the oil-cooler
8 above 60°C
9 oil filter
10 minimum-pressure non-return valve
11 compressed air cooler.

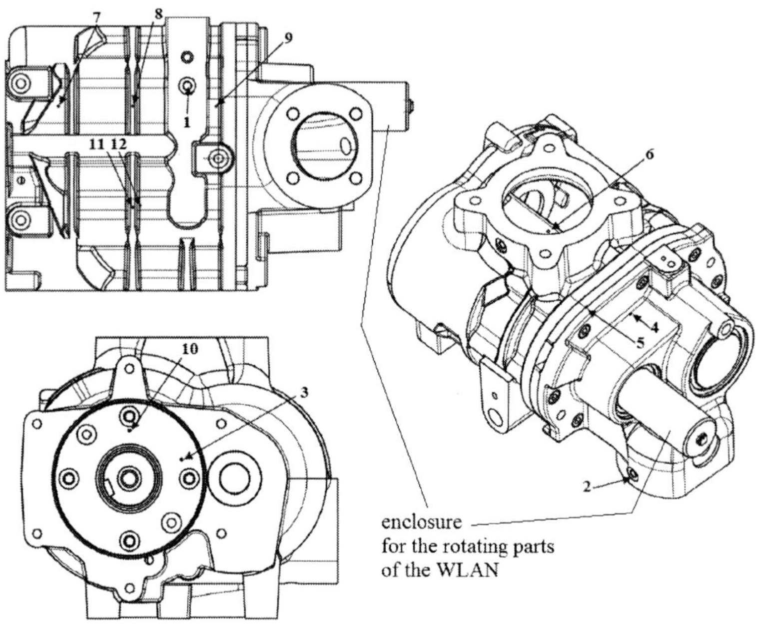

Figure 4. Position of the thermocouples in the casing.

3. Way of measurement

The thermal behaviour of the casing was measured in the standard way, with small thermocouples, embedded in tiny holes drilled into the casing, Figure 4, and fixed with adhesive.

The measurement of the temperature in the gate rotor teeth needs small temperature probes, in order not to influence the heat flow. So small Pt1000-probes in small drilled holes, Figure 5, fixed with thermally conductive (0,1 W/m*K) adhesive are suitable. The thermo-signal voltage is small and needs transformation into a rugged signal before being sent from the rotating rotor to the recording device. After comparing different methods the use of an electronic device, rotating with the rotor and sending the temperature signal in digital form by WLAN to the recording device, was chosen. This resulted in low costs (almost independent of the number of temperatures) and reliable transmission for

IOP Conf. Series: Materials Science and Engineering **425** (2018) 012010 doi:10.1088/1757-899X/425/1/012010

several temperatures (If the transmission is disturbed, the results are not falsified; either they are transmitted correctly, or there is no signal). The supply voltage is transferred with the single contact pin and a carbon brush, Figure 6.

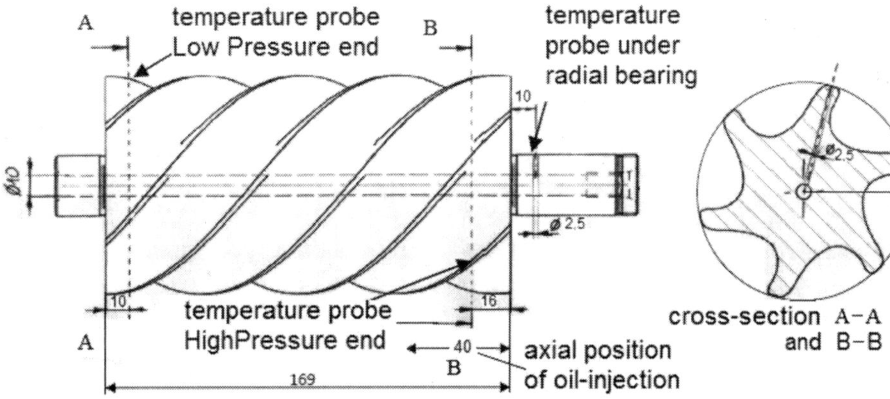

Figure 5: position of the temperature probes in the gate rotor.

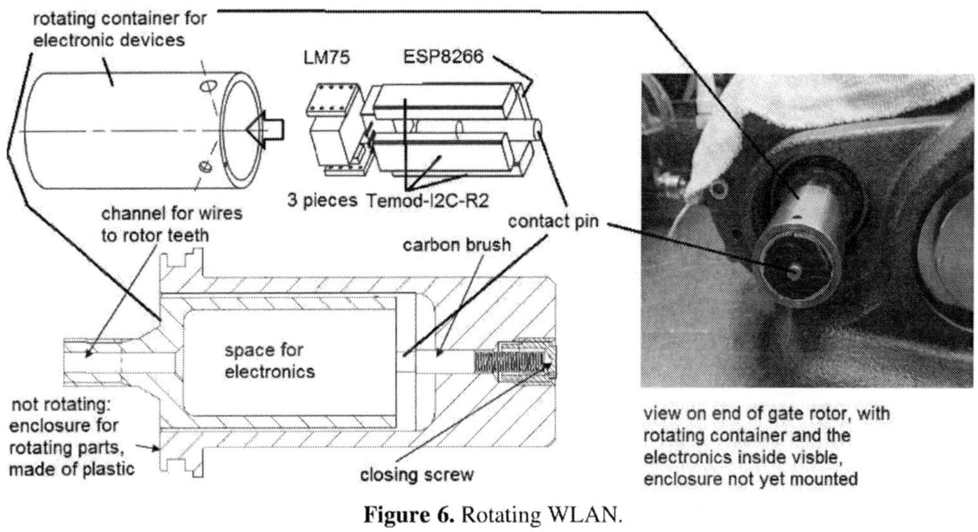

Figure 6. Rotating WLAN.

The rotating parts are enclosed with a plastic part (manufactured as a turning workpiece), keeping the pressure and the oil within and allowing the WLAN signals to be transmitted. The WLAN-signals were received and recorded with a standard personal computer. The signal transmission rate was 1 record of temperature data per second. A higher transmission rate is possible, but wasn't necessary in this case. The supply of electric energy through the contact pin (with closing the electric circuit through the rotor ball bearings and the casing) is not suitable for long-time measurements because of wear; in that case a supply through induction would be better. The monitoring of the temperature at the electronic devices, rated for a maximum 85°C, showed a temperature more than 10 K lower. The number of temperature measuring points in the rotor is limited only by the number of holes, which can be drilled without weakening the rotor, so 6-20 would be possible.

4. Results

The maximum temperature differences between rotor tooth and casing were recorded when starting the cold compressor. The oil-injection temperature shows a spike at 13:23 (figure 7), caused by a delay in the thermostat. When the oil temperature reaches 60°C, the thermostatic valve does not open immediately; it takes about half a minute, until the wax is molten and the valve open. Before the thermostatic valve opens, the temperatures increase quickly, after that, they are rising slower, and after a quarter of an hour, they almost reach a steady state. Only the casing temperature at the low pressure end of the rotors and the gate rotor tooth temperature at the low pressure end show a much slower reaction. Because the oil is injected in an already closed working chamber, oil reaches these points only indirectly and in low quantity. After stopping the compressor, the temperatures are converging and then returning to ambient temperature slowly.

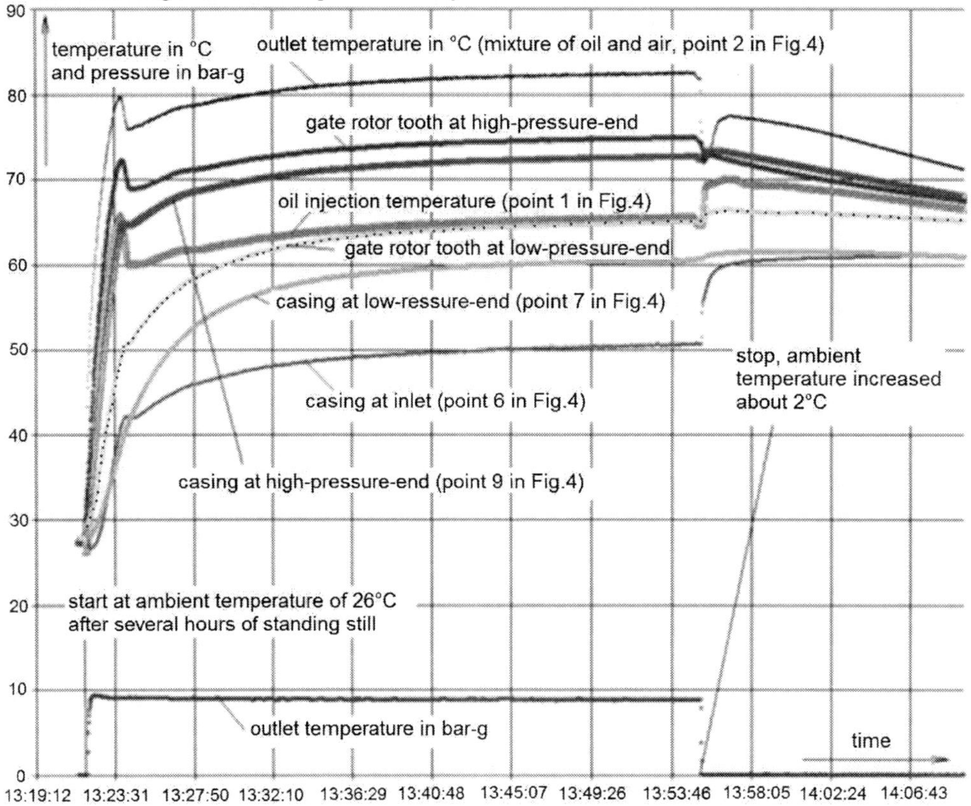

Figure 7. Temperatures recorded during star-stop-cycle (here and in Figure 9 and 10: horizontal axis – time hh:mm:ss, vertical axis – temperature in °C).

The temperature spike at 13:23 is in the outlet temperature and the gate rotor temperature at the high pressure end, too, what shows a very strong influence of oil inlet temperature on the whole screw compressor block. The casing temperatures show a bigger inertness (slower reaction / more delay) compared with the corresponding rotor tooth temperature, but the difference is lower than between high-pressure-end and low-pressure-end.

The test runs were repeated several times, and the repeatability was < 2 K.

The temperature differences between gate rotor tooth and casing reach their maximum at about 1 to 2 minutes after starting, during the temperature spike, with some 12 K. At steady state the difference is only 2 K (high-pressure end) or 4 K (low-pressure end).

When computing the diameter change and the gap change resulting from temperature differences (gate rotor - casing) at the high-pressure-end and at the low-pressure-end, the maximum gap change is only approximately 8 µm.

Conclusion: The temperature difference between rotor tooth and casing is lower than we expected for medium-size oil-injected screw compressors, and the main influence on temperatures is from the oil thermostat.

5. Extrapolation:

Temperature differences are important for bigger compressors, too. Here we find a dependence on length L. When the lengths L of a screw compressor are redoubled, the heat flux through a surface is 4-times higher, being proportional to L^2, the area. The heat flux through a volume, like for example a cube-like block is 2-times higher (because the surface is 4-times higher, and the gradient of temperature, $\Delta T/L$, is halved), proportional to L. The thermal capacity is 8 times higher, being proportional to the mass, and therefore proportional to L^3. As a result, the temperature influence on gaps might be more critical with bigger machines. So the test should be repeated with a big screw – but the costs for an experiment with a 200 kW-machine are prohibitive.

An extrapolation method for temperatures could be helpful, even without an experimental proof. Using words like "thermal capacity" and "heat conductivity" points to an analogy with electric circuits with capacitors and resistors, figure 8. The temperature (corresponding to the voltage U in the model) at a rotor surface spot varies during rotation, as it is exposed to different states of the working chamber: hot when the content of the working chamber is near to full pressure, cold when the working chamber is drawing in fresh air. So an effective medium temperature Teff has to be found which can be modelled as, for example, x % of intake temperature, y % of oil injection temperature and (100-x-y) % of outlet temperature.

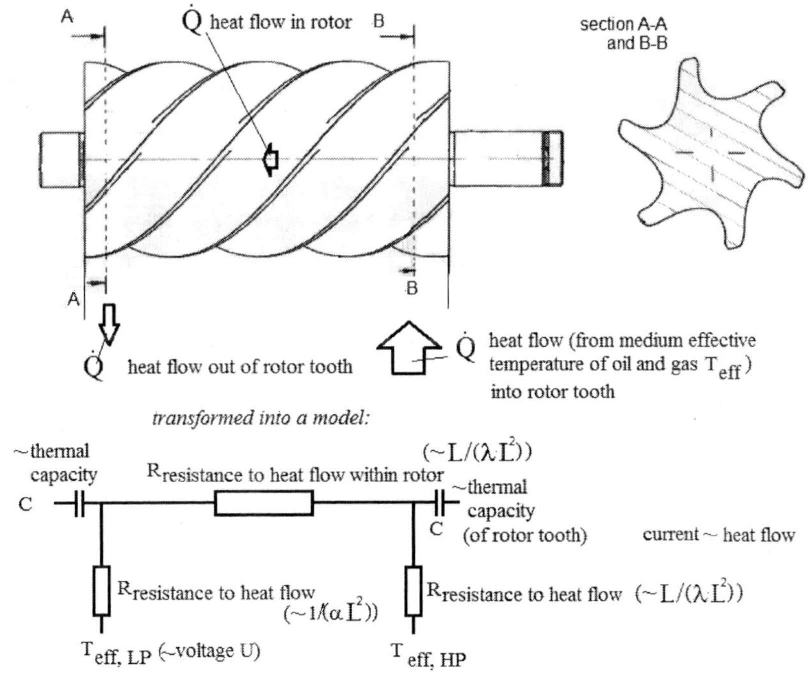

Figure 8. Model for extrapolation of rotor tooth temperature.

Factors for effective temperature and heat flow resistance were found by an iterative process: comparing the results of the model with measured results in a diagram, changing factors until the model's results are getting better. The factors can be seen in [1].

Figure 9 shows the results of the model with those factors compared with the measurements. The temperature of the high-pressure gate-rotor tooth, when running, is almost the same in the model and in the measurement. Other temperatures show a deviation up to 3 K. These deviations can be explained by the fact that heat transfer is not constant; for example caused by changes in oil flow through gaps, when oil viscosity is lowered by higher temperature.

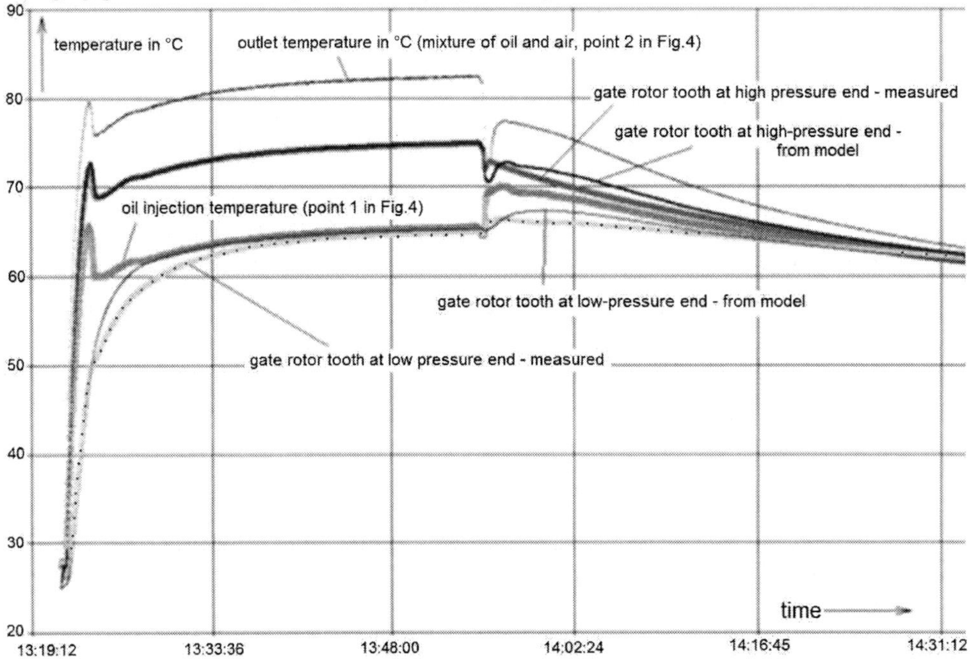

Figure 9. Comparison of measured rotor tooth temperature with model.

This model can be used for an extrapolation. When increasing all lengths with a factor 3, the heat capacity is 27-times higher. Heat-flow-resistance within the rotor and most probably at the high-pressure-end tooth is dominated by heat-flow through a volume, so these resistances are reduced by a factor 3. The heat-flow-resistance at the low-pressure side of the rotor is probably dominated by the heat transfer through the surface and so is reduced with a factor of 9. The power of a compressor of that size would be some 200 kW.

Figure 10 shows the differences in gate-rotor tooth temperatures for that case. The spike at the high-pressure-end rotor-tooth-temperature, when starting, practically disappears, the low-pressure-end rotor-tooth- temperature has a much bigger delay.

This model needs more calibration in different machines before being used as a standard, but it gives a first hint for thermal expansion of rotor teeth under the conditions:

- liquid-injected screw compressor, with "thermal capacity of injected liquid flow" > 2 x "thermal capacity of compressed gas flow"
- the liquid is injected into a closed working chamber, at a place that is not near to the suction-side rotor-end

The question might arise: Why not extrapolating the casing temperature as well ? In order to evaluate the gap changes when starting, both temperatures, of rotor tooth and of casing, are needed.

One answer is: The shape of a casing – wall thickness, stiffening ribs, oil channels - can differ very much, so an extrapolation for a casing is more uncertain than for a rotor. The second answer is: You can measure the heating-up of a casing rather easily, with a handheld thermo-element for example. That gives casing temperature curves. And this extrapolation (not forgetting the influence of the oil-thermostatic-valve, so the oil injection temperature curve should be measured, too) helps you with the rotor tooth temperature curves.

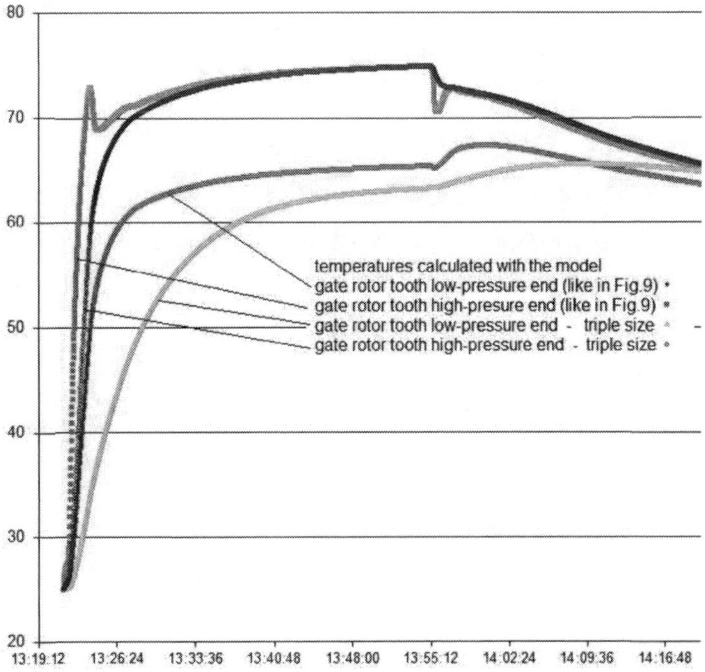

Figure 10. Result of extrapolation to triple size

6. Outlook

This consideration is not exhaustive - there exists not just a thermal expansion of diameters, but also some distortion of screws and casing, which may be evaluated with the help of a more complex model.

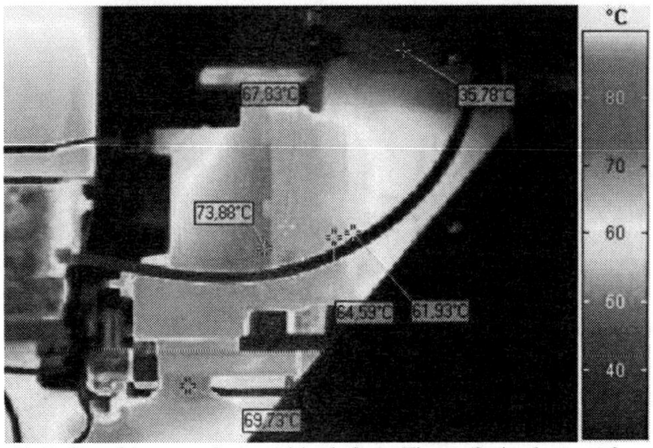

Figure 11. Infrared image of screw block from Figure 2; 4 minutes after starting

These measurements (including a series of IR-photographs at different instants of time, like in figure 11) could be used for the calibration of detailed computational models of liquid-injected screw compressors including heat transfer, thermal modelling, and thermal deformation for gap calculation. Boge does not intend that at the moment. Giving results of the measurements to other researchers is not planned, but possible.

This work was done in cooperation with the Ostwestfalen-Lippe-University-of-Applied-Sciences.

References

[1] Thermal expansion in liquid-injected screw compressors; pages 291-01
in: Proceedings Compressor Users International Forum 2016 (International Rotating Equipment Conference 2016 Düsseldorf), VDMA-Verlag, Frankfurt

Appendix

Heat capacity: $Q = V * \Delta T * \rho * c = L^3 * \Delta T * \rho * c$

heat flow through volume: $\frac{Q}{t} = A * \frac{\Delta T}{L} * \lambda = L^2 * \frac{\Delta T}{L} * \lambda = L * \Delta T * \lambda$

resistance to heat flow: $\frac{\Delta T}{\frac{Q}{t}} = \Delta T/(L * \Delta T * \lambda) = 1/(L * \lambda)$

(when dominated by resistance in the volume)

heat flow through surface: $\frac{Q}{t} = A * \Delta T * \alpha = L^2 * \Delta T * \alpha$

resistance to heat flow: $\frac{\Delta T}{\frac{Q}{t}} = \Delta T/(L^2 * \Delta T * \alpha\ ) = 1/(L^2 * \alpha)$

(when dominated by resistance of the heat transfer at the surface)

V volume ; Q heat; t time ;
ΔT temperature difference (temperature at same place and different moments for heat capacity; temperature at different sides of the volume for heat flow) ;
ρ density ; c specific heat capacity ; L edge length of the cube (as an example for a geometric body);
λ heat conductivity (of the cast iron in this case) ; A area of the surface at one side of the cube

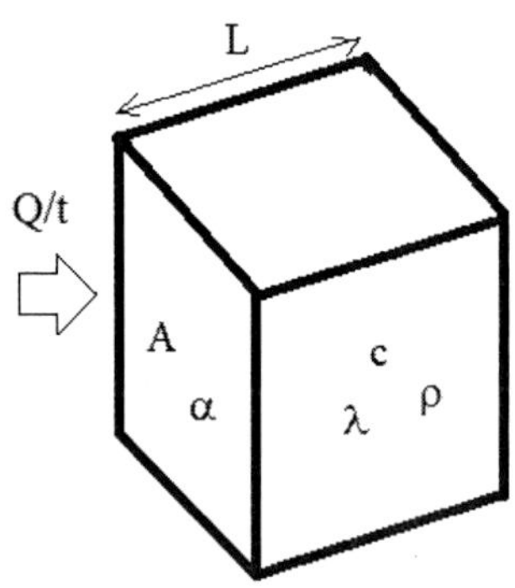

IOP Conf. Series: Materials Science and Engineering **425** (2018) 012011 doi:10.1088/1757-899X/425/1/012011

Novel approach to single-screw compressors and expanders design*

Davide Ziviani[1], Patrick J. Goeghegan[2] and Eckhard A. Groll[1]

[1]Ray W. Herrick Laboratories, School of Mechanical Engineering, Purdue University, West Lafayette, 47907-2099, USA

[2]Oak Ridge National Laboratory (ORNL),Oak Ridge, TN, USA

E-mail: dziviani@purdue.edu; geogheganpj@ornl.gov; groll@purdue.edu

Abstract. Single-screw machines are currently employed as both compressors and expanders in vapor compression systems and organic Rankine cycles (ORCs), respectively. The working principle of single-screw machines is based on the simultaneous meshing of two starwheels with one central grooved rotor. The performance of the machine is heavily affected by the wear of the tooth meshing profile after several hours of running. In order to improve the wear resistance of the conventional straight-line profile, multi-column tooth flank designs have been introduced and investigated by several researchers. Cylindrical multi-column envelope profiles are able to distribute the local contact over a larger surface area reducing the wear of the tooth flank profile. Nevertheless, the manufacturability of such profiles is directly linked to the accuracy and limitations of CNC machines.

In this work, an attempt is made to overcome the aforementioned limitations. To push the boundaries of possible tooth and rotor profile designs, 3D printing is utilized as manufacturing technique. Such approach allows the investigation of more complex designs that improves the sealing lines during the meshing process.

A detailed geometry model based on polygon intersections is utilized to calculate the sealing lines, groove volume curves, and surface areas of the different designs. As a result of this study, a novel rotor design has been manufactured.

1. Introduction

The single-screw concept was first introduced by Bernard Zimmern in the early 1960s [1] and consisted of a single six helically grooved central rotor with a cylindrical periphery and a globoid core and two identical 11-teeth starwheels engaging with the rotor [2, 3]. After being rejected several times from French and American companies, a first prototype of single-screw compressor for compressed air applications was built in 1963 and it was licensed two years later. After that point, the development of the single-screw continued and its application was extended to refrigeration within a decade [4, 5, 6, 7].

*This manuscript has been authored by UT-Battelle, LLC, under contract DE-AC05-00OR22725 with the US Department of Energy (DOE). The US government retains and the publisher, by accepting the article for publication, acknowledges that the US government retains a nonexclusive, paid-up, irrevocable, worldwide license to publish or reproduce the published form of this manuscript, or allow others to do so, for US government purposes. DOE will provide public access to these results of federally sponsored research in accordance with the DOE Public Access Plan (http://energy.gov/downloads/doe-public-access-plan).

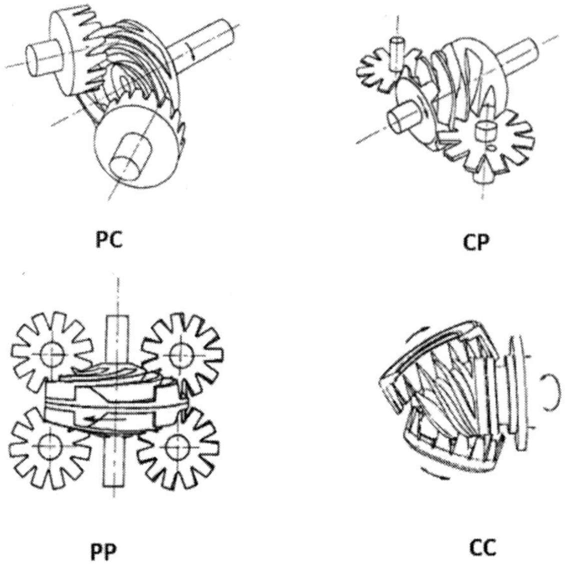

Figure 1. Classification of single screw machines depending on the meshing conditions. Adapted from Zimmern B. [7].

The central position of the rotor in the single-screw potentially leads to different configurations depending on the shape of the grooved rotor and the position of the starwheels. In particular, four main configurations can be distinguished at the moment, as shown in Figure 1 [8]:

- Plate-Cylindrical (PC): planar main screw rotor and two cylindrical starwheels. The working process occurs only on the front face of the screw rotor. The starwheel shafts are oriented at opposing angles from the axial face of the main screw.
- Cylindrical-Plate (CP): cylindrical main screw rotor and two planar starwheels. The starwheel shafts are parallel to each other with their rotation axes perpendicular to the rotation axis of the main screw. The working process occurs within the volume created when the starwheels mesh with the main screw rotor.
- Plate-Plate (PP): planar screw rotor and planar starwheels. The grooves are located on the front and back faces of the main screw rotor. Two starwheels are employed for each side or the rotor and the rotation axis of their shafts are parallel to each other.
- Cylindrical-Cylindrical (CC): cylindrical starwheels and cylindrical screw rotor. The two starwheels are engaged on the top and bottom face of the main screw, respectively. As with the PC design, the starwheel shafts are oriented at opposing angles.

The most common single-screw configuration is the CP-type, which includes as primary components a screw rotor and two starwheels. One of the main reasons is the higher complexity of the manufacturing process of the other configurations that leads only to a limited increase in the efficiency and performance with respect to the CP-type.

The improvement of the single-screw design is directly related to advancements in the manufacturing techniques. Over the years, the accuracy of the Computer Numerically Controlled (CNC) machines has increased allowing for a better control of the machining process of the main rotor. Initially, specially designed cutting tools that mimicked the shape and motion of the starwheel were employed to manufacture the main rotor grooves, resulting in long machining times and limited range of sizes [9]. The introduction of 5-axis CNC machine centers enabled advancements in the manufacturing process of the rotors. In fact, each groove became a ruled surface created by the milling tool while sweeping through the body of the main rotor [10, 11]. However, difficulties in the manufacturing process of the grooves still exist, especially with respect to the root of the grooves [12].

The machining accuracy of the meshing pair is an important aspect and according to Li et al. [13], the average maximum machining error is on the order of two times the design precision required of the single screw rotor. Furthermore, the indexing accuracy of a common CNC machine tool can reach 10 seconds, which means the maximum machining error is around 0.01 mm for a 200 mm diameter, while the design precision of the meshing pair in a single-screw is around 0.05 mm. The uncertainties related to the manufacturing process of the meshing pair and the type of the generating profile yield a potential rapid wear of the starwheel tooth flank surfaces meshing with the screw groove flanks. Ultimately, the wear leads to larger gaps which increase the leakages and decrease the volumetric efficiency.

The shape of the tooth flanks, the materials and the machining accuracy of the meshing pair represent the important aspects with respect to improving the wear resistance. Different geometric meshing conditions have been proposed to achieve a smoother contact during the meshing. The original profile proposed for the single-screw compressor meshing pair was substantially a straight-line envelope mashing pair (SEMP or LEMP) [14]. The contact area between the starwheel tooth flank and the screw rotor is a fixed straight line on the starwheel tooth flank, resulting in a low wear resistance because the contacts are concentrated on a line. The detailed model for LEMP was presented by Yang S.-C. [15]. In order to better distribute the contact area and reduce the progressive wearing, a column envelope meshing profile (CEMP) was proposed by Zimmern [3]. A more comprehensive mathematical model of CEMP with design purposes was presented by Wu et al. [16], but due to limitation in machining capabilities, the solution was not suitable for industrial applications. Successively, Wu and Feng [17, 18] improved the feasibility of the column profile by introducing a multi-column (or multi-straight line) approach as well. The proposed approach allows the contact line between the tooth flak and the groove flank to move back and forth in the whole area of the tooth flank, enhancing the lubrication between tooth and groove flanks and improving the service life. Basically, the meshing area of a multi-column envelope is not continuous but characterized by jumps between one cylindrical segment to another. Furthermore, different curved profiles of the gate-rotor teeth, e.g., elliptical, hyperbolic, or involute, were analyzed by Li et al. [13] in order to enhance the lubrication between tooth and groove flanks. It is possible to find an overview of the different meshing pair profiles in a recent work published by Wu et al. [19]. Conical teeth were also proposed by Yang and Liang [20].

Novel manufacturing techniques such as 3D printing would open the possibility of designing novel types of single-screw meshing configurations that are not cost effective with the current manufacturing processes, such as CC single screw compressor described by Heidrich and Yang [21, 22].

In this work, a non-conventional design of the starwheel tooth profile has been explored to improve the manufacturing control over the meshing pair and the sealing lines. In order to verify the feasibility of such profile, 3D printing has been employed to prototype the main rotor and the starwheel. Such methodology enables advancements in the design of single-screw compressors and expanders which are not suitable to be manufactured with conventional CNC machines.

2. Novel tooth design

As described by Ziviani et al. [23], the geometry model of single-screw machine is based on three groups of parameters denoted as independent variables (such as the number of grooves/teeth,

IOP Conf. Series: Materials Science and Engineering **425** (2018) 012011 doi:10.1088/1757-899X/425/1/012011

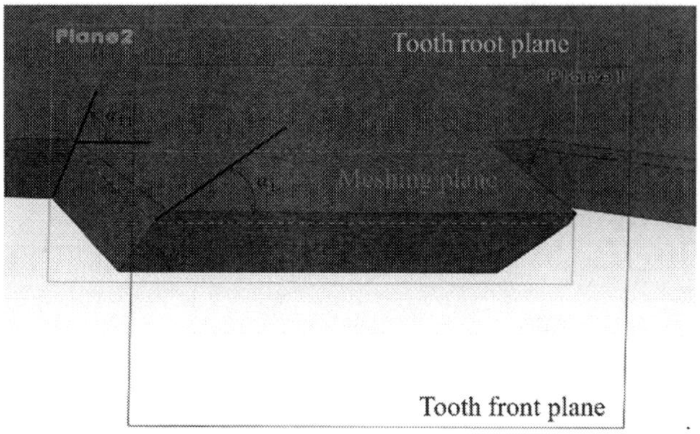

(a)

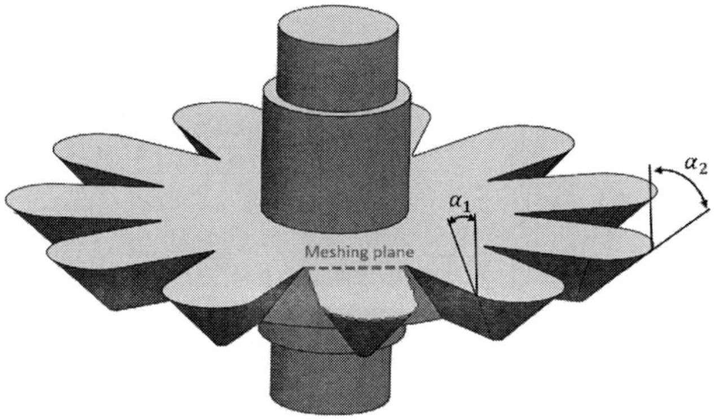

(b)

Figure 2. (*a*) Conventional tooth design with straight-line meshing profile; (*b*) new tooth design with curved tip and drafted flank surfaces.

number of starwheels, diameters of rotor and starwheels), dependent variables (such as engaging angles, length of the rotor), and limit-set variables that enforce a realistic design space for the meshing pair. The selection of the generating profile uniquely defines the meshing pair. In this work, a single straight line envelope meshing pair (SSLEMP) is considered to have a direct comparison with the conventional tooth design, shown Figure 2(a). The proposed tooth profile consists of two straight lines and a circular arc, shown in Figure 2(b). Furthermore, new tooth design does not feature an upper and lower surfaces that contribute to the meshing (and sealing) process. In the conventional tooth profile, the cross section of the tooth tip is characterized by two design angles, α_1 and α_2, that merge to a single flank angle, α_{11}, at the root of the tooth. Whereas, in the new design, the upper surface of the tooth is the meshing plane facilitating the alignment and mating between central rotor and starwheels. The tooth flanks are drafted accordingly to avoid interference with the groove flanks.

3. Kinematics of the meshing profile

From the gear kinematics theory [24], the meshing process of the screw rotor (*sr*) with two starwheels (*sw*) can be described as an envelope of two-parameter family of surfaces with a cylindrical-plate (CP) configuration. In particular, the generating surface is represented by the

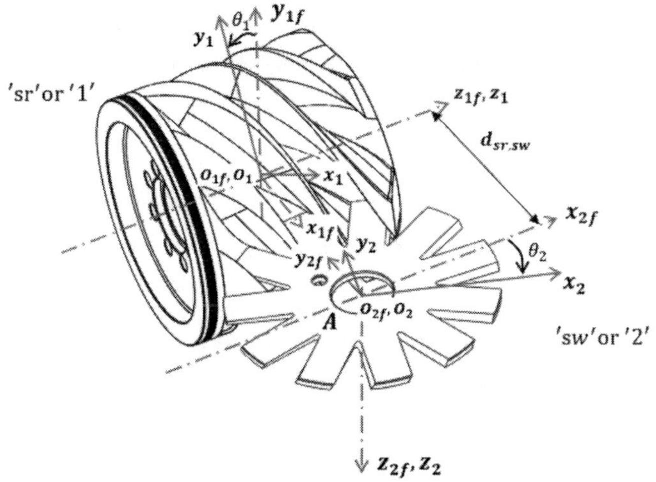

(a)

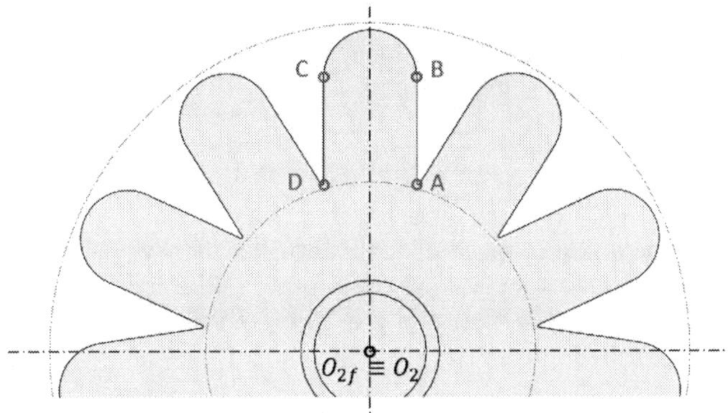

(b)

Figure 3. (*a*) Fixed and rotating reference systems of the meshing pair; (*b*) representation of the new tooth design.

starwheel, i.e. plate, (direct problem). The profile of the starwheel is obtained by applying the inverse envelope concept. A right-handed fixed, $O_{\mathrm{i,f}}(x_{\mathrm{if}}, y_{\mathrm{if}}, z_{\mathrm{if}})$, and a rotating, $O_{\mathrm{i}}(x_{\mathrm{i}}, y_{\mathrm{i}}, z_{\mathrm{i}})$, coordinate system have to be defined for both screw rotor ($i = 1$) and starwheel ($i = 2$), as shown in Figure 3(a).

The starwheel profile is expressed in general terms by:

$$\mathbf{R}_{2\mathrm{j}}(u,t) = [\mathbf{R}_{2\mathrm{j,x}}(u,t)\,\mathbf{R}_{2\mathrm{j,y}}(u,t)\,\mathbf{R}_{2\mathrm{j,z}}(u,t)]^T\,, \tag{1}$$

where $j = \overline{AB},\ \widehat{BC}\ ,\overline{CD}$. The corresponding family of curves on the main rotor, $\mathbf{R}_{1\mathrm{j}}(u,t,\theta_1)$ are obtained through a rotation matrix $\mathbf{M}_{12}$ and ensuring the meshing condition at each angular position. The meshing equation imposes the sliding velocity between the two bodies in contact as zero, i.e. the cross product between the normal vector to the contact surface, $\mathbf{N}_f^1$, and the relative velocity $\mathbf{v}_f^{(12)}$ is zero. Mathematically, the system of equations that has to be solved simultaneously is given by:

$$\begin{cases} \mathbf{N}_f^{(1)} \cdot \mathbf{v}_f^{(12)} = \left(\frac{\partial \mathbf{R}_{1j}}{\partial u} \times \frac{\partial \mathbf{R}_{1j}}{\partial t} \right) \cdot \mathbf{v}_f^{(12)} = 0 \\ \mathbf{R}_{1j} = \mathbf{M}_{12}\, \mathbf{R}_{2j} \end{cases} \tag{2}$$

where $j = \overline{AB},\ \overparen{BC}\ , \overline{CD}$. In particular, the rotation matrix can be expanded as follows:

$$\mathbf{R}_{1j}(u, \theta_1) = \mathbf{M}_{1,1f}\mathbf{M}_{1f,2f}\mathbf{M}_{2f,2}\mathbf{R}_{2j}(u, \theta_1). \tag{3}$$

Usually, to perform coordinate transformations, mixed matrix operations are needed. Homogeneous coordinates can be adopted to include only multiplications [24].

By considering the reference system O_1 on the screw rotor as in Figure 3(a), the system of equations for meshing is given by:

$$\begin{cases} x_1 = -x_2 \sin\theta_{sw} \sin\theta_{sr} - y_2 \cos\theta_{sw} \sin\theta_{sr} + z_2 \cos\theta_{sr} + d_{sr,sw} \sin\theta_{sr} \\ y_1 = -x_2 \sin\theta_{sw} \cos\theta_{sr} - y_2 \cos\theta_{sw} \cos\theta_{sr} - z_2 \sin\theta_{sr} + d_{sr,sw} \cos\theta_{sr} \\ z_1 = x_2 \cos\theta_{sw} - y_2 \sin\theta_{sw} \end{cases} \tag{4}$$

where $d_{sr,sw}$ is the distance between rotation axes of the main rotor and a starwheel, the relationship between θ_{sw} and θ_{sr} is given by:

$$i = \frac{z_{\text{sw}}}{z_{\text{sr}}} = \frac{\theta_{\text{sr}}}{\theta_{\text{sw}}} = \frac{\omega_{\text{sr}}}{\omega_{\text{sw}}} = \frac{11}{6}. \tag{5}$$

where ω_{sr} and ω_{sw} are the angular speed of main rotor and starwheel, respectively.

By introducing the parametric curves of the tooth profile, it is possible to obtain a family of meshing curves. To be noted is that the meshing equations refer the plain containing the straight-line profile, shown in Figure 2(b).

Table 1. Geometry details of the designed single-screw machine.

Parameter	Value	Description
D_{sr}	140 mm	Main rotor diameter
D_{sw}	170 mm	Starwheel diameter
$d_{sw,sr}$	119 mm	Distance main rotor and starwheel
w_{tooth}	25 mm	Tooth width
t_{tooth}	15 mm	Tooth thickness
L_{rotor}	134.9 mm	Rotor length suction side

4. Results and discussion

In order to prove the feasibility of the concept tooth profile, a rotor with a diameter of 140 mm has been chosen as reference case. The main dimensions of the meshing pair are listed in Table 1. The meshing equations have been implemented in SOLIDWORKS to generate the groove profiles starting from the new tooth design. Then, a 3D printer has been employed to manufacture the assembly in a plastic material, as shown in Figure 4(a) and in Figure 4(b). A second CAD model has been obtained from a conventional tooth profile by adopting the same rotor diameter and the meshing pair can be seen in Figure 5(a). Such conventional rotor profile has been previously studied by Ziviani et al. [25] as an expander for organic Rankine cycle applications. For completeness, a view of the actual rotor is provided in Figure 5(b).

By looking at the two designs a number of observations can be made. In the conventional design, the manufacturing process is done by CNC machines and entails the control over the flank surfaces and the bottom surface of the groove. Inaccuracies of the tolerances during the manufacturing lead to leakage gaps as well as abnormal wear of the tooth profile. In the new design, the groove profile does not present discontinuities yielding to a higher control of the meshing process as well as the sealing lines, as shown in Figure 4. The continuity of the meshing helps establishing hydrodynamic lubrication conditions which can potentially reduce the tooth wear and leakage flows. Furthermore, due to the fact that the upper surface of the new tooth coincides with the meshing plane, the correct alignment of the starwheels can be facilitated. Whereas in the conventional tooth profile (see Figure 2(a)), upper and lower tooth surfaces need to be considered during the machining process to ensure the proper sliding of the tooth inside the mating groove.

Additional considerations can be made with respect to the usability of 3D printing.In fact, on the one hand, 3D printing enables new families of meshing profiles that optimize the single-screw geometry as well as allows to have higher uniformity of tolerances on the groove surfaces (both flanks and bottom). On the other hand, the current state-of-the-art of 3D printing does not allow for high-volume production required in industry. Moreover, the printed surfaces still require post treatment to ensure the correct surface roughness. Nevertheless, it is believed that such manufacturing technique will reduce current limitations in rotor profiles and research work in ongoing to evaluate different aspects of the manufacturing process.

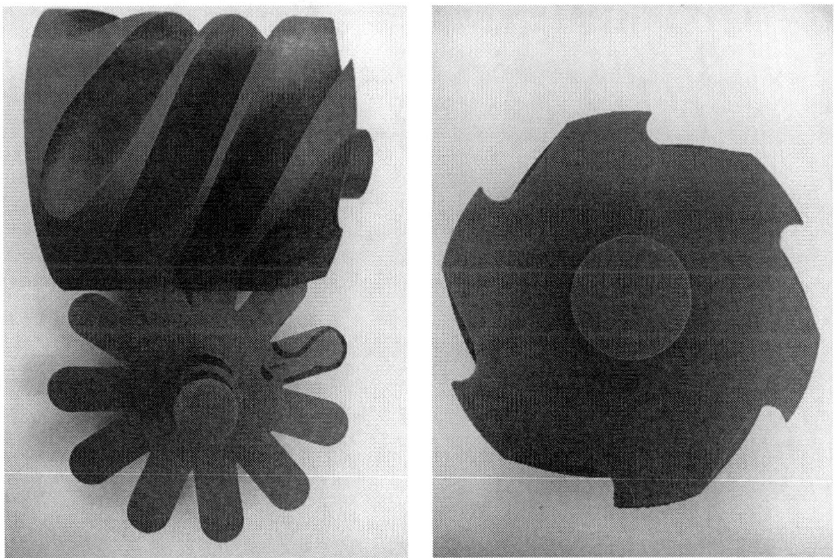

(a) (b)

Figure 4. (*a*) View of the 3D printed meshing pair; (*b*) View of the main rotor end face.

5. Conclusions

In this paper, a non-conventional tooth profile for single-screw machines has been developed. The new profile is believed to enable higher control over the sealing lines reducing leakage flows and wear of the starwheel teeth. The proposed design has been manufactured by using 3D

(a)

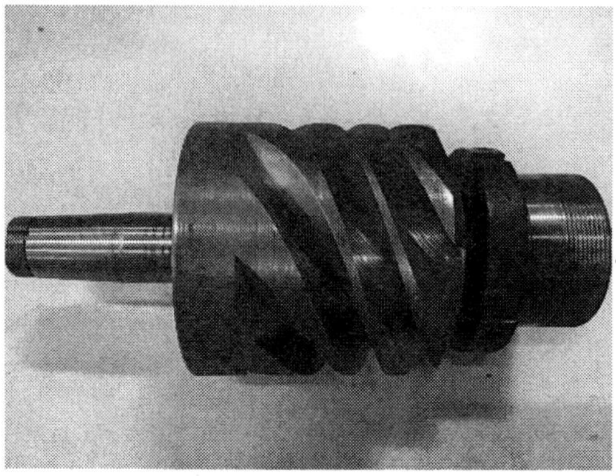

(b)

Figure 5. View of a conventional straight line meshing pair: (*a*) CAD model; (*b*) Example of single-screw expander rotor [23].

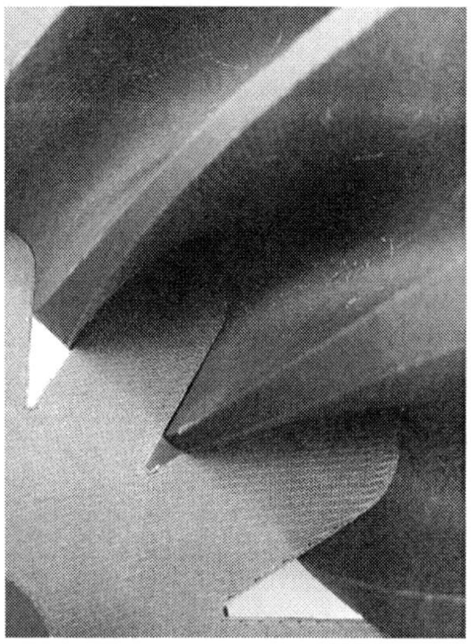

Figure 6. Close-up view of the tooth sealing line while engaged in the mating groove.

printing technique. Such methodology opens up more complex single-screw meshing profiles that could improve the performance of such machines over a wide range of rotor sizes.

As a next step, a metallic rotor and a PEEK (polyetheretherketone) starwheel will be manufactured and tested to validate the performance improvements and durability of the new profile.

Acknowledgments
The authors would like to thank Bob Gemmer, Technology Manager, Research and Development for the Department of Energy Advanced Manufacturing Office.

References

[1] Vrinat G 1984 *International Journal of Refrigeration* **7** 107–114
[2] Zimmern B 1965 Worm rotary compressors with liquid joints
[3] Zimmern B 1976 Rotary injection worm and worm wheel with specific tooth shape
[4] Zimmern B and Patel G C 1972 *International Compressor Engineering Conference* paper 16
[5] Clark J R, Hodge J M and Hundy G F amd Zimmern B 1976 *Institute of Refrigeration*
[6] van Male J 1978 *International Journal of Refrigeration* **1** 242–248
[7] Zimmern B 1984 *Int. Compr. Eng. Conf. at Purdue*
[8] Masuda M, Ueno H, Inoue T, Hori K and Hossain M 2013 *Proceedings of 8th Int. Conf. on Compressors and their Systems, City University of London, London* pp 257–264
[9] Jensen D 1998 *Int. Compr. Eng. Conf. at Purdue* Paper 1306
[10] Zhou S, Li X, Zhang Y and Zhou D 2006 *Int. Compr. Eng. Conf. at Purdue* Paper 1815
[11] Wu W, Feng Q and Yu X 2009 *Journal of Mechanical Design* **131** 1–5
[12] Wu W, Li J and Feng Q 2011 *Computer-Aided Design* **43** 67–71
[13] Li J, Liu F, Feng Q and Wu W 2012 *International Compressor Engineering Conference at Purdue* Paper 1209
[14] Sun S, Wu W, Yu X and Feng Q 2010 *International Compressor Engineering Conference at Purdue* Paper 2024
[15] Yang S C 2002 *Proceedings of IMechE, Part C: J. of Mechanical Engineering Science* **216** 343–351
[16] Wu W, Feng Q, Xu J and Feng X 2010 *International Compressor Engineering Conference at Purdue* Paper 1863
[17] Wu W and Feng Q 2009 *Journal of Zhejiang University* **10** 31–36
[18] Wu W and Feng Q 2009 *Journal of Mechanical Design* **131** 1–4
[19] Wu W, Hao X, He Z and Li J 2014 *Journal of Mechanical Design* **136** 1–5
[20] Yang S C and Liang T L 2008 *Transactions of the CSME* **32** 333–352
[21] Heidrich F L 1996 *International Compressor Engineering Conference at Purdue* pp 145–150
[22] Yang S C 2004 *IMechE Part C: J. Mechanical Engineering Science* **218** 437–448
[23] Ziviani D, Braun J E, Groll E A and De Paepe M 2018 *International Journal of Refrigeration* **92** 10–26
[24] Litvin F L 1994 *Gear Geometry and Applied Theory* (Prentice-Hall, New Jersey)
[25] Ziviani D, Sergei S, Lecompte S, Groll E A, Braun J E, Horton W T, van den Broek M and De Paepe M 2016 *Applied Energy* **181** 155–170

IOP Conf. Series: Materials Science and Engineering **425** (2018) 012012 doi:10.1088/1757-899X/425/1/012012

The influence of profile geometric parameters on characteristics of rotor-gearing compressor

A A Kotlov[1], I A Maksimenko[1] and Y L Kuznetsov[2]

[1] R&D Laboratory "Gas dynamics of turbo machines" Peter the Great St. Petersburg Polytechnic University, Polytechnical st. 29, St. Petersburg, Russia

[2] JSC «Compressor», Bolshoy Sampsonievskiy prospect, 64, St. Petersburg, Russia.

Email: kotlov_andrej@mail.ru, office@compressor-gaz.ru.

Abstract. Dry (without lubrication) claw compressors have a promising future because they feature both advantages of piston and rotary compressors. The rotor profiles have sophisticated form including several curves of similar type, combined so, that the line of contact is always parallel to the rotor axis. The effect of the main geometric parameters of the profile ($\overline{R}$, α_R and r) on the rotor-swept area value is considered. The research was carried out by the developed computer program intended for selection of the optimum variant of a ratio of profile geometrical parameters. In this paper the results of theoretical studies of influence of profile geometric parameters on integrated characteristics of rotor-geared compressor are given. Optimal relations between parameters are given, defining geometry of gas distribution elements.

1. Object

Oil-free claw compressors are very promising since they combine advantages of piston pumps and those of rotary compressors. In a claw compressor, the relative length of rotors may be shorter than in other types of rotor machines. With that optimal tip speeds along the rotor periphery may be obtained at a lower speed of rotation. A larger diameter allows greater ratios of suction port areas and working space volumes, thus reducing gas-dynamic losses at a higher speed. Rotor profiles are of complex shape and are so designed that the contact line is always parallel to rotor axes [1].

The rotor profile shall meet the process requirements, that is shall allow easy machining [2]. That is why profiles are made with mainly straight sections and arcs of a circle are preferred. As a result, using similar single-tooth rotors with a straight back of the tooth would be efficient in the compressor design.

2. Profile geometry

A profile with a straight back of a tooth shall be developed basing on the following parameters (see figure 1): pitch radius r, relative radius of addendum circle $\overline{R} = R/r$ and angular tooth thickness over addendum circle α_R [3].

The profile consists of seven curves fitting one another in such a manner that almost backlash-free mesh of rotors is obtained during their rotation. Analytical description of curves looks like a system of parametric equations. To derive curve equations, mutual arrangement of profile sections was taken into account, with these sections matched (see figure 1). Equations in their general presentation were expressed via the parameters that determine profile geometry $\overline{R}$, α_R and r:

IOP Conf. Series: Materials Science and Engineering **425** (2018) 012012 doi:10.1088/1757-899X/425/1/012012

- *AB curve* – original straight section described by equations (1)

$$X = r\cdot\cos\left[-\left(\arccos\left(1/\bar{R}\right)+\alpha_R\right)\right]-r\cdot tg\varphi\cdot\sin\left[-\left(\arccos\left(1/\bar{R}\right)+\alpha_R\right)\right],$$
$$Y = r\cdot\sin\left[-\left(\arccos\left(1/\bar{R}\right)+\alpha_R\right)\right]+r\cdot tg\varphi\cdot\cos\left[-\left(\arccos\left(1/\bar{R}\right)+\alpha_R\right)\right], \quad (1)$$
$$0^\circ < \varphi < \arccos\left(1/\bar{R}\right);$$

- *BC curve* – an arc of a circle described by equations (2)

$$X = r\cdot\bar{R}\cdot\cos\varphi\cdot\cos\left(-\alpha_R\right)-r\cdot\bar{R}\cdot\sin\varphi\cdot\sin\left(-\alpha_R\right),$$
$$Y = r\cdot\bar{R}\cdot\cos\varphi\cdot\sin\left(-\alpha_R\right)+r\cdot\bar{R}\cdot\sin\varphi\cdot\cos\left(-\alpha_R\right), \quad (2)$$
$$0^\circ < \varphi < \alpha_R;$$

- *CD curve* – epicycloidal oblong section described by equations (3)

$$X = 2\cdot r\cdot\cos\varphi - r\cdot\bar{R}\cdot\cos(2\varphi),$$
$$Y = 2\cdot r\cdot\sin\varphi - r\cdot\bar{R}\cdot\sin(2\varphi), \quad (3)$$
$$\arccos\left(1/\bar{R}\right) > \varphi > 0^\circ;$$

- *DE curve* – an arc of a circle described by equations (4)

$$X = r\cdot\left(2-\bar{R}\right)\cdot\cos\varphi,$$
$$Y = r\cdot\left(2-\bar{R}\right)\cdot\sin\varphi, \quad (4)$$
$$0^\circ < \varphi < \alpha_R;$$

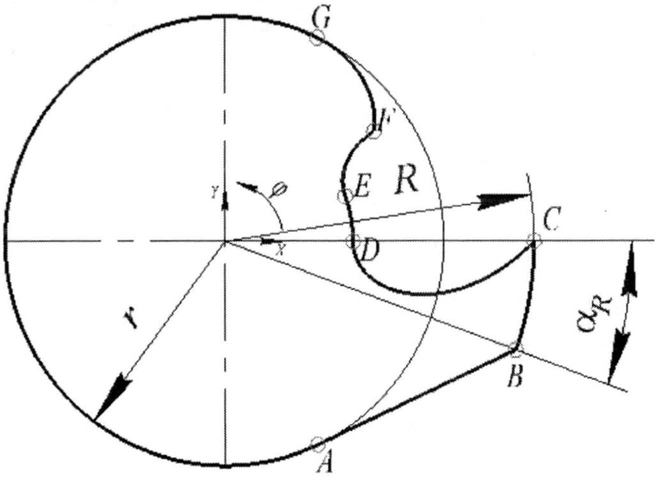

Figure 1. Profile with a straight back of a tooth.

- *EF curve* – epicycloidal oblong section described by equations (5)

$$X = \left[2\cdot r\cdot\cos\varphi - r\cdot\bar{R}\cdot\cos(2\varphi)\right]\cdot\cos\alpha_R + \left[2\cdot r\cdot\sin\varphi - r\cdot\bar{R}\cdot\sin(2\varphi)\right]\cdot\sin\alpha_R,$$
$$Y = \left[2\cdot r\cdot\cos\varphi - r\cdot\bar{R}\cdot\cos(2\varphi)\right]\cdot\sin\alpha_R - \left[2\cdot r\cdot\sin\varphi - r\cdot\bar{R}\cdot\sin(2\varphi)\right]\cdot\cos\alpha_R, \quad (5)$$
$$0^\circ < \varphi < \varphi^*;$$

- *FG curve* – curvilinear section described by equations (6)

$$X = r\cdot\left[2\cdot\left(1-tg^2\varphi\right)^{\frac{3}{2}}-\left(1-2\cdot tg^2\varphi\right)\right]\cdot\cos\left[\arccos\left(1/\bar{R}\right)+\alpha_R\right]+$$
$$+r\cdot\left[2\cdot tg\varphi\cdot\left(1-\sqrt{1-tg^2\varphi}\right)+tg\varphi\cdot\left(1-2\cdot tg^2\varphi\right)\right]\cdot\sin\left[\arccos\left(1/\bar{R}\right)+\alpha_R\right],$$
$$Y = r\cdot\left[2\cdot\left(1-tg^2\varphi\right)^{\frac{3}{2}}-\left(1-2\cdot tg^2\varphi\right)\right]\cdot\sin\left[\arccos\left(1/\bar{R}\right)+\alpha_R\right]- \qquad (6)$$
$$-r\cdot\left[2\cdot tg\varphi\cdot\left(1-\sqrt{1-tg^2\varphi}\right)+tg\varphi\cdot\left(1-2\cdot tg^2\varphi\right)\right]\cdot\cos\left[\arccos\left(1/\bar{R}\right)+\alpha_R\right],$$
$$\varphi^{**} > \varphi > 0^{\circ};$$

- *GA curve* – an arc of a circle described by equations (7)

$$X = r\cdot\cos\varphi\cdot\cos\left[-\left(\arccos\left(1/\bar{R}\right)+\alpha_R\right)\right]-r\cdot\sin\varphi\cdot\sin\left[-\left(\arccos\left(1/\bar{R}\right)+\alpha_R\right)\right],$$
$$Y = r\cdot\cos\varphi\cdot\sin\left[-\left(\arccos\left(1/\bar{R}\right)+\alpha_R\right)\right]+r\cdot\sin\varphi\cdot\cos\left[-\left(\arccos\left(1/\bar{R}\right)+\alpha_R\right)\right], \qquad (7)$$
$$\left[2\cdot\left(\arccos\left(1/\bar{R}\right)+\alpha_R\right)-360^{\circ}\right] < \varphi < 0^{\circ};$$

A cross point of *EF* and *FG* curves determines values of φ^*. A cross point of *FG* and *GA* curves determines values of φ^{**}. These angles are derived by the numerical technique.

Gas pressure and temperature in the working chamber at a random moment of time, is determined by dependence of the chamber volume on the angle of rotation of rotors. Since rotors are straight-line, of gas compression in the compressor be dependence of the working chamber end area $S(\varphi)$.

The current value $S(\varphi)$ is determined by the profile geometry and depends on the rotor-swept area, which is the difference of the standard cross-sectional area of the housing bore and standard cross-sectional area of rotors. The rotor-swept area may be derived by equation (8)

$$S = 2\cdot\left(r\cdot\bar{R}\right)^2\cdot\left[\pi-\arccos\left(1/\bar{R}\right)+0{,}5\cdot\sin\left(2\cdot\arccos\left(1/\bar{R}\right)\right)\right]-2\cdot S_r, \qquad (8)$$

The study of profile geometry influence on the compressor performance employed an algorithm that allows calculating theoretical profile coordinates at any physically sound values of $\bar{R}$, α_R and r for a random angle of rotor rotation, mutual arrangement of rotors relative each other, standard cross-sectional area of rotors and rotor-swept area.

The computer program developed allowed to obtain and approximate dependences of the rotor-swept area value on each of the three profile parameters ($\bar{R}$, α_R and r) with the other parameters remaining unchanged. The profile with r=40 mm, $\bar{R}=\sqrt{2}$, α_R=20° (S_{40}) was assumed as a reference variant. Results are shown in figures 2 – 4.

Results of theoretical calculations shown in Figures 2-4 demonstrate that dependence of the rotor-swept area on pitch radius r is virtually parabolic. Dependence of the rotor-swept area on relative depth and angular thickness α_R of a tooth were derived in a relative form as compared to the reference profile variant and are virtually linear. After approximating the given dependencies, a simplified equation (9) is obtained for calculating the rotor-swept area

$$S = 5{,}3\cdot r^2\cdot\left(2{,}758\cdot\bar{R}-2{,}895\right)\cdot\left(1{,}023-0{,}406\cdot\frac{\alpha_R^{\circ}}{360^{\circ}}\right), \qquad (9)$$

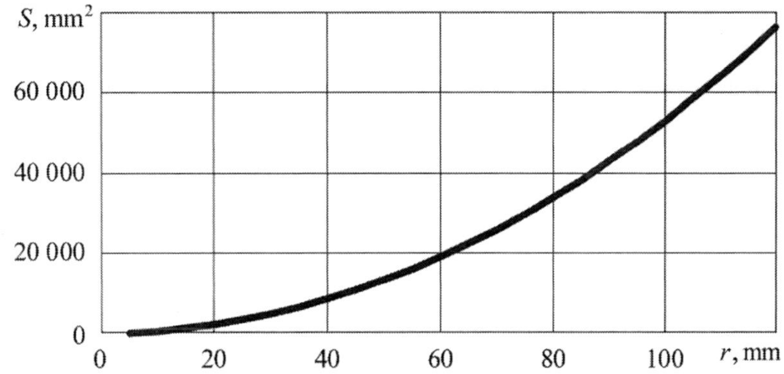

Figure 2. Dependence of $S(r)$.

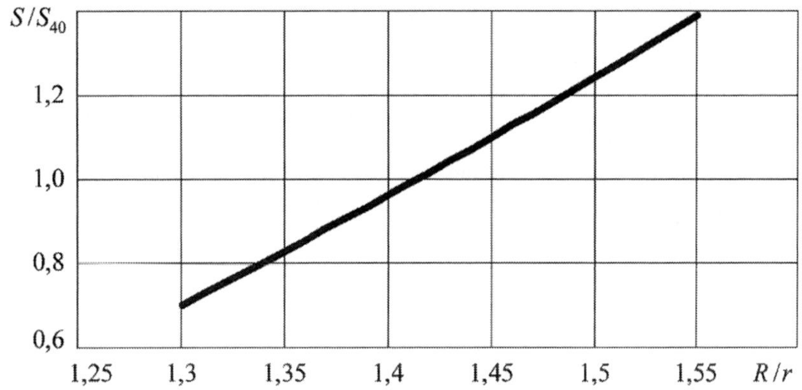

Figure 3. Dependence of $S(R/r)/S_{40}$.

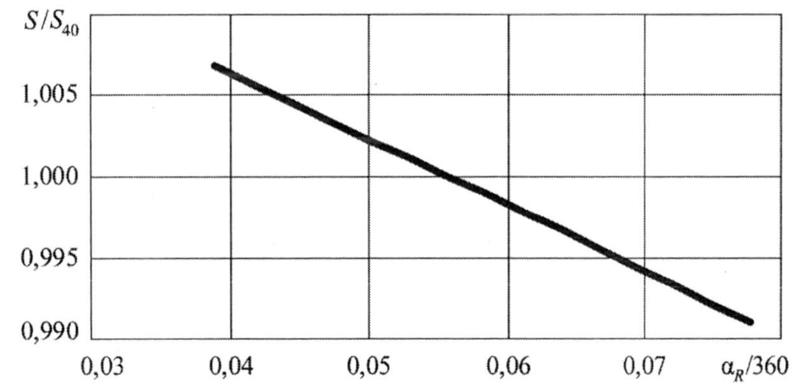

Figure 4. Dependence of $S(\alpha_R/360°)/S_{40}$.

A value of the rotor-swept area found by the proposed formula is used to calculate the current value of the working chamber end cross-section area $S(\varphi)$. To determine the working chamber volume, it is enough to multiply the calculated area by the rotor working length.

Figure 5 shows calculation dependence $V/V_{max}(\varphi)$ for the profile with parameters $\bar{R} = \sqrt{2}$, α_R=20°. Working chamber maximum volume V_{max} is determined as a product of the rotor working length and the rotor-swept area.

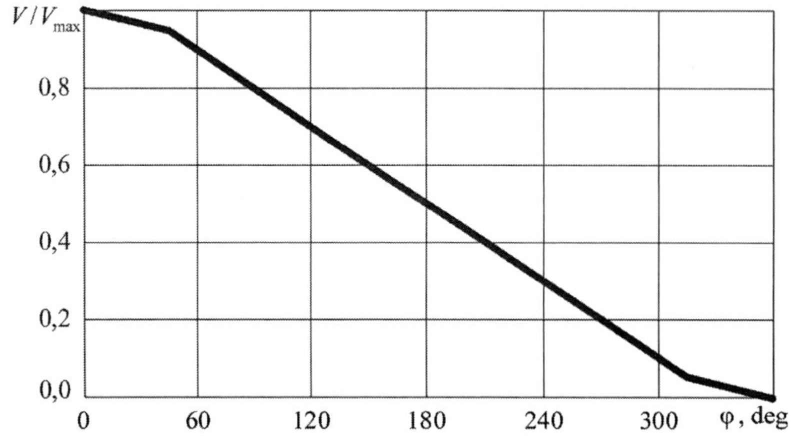

Figure 5. Dependence $V/V_{max}(\varphi)$ for the profile with parameters $\bar{R} = \sqrt{2}$, α_R=20°.

Suction and delivery ports are located on end covers and are so designed that rotation of one rotor causes suction ports to open and close, and rotation of the other rotor causes delivery ports to open and close.

Knowledge of the working chamber volume law of variation allows determining a delivery port angle of opening φ_3 (see figure 6)

$$V(\varphi_1)/V_{max} = \left(p_1/p_3\right)^{1/n}, \tag{10}$$

where p_1 – gas pressure in the working chamber at the starting moment of geometric compression; p_3 – gas pressure in the working chamber at the moment of delivery port opening.

Delivery port closing angle φ_4 (see figure 6) is selected subject to less flow-overs and working chamber leak-tightness. This angle determines the entrapped volume value which leads to indicated power growth.

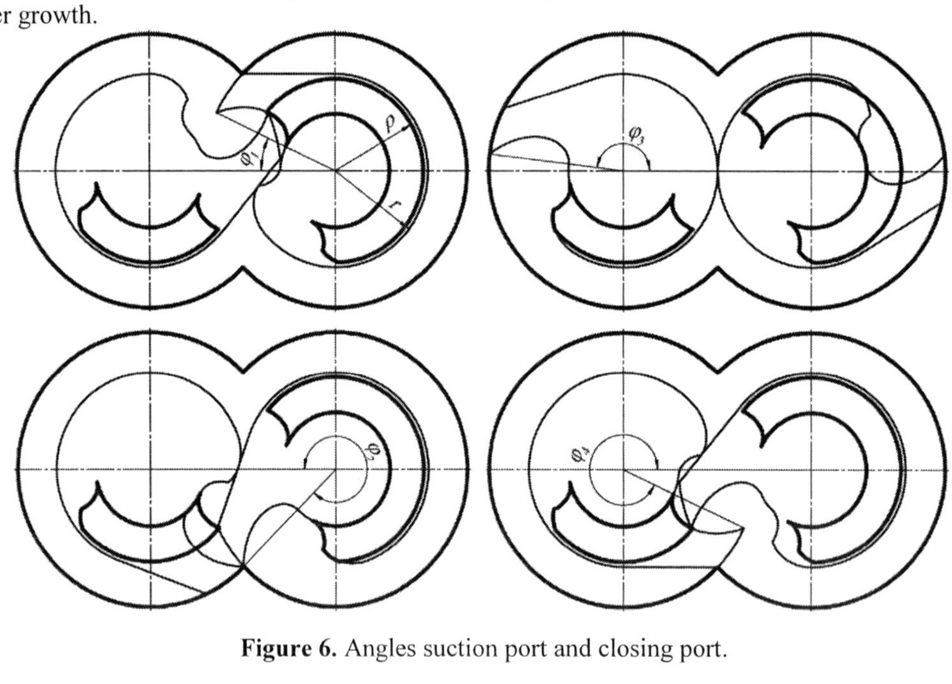

Figure 6. Angles suction port and closing port.

Suction port opening angle φ_1 (see figure 6) is selected subject to provision of the processable suction port section, which is well developed along the entire depth of the space. Angular length of the suction port influences the relative value of gas flow-overs and the rate of underutilization of the working chamber volume, which determines suction port closing angle φ_2 (see figure 6).

Arrangement of suction and delivery ports that determines φ_1, φ_2 and φ_4 is justified geometrically. A theoretical outline of gas distribution ports results from crossing of opening and closing profile edges. An actual outline of gas distribution ports has edges easier in manufacturing and slightly different sizes, which has almost no influence on compressor performance because narrow and low-efficient parts of gas distribution ports are cut off. Gas flow-overs from the compression chamber may be decreased by changing the relative height of the top edge of gas distribution ports ρ/r (see figure 6).

3. Calculation method

To study the influence of profile geometry ratios and other determinants of gas- and thermodynamics of the compression process on the claw compressor performance, a program to calculate integral characteristics was developed to select the optimum ratio of geometry parameters of the profile and gas distribution items.

The thermodynamic calculation was performed according to the procedure of Saint Petersburg Polytechnic University, designed for calculation of the claw compressor [4]. The procedure is based on compressor efficiency estimation with the use of the coefficient of efficiency [5]:

$$\lambda = \lambda_p \cdot \lambda_t \cdot (1 - \nu_u) - \nu_l - \nu_{el}, \tag{11}$$

Pressure factor is a factor taking into account pressure losses at gas flow through the suction port, which is determined by equation (12)

$$\lambda_p = 1 - \frac{\Delta p}{p_s}, \tag{12}$$

where Δp is pressure losses at flow through the suction port.

Temperature factor λ_t considers the increase of suction gas temperature, i.e. $\lambda_t = T_s/T_1$, where T_s is suction temperature, T_1 is the temperature in the working cavity at the end of the suction process. A rise of temperature leads to a decrease of the intake gas density and, as a consequence, a decrease of capacity.

Utilization factor ν_u shows how much of the theoretically possible volume is used in the suction process, i.e. $\nu_u = 1 - S_s/S$, where S_s the end area of the chamber at the time of closing the suction port.

Leakage factor ν_l considers gas leakage from the high-pressure cavities in the low-pressure cavity. Leaks reduce the volume of the suction gas, as they themselves occupy part of the suction cavity volume and heat the intake gas.

External leakage factor ν_{el} considers the volume of gas that flows from the compressor's working cavity to the environment.

Inflows and outflows were calculated by the S. E. Zakharchenko method [4]. Mass rates of inflows and outflows at each fixed angle of rotor rotation are found from the formula (flows from pressure space p_2 to pressure space p_1):

$$\dot{m}_\varphi = \mu \cdot \delta \cdot l \cdot \left(\frac{\rho_1^* \cdot p_1 \cdot (\varepsilon^2 - 1)}{2 \cdot \ln \varepsilon + \xi + \Sigma \cdot \lambda_r} \right)^{1/2}, \tag{13}$$

where $\rho_1^* = p_1/(R \cdot T_2)$; $\varepsilon = p_2/p_1$; $\Sigma = [b/(2 \cdot \delta)] \cdot [1 + \delta/l]$.

The flow value from equation (13) is found by the iteration method. The first step specifies the zero approximation $\dot{m}_0$. In the second step, according to the value of $\dot{m}_0$ found, the number Re_0 is calculated from equation (14).

$$\mathrm{Re}_0 = \frac{2\dot{m}_0}{\mu_v(\delta + L)}, \tag{14}$$

where μ_v is dynamic viscosity of gas.

Depending on Re, the coefficients of friction are found, and the value found is substituted in (13). Thus, a first approximation of the mass flow rate $\dot{m}_1$ is found. Substituting $\dot{m}_1$ into equation (14), is found a new value of Re and friction coefficient. The calculation is repeated until the condition $|\dot{m}_{i+1} - \dot{m}_i|/\dot{m}_i \geq \Delta$, where Δ is given accuracy.

A value of the coefficient of local resistances depends on the gap shape: for gaps with abrupt narrowing and expansion of flow ξ=2.5; for gaps with smooth narrowing and expansion of flow ξ=1.418.

A value of the roughness coefficient is derived from empirical dependences: for Re< 1.200 by equation $\lambda_r = 189.2 \cdot \mathrm{Re}^{-1.127}$; for Re≥ 1.200 – $\lambda_r = 3.6 \cdot \mathrm{Re}^{-0.566}$.

4. Calculated data representation

Using the developed computer program, let's estimate the profile geometry influence on performance of a single-stage compressor with the following parameters: 2 m³/min capacity, 0.25 MPa final pressure, 3,000 rpm synchronous speed of rotors.

There are two ways to design the claw compressor: 1) by determining the optimal working length of the rotor and addendum geometry, with the rotor profile pitch radius remaining unchanged; 2) by determining the optimal working length of the rotor and addendum geometry, with the housing bore radius remaining unchanged. The selected design method does not influence dependence patterns of efficiency-related parameters of compressor performance. The developed program allows calculation for both design methods. The profile geometry influence on compressor performance designed by the first method is given below. The profile pitch radius was assumed to be *r*=55 mm from design considerations.

Figures 7-12 shows results of the study. The profile geometry determines the rotor working length required for the capacity specified at the selected speed of compressor rotors (see figure 7). Herewith, profile parameters and rotor working length determine the geometry of end and longitudinal gaps where the air overflows thus influencing the compressor efficiency.

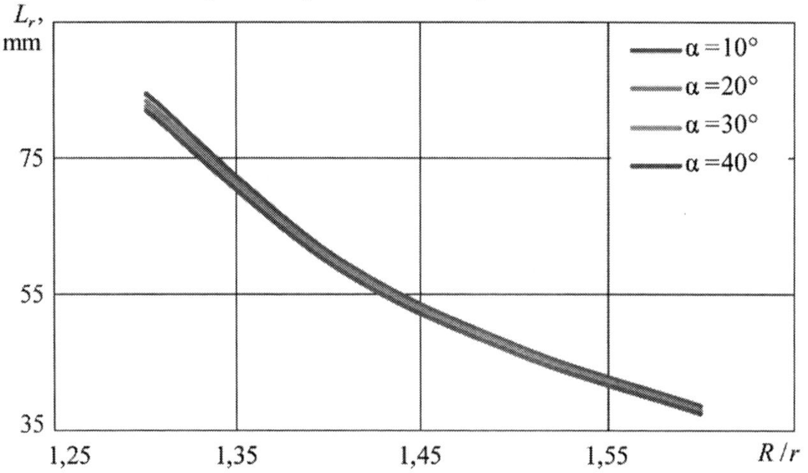

Figure 7. Axial working length of rotors depending on relative addendum and angular tooth thickness.

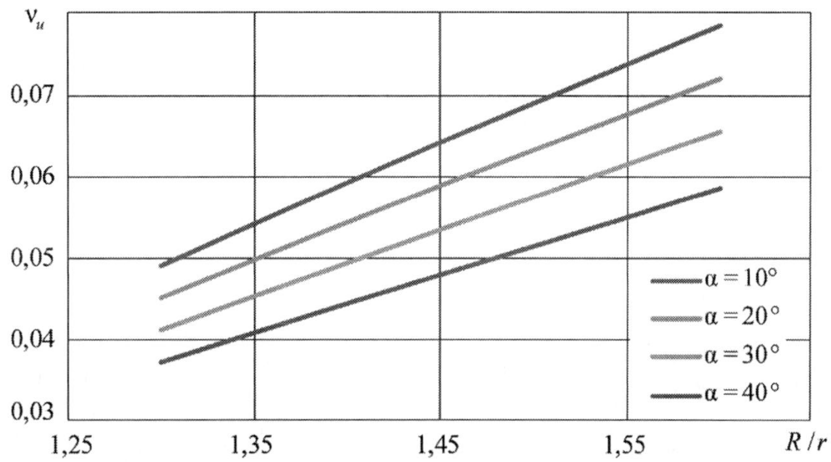

Figure 8. Utilization factor of the steam space volume.

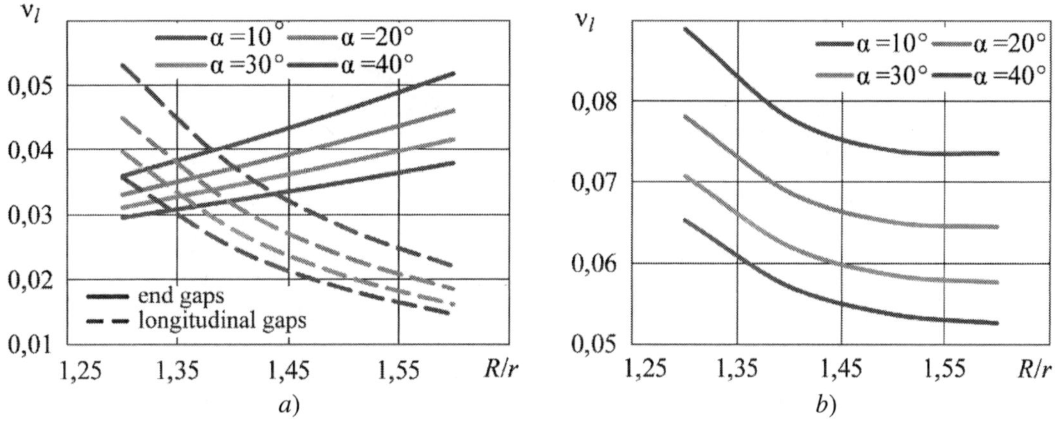

Figure 9. Relative value of internal inflows: *a*) depending on the gap type; *b*) total.

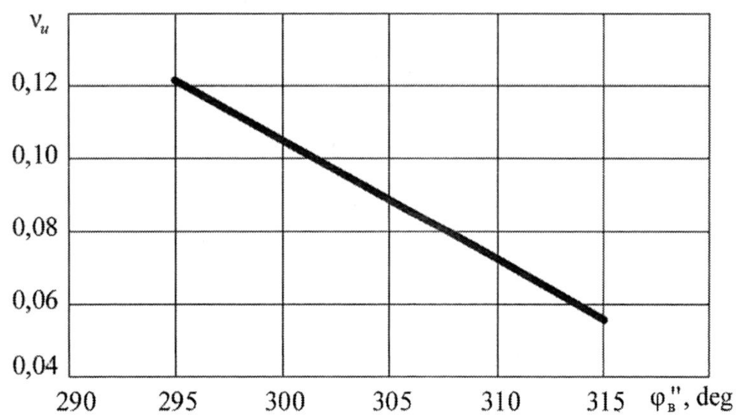

Figure 10. The rate of underutilization of the steam space volume, depending on the axial length of the suction port (α_R=20°).

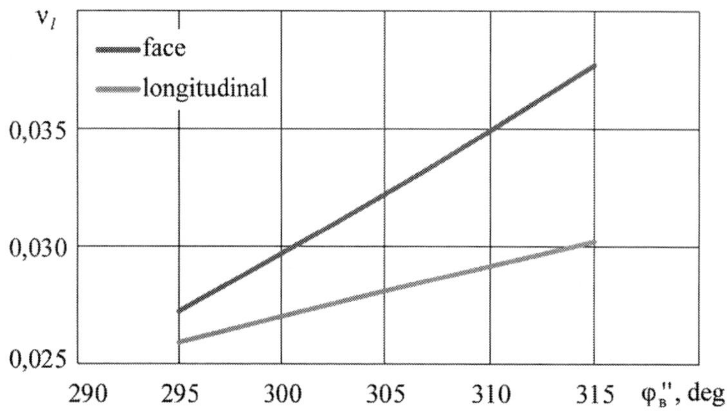

Figure 11. Relative value of internal inflows depending on the axial length of the suction port (α_R=20°).

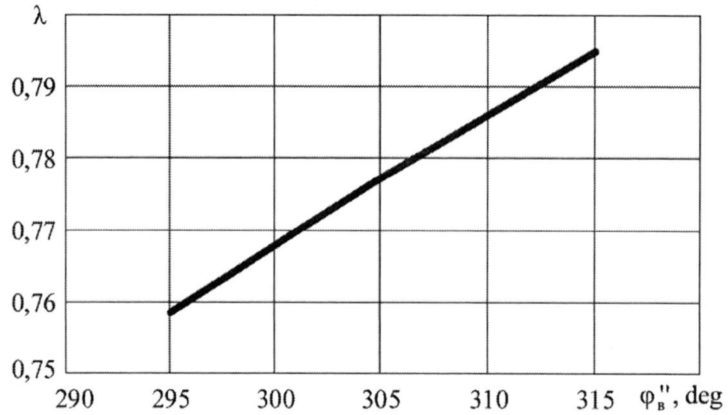

Figure 12. Variation of the coefficient of efficiency depending on the axial length of the suction port (α_R=20°).

Variation of profile parameters determines the change in extent of utilization of the steam space theoretical volume while the working chamber is being filled during suction (see figure 8) and leads to re-distribution of inflows and outflows between end and longitudinal gaps because of changing depth and thickness of the tooth addendum and rotor working length (see figure 9).

The rate of underutilization and the rotor working length are also influenced by arrangement of edges and angular length of the suction port (see figure 10). As the angular length of the suction port decreases, the relative value of inflows through both end and longitudinal gaps is reduced (see figure 11), which, however, does not compensate the growing rate of underutilization and leads to drop in the coefficient of efficiency (see figure 12).

5. Conclusion

Claw compressors belong to the rotating machinery class where the prevailing determinant of the performance efficiency is gas flows through gaps. Reducing gas flows from one working space to another and to the environment is an urgent task for rotary compressors. Capacity and power losses are directly dependent on amount of gas flowing through gaps.

The developed program allows estimating the claw compressor performance at various geometry parameters of the rotor profile and for different design solutions. The program is intended for calculation of the rotor profile geometry and working space volume, as well as integral characteristics of the compressor under design. Since the compressor analysis performed has demonstrated that the compressor capacity is influenced by internal inflows through gaps, the first task to accomplish is to minimize the end clearance from the suction port end. According to the analysis of the suction port length, decrease thereof leads to drop in the coefficient of efficiency.

6. Acknowledgement
The article is written in the framework of the program "5-100-2020".

List of symbols and abbreviations

Symbols

b	reduced length of gas travel in a gap	(m)
l	reduced length of a gap	(m)
n	temperature indicator of compression polytrope based on terminal parameters	(-)
p	pressure	(N m^{-2})
R	gas constant	m^2 s^{-2} K^{-1}
S_r	standard cross-sectional area of the rotor	(m^2)
V	volume	(m^3)
Σ	coefficient of form	(-)
δ	reduced width of a gap	(m)
λ_r	roughness coefficient	(-)
λ_p	pressure factor	(-)
λ_t	temperature factor	(-)
μ	flow rate factor	(-)
ξ	coefficient of local resistances	(-)
ν_u	utilization factor	(-)
ν_l	leakage factor	(-)
ν_{el}	external leakage factor	(-)
ρ	density	(kg m^{-3})

References
[1] Northey A J 1937 *Rotary compressor or vacuum pump* (United States Patent 2097037).
[2] Ksayer E B and Clodic D 2010 Analysis and Development of a Turbivo Comoressor for MVR Applications *International Compressor Engineering Conference* 1275 p. 8
[3] Yaminsky V V 1960 *Rotary Compressors* (Moscow: Mashgiz) p 223
[4] Dimentov Y I and Prilutsky I K 1978 *Screw compressors* (Leningrad: Leningrad polytechnic institute of M.I. Kalinin) p 69
[5] Grieb M and Brümmer A 2014 *International Conference on Screw Machines – VDI Bericht* 2228 pp. 197-210

IOP Conf. Series: Materials Science and Engineering **425** (2018) 012013 doi:10.1088/1757-899X/425/1/012013

Mini Screw: the development of a high-CFM compact compressor for LGWP A1 low pressure refrigerant

Masao Akei, Vishnu Sishtla and Scott MacBain

UTC Climate Controls & Security (Carrier), Syracuse, New York, USA
E-mail: Masao.Akei@carrier.utc.com

Abstract. Addressing global warming concerns is one of the largest challenges facing the HVAC industry currently. Although A2L refrigerants are important options to consider for certain sizes of HVAC units, there remain many systems that require A1 category refrigerants. Among the several available refrigerants, we looked at DR-12 as one of the attractive refrigerants. Along with it being in the A1 category, its very low GWP value (32) and high cycle performance make it an excellent long-term solution if successfully applied. A significant challenge to overcome with this refrigerant is its low density. DR-12 will require almost 7 times the volumetric flow for the same cooling capacity compared to R410A. The current scroll or rotary technologies are not realistic to be applied in such a low density refrigerant system. The development of a new compact large CFM compressor is key to realize the scope of the project successfully. The authors have explored various types of compression mechanisms and identified screw as one of the most suited technologies for this refrigerant. In order to obtain such large volume flow (19.8L/s, 42CFM), a unique screw rotor design and a high speed PM motor were employed. Innovative bearing design and new compressor layout also enabled the compressor to be compact (ϕ145mm) and cost effective. A prototype compressor was built and tested. This concept delivered 17.6kW (5 ton) capacity at 11,000rpm and 4.4kW (1.25 ton) at 2,500rpm, meeting its target requirements. This paper will introduce details of the compressor concept, along with the test results from a physical prototype.

1. Introduction

Global warming potential of refrigerants is one of the largest challenges that the HVAC industry is currently facing. Various low global warming potential (LGWP) refrigerants have been proposed and some of them have already been deployed in applications. However, most of these recently proposed LGWP refrigerants are categorized as A2L (mildly-flammable refrigerants). This requires special hardware for leak detection and ventilation especially when it is used in the systems with large amount of refrigerant charge. Neither of the LGWP refrigerants offers the lowest global warming potential option. For example, R32 and R452B (DR-55), which are two of the R410A substitute candidates, still have a GWP of 600-700 [1]. Although these are excellent drop-in candidates, they are considered as interim solutions since considerable GWP reduction potential remains. At this time, there have been no drop-in refrigerants developed with very-low GWP and A1 (non-flammable) property for replacement of R410A.

When a long term solution such as A1 and GWP<100 is considered, options are currently very limited. Carbon dioxide (CO_2) has been used in some applications such as water heating and refrigeration (cascade) systems. It is a vital option for the long term yet would require high strength components in the A/C system due to its extreme pressure. This would likely be more challenging when a larger size system is considered. In addition, the cycle efficiency with CO_2 as a refrigerant for typical A/C system duty is at a disadvantage to other replacement options. Another path for long term LGWP A1 refrigerant is to move to a lower pressure refrigerant option such as R1233zd(E). This type of refrigerant has been applied in some large chillers using centrifugal compressors. Since it is a low pressure refrigerant, compressors need to be large to generate the large volumetric flow required. In fact, when cycle capacities are calculated at the industry's cooling rating condition [2], the required volumetric flow for the same cooling capacity reaches almost ten times with R1233zd(E) compared to R410A. The current compression technologies used in R410A systems such as scroll and rotary are not realistic to apply in such low pressure systems.

DR-12 is a relatively new refrigerant that has been proposed for achieving a LGWP A1 solution [3]. It is still a low pressure refrigerant with the required volumetric flow of approximately seven times that of R410A. Although the GWP and flammability issues can be mitigated with DR-12, the challenge still remains with the compression technology of how to create seven times the volumetric flow with a single compressor. An "out-of-the-box" solution is anxiously desired when considering low pressure refrigerants.

Screw compressors are widely used in larger applications above the capacity of typical R410A products. The compressor is well-suited for large volumetric flow, therefore a practical compressor could be constructed for a low pressure system. Since typical R410A applications are very different from screw chillers, the compressor should be re-designed entirely to accommodate the requirements of product styles leveraging R410A compressors. In this paper, a new compressor concept will be introduced using a novel screw mechanism. With this concept, a very large volumetric flow can be obtained without an increase of compressor shell size versus R410A. Also, as it is critical to typical R410A applications, various design elements were put into the concept in order to produce the compressor more economically. A comprehensive design study was carried out to get the optimized screw design for 17.6kW (5 ton) cooling capacity with DR-12. A prototype compressor was made and the performance was verified. More details will be described in the main article below. It should be noted that the developed compressor has been named "Mini Screw" as it is literally a mini size compared with typical screw compressors.

2. Mini Screw Concept

2.1. Compression mechanism comparison

The study started with a comparison of how the screw compressor had an advantage over other compression mechanisms. Figure 1 shows the comparison with scroll. Both screw and scroll models shown in the figure have the same displacement (swept volume), same volume ratio (Vi), and the same discharge port area. Since the compression volume shrinks inwardly in the scroll compressor, the compressor tends to grow in the radial direction when a larger displacement is required. Increasing vane height is limited in the scroll compressor due to its own overturning moment. Large contact force (thus large friction loss) between the two mating scrolls is unavoidable in a tall scroll to prevent its overturning during operation. On the other hand, the screw volume can be extended in the axial direction, therefore the same displacement can be achieved in smaller shell diameter with screw. It should be noted that too long of a screw shaft would cause excess deflection in the screw compressor. However, the pressure differential that causes the shaft deflection is expected to be small with the low pressure refrigerant, therefore the limitation is not overly restrictive with DR-12. The overturning moment issue will still exist in the scroll compressor regardless of the pressure level of the refrigerant.

Another advantage of the screw compressor is that it is a perfectly balanced machine. Unlike other mechanisms, there are no unbalancing components in the compressor such as pistons, counter weights, discharge valves, Oldham rings, vanes, etc. This will be very helpful for constructing a high speed compressor without facing typical speed-related issues like valve dynamics, vibration, wear, etc. In fact, the optimum speed of the screw compressor is usually higher than other type of compressors due to these characteristics of the compressor. By setting the compressor speed higher than scroll, the screw compressor could output even higher volumetric flow within the limited size of the compressor.

IOP Conf. Series: Materials Science and Engineering **425** (2018) 012013 doi:10.1088/1757-899X/425/1/012013

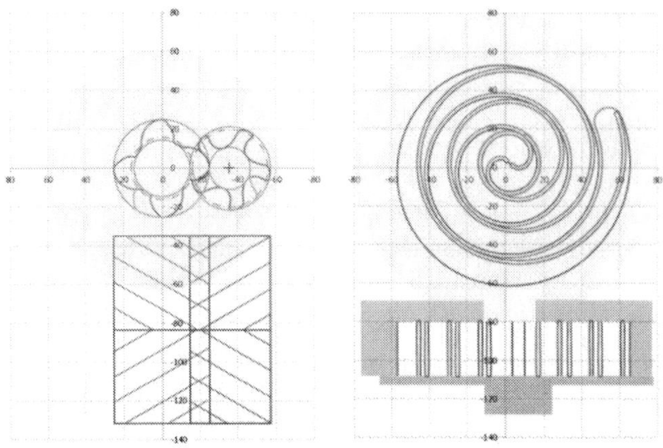

Figure 1. Compression mechanism comparison between screw and scroll.

2.2. Screw compressor design for small capacity application

R410A compressors are typically used in residential and light commercial units in the capacity range of approximately 3-100 kW (1-30 ton). With these smaller capacity applications, production volumes (number of units) are considerably large compared to typical screw products. Most of the R410A compressors are made in a welded can (hermetic) configuration for mass production. Shell diameter (footprint) is always an important factor because it tends to drive overall cost of the compressor. The development of a Mini Screw compressor needs to accommodate these requirements. Below are some of the design considerations:

2.2.1. Opposed rotor screw compressor. Figure 2 shows the opposed rotor screw compressor concept. This configuration enables an economical bearing design without the use of rolling element bearings or special bushings. The concept itself has been known since 1923 [4]. A pair of helically opposed twin screw rotors are connected on a common male and a female shaft. Refrigerant suction takes place at the mid portion of the connected rotors and the gas flow splits towards the opposite ends. Gas compression is balanced and therefore thrust bearings can be omitted. In our configuration, screw rotors are machined separately from the shafts. Each male and female rotor has an axial bored hole within the rotor. Two male rotors (the opposed helical pair) are securely attached to a drive shaft through their rotor bores (tight fit). The drive shaft will be supported by journal bearings at the upper and lower ends of the attached rotors. The female rotors are made to have small clearance between their bore diameters and its supporting shaft (loose fit) so that the rotor bore itself can be utilized as a cost effective journal bearing. This loose-fit configuration also provides a self-aligning function to the female rotors. Since each female rotor can spin independently on a stationary shaft, it can mesh to the respective male rotor regardless of angular positioning (clocking) between the two male rotors on the shaft. This design can greatly reduce the level of complexity in the screw rotor assembly and make the opposing concept practical.

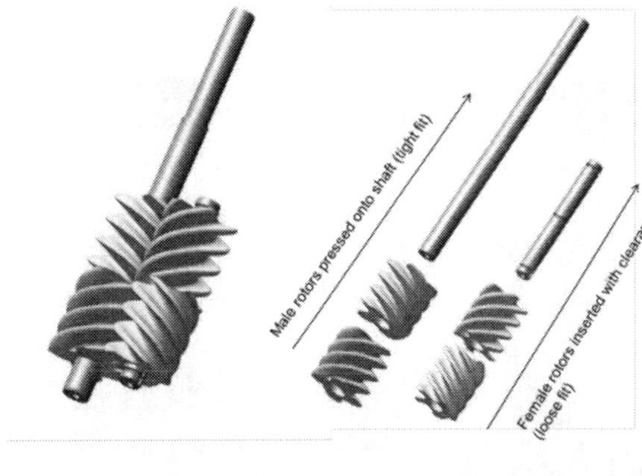

Figure 2. Opposed rotor screw compressor concept.

2.2.2. High side motor shell. The high-side motor shell design is widely used in small capacity compressors. It has an advantage of simplifying the lubrication system, especially for compressors using variable frequency drives. Pressure differential is utilized to supply oil to the bearings eliminating the need for a dedicated oil pump. Figures 3 and 4 show the concept of the high-side design. Suction gas is directly introduced to the compression mechanism through the inlet pipe at the bottom side. Discharge gases from the opposed screw rotors are routed along the channel provided within the rotor housing, and flows out of the slots at the top of the rotor housing. The gas flows up through the motor/stator and provides the cooling effect to the motor. Gas pulsation could be dampened and reduced within the compressor shell before it is discharged to the outlet. Oil that mixed with the refrigerant gas is separated in the space above the motor, which then drips down to the bottom sump by gravity. The oil is supplied back to the bearings via the holes within the male/female shafts. Internal tight screw rotor discharge end clearances are utilized to regulate oil flow and therefore excessive oil circulation is prevented. A discharge outlet is provided at the top of the compressor shell. Inlet/outlet pipe sizes are chosen to accommodate the large volumetric flow from the compressor. Expected overall compressor dimensions (approximate) are also shown in the figure 3. It can be seen that the compressor size is maintained compact (ϕ145mm) even though it provides seven times the volumetric flow (19.8L/s, 42CFM) of the equivalent size R410A compressor.

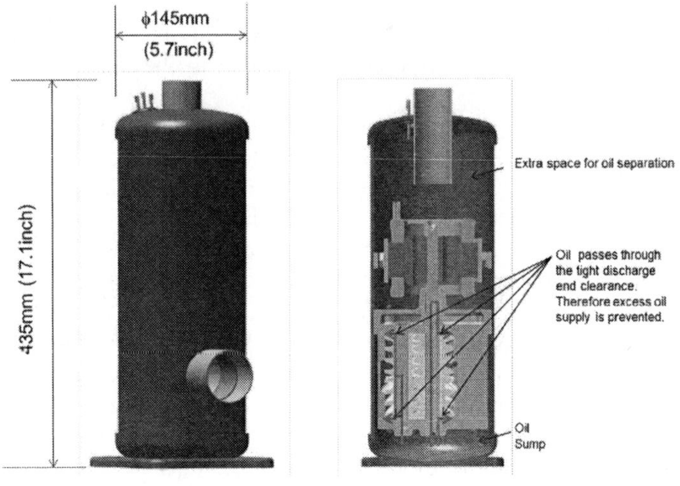

Figure 3. Mini Screw compressor concept in high side shell (19.8L/s, 42CFM).

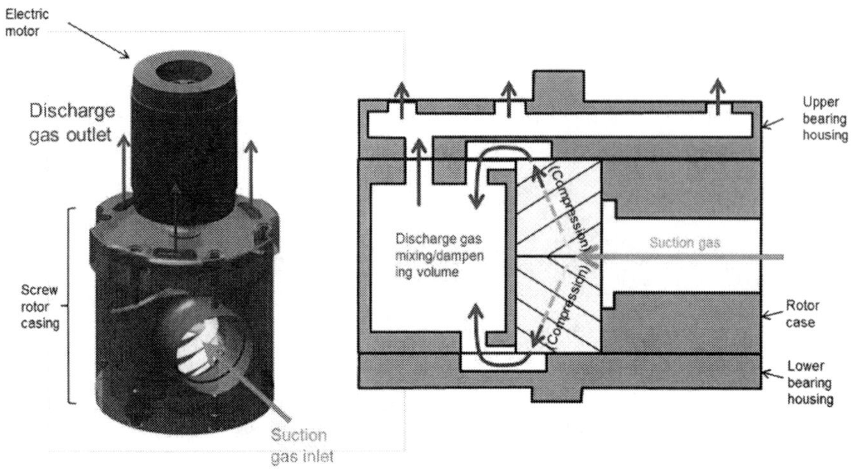

Figure 4. Mini Screw concept showing the internal gas passage

2.3. Design optimization

In positive displacement machines, compressor output (capacity) is determined by its displacement and shaft speed. Variable frequency drives (VFD) are vital options, not only for adding capacity modulation capability, but also to help reduce the compressor size by operation at a higher frequency than utility supply power. Although modern VFDs are capable of handling frequencies as high as 1 kHz, the associated small displacement compressor may not be efficient due to increase in the various internal losses. In order to design the right size compressor with a VFD, it is necessary to understand the effect of speed on the compressor efficiency.

In this study, an in-house screw compressor performance simulation program (1-D model) was used to determine the optimum speed of the compressor. This simulation tool was validated against test data for larger compressors. The 1-D program includes comprehensive models to account for most of the physical effects such as leakages, porting losses, bearing losses, deflection, oil dilution, motor cooling, etc. Modifications were made to model the Mini Screw concept for its unique design considerations. A hydrodynamic fluid film loss model based on Reynolds equation was added to handle the journal bearings. Thrust bearing loss was removed since it is absent in the compressor. The oil dilution (PVT characteristics) was calculated based on the discharge condition of the compressor. The motor loss was also calculated based on the discharge temperature due to the high-side design of the Mini Screw compressor.

Speed sweep analyses were performed on five different screw displacement designs (14, 28, 57, 113 and 227cm^3/rev). Table 1 shows basic screw rotor information used in the simulation. Screw designs were kept similar among all five displacements. Clearance was increased proportionally with the rotor diameter (i.e. constant clearance ratio). Figure 5 shows the effect of speed on the compressor efficiency (vertical axis) and the capacity output (horizontal axis). With each displacement, the compressor efficiency peaked during the speed sweep. For example, the 14cm^3 compressor had an efficiency peak at 29,000rpm for generating 5.3kW (1.5 ton) cooling capacity, and 28cm^3 had it at 24,000rpm for 8.4kW (2.4 ton) capacity. There are two trends worth noting that peak efficiency reduces with size and large efficiency change with speed in smaller sizes. Efficiency levels of these small displacement compressors were low since the size of the screw rotors were too small, resulting in relatively high internal leakage. On the other hand, larger size compressors had more favorable simulation results. Due to less leakage per volume, the compressor efficiency improves with the displacement. According to the simulation, 113cm^3 and 227cm^3 were expected to be superior by 18% and 21%, respectively, compared to the smallest 14cm^3 compressor. Table 2 shows the summary of the analysis.

Table 1. Screw rotor information

Lobe counts (male/female)	4/5
Wrap angle	330deg
L/D (male rotor)	0.96

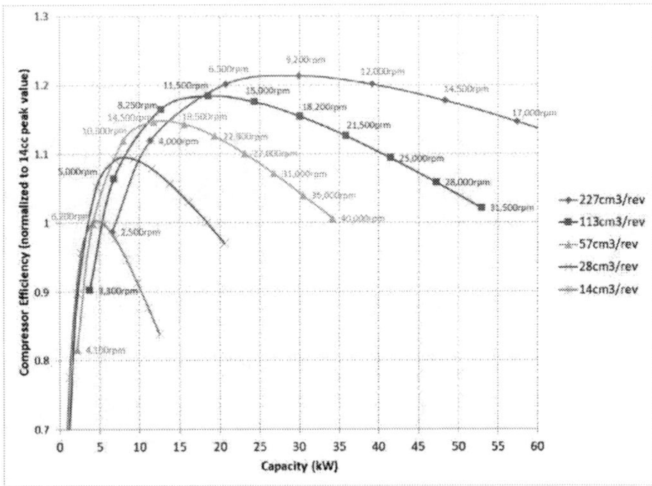

Figure 5. Compressor speed sweep analysis using 1-D simulation tool.

Table 2. Speed sweep analysis summary.

Displacement (cm^3/rev)	Optimum Speed (rpm)	Optimum Capacity (kW)	Peak efficiency (normalized to 14 cm^3)
14	29,000	5.3	1.00
28	24,000	8.4	1.09
57	14,000	12.7	1.15
113	11,000	18.3	1.18
227	9,000	28.1	1.21

This result can be used to determine the compressor speed and displacement for a given cooling capacity. For example, 227cm^3 is indicated to be the best for 25-30 kW (7-9 ton) capacity and 113cm^3 is shown as optimum for 15-20 kW (4-6 ton). Although 10kW (3 ton) and below capacities can be possibly pursued with 57cm^3 and lower displacement, efficiency levels may not be as attractive as the larger models. In this study, 113cm^3/rev was selected for 17.6kW (5 ton) target full load capacity. The compressor is planned to be operated at 11,000rpm for this capacity.

2.4. Prototype compressor

Based on the above optimization, the screw rotor design was finalized, and associated motor torque and bearing load calculations were made. Volume index (Vi) was further optimized based on the part load conditions of an expected application. A permanent magnet motor and VFD were selected which could accommodate the required range of speed and torque of the compressor. Figure 6 shows pictures of the actual prototype built. Instead of a welded canned shell, a bolted construction with increased thickness was used. The prototype design allows adding instrumentation connectors and facilitates periodic inspections of the rotor and bearings. The critical design elements of the compressor such as screw rotors, shafts, suction/discharge ports, clearances, etc. were maintained as proposed in the original hermetic concept, therefore there should be no influence of the modifications on the compressor performance. In addition, Coriolis flow meter was connected to measure the amount of oil flow to the bearings during the test.

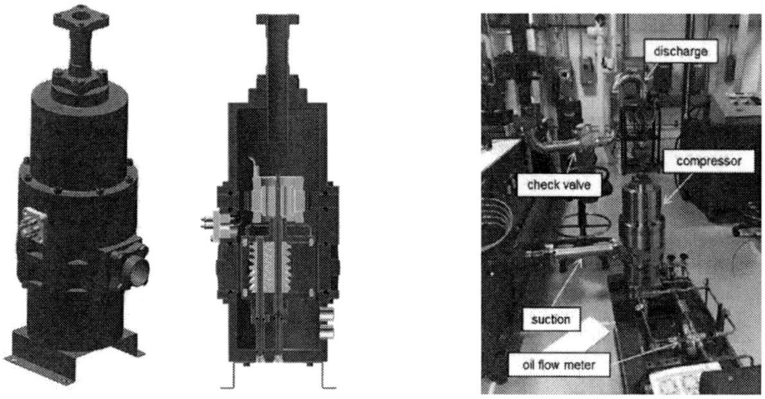

Figure 6. Mini Screw compressor prototype.

3. Prototype Performance

Compressor tests were carried out on an automatic calorimeter and tests were performed in accordance with ASHRAE Standard 23.1. Electric power was measured before the VFD, therefore power loss of the drive is included in the efficiency data. Oil circulation rate (OCR) was measured by sampling a small amount of condensing liquid from the calorimeter and then measuring the weight of non-evaporated liquid (oil) from the sample. DR-12 was used for refrigerant. VG32 POE was used for oil. Figure 7 shows the schematic diagram of the calorimeter setup.

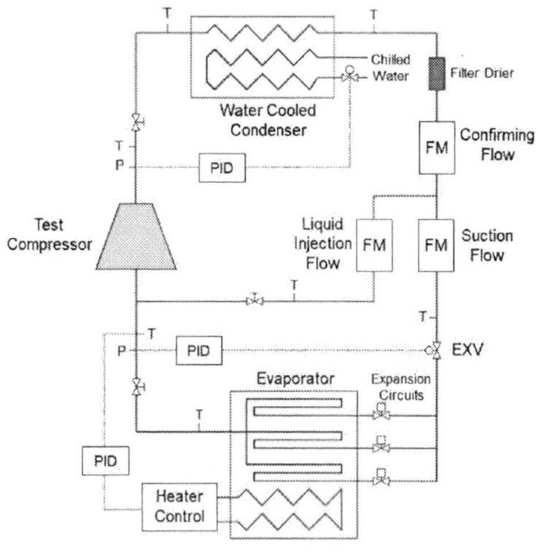

Figure 7. Calorimeter piping diagram.

Compressor performance was measured at four different rating conditions (named here as A, B, C and D). Condition A is for the full load operation, where the compressor should output 100% of maximum capacity i.e. 17.6kW (5 ton) in our case. The other conditions, B, C and D, are part load at capacities corresponding to 75%, 50% and 25%. In the present case, the compressor speed was reduced from 11,000rpm to 2,500rpm to achieve the 25% (4.4kW, 1.25 ton) unloaded capacity. All four of these conditions constitute a system efficiency rating, therefore it is important that the compressor should achieve near optimum performance at all conditions. Table 3 shows the test conditions. It should be noted that discharge pressure decreases with reduced capacity due to low refrigerant mass flow and reduced heat rejection in an A/C system.

Table 3. Test conditions.

Test Condition	Target Capacity (kW)	Compressor Speed (rpm)
A	17.6 (100%)	11,000
B	13.2 (75%)	7,200
C	8.8 (50%)	4,500
D	4.4 (25%)	2,500

The Mini Screw prototype compressor was run at all four conditions without any issues during the tests. The initial performance data is summarized in figure 8 (red line). It can be seen on the horizontal axis that the 17.6kW full load capacity and respective 75% (13.2kW), 50% (8.8kW) and 25% (4.4kW) capacities were all achieved per the targets. However, the compressor efficiency (EER; vertical axis) did not achieve the targeted results, especially at the reduced capacity conditions. As shown in the figure, EER was inferior by 27% at 4.4kW unloaded capacity when compared to a current R410A compressor (shown in the black line).

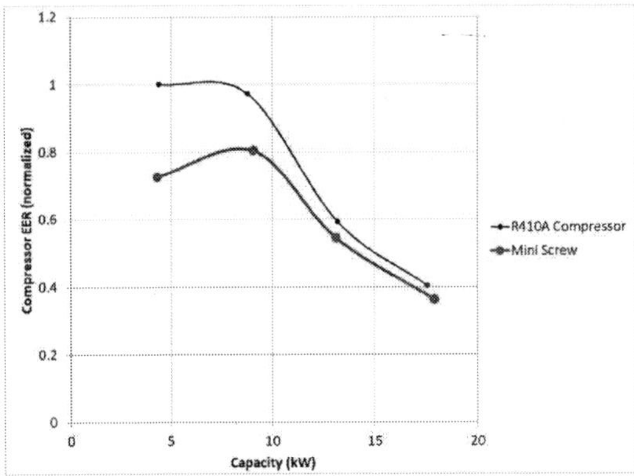

Figure 8. Mini Screw calorimeter performance (first result).

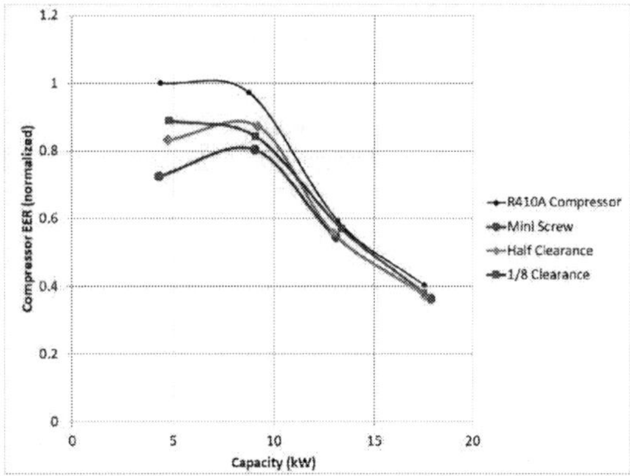

Figure 9. Mini Screw calorimeter performance (after clearance tightened).

The main reason of the lower performance was the leakage within the screw rotors. In order to take a conservative approach for initial tests, the first prototype was made with large clearances between the screw rotors and housing. After the first test, the rotor housing was re-machined and components of the clearance were tightened in several steps. Figure 9 shows the performance results for the tightened clearance compressors. The clearance was reduced to half (shown in green line) and then to one-eighth (purple line). It can be seen that EER improved at all four conditions and the performance gap versus the R410A compressor had been narrowed. With one-eighth the clearance, the gap decreased to 4% at 13.5kW and 11% at 4.5kW unloaded capacity, respectively.

IOP Conf. Series: Materials Science and Engineering **425** (2018) 012013 doi:10.1088/1757-899X/425/1/012013

Table 4 shows a summary of the performance tests along with the OCR results. It can be seen that the OCR maintained levels less than 1% at all four conditions. The compressor was run for more than 800 hours without any issues. Figure 10 shows pictures of the internal components after the performance test.

Table 4. Mini Screw performance data summary (DR-12 refrigerant).

Condition	Clearance			R410A compressor
	Original	Half	1/8	
A	Cap[a]: 17.9 kW EER[b]: 0.90	Cap: 18.0 kW EER: 0.92 OCR[c]: 0.6%	Cap: 18.3 kW EER: 0.93 OCR: 0.9%	Cap: 17.6 kW EER: 1.00
B	Cap: 13.1 kW EER: 0.92	Cap: 12.9 kW EER: 0.94 OCR: 0.2%	Cap: 13.4 kW EER: 0.96 OCR: 0.3%	Cap: 13.2 kW EER: 1.00
C	Cap: 9.1 kW EER: 0.83	Cap: 9.2 kW EER: 0.90 OCR: 0.1%	Cap: 9.1 kW EER: 0.87 OCR: 0.6%	Cap: 8.8 kW EER: 1.00
D	Cap: 4.3 kW EER: 0.73	Cap: 4.7 kW EER: 0.83 OCR: 0.2%	Cap: 4.9 kW EER: 0.89 OCR: 0.6%	Cap: 4.4 kW EER: 1.00

[a] Capacity
[b] EER (Energy Efficiency Ratio) normalized to R410A compressor
[c] Oil circulation rate [weight ratio: oil / (oil + refrigerant)]

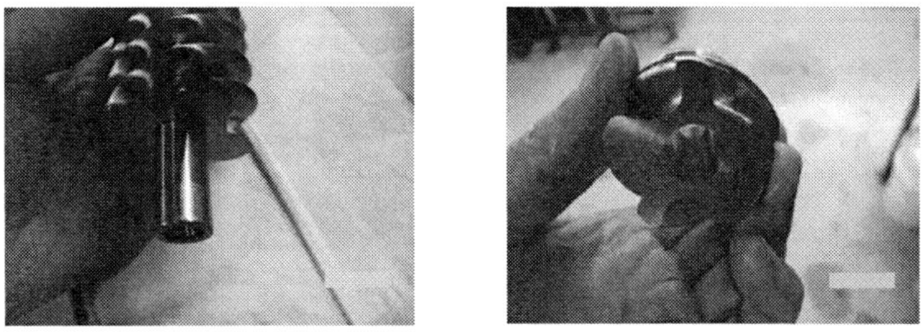

Figure 10. Screw rotors after the test.

After the calorimeter test, the test results were compared with the 1-D simulation data. Table 5 shows the comparison summary. Although the test data agreed fairly well with the 1-D tool at the high speed condition such as A (11,000rpm), the data deviated more at lower speeds, especially 2,500rpm. Also, the prediction error was more pronounced with the original larger clearance compressor. This result indicates that the 1-D tool was not capturing well the increased leakage with large clearance and/or slow speed operation. This result will be used for refining the physics model in the program. The simulation tool will be utilized for further improvement of the Mini Screw compressor.

Table 5. Test data versus 1-D simulation tool prediction (test result / simulation data)

Condition	Clearance		
	Original	Half	1/8
A (11,000rpm)	Cap: 1.13 EER: 1.03	Cap: 1.06 EER: 1.00	Cap: 1.02 EER: 0.93
B (7,200rpm)	Cap: 1.21 EER: 1.10	Cap: 1.08 EER: 1.03	Cap: 1.05 EER: 0.98
C (4,500rpm)	Cap: 1.31 EER: 1.12	Cap: 1.16 EER: 1.05	Cap: 1.05 EER: 0.95
D (2,500rpm)	Cap: 1.97 EER: 1.55	Cap: 1.48 EER: 1.18	Cap: 1.22 EER: 1.04

4. Conclusion

A unique screw compressor concept has been developed for LGWP A1 low pressure refrigerant. Seven times the volumetric flow was achieved in an equivalent size to a typical R410A compressor. Various economical design aspects were introduced in the compressor to accommodate mass production in the future. A prototype was built and the concept was validated. Although the initial performance did not completely reach that of the current R410A compressor, the gap was only 4-11%. At this point in development, there are several areas that have not yet been fully optimized in the prototype such as bearings, motor, oil, gas passages, etc. Therefore, there is a high degree of confidence that the compressor will outperform the existing R410A compressors. It also should be noted that the refrigerant does not have to be limited to DR-12. Any low-to-medium pressure refrigerant that requires large volumetric flow per unit of capacity could be applied with this compressor concept. Further development is currently in progress.

5. References

[1] Brown R and Obosu C 2016 AHRI Low-GWP AREP Test Report #60
[2] AHRI Standard 540
[3] Kontomaris K 2014 *ASHRAE Annual Conference (Seattle Washinton)* Seminar 35
[4] Lithgoe J 1925 *United States Patent Office* 1534182

Acknowledgements

The authors would like to express their appliciation to Rick Pandzik for his essential advise on the compressor design.

IOP Conf. Series: Materials Science and Engineering **425** (2018) 012014 doi:10.1088/1757-899X/425/1/012014

Influence of suction port parameters on integral characteristics of screw-type compressor

A A Kotlov

R&D Laboratory "Gas dynamics of turbo machines" Peter the Great St. Petersburg Polytechnic University, Polytechnical st. 29, St. Petersburg, Russia.

Email: kotlov_andrej@mail.ru.

Abstract. At engineering screw-type machines different tasks on search for more effective and reliable design solutions arise connected with: organization of work process; selection of design shapes of the flow channel; strength calculation of machine components and selection of material. Considered in the work is enhancement of suction process efficiency by the use of dynamic head of the suction gas flow. Influence of rotor length to diameter ratio and suction port opening angle on integral characteristics of the compressor is considered. Calculation of work process in the compressor is performed by means of mathematical simulation. According to the work results it was determined that change in performance coefficient may reach 5%. At increase in suction port opening angle the leakage from compression cavities to a suction chamber increases and may reach 20%.

1. Object

Screw-type compressors (SC) are widely used in different spheres of industry. This is conditioned by high technical and economical parameters of SC. Screw-type compressors are constantly improved, their volumetric and power characteristics, reliability and service life, automation degree are increased, and overall dimensions and specific amount of metal, noise and vibration level are decreased. All this leads to a high demand for SC as compared to other displacement compressors.

At engineering screw-type machines different tasks on search for more effective and reliable design solutions arise connected with: organization of work process [1, 2, 3]; selection of design shapes of the flow channel [4, 5]; strength calculation of machine components and selection of material [6]. The SC application range is expanded by capacity and end pressure, new screw shapes are created, design and manufacturing technology are improved.

The purpose of the present work is:

- increase in SC working process efficiency by means of suction port opening angle optimization;
- derivation of dependence of SC performance coefficient change on suction port opening angle;
- elaboration of recommendations on suction port designing.

Organization of the effective suction process in screw-type compressors is a challenging task [3, 7]. Gas flow from the compressor connecting branch flange to the suction port makes a turn, gas acquires a significant speed in the suction port, at that, gas speed becomes irregular due to discrete supply. Poor stream line of screw teeth at the cavity inlet and absence of the necessary directivity of gas flow leads to shocks and vortex. Gas motion through the screw channels leads to its throw back to the circumference by centrifugal forces. Simultaneously, which is quite significant, substantial volume of heated gas flowing through slots comes to the suction cavity. All this adversely affects the compressor characteristics and its power (economical) parameters.

2. Calculation method

The actual mass $\dot{m}_a$ capacity of the compressor is less than the theoretical one $\dot{m}_t$, i.e. $\dot{m}_a = \lambda \dot{m}_t$, where $\lambda<1$ — performance coefficient [8, 9].

The performance coefficient for SC can be represented as (1) [8]

$$\lambda = \lambda_p \cdot \lambda_t \cdot (1 - \nu_u) - \nu_l - \nu_{el}, \tag{1}$$

where λ_p is pressure factor, λ_t is temperature factor, ν_u is utilization factor, ν_l is leakage factor, ν_{el} is external leakage factor.

Pressure factor is a factor taking into account pressure losses at gas flow through the suction port, which is determined by equation (2)

$$\lambda_p = 1 - \frac{\Delta p}{p_s}, \tag{2}$$

where Δp is pressure losses at flow through the suction port.

Temperature factor λ_t considers the increase of suction gas temperature. A rise of temperature leads to a decrease of the intake gas density and, as a consequence, a decrease of capacity.

Utilization factor ν_u shows how much of the theoretically possible volume is used in the suction process.

Leakage factor ν_l considers gas leakage from the high-pressure cavities in the low-pressure cavity. Leaks reduce the volume of the suction gas, as they themselves occupy part of the suction cavity volume and heat the intake gas.

External leakage factor ν_{el} considers the volume of gas that flows from the compressor's working cavity to the environment.

Pressure losses can be determined by equation (3)

$$\Delta p = \xi_s \frac{\rho_s \cdot c_s^2}{2}, \tag{3}$$

where $\xi_s = \xi_s(\mathrm{Re})$ is coefficient of hydrodynamic resistance at suction, determined by approximation of graphs [10].

Average speed of gas directed motion in the cavity can be found from continuity equation (4) [7].

$$c_s = \frac{2\pi \cdot l \cdot n}{\alpha_{1s}} \tag{4}$$

where α_{1s} is central angle of suction port (see figure 1).

These formula indicate that in order to increase the pressure factor it is necessary to increase the suction port angle or decrease the rotor length and number of turns. At that, it is necessary to take into account that l and n values cannot be used freely at any time. It should be noted that the suction port increase leads to decrease in gas dynamic losses and, consequently, to increase in pressure factor. On the other hand, this leads to suction process delay and increase in mass exchange in the work cavities, which shall lead to gas heating and decrease in suction gas fresh portion. On the one hand, decrease in suction port leads to reduced leakages, and on the other hand to increase in volume underutilization coefficient. This paper is devoted to the optimization of the central angle of suction port α_{1s}.

Besides, it is advisable to design the suction port so that the time within which the suction port connects the suction chamber with the double cavity after its complete disengagement from the double screw tooth is equal to the time for shock wave passing from the discharge end to the suction end. The dynamic head of suction gas flow thus created shall lead to improvement of cavity filling with fresh gas.

Shock wave creation can be represented as follows. Gas moving in the suction cavity meets a fixed compressor housing - end on the discharge side. Flow disturbance thus created propagates in the double cavity of screws towards suction side. As is known, a speed of small disturbances propagation equals to local sound velocity. Consequently, when gas flow runs in the housing end wall the combination of sound waves constantly following each other creates a compression wave which is called the shock wave.

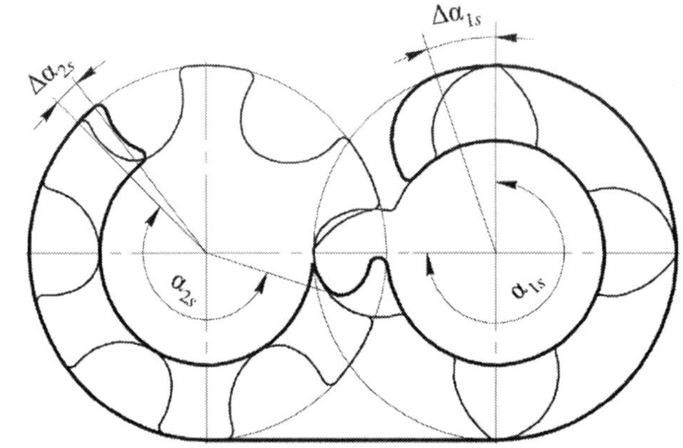

Figure 1. Suction port angles.

Thus, it is necessary to solve the optimization task which makes it possible to find the most rational parameters of the suction port. To this end, a mathematical model of work process and the software program were developed.

In order to simplify calculations the following assumptions have been made:

1. gas is considered as ideal one;
2. process is quasi-static, i.e. the parameters are the same at any time and in each point of the double cavity;
3. pressure and temperature in the suction and discharge branches remain constant;
4. dry gas compression process without internal cooling is considered. The amount of heat removed through the compressor housing in the process of compression shall be ignored.

At the ends of the calculation area the following boundary conditions are set:

1. at inlet – gas pressure in front of the suction port is equal to gas pressure in the suction branch;
2. at outlet – gas discharge process takes place under internal compression pressure equal to the pressure in the discharge branch (nominal mode).

In order to describe the work process in the screw double cavity the following equations shall be used: state, first law of variable-mass body thermodynamics, adiabatic egress.

The finite equation of parameter change in the double cavity looks as follows

$$\frac{dT}{d\tau}=\frac{T(k-1)}{p\cdot V}\left[\frac{dQ_0}{d\tau}+i_{\text{in}}\left(\dot{m}_{\text{in}}-\dot{m}_{\text{out}}\right)-\frac{dL_0}{d\tau}-i_{\text{out}}\frac{k-1}{k}\left(\dot{m}_{\text{in}}-\dot{m}_{\text{out}}\right)\right],$$
$$\frac{dp}{d\tau}=\frac{k-1}{V}\left(\frac{dQ_0}{d\tau}-\frac{k-1}{k}\frac{dL_0}{d\tau}+i_{\text{in}}\cdot\dot{m}_{\text{in}}-i_{\text{out}}\cdot\dot{m}_{\text{out}}\right). \tag{5}$$

where dQ_0 is heat supplied from surfaces limiting the working cavity, dL_0 is work performed by heat.

The current volume value V is a function of drive rotor turning angle and is determined by analytical, graphical or numerical method [7, 11]. The analytical method stated in [8] is used in the work.

Gas leakage through gaps which make significant part of the useful capacity takes place in the screw-type compressors [12, 13, 14]. Increase in the number of rotor turns leads to reduction of the relative leakage value, however it leads to increase in losses at suction. Therefore, decrease in the absolute value of gas losses through slots is one of the main factors of SC efficiency improvement. The following issues are of great significance in SC theory:

1. study of qualitative influence of leakages on the machine parameters and their classification;
2. determination of gas amount flowing through slots of different shape [15];
3. determination of quantitative relations between the total leakage value and the most important machine parameters, as well as analysis of the obtained dependences.

Mass gas flow rate through slots shall be determined by the Saint-Venant - Wantzel equation:

– for pre-critical egress by equation (6)

$$\dot{m} = \mu \cdot f \cdot \frac{p_2}{T_2} \sqrt{\frac{2k}{k+1} \frac{T_1}{R} \left(1 - \left(\frac{p_2}{p_1}\right)^{\frac{k-1}{k}}\right)}, \tag{6}$$

– for critical egress by equation (7)

$$\dot{m} = \mu \cdot f \cdot p_2 \sqrt{\frac{k}{R \cdot T_2} \left(\frac{2}{k+1}\right)^{\frac{k+1}{k-1}}}. \tag{7}$$

In order to determine leakages it is necessary to know the change in sealing lip length along the contact lines and screw edges for one double cavity and the gap between screws. Figure 2 contains projections of the engagement lines and screw contact for two-sided asymmetric profile of teeth [1, 5].

The compressor screws actually do not contact each other, there is some gap between them all the time [16] which is necessary to ensure compressor safe operation. Meant by the contact lines are the imagined lines used to set minimum gaps between the operating screws. It is not simple to determine them by calculation. In the practice of calculations and investigations the contact lines are the tangent lines of screws made in strict accordance with theoretical sizes.

The contact line length is an important parameter of screws engagement since this makes it possible to calculate slot cross-section by screws.

The contact lines must be continuous to ensure safe separation of areas with the increased gas pressure and the suction area. The task of contact lines determination is a spatial one. The suction port must provide maximum filling of cavities with fresh gas. The suction port dimension is characterized by central angles α_{1s} and α_{2s} (see figure 1).

Usually, the double cavities of the driving and driven screws are disconnected from the suction chamber simultaneously. For screws with asymmetric shape of teeth [7] by equation (8)

$$\alpha_{2s} = \frac{\alpha_{1s} + \frac{2\pi}{m_1}}{i_{12}} - \left(\theta_{III} + \theta_{IV}\right), \tag{8}$$

where i_{12} is gear ratio, θ_{III} and θ_{IV} is profile angles.

Connection of suction angles of driving and driven screws makes it possible to bring the task of rational port sizes determination to search optimal sizes of suction angles for one of screws. This is convenient to be done for the driving screw (angle α_{1s}).

With account of the overlap angle the driving rotor suction angle will be equal to

$$\alpha_{1s} = \tau_{1sa} + \beta_{01} + \Delta\alpha_{1s}, \tag{9}$$

where τ_{1sa} is swirl angle, β_{01} is angle between the center line and the beam drawn through the rotation center of the driving screw and the tooth top in position of the gas compression start in the double cavity (compression start angle).

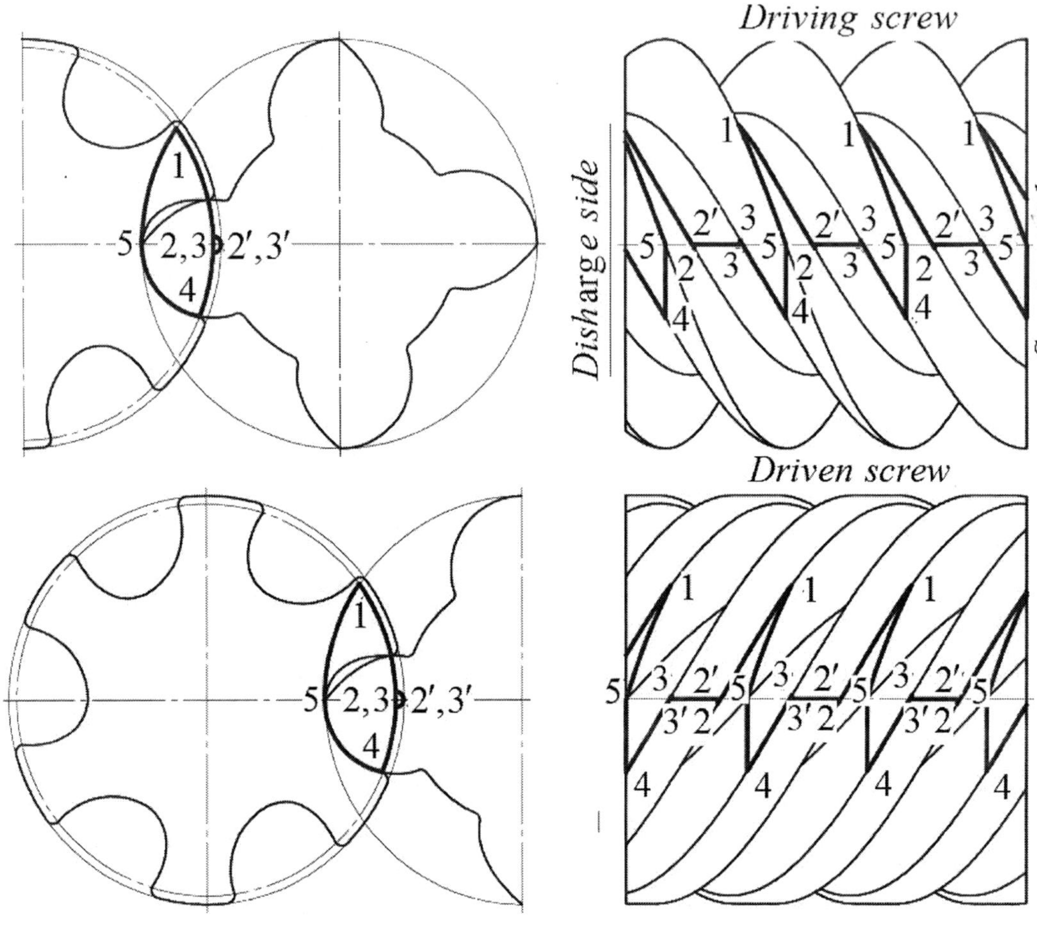

Figure 2. Engagement and contact lines of screws with two-sided asymmetrical shape of teeth.

The time within which the double cavity remains connected with the suction branch after complete disengagement of the cavity from the tooth is equal to [7]

$$t = \frac{\Delta\alpha_{1s}^{\circ}}{2\pi n_1}, \tag{10}$$

where $\Delta\alpha_{1s}$ is additional value of suction angle (overlap angle) (see figure 1), n_1 is number of driving screw turns.

The overlap angle is equal to the rotation angle to which the screw will turn after complete disengagement of the cavity under consideration till its disconnection from the suction chamber. The time of shock wave passing from the discharge end to the section end

$$t = \frac{l}{c^*}, \tag{11}$$

where l is screw length, c* is shock wave propagation speed.

By equating the time (10) and (11) we obtain

$$c^* = \frac{2\pi \cdot l \cdot n_1}{\Delta\alpha_{1s}^{\circ}}. \tag{12}$$

Then the additional value of the suction angle is determined by equation (13)

$$\Delta\alpha^{\circ}_{1s} \approx \frac{2\pi \cdot l \cdot n_1}{\sqrt{kRT_1}}. \quad (13)$$

Solution of the differential equations represented above is not possible in general terms since the equation system is not immediately differentiable. Thus, the special software program was developed for solution of these equations. The differential equations shall be solved by numerical method of step-by-step approximations.

The initial parameters in each chamber must be set in the first approximation for numerical calculation. They can be selected randomly, but this will increase a number of iterations necessary to obtain the required accuracy. That is why the initial parameters in each chamber shall be determined by the adiabatic equation. With a knowledge of initial parameters and having selected the count length the main differential equations can be solved and the final parameters in each chamber and all intermediate values of rotation angle can be found, i.e. the indicator diagram of the first approximation can be obtained. This diagram is not actual yet.

Upon its obtaining the calculation of the second approximation begins; at that, the initial parameters are values obtained at calculation of the first approximation, except for the first compression chamber. The initial parameters in this chamber for the second and subsequent approximations shall be taken as follows: the compression start pressure in each approximation on the basis of the accepted assumptions shall be taken equal to the pressure in the suction branch; the compression start temperature shall be determined according to the calculation results of the previous approximation and shall take into account the gas heating due to leakage.

The program makes it possible to obtain the following dependences: end area of rotor cavities (depending on the driving rotor turning angle), volume of double cavity (depending on the driving rotor turning angle), change in pressure and temperature due to the double cavity volume, change in contact lines (depending on the driving rotor turning angle), change in mass leakage through the contact lines (depending on the driving rotor turning angle).

3. Calculated data representation

The basic variant in calculations is a dry SC with asymmetric shape (see figure 2) of teeth according to 4+6 scheme (four teeth in the driving rotor and six teeth in the driven rotor) [7]. The suction gas - air, suction pressure 0.1 MPa, discharge pressure 0.3 MPa, compressor volumetric capacity 20 (m^3/min), rotor length to diameter ratio l/d=1, rotor turning angle 300 deg. The geometric compression ratio for all variants is constant. Calculations to reveal dependencies of the performance coefficient and its component parts at change in geometrical and mode parameters has been carried out.

The following parameters have been selected as variable ones: ratio l/d (rotor length to diameter); discharge pressure within 0.25 to 0.35 MPa; suction temperature within 293 to 313 K. Change in test pressure, suction temperature has taken place at different values of l/d. Given below are graphs of change in performance coefficient and its component parts at change in ratio l/d.

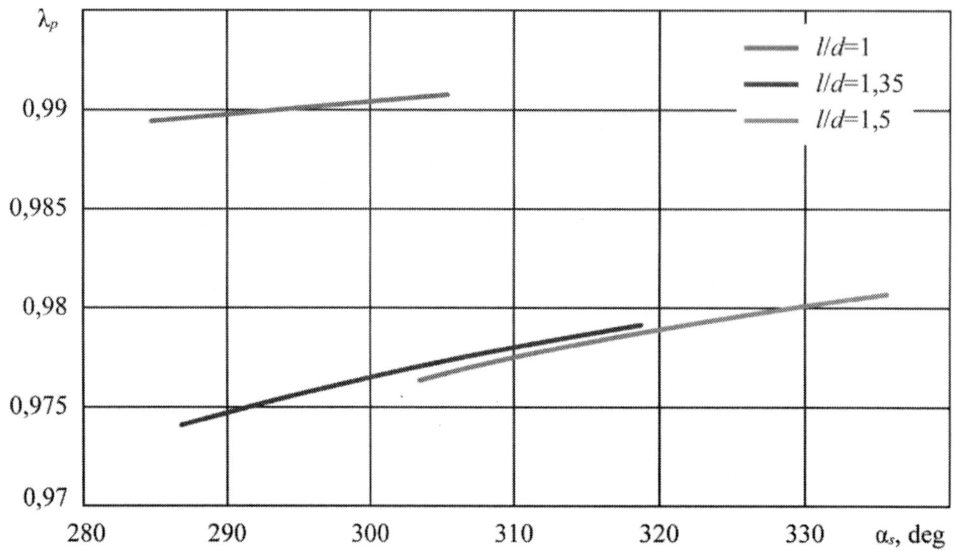

Figure 3. Change in pressure coefficient depending on suction port opening angle.

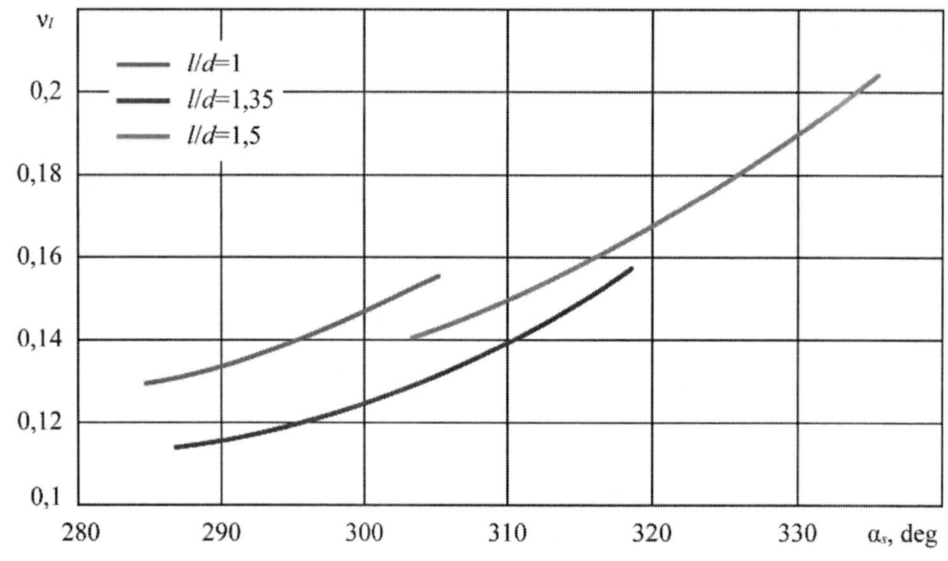

Figure 4. Change in leakage coefficient depending on suction port opening angle.

IOP Conf. Series: Materials Science and Engineering **425** (2018) 012014 doi:10.1088/1757-899X/425/1/012014

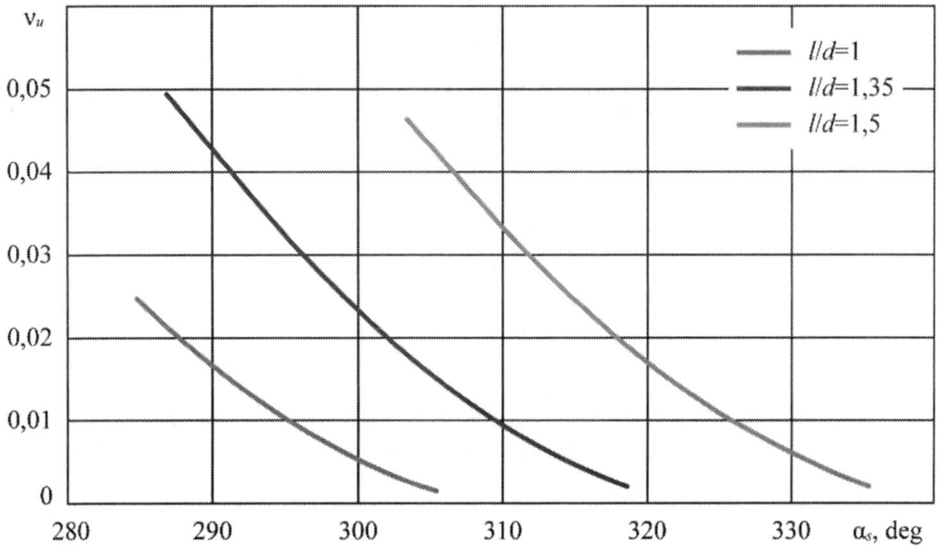

Figure 5. Change in utilization coefficient depending on suction port opening angle.

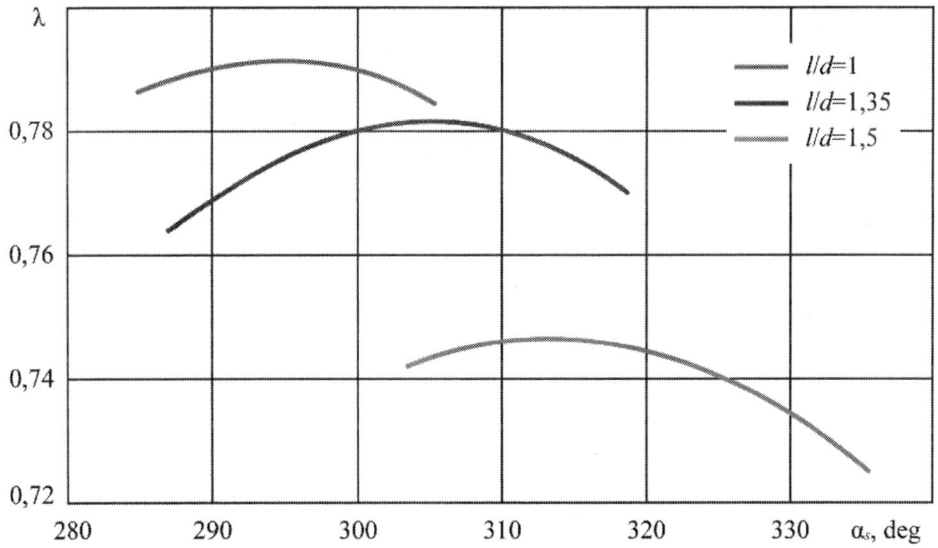

Figure 6. Change in performance coefficient depending on suction port opening angle.

In the process of gas suction to the compressor machine the gas pressure decrease takes place and, as a consequence, slight decrease in its temperature. But simultaneously heat exchange between gas and compressor parts, which are more heated, as a rule, takes place. Therefore, gas temperature is slightly increased. For different values of the central angle of suction port, the components influence of the performance coefficient is different.

The results of the calculations show that changing of the central angle of suction port divides the graph of the performance coefficient function into three zones (figure 6). On the left - the effect of gas-dynamic resistances (figure 3) and underutilization of the volume (figure 5), on the right - the effect of mass leakage (figure 4), the middle part depends on both the first and the second factor.

4. Conclusion

According to work results it was established that the performance coefficient (see figure 6) has an expressed maximum depending on the suction port parameters. Depending on ratio *l/d* the maximum value shall be shifted to the right (*l/d*=1.35) or to the left (*l/d*=1.5). Change in the performance coefficient at different modes in the considered intervals reaches 3-5%.

Influence on the performance coefficient constituents:

1. influence on pressure factor (see figure 3): change in pressure factor reaches 1÷3% on the average, the less values correspond to *l/d*=1, the more – to *l/d*=1.5;
2. influence on utilization factor (see figure 4): At increase in the suction port opening angle the utilization factor is decreased, tending to zero;
3. influence on leakage factor (see figure 5): at increase in the suction port opening angle the leakage factor is increased due to higher leakage from the compression cavities to the suction chamber. Change in the considered ranges reaches 20%.

It should be noted that calculations have been performed at constant volumetric capacity. At that, the compressor geometrical parameters with different *l/d* did not coincide. Difference between the pressure factors at *l/d*=1 and other values (see figure 3) is connected exactly with this fact.

The recommended values of α_{1s}: for ratio *l/d*=1 the value α_{1s} can be equal to 295 deg; *l/d*=1.35 – α_{1s}=305; *l/d*=1.5 – α_{1s}=315. The given values are true for dry compression compressors with a capacity of up to 100 m^3/min and ratio π up to 4.

5. Acknowledgement

The article is written in the framework of the program "5-100-2020".

List of symbols and abbreviations

Symbols

c	speed	($m\ s^{-1}$)
d	diameter	(m)
i	enthalpy	($J\ kg^{-1}$)
f	area	(m^2)
k	adiabatic exponent	(-)
l	length	(m)
m	mass	(kg)
n	rotational speed	(s^{-1})
p	pressure	($N\ m^{-2}$)
R	gas constant	$m^2\ s^{-2}\ K^{-1}$
T	temperature	(K)
V	volume	(m^3)
ρ	density	($kg\ m^{-3}$)
τ	time	(s)
μ	flow rate factor	(-)

Subscripts

1	gas parameters upstream the port
2	gas parameters downstream the port
in	incoming gas
out	outgoing gas
s	suction

References

[1] Khisameyev I G and Maksimov V A 2000 *Two-rotor screw-type and straight-toothed compressors. Theory, calculation and design* (Kazan, FEN) p 638 [Russian]

[2] Hütker J and Brümmer A 2010 Comparative examination of steam-powered screw motors for specific installation conditions. *8. VDI-Fachtagung Schraubenmaschinen, Kompressoren, Lader, Motoren, Vakuum- und Flüssigkeitspumpen, VDI-Gesellschaft Energie und Umwelt* (Dortmund, Germany) pp. 109-123

[3] Kovacevic A, Arjeneh M, Rane S, Stosic N and Gavaises M 2014 Flow visualization at suction

of a twin screw compressor *International conference, Screw machines* (Dortmund, Germany) pp. 309-318

[4] Corneli T, Preuß N and Troßmann O 2014 Experimental studies on the volumetric efficiency of triple screw pumps *International Conference on Screw Machines* (Dortmund University, Germany)

[5] Herlemann S, Hauser J and Henning N 2014 Optimization of rotor profiles for energy efficiency by using chamber-based screw model *International conference, Screw machines*; (Dortmund, Germany) pp. 211-224

[6] Schuhmann R, Ochs H, Missal R and Schmitt T 2014 Large industrial screw compressors for refrigeration units. Experiences out of operation and maintenance *International Conference on Screw Machines* (Dortmund, Germany) pp. 15-28

[7] Sakun I A 1970 *Screw-type compressors* (Leningrad: Mashinstroienie) p 400 [Russian]

[8] Andreyev P A 1961 *Screw-type compressor machines* (Leningrad: Sudpromgiz) p 252 [Russian]

[9] Grieb M and Brümmer A 2014 Design and examination of a small-scale screw expander for waste heat recovery *International Conference on Screw Machines* (Dortmund, Germany) pp. 197-210

[10] Idelchik I E 1975 *Reference book on hydraulic resistance* (Moscow: Mashinstroienie) p 559 [Russian]

[11] Avtonomova I V, Bratus A V and Sorokin S G 2014 Method of screw-type compressor double cavity volume determination. *Bulletins of higher educational establishments* **7** pp. 11-19 [Russian]

[12] Dokukin V N and Pronin V A 2014 To calculation of leakage in screw-type compressor working part *Scientific magazine of NIU ITMO. Series "Refrigerating appliances and conditioning"* **3** pp. 17-23 [Russian]

[13] Chernov G I, Panyutich A A, Yusha V L, Vasilyev V K, Zinovyeva A V and Mikhailets S N 2015 Theoretical analysis of working process economical efficiency for screw-type oil-filled compressor at the modes of refrigerating capacity frequency regulation *Omsk scientific bulletin* **3** pp. 100-103 [Russian]

[14] Kauder K and Fost C 2002 Improving chamber filling in screw-type engines *Schraubenmaschinen 2002, Tagung, VDI-Gesellschaft Energietechnik* (Dortmund, Germany) pp. 179-196

[15] Zakharenko S E 1965 Gas flow rate through narrow slots *Tr. of Leningrad polytechnic institute of M.I. Kalinin* **249** pp. 69-75 [Russian]

[16] Noskov A N and Tarasenkov D S 2017 Selection of shaped gaps for refrigerating oil-filled screw-type compressor *Bulletin of International Frost Academy* **4** pp. 41-45 [Russian]

IOP Conf. Series: Materials Science and Engineering **425** (2018) 012015 doi:10.1088/1757-899X/425/1/012015

Identification and analysis of screw compressor mechanical losses

S Abdan[1,2], N Stosic[1], A Kovacevic[1], I Smith[1] and P Deore[2]

[1] Centre for Compressor Technology, City, University of London, London, U.K.
[2] Kirloskar Pneumatic Company Limited, Pune, India.

E-mail: suraj.abdan@city.ac.uk

Abstract. Screw compressors are compact machines, used for a wide range of applications where gases or vapours are required to be delivered at moderate pressures with high efficiency and reliability. They are most effective when the compressed medium requires power inputs, approximately in the 10 kW - 1-2 MW range. At lower inputs alternatives such as reciprocating and scroll compressors are preferable and at higher inputs turbo-compressors are more suitable.

In industrialised countries, compressors absorb 15-20% of the total electrical power generated. Hence there is a continuing demand to improve their efficiency. This is normally expressed as the specific power consumption, which is the power required to compress unit mass of gas delivered.

There already exist mathematical models to assist in the design of such machines and to estimate their performance, which include the estimation of the dynamic loads acting on the rotors and bearings and these loads determine their mechanical efficiency. However, these models do not estimate the magnitude of the mechanical losses, which are only guesstimated as an additional increment to the power required to compress the gas. Such an approximation does not enable the optimum selection of bearings and lubricating oil to minimise the frictional power losses.

The aim of the study, described in this paper, was to estimate the effect of the individual parameters responsible for mechanical power loss in oil injected screw compressors and is focussed on the losses incurred in the gear box, bearings and shaft seals.

It was found that in the gearbox, meshing, bearing and seal losses all increase both with speed and gear ratio. In the main rotors, it was found that sliding friction losses in the bearings are not significantly affected by the radial load, nor are rolling friction losses significantly dependent on the axial load. However, both axial and radial loads have a significant effect on the total frictional power loss. Lubricant viscosity affects the frictional power losses but the oil level does not.

Notation

A	centre distance	mm
B	bearing width	mm
b	face width	mm
C_1	bath lubrication constant	
C_2	bath lubrication constant	

IOP Conf. Series: Materials Science and Engineering **425** (2018) 012015 doi:10.1088/1757-899X/425/1/012015

C_{Sp}	factor of oil spraying	
C_W	variable	
d_m	bearing mean diameter	mm
d_{sh}	shaft diameter	mm
f_t	variable	
F_t	reference circle peripheral force	N
G_{rr}	variable for rolling frictional moment	
G_{sl}	variable for sliding frictional moment	
H	oil level inside the bearing	mm
H_V	factor of power losses in the mesh	
i_{rw}	number of ball rows	
K_A	application factor	
$K_{B\alpha}$	transverse load factor	
$K_{B\beta}$	face factor	
$K_{B\gamma}$	helix angle factor	
K_{ball}	constants for ball bearing	
K_{roll}	constants for roller bearing	
K_V	internal dynamic factor	
M	total frictional moment	N-mm
M_{drag}	frictional moment of drag losses, churning, splashing etc.	N-mm
M_{rr}	rolling frictional moment	N-mm
M_{seal}	frictional moment of seals	N-mm
M_{sl}	sliding frictional moment	N-mm
N	rotational speed	rpm
P	transmitted power	W
P_{loss}	power losses	W
P_S	power losses in seals	W
P_{Z0}	power losses in idle motion	W
P_{ZP}	power losses in mesh, under load	W
R_a	arithmetic mean roughness of meshed teeth	µm
R_S	variable	
T_H	hydraulic moment of power losses	N-m
T_{oil}	operating temperature of oil	°C
U	gear ratio	
V_M	drag loss factor	
$v_{\Sigma c}$	sum of the peripheral speeds in the pitch point	m/s
v_{t0}	constant, 10 m/s	m/s
w_{Bt}	specific load in transverse plane	N/mm
z_1	pinion number of teeth	
z_2	gear number of teeth	

Greek letters

α_w	pitch circle pressure angle	degrees

α_t	transverse plane pressure angle	degrees
β_b	base circle helix angle	degrees
ε_1	partial contact ratio of pinion	
ε_2	partial contact ratio of gear	
ε_α	transverse plane contact ratio	
Φ_{ish}	inlet shear heating reduction factor	
Φ_{rs}	kinematic replenishment/starvation reduction factor	
v	actual operating viscosity of oil or the base oil of the grease	mm^2/s
μ_{sl}	sliding friction coefficient	
v_{40}	ISO VG number of oil	
μ_{mz}	mean coefficient of friction in mesh	
η_{oil}	dynamic viscosity of oil at operating temperature	mPa.s
ρ_{cn}	equivalent radius of meshed profiles curvature in the pitch point in normal plane	mm

1. Introduction

Due to the cooling effect of the entrained oil in screw compressors, these machines can attain very high isentropic efficiencies. However, mainly because the flow through them is not axi-symmetric, bearing loads within them are much larger than in turbo-compressors and these, together with other sources of frictional loss, can reduce their overall adiabatic efficiencies by as much as 10%. Currently, although isentropic power input, efficiencies and the magnitude of the bearing loads can be predicted very accurately, [3], there are no established methods to predict frictional losses and these are normally only accounted for assuming them to be some percentage of the total input power. A better understanding of the nature of the sources of loss and an improved ability to predict them, is, therefore essential if these relatively large losses are to be minimised.

In general, it is difficult mechanical losses, but it has been shown [1] that these increase at much lower rate than the compression losses at higher compressor duties. These include losses which happen before the work is transferred to the fluid, in the bearings and seals. In most publications they are assumed to be 7-10% of the shaft power but it has been shown that this value decreases with the increase in compressor size [2].

This study includes the estimation of losses in the gear boxes, usually contained in these machines, with account taken of meshing losses both in idling and under load, as well as losses in the seals, while the model for frictional losses in the bearings includes estimates of the effect of rolling friction, sliding friction, friction in the seals and drag losses. The model also accounts for the type of lubricant used and the effect of its viscosity on frictional losses. A parametric analysis of the type and size of bearings selected and how both axial and radial loads and drive speed act on them to affect the compressor overall performance is also included. However, mechanical losses caused by rotor contact and due to shear stresses within the oil contained in the clearances were not considered.

By including such predictive methods performance estimates should be closer to the experimental results and these can be used to help design machines with lower mechanical power losses.

2. Elements of mechanical power loss in compressor plant

A typical oil-injected air screw compressor package consists of a drive system; where it can be either driven by an electric motor or by an internal combustion engine, a transmission system if required, and the compressor. There are downstream components from compressor as well, like oil separator tank, heat exchangers for oil and for air, and control valves to regulate the discharge pressure. This paper focuses on the mechanical losses of the components which are upstream of the compressor including the compressor as well.

2.1. Drive system

Typically small to medium pressure air screw compressors are driven by an electric motor. The mechanical elements of the electric motor are those which contribute to the mechanical losses inside the electric motor. These elements are bearings and seals. The loading of the bearings and type of lubrication used can be used to calculate the losses inside the electric motor bearing. The diameter of the motor shaft and the speed of rotation define the seal losses which are explained in further sections.

2.2. Transmission system

The transmission system is used to transfer mechanical power from the driver to the driven. The drive in this case can either be an electric motor or an engine while the driven is compressor. The power in small pressure screw compressor system is transmitted by a belt, while for medium to high pressure compressor system, it is transmitted through gear box system.

In a gear box system, the power is transferred through rotational motion at different speeds, torque and direction. A part of power is lost due to friction between the geared elements during mesh and also in other mechanical elements like bearings and seals [4]. The mechanical losses in a transmission system are explained in subsequent sections.

2.3. Compressor

The major contributors to mechanical losses in a screw compressor are bearings and seals. A usual approximation of these losses is if they are considered to be 5-12% of the shaft power. Different types of bearings are used inside of a screw compressor machine to take different loads. Generally, the cylindrical roller bearings are used to carry radial loads while the ball bearings are used to transfer axial loads from the rotors to the housing. In an oil-injected screw compressor machine, the viscosity and level of oil occurring in the bearing housing also affect the frictional losses.

The effects of loads, type of bearing, size of bearing, type and level of oil are analysed and compared in the power loss section.

3. Power loss in transmission system

3.1. Gearbox

As explained in the previous section, a gear box is one of the types of mechanical power transmission devices. Power losses in the gear box consist of the power losses in a gear mesh under load, the power losses during idle motion, the power losses in the gearbox bearings and seals. The power losses in mesh, under load, for a single gear pair in a cylindrical gear drive are obtained by the following equation [4].

$$P_{ZP} = \mu_{mZ} P H_V \tag{1}$$

The mean coefficient of friction in the mesh is approximately equal to the coefficient of friction at the gear pitch point, and is given by:

$$\mu_{mZ} \approx \mu_c = 0.12 \times \sqrt[4]{\frac{w_{Bt} R_a}{\eta_{oil} v_{\Sigma C} \rho_{cn}}} \tag{2}$$

Specific load in the transverse plane w_{Bt} is calculated by following equation

$$w_{Bt} = K_A K_V K_{B\alpha} K_{B\beta} K_{B\gamma} \frac{F_t}{b} \tag{3}$$

The sum of peripheral speeds in the pitch point is equal to

$$v_{\Sigma c} = 2v \tan \alpha_w \cos \alpha_t \tag{4}$$

The equivalent radius of meshed profiles curvature at pitch point in the normal plane is given by:

$$\rho_{cn} = a\frac{u}{(u+1)^2}\frac{\sin\alpha_w}{\cos\beta_b} \tag{5}$$

The dynamic viscosity of oil at operating oil temperature is determined by the following equation:

$$\eta_{oil} = 0.1282 \times \eta_{50}^{\frac{174}{\upsilon_{oil}+95}-0.198\frac{298}{e^{\upsilon_{oil}+95}}} \tag{6}$$

where, η_{50} is the dynamic viscosity of oil at operating temperature of 50°C and υ_{oil} is the operating temperature of oil (in °C).

The power loss factor of the gears in mesh is determined by the following equation:

$$H_V = \pi\left(\frac{1}{z_1}+\frac{1}{z_2}\right)\frac{1}{\cos\beta_b}\left(1+\varepsilon_1^2+\varepsilon_2^2-\varepsilon_\alpha\right) \tag{7}$$

It is also necessary to calculate the power loss when the gears are not under load but are still in motion during the compressor unloading conditions. The power loss of idle motion can be expressed as follows:

$$P_{Z0} = \sum_{i=1}^{k} T_{Hi}\frac{\pi n_i}{30} \tag{8}$$

where the hydraulic moment of power loss is given as:

$$T_H = C_{Sp}C_1 e^{C_2(v/v_{t0})} \tag{9}$$

The oil spraying factor C_{Sp} and, C_1, C_2 factors for gear width and depth of their position in the oil bath are considered while calculating the power loss.

3.2. Power losses in seals

The power loss in shaft seals is function of the shaft speed only. Hence the variation of input power has no effect on power loss in seals. The power losses in the shaft seals can be calculated using following equation [4]:

$$P_{Si} = \left[145 - 1.6T_{oil} + 350\log\log\left(v_{40} + 0.8\right)\right]d_{sh}^2 n.10^{-7} \tag{10}$$

Figure 1 shows the effect of input power upon the power losses in seals, power losses during the idle motion and power losses during the gear mesh under load. The helical gears are presented in Figure 1 have a gear ratio of 2.23 with a centre distance 125 mm, helix angle 12°, pressure angle 20° and a module of 2.5.

The power losses in idle motion are mainly function of peripheral speed, position of the gears in oil and the direction of rotation of mating gears. Since the input power does not affect any of the parameters mentioned above, the power loss in idle motion for various input power remains constant. Similarly, the power losses in seals are function of the shaft diameter, shaft speed and oil temperature. Hence, the input power does not influence the power losses in the seals.

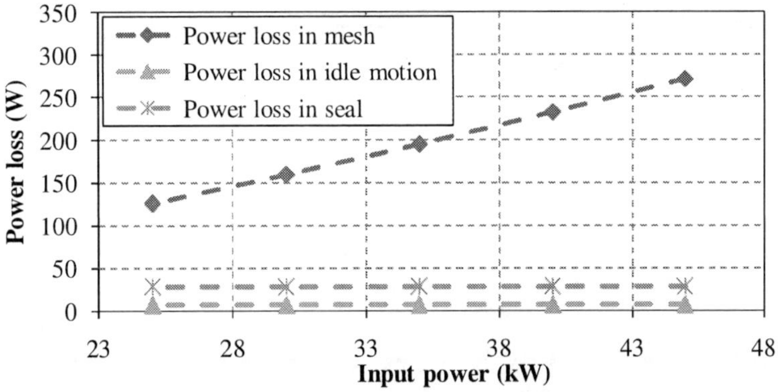

Figure 1. Power loss v/s input power.

The power loss is also a function of gear ratio. Figure 2 shows the effect of gear ratio on power losses. As the gear ratio increases, the power loss during mesh under load increases. The reason behind this is due to increase in gear ratio, the contact ratio between mating gears increases which causes increase in the power loss. However, the power losses in seals and power losses in idle motion do not change, as explained in the earlier section. The same helical gear specification as mentioned above with the power input of 24 kW is used for analysis of Figure 2 and Figure 3.

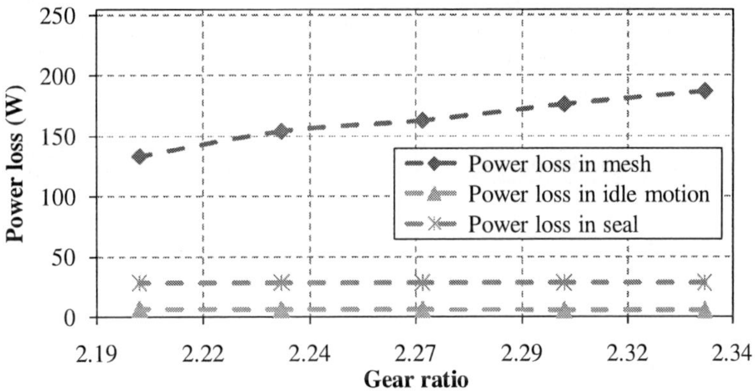

Figure 2. Power loss v/s gear ratio.

Figure 3 shows that the transmission efficiency decreases with the increase of gear ratio. As gear ratio increases, the efficiency drops because of the increased power losses for the gears in mesh under load.

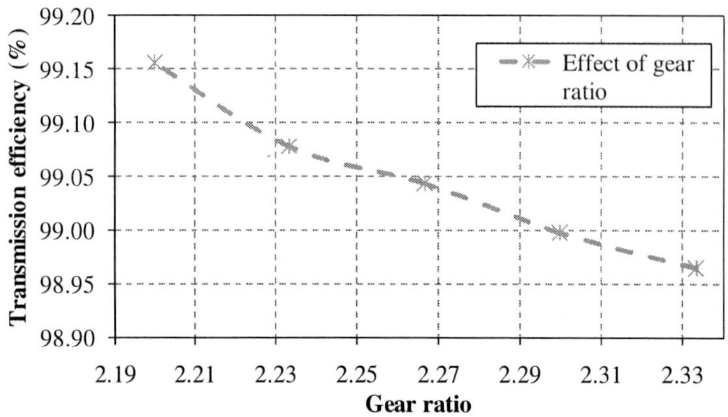

Figure 3. Transmission efficiency v/s gear ratio.

The parametric study shows that total power loss increases with the increase of input speed, increase in gear ratio and increase in input power. The major contributor to the total power loss is the power loss in mesh under load, while the power loss in seals and bearings also contribute by considerable amount.

3.3. Types of bearing frictional loss

A number of operational and non-operational factors affect the bearing friction. The friction between interacting surfaces which are in relative motion inside the bearings always changes. SKF model for calculating the frictional moment [5] presents how the friction changes with the bearing speed, as shown in Figure 4.

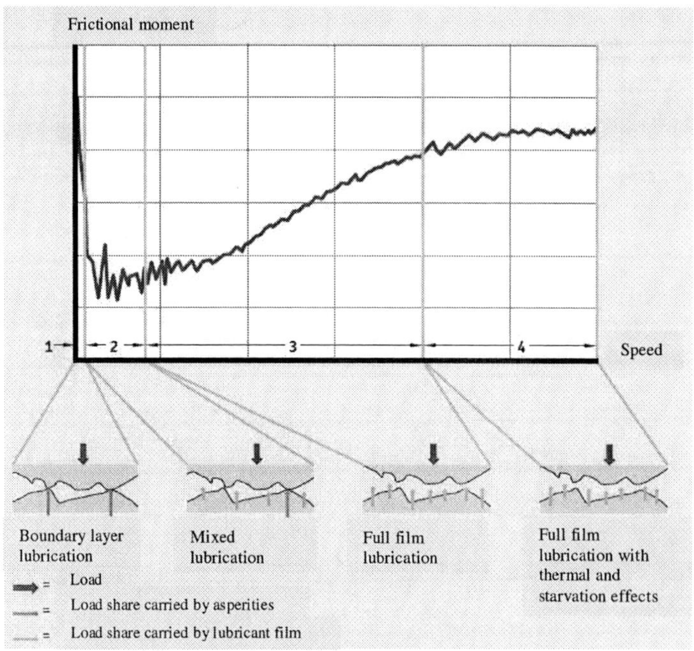

Figure 4. Bearing frictional moment v/s speed. [5]

The total frictional moment in a rolling bearing consists of four elements; the rolling frictional moment, sliding frictional moment, frictional moment of the seals and frictional moment caused by the drag losses, churning and splashing. It is defined as follows:

$$M = M_{rr} + M_{sl} + M_{seal} + M_{drag} \tag{11}$$

The rolling frictional moment is calculated by using the following equation:

$$M_{rr} = \phi_{ish}\phi_{rs}G_{rr}\left(\nu n\right)^{0.6} \tag{12}$$

The equation indicates that the rolling frictional moment is a function of the inlet shear heating reduction factor, kinematic reduction factor caused by replenishment/starvation, viscosity and speed, together with several other geometrical and bearing load variables.

A reverse flow of lubricant just behind the rolling element shears the lubricant which generates heat and consequently reduces oil viscosity. This results in reduction of the lubricant film thickness and rolling friction. The inlet shear heating factor takes into account this phenomenon. The kinematic replenishment/starvation factor estimates reduction of the lubricant film thickness because the oil starvation. The continuous rolling displaces lubricant from the racer ways not allowing sufficient time to replenish the lubricant at higher speeds. The influence of the rolling frictional moment depends upon the bearing mean diameter and its radial and axial loads.

The sliding frictional moment is a function of sliding friction coefficient and sliding frictional moment and it can be expressed as follows:

$$M_{sl} = G_{sl}\mu_{sl} \tag{13}$$

The effect of the full-film lubrication and mixed lubrication conditions can be approximated by evaluation of the sliding friction coefficient.

The drag losses caused by rotation of the bearings inside the oil bath can influence total frictional moment. Not only the bearing speed, but also the oil viscosity, oil level, size and shape of the oil reservoir affect the drag loss. The drag loss in the oil bath of the ball bearings can be calculated as follows:

$$M_{drag} = 0.4V_M K_{ball} d_m{}^5 n^2 + 1.093\times10^{-7} n^2 d_m{}^3 \left(\frac{n d_m{}^2 f_t}{\nu}\right)^{-1.379} R_s \tag{14}$$

For roller bearing, the same can be estimated as follows

$$M_{drag} = 4V_M K_{roll} C_w B d_m{}^4 n^2 + 1.093\times10^{-7} n^2 d_m{}^3 \left(\frac{n d_m{}^2 f_t}{\nu}\right)^{-1.379} R_s \tag{15}$$

The bearing frictional power loss can be calculated using the following equation [6]

$$P_{loss} = 1.05\times10^{-4} M\, n \tag{16}$$

The mathematical model of compressor mechanical losses is developed by using MATLAB software [7], and a parametric analysis of the bearing frictional losses is carried out. The results of this parametric analysis are presented in the next section, where the effects of shaft speed, bearing loads, lubricant viscosity, its level and the size of bearing are analyzed.

An oil-injected air screw compressor with the input power of 145 kW and rotor bearing loads calculated for 8.5 bar absolute discharge pressure is considered as a calculation case for the bearing frictional power loss. The radial loads of the rotors at the suction as well as at the discharge side are carried by cylindrical roller bearings, while the axial loads of the rotors on discharge side are taken by angular contact ball bearings.

3.3.1. Effect of speed. Figure 5 shows the effect of speed upon the bearing frictional power loss for the selected compressor set of bearings. The frictional power loss is directly proportional to the shaft speed; and hence as the speed increases, the frictional power loss also increases. The frictional power loss increases with the increase in the speed for all bearings; but for a given speed, it is noticed that different bearings have different frictional power loss.

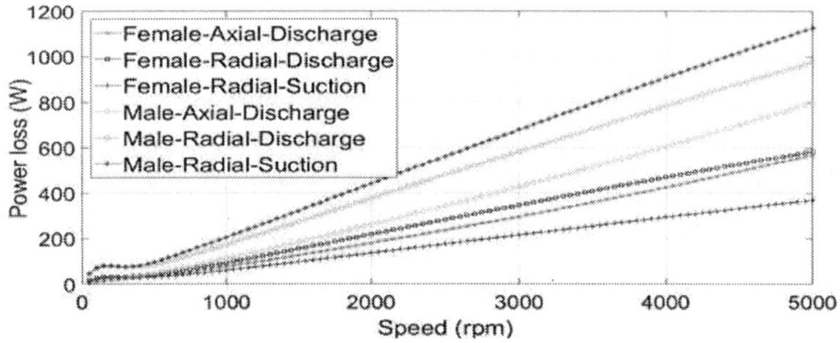

Figure 5. Bearing power loss v/s shaft speed.

For a given speed, the highest frictional power loss is noticed for the male radial suction bearing which is a cylindrical roller bearing. This can be attributed to the fact that it is the biggest bearing in the bearing set. The bigger the size means the larger the contact area in the bearings which increases the frictional power loss. The female suction radial bearing is the smallest bearing, and hence it causes the lowest frictional power loss.

3.3.2. Effect of bearing loads. As previously discussed, the bearing frictional loss consists of the components which represent the rolling, sliding and drag losses. The effect of radial loads upon these components along with total frictional moment and power loss is shown in Figure 6 where (a) shows effect of the radial load on roller bearing (male radial suction) and (b) shows effect of the radial load on ball bearing (male axial discharge).

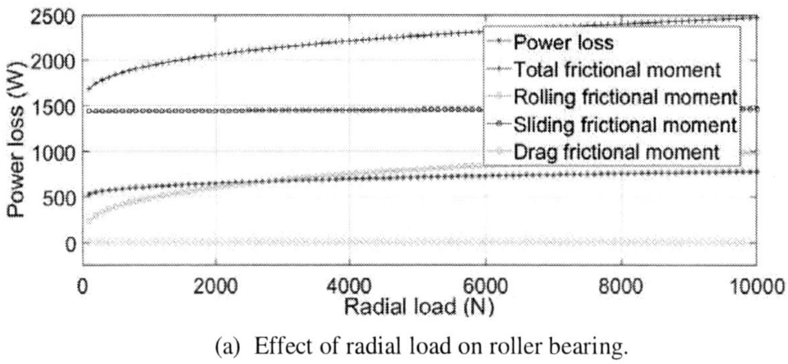

(a) Effect of radial load on roller bearing.

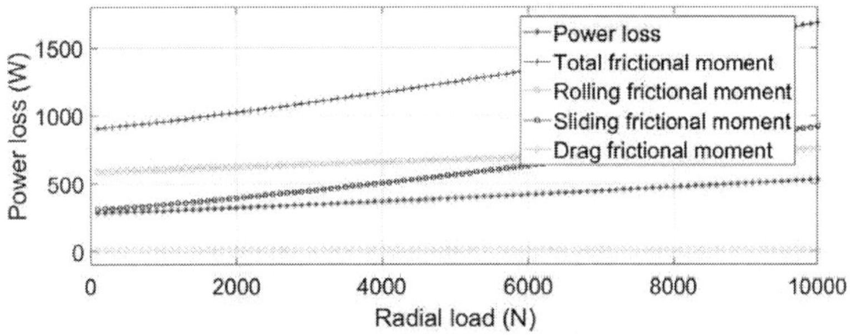

(b) Effect of radial load on ball bearing.

Figure 6. Effect of radial load on bearing loss.

For the roller bearing at the male rotor suction, the effect of the radial load is significant in the rolling frictional component, while the change in the sliding frictional componentis negligible and can be ignored. The frictional drag component is constant and the radial load has no effect upon. Since the total frictional moment is added for all components, the total frictional moment as well as power loss of the radial bearings increases substantially with the increase of the radial load.

In case of the axial ball bearing at the male rotor discharge, along with the rolling frictional component, the sliding frictional element also increases with the increase of the radial load. The rolling and sliding variables which are dependent upon the geometry and loads are G_{rr} and G_{sl}. The roller bearings are considered to take purely radial load and no axial load; while the ball bearings are considered to take the axial load and some part of the radial load. When the axial load is zero, the effect of radial load upon the sliding variable G_{sl} is negligible, and this is the reason of the nearly constant sliding frictional loss of the roller bearing.

3.3.3. Effect of axial load on bearing frictional loss. Figure 7 (a) shows the effect of axial load on roller bearing (male radial suction), and Figure 7 (b) shows the effect on ball bearing (male axial discharge). As the axial load on roller bearing increases, only sliding component of frictional loss increases, while the rolling and drag component remains constant. In the case of ball bearing, the rolling component along with the sliding component of the frictional loss increases with the increase of the axial load.

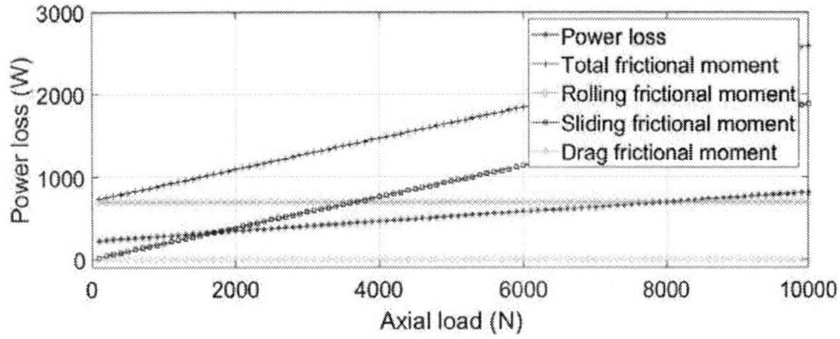

(a) Effect of axial load on roller bearing.

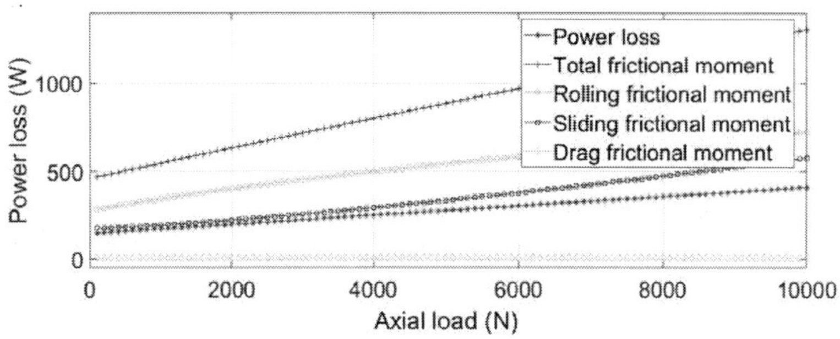

(b) Effect of axial load on ball bearing.

Figure 7. Effect of axial load on bearing loss.

The effect of axial load is not significant for the rolling frictional loss, while the effect of radial load is not significant for the sliding frictional loss, as can be seen in the Figure 6 and Figure 7.

3.3.4. Effect of viscosity. The effect of viscosity upon the bearing frictional power loss is presented in Figure 8. The frictional power loss is the largest for the roller bearing at the male radial suction and it is lowest for the female radial suction. Similar explanation is applicable here, that with the bigger bearing, the contact area is larger and hence the viscosity effect is higher.

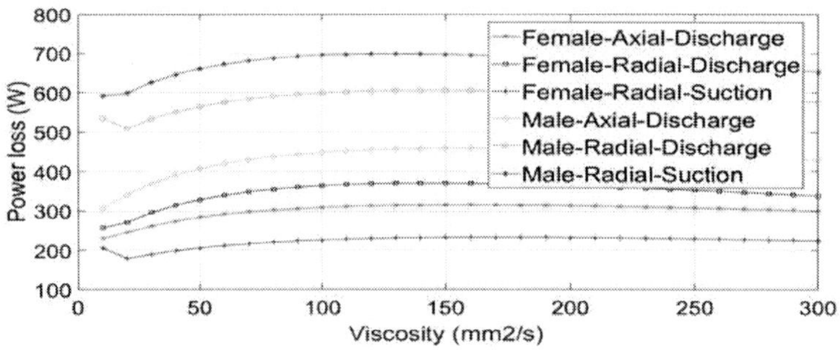

Figure 8. Frictional power loss v/s Viscosity.

3.3.5. Effect of oil level. One of the factors that influences the drag losses is oil level. The oil level is defined here as a level of oil above the inner diameter of the outer bearing racer. As can be seen from Figure 9, the drag loss does not change significantly with the oil level. The drag loss is higher for the bigger bearing compared to the smaller bearing.

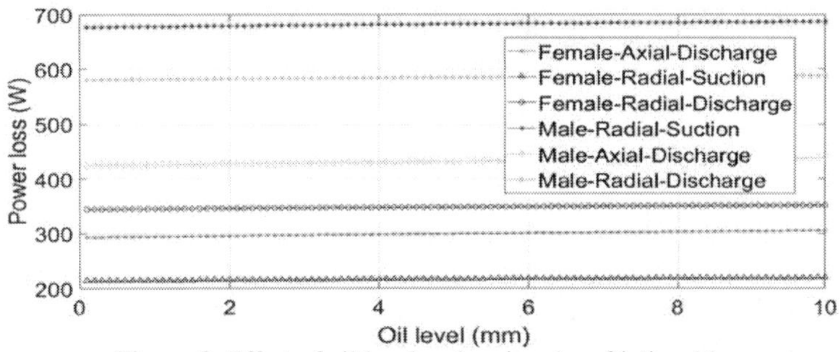

Figure 9. Effect of oil level on bearing drag frictional loss.

3.3.6. Effect of size of bearing. An analysis is carried out to see the effect of the bearing size upon the frictional power loss. The analysis is carried out for roller as well as for ball bearings. For this analysis, the radial and axial loads are considered to be similar for different bearing sizes, and the calculated values of frictional losses are presented in Table 1 and Table 2. It is noticed that the frictional power loss increases with the bearing size. This behaviour is observed for the roller as well as for the ball bearings.

Table 1. Effect of radial bearing size on power loss.

Bearing	Size (all in mm)	Mrr (N-mm)	Msl (N-mm)	Mdrag (N-mm)	Ploss (W)
NU 2212 ECP	60x110x28	754.37	1343.64	8.35	663.50
NU 2213 ECP	65x120x31	897.69	1449.96	9.60	742.54
NU 2214 ECP	70x125x31	991.78	1520.36	11.02	794.79

Table 2. Effect of axial bearing size on power loss.

Bearing	Size (all in mm)	Mrr (N-mm)	Msl (N-mm)	Mdrag (N-mm)	Ploss (W)
7410 BEP	50x130x27	745.423	666.517	5.31	446.434
7411 BEP	55x140x33	854.805	709.578	5.72	494.582
7412 BEP	60x150x35	971.374	761.412	5.843	547.668

The bigger size of the bearing makes larger contact surfaces and hence it contributes more towards the frictional power loss. From Table 1 and Table 2, it is observed that the rolling element of frictional power loss is affected more than the sliding and drag components. For 5 mm increase in the shaft diameter or bearing bore diameter, the increase in rolling frictional moment is about 10-20%; while for the sliding frictional element, it is 3-8%, and for the drag frictional element it is 2-14%.

4. Conclusions

A discussion on the effect of individual parameters upon the mechanical power losses is presented in this paper. It is noticed that in the gearbox system, the power losses of the gear engagement under load and under idle motion, as well as the power losses in the bearings and seals increase with the gear ratio and input speed. Consequently the gearbox efficiency decreases. The power losses during the idle motion are only functions of the peripheral speed, oil bath depth of gears and direction of rotation, while the power losses in seals are functions of the shaft diameter, shaft speed and oil temperature.

The power losses in the bearings are affected by several parameters. The frictional losses rise with the shaft speed and also with the bearing size. The frictional power loss increases with the increase of the lubricant viscosity. The sliding frictional component is not significantly changed by radial load, while the rolling frictional element is not significantly influenced by the axial load. However, both loads, the radial, as well as the axial, individually have significant effect upon the total frictional power loss. The increased oil level does not significantly affect the drag frictional losses in the bearings.

References

[1] O'Neill P 1993 *Industrial Compressors: Theory and Equipment* (Oxford: Butterworth-Heinemann)
[2] Brown R 2005 *Compressors: Selection and Sizing* (Elsevier Science & Technology Books)
[3] Stosic N, Smith I and Kovacevic A 2005 *Screw Compressors: Mathematical Modelling and Performance Calculation* (Springer)
[4] Jelaska D 2012 *Gears and Gear Drives* (John Wiley & Sons Ltd.)
[5] The SKF model for calculating the frictional moment (2017, December) Retrieved from http://www.skf.com/binary/110-299767/The%20SKF%20model%20for%20calculating%20the%20frictional%20moment_tcm_12-299767.pdf
[6] Bearing friction, power loss and starting torque (2017, December) Retrieved from http://www.skf.com/sg/products/bearings-units-housings/principles/bearing-selection-process/operating-temperature-and-speed/friction-powerloss-startingtorque/index.html
[7] MATLAB version 7.10.0. Natick, Massachusetts: The MathWorks Inc., 2010.

IOP Conf. Series: Materials Science and Engineering **425** (2018) 012016 doi:10.1088/1757-899X/425/1/012016

Raman scattering study of micrometer-sized spots of magnetite and hematite formed at 18CrNiMo7-6 screw rotor surfaces due to liquid-free, unsynchronized operation

Henning Moldenhauer, Manfred Bayer, Jörg Debus
Experimental Physics 2, TU Dortmund University, 44227 Dortmund, Germany

Alexander Nikolov, Andreas Brümmer
Chair of Fluidics, TU Dortmund University, 44227 Dortmund, Germany

E-mail: `joerg.debus@tu-dortmund.de`

Abstract. We analyze differently worn surface positions of a screw rotor's contact zone using confocal Raman spectroscopy with 2 μm spatial resolution. The Raman scattering spectra demonstrate signals from magnetite Fe_3O_4 and hematite α-Fe_2O_3 whose intensities and energy shifts are associated with the degree of wear of the surface locations. While magnetite is identified at the worn surface areas studied, excluding a possible white etching domain, scattered light signatures from hematite are only detected at surface positions, at which high temperatures and/or high pressure inputs were proposed to be present during the liquid-free wear process of the screw rotors. The findings may be used to gain a more general understanding of tribological layers and to develop microscopic indicators of wear.

1. Introduction

The ideal product designed for mechanical-engineering purposes remains its functionality during an infinite long lifetime. For the conditioning of the functions as well as the product's operation, surfaces being in contact with each other are of major importance. The understanding of how the surface properties are changed and controlled needs to be developed [1]. It is well known that a tribological layer of a few nanometers thickness is formed at the surfaces [2]. This layer determines the short- and long-term behavior of the engineered product [3]. Predicting the properties of the layer by monitoring changes in the atomic and molecular structures is highly desirable, but it was up to now only realized within the framework of fundamental theoretical [4] or to some extent in complex ex-situ studies [5]. In the latter case, detailed information about the surface tribological properties on the atomic scale are drawn from x-ray scanning and transmission electron microscopy experiments. An in-situ and a non-destructive tool for characterizing the tribological layers ideally during their formation process will be the key to understand and control the functionality and lifetime of an engineered product.

Such a key could be provided by Raman spectroscopy, which is a powerful optical technique used for a wide range of applications [6]. It was exploited to characterize and identify materials, to discover novel effects in physics and chemistry and to enhance the understanding of elementary mechanisms in fundamental, technological and industrial research studies [7, 8]. The main advantage of the Raman spectroscopy is based on its principle mechanism, which is mediated predominantly by a quantized lattice vibration (phonon), whose energy is directly measured through the difference in energy of the incident and scattered light [6]. Combining the Raman spectroscopy with confocal microscopy adds high spatial resolution to its large set of surface-sensitive features. Confocal Raman spectroscopy could be the key technique to reveal the atomic properties of heterogeneous surfaces, such as tribological layers.

Tribological layers play a major role in the wear process of the sliding surfaces of oil-injected twin-screw rotors. Due to their cost-effective design twin-screw machines are used for the compression and expansion of gases [9]. To reduce the degree of wear and its impact on the rotor functionality by preventing a direct rotor-rotor contact, oil is usually injected into the twin-screw-rotor system. The usage of oil as lubricant is a drawback, since it has to be separated from the processed gas, which enhances significantly the total cost of ownership [9, 10]. On the other hand, dry-running screw compressors or expanders are always equipped with an oil flooded timing gear. Therefore, during operation there is no contact between both rotors. The main drawback of this type of screw machines is their reduced efficiency and the required sealing between the timing gear and the working chambers of the rotors. Hence, they are more expensive in manufacturing and operation. In order to protect the environment and to lower the manufacturing and operating costs it is desirable to design dry-running (liquid-free) screw rotor machines. Short-time experiments on twin-screw rotors with WC/C coating for different applications (supercharger, expander, and vacuum blower) have already been performed by achieving 200 to 400 hours of operation. The wear process at the rotor surfaces in the contact zones is not understood yet; however, first confocal Raman spectroscopy experiments have gained insight into the microscopic properties of the tribological layer [11].

Here, we study worn surfaces of uncoated twin-screw rotors made from 18CrNiMo7-6 steel using the confocal Raman spectroscopy. We identify magnetite Fe_3O_4 and hematite α-Fe_2O_3 Raman scattering signals in dependence on the degree of wear of the surface locations. The spatial dimensions of surface areas containing hematite and magnetite are a few micrometers. Spatial Raman scattering patterns suggest that hematite is only present at surface locations, which were exposed to high temperatures and/or high pressures during the liquid-free operation. Moreover, the surface positions without heavy wear history contain primarily magnetite as iron oxidation product. Since the respective Raman line widths are not broadened, the molecular bond lengths and strengths are homogeneously distributed. Our spectroscopic findings of iron oxide compounds, which play a fundamental role in corrosion processes of surfaces [12, 13], may provide a step forward to understand wear mechanisms and properties of tribological layers at uncoated twin-screw rotors.

2. Experimental Details

Within the following investigation, the effects of the tribological interaction between the female and male rotor in the interlobe contact of a dry-running, unsynchronized twin-screw expander are analyzed. The twin-screw expander was operated at a constant inlet pressure of 4 bar (absolute), an inlet temperature of 90°C, and a rotational speed of 9500 min^{-1} corresponding to a male rotor tip-speed of 35.7 ms^{-1}. At the expander outlet an ambient pressure of 1 bar (absolute) was applied. The rotors with an asymmetric SRM profile were in use for 40 minutes with 4×10^5 changes of load under liquid-free conditions (dry air). In this context, we studied the surface area of a female screw rotor, which was in contact with its male counterpart. Both screw rotors were made of chromium steel 18CrNiMo7-6.

A photography of the part, which was mechanically excavated from the female screw rotor, is shown in figure 1(a). A horizontal arrow indicates the contact area which clearly demonstrates

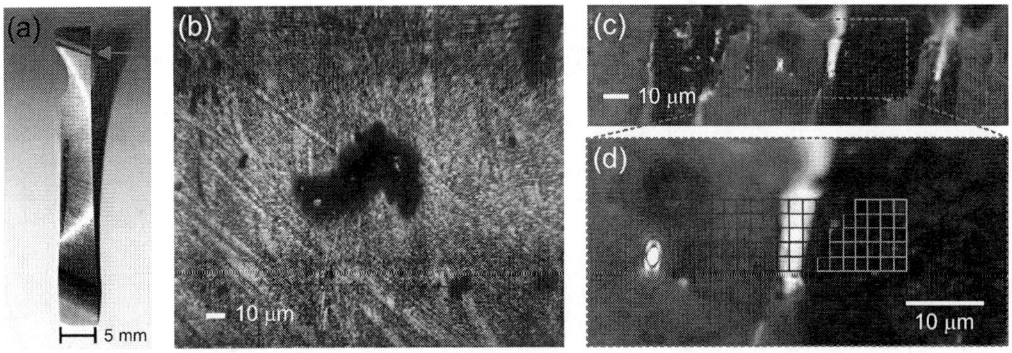

Figure 1. (a) Photography of the excavated part of the female screw rotor. The red arrow indicates the line contact zone. (b) Video-camera image with corrected height-profile focusing. (c) Another contact-zone image illustrates two heavily worn domains. The area marked by the red dashed lines is shown magnified in panel (d). The studied surface area is separated in 80 squares of 2×2 μm^2. The bright spot on the left side stems from the laser beam.

wear traces resulting from the running process. In figure 1(b) a monochrome video-camera image shows a 150 μm $\times$ 150 μm area which was obtained after a focus stacking. The surface is mainly characterized by shallow strips and several dark discolored domains having different dimensions. Surfaces without wear history do not exhibit such black shaded spots. Hence, one can assume that the dark discolored domains were formed during the wear process. Using the vertical scanning option (perpendicular to the rotor surface) of our microscope system, we could determine the depths of the dark areas; they typically vary between 5 and 20 μm. In figure 1(c) another example of a surface area, which was inside the contact zone of the screw rotors, is depicted. In the center of the image the laser spot can be seen as bright spot with a few interference rings. At the left and right side of the image heavily worn surface domains can be identified. For the confocal Raman scattering experiments, we investigated the area marked in figure 1(d), which shows a part of panel (c) in a magnified way. This surface area of 320 μm^2 contains three separated domains with different degrees of wear; a worn surface domain (purple), a transition zone (dark blue) and a heavily worn domain (light blue). At every 2 μm in the x- and y-direction the surface was illuminated with a laser beam and a Raman spectrum was measured. In that way, 80 Raman spectra in the corresponding 16×5 scanning matrix, as presented in figure 1(d), were detected. Every spectrum was recorded during an exposure time of 600 s.

The spatial mapping of the surface area marked in figure 1(d) was performed at room temperature using a confocal Raman spectrometer (S&I MonoVista Raman Microscope System CRS+ 750/BX51). A scheme of the Raman scattering setup is presented in figure 2(a). A 532 nm single-frequency laser (Cobolt Samba) was used in the experiments. The laser power was set well below 1 mW through neutral density filters. For this laser power, a sample degradation was not detected. After each measurement, the surface was visually checked using white light illumination and a video camera installed at the microscope. The laser beam was focused with a 50×microscope objective (Olympus LMPLFLN50x, NA=0.5) on the screw rotor surface. The laser spot size did not exceed 2 μm. The scattered light was examined using a 0.75 m monochromator (Acton SpectraPro SP2750) with a liquid-nitrogen cooled charge-coupled-device (CCD) camera (Princeton Instruments PyLoN 2KF). An edge filter suppressed the laser light, so

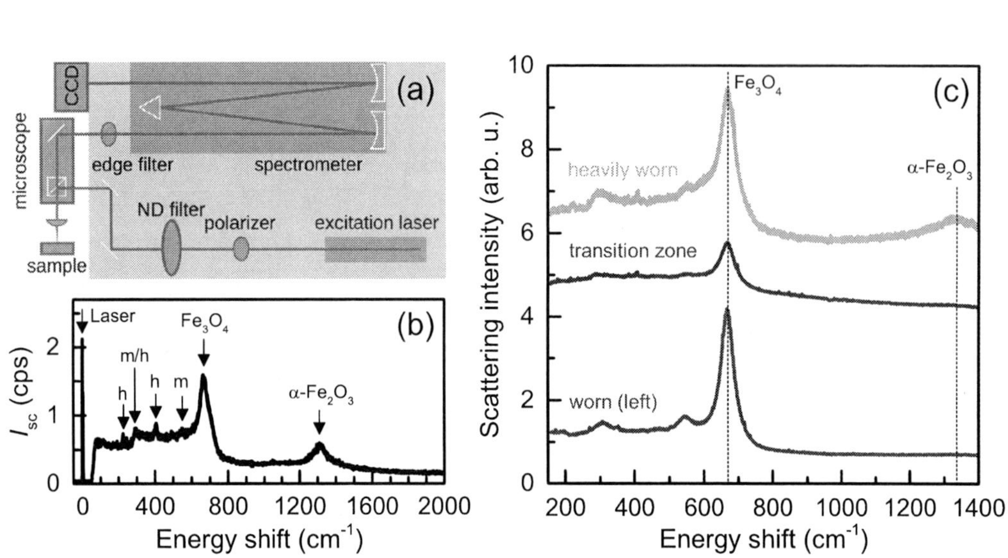

Figure 2. (a) Scheme of the confocal Raman scattering setup. (b) Exemplary Raman scattering spectrum shows the scattered light intensity I_{sc} as function of the energy shift with respect to the laser light energy. It was measured at a heavily worn surface position of the female screw rotor. The hematite (h) and magnetite (m) signals as well as the laser line are marked by arrows. (c) Raman spectra integrated over the three different spatial domains, which are marked by the light blue, dark blue and purple patterns in figure 1(d). The Raman line at 668 cm^{-1} of magnetite and the two-magnon line at 1310 cm^{-1} of hematite are selected for the mapping images shown in figure 3 and figure 4.

that scattered light signals with energy shifts of 50 cm^{-1} with respect to the laser line position at zero energy shift could be resolved. The excavated part of the female screw rotor was placed on a microscopy table, whose position could automatically be changed with sub-micrometer spatial scanning resolution.

3. Results and Discussion

In a typical Raman scattering experiment the incident laser light interacts with a molecular vibration at the surface leading to scattered light whose energy is reduced or enhanced. The energy difference between the incident and scattered light is given by the energy of the elementary excitation, which has mediated the scattering process [6]. The energy shift, which provides information about the vibrational modes, indicates the type of the bonding; thus, the Raman shift can be used to identify chemical compositions. Light-weighted atoms give rise to large Raman shifts, while a weak bonding of atoms results in small Raman shifts. One can assume that the Raman shift ΔE is in good approximation proportional to the coupling strength $\sqrt{W}$ of the respective atoms and inversely proportional to the length d of the bonding and effective mass $\sqrt{m}$ of the interacting atoms: $\Delta E \propto \frac{1}{d}\sqrt{\frac{W}{m}}$ [14]. Each type of bonding is characterized by unique motions of the atoms or molecules, respectively. One can distinguish between translational, rotational and vibrational motions. Accordingly, the molecules and chemical bondings can be described. Moreover, in amorphous samples the Raman lines are typically inhomogeneously broadened, since the chemical composition, molecular chain length and, for example, morphology underlie considerable variations at different sample positions. Since each variation causes slightly different Raman line energies, the integrated spectrum of the ensemble of molecular vibrations can be broadened significantly. Thus, the width of a Raman line is an essential parameter to describe local changes in the molecular vibrations and, in turn, to identify chemical and mechanical property changes across the surface.

To begin with, the experimental Raman scattering data for the excavated part of the female screw rotor are presented. In figure 2(b) an exemplary Raman spectrum obtained from a heavily worn surface position is shown. Six Raman lines are typically observed, their Raman shifts are approximately 230, 292, 405, 544, 668, and 1310 cm^{-1}. Here, the error tolerance in the absolute spectral position is set to approximately ± 10 cm^{-1}. Within the scope of this article we do not want to go into details on line shifting and broadening. These lines can be attributed to molecular vibrations of iron oxide being present in different oxidation stages. Relying on the analyses of de Faria et al. and Shebanova et al. [15, 16], we assign the Raman lines detected at 544 and 668 cm^{-1} to magnetite, Fe_3O_4. The lines with energy shifts of 230, 405, and 1310 cm^{-1} stem from hematite, α-Fe_2O_3. The Raman line with an energy shift of 292 cm^{-1} cannot be related definitely to an iron-oxide vibrational mode. However, the integrated Raman spectra evaluated from the spatial mapping experiments, see figure 2(c), indicate that this line is most probably due to a magnetite molecular vibration.

Magnetite, which has a spinel structure at room temperature and contains both Fe^{2+} and Fe^{3+} cations [16], possesses several Raman bands. The two Raman peaks assigned above belong to the T_{2g} and, respectively, A_{1g} mode: The T_{2g} mode at 544 cm^{-1} is an asymmetric stretching of Fe and O, the A_{1g} mode is a symmetric stretch of the oxygen atoms along the Fe-O bondings [17, 18]. Apart from magnetite, maghemite γ-Fe_2O_3, which also has a spinel structure with iron cations only in the trivalent state and is the second most stable polymorph of iron oxide [19], could give rise to the Raman lines observed. Nevertheless, the Raman signatures owing to vibrations of maghemite have a doublet-like structure thus being much broader than the Raman lines of magnetite [15]. Since the detected Raman lines are quite narrow, we rule out the presence of maghemite at the worn surface areas of the female screw rotor. In the following, the intensive A_{1g} Raman mode with an energy shift of about 668 cm^{-1} is considered as indicator of magnetite.

In comparison to that, hematite belongs to the rhombohedral (hexagonal) crystal space group. The Raman line at 230 cm^{-1} is an A_{1g} mode, which reflects a symmetric stretching of the iron and oxygen ions, the Raman line at 405 cm^{-1} describes a symmetric bending of the iron and oxygen ions (E_g mode), while the 1310 cm^{-1} line results from a two-magnon scattering process [15, 20]. In the antiferromagnetic hematite two magnons, which are collective spin waves, are excited at adjacent lattice sites with antiparallel spin orientations [20]. This two-magnon scattering line is not a feature of magnetite [15]. Accordingly, this line in the scattering spectra shall be exploited in the following to designate hematite.

The presence of wustite FeO and a participation of hydrogen in the iron oxide bondings forming oxyhydroxides cannot clearly be identified. However, the focus of future Raman scattering studies shall be drawn on the possible impact of those compounds on the tribological properties of the screw rotor contact zones. Moreover, we have carefully chosen the laser power to avoid a degradation of the laser illuminated surface domains. As underlined by de Faria et al. [15], we also observe changes of the Raman line features depending on the laser light power applied to the screw rotor surface. For high laser powers, we obtain Raman scattering spectra containing only signatures of hematite; we detect the laser-induced transformation called martitization of magnetite into hematite [21]. For future studies, it would be interesting to study the dependence of this transformation on the surface morphology of the screw rotors. A detailed laser wavelength dependence is also out of the scope of the present article, which will however be essential to identify the photochemical behavior of the iron oxides as well as oxyhydroxides.

In figure 2(c) the integrated Raman spectra characteristic for the heavily worn (light blue), transition zone (dark blue) and worn (purple) surface areas are shown. For the three differently worn domains, the Raman spectra of each 2×2 μm^2 pattern square are summed up to provide an averaged spectrum. At the heavily worn surface positions hematite is present, while it is absent

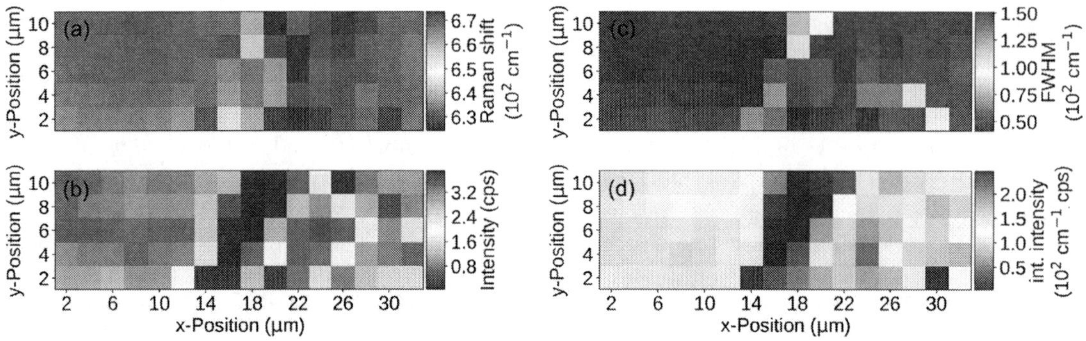

Figure 3. Confocal mapping of the A_{1g} Raman mode of magnetite at about 668 cm^{-1} across the three surface areas marked in figure 1(d). The energy shift (a), intensity (b), full width at half maximum (c), and integral intensity (d) are shown. These parameters result from Lorentzian peak fittings.

at the two other areas. Raman signals from magnetite are detected for all surface domains, while their intensities are considerably weaker at the transition zone. The change in the energy of the T_{2g} magnetite Raman mode is interesting: Its line shifts from large energies at the heavily worn positions, 580 cm^{-1}, to small Raman shifts of 572 and 547 cm^{-1} at the transition and worn areas, respectively. The position of the Raman line at about 300 cm^{-1} also depends on the degree of wear, but it demonstrates an opposite trend (299 cm^{-1} $\rightarrow$ 307 cm^{-1}) compared to the previously described Raman line. The shift to high energies, as demonstrated for the T_{2g} magnetite Raman line, could be caused by compressive pressure leading to a reduction of the bonding length and an enhancement of the interaction strength [17]. Since a hematite Raman signal is less probable at the worn surface positions, one can propose that the Raman signals observed at the worn positions including the Raman line at about 300 cm^{-1} belong to magnetite.

The properties of the magnetite-related Raman line with an energy shift of 668 cm^{-1} evaluated from the mapping across the three surface domains, which are marked in figure 1(d) by the light blue, dark blue and purple colors, are shown in the panels of figure 3. While the changes in the Raman shift do not exceed considerably the ± 10 cm^{-1} error tolerance, the line intensity varies significantly. At worn positions (left area) the intensity is homogeneously distributed. At the transition zone the magnetite-related Raman line practically vanishes, and at the heavily worn domain the intensity fluctuates remarkably. The same behavior can also be observed for the full width at half maximum (FWHM) of the line, see figure 3(c), and the integral intensity, see figure 3(d).

For the two-magnon scattering line with an energy shift of approximately 1310 cm^{-1}, which is clearly attributed to α-Fe_2O_3, the spatial mapping yields distinct differences, see panels of figure 4: At the worn and transition areas, the two-magnon line is not observed. It is only present at the heavily worn surface positions, as concluded from figure 2(c). At the latter region, the line parameters furthermore demonstrate strong fluctuations. The energy shift ranges from 1300 to 1400 cm^{-1} at several positions. The relatively weak intensities and large line widths indicate a great distribution of different energies of the magnon scattering process and a small amount of hematite across the surface area selected.

The appearance of hematite at the heavily worn surface positions of the female screw rotor may result from high temperatures being present during the liquid-free wear process at the contact zone of the screw rotors. Temperatures of about 400°C are necessary to initiate the transformation of magnetite to hematite [15]. Due to the significant spatial variations in the hematite signal parameters one may propose that such high peak temperatures were present and

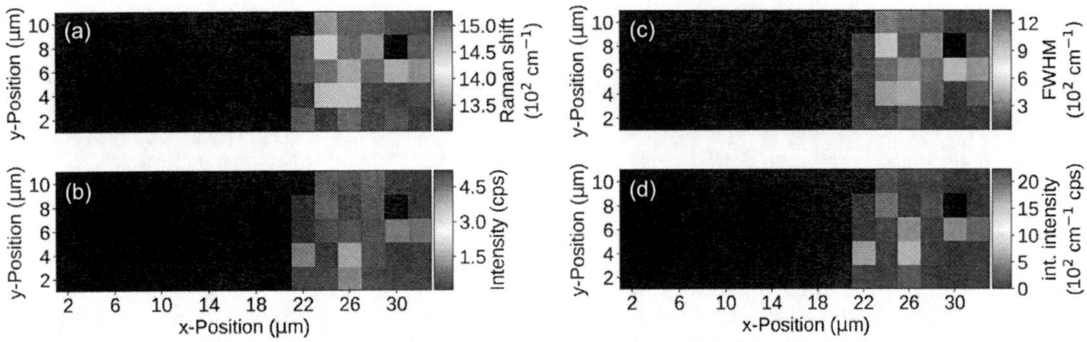

Figure 4. Confocal mapping of the two-magnon scattering line of hematite at about 1310 cm^{-1} across the three surface areas marked in figure 1(d). The energy shift (a), intensity (b), full width at half maximum (c), and integral intensity (d) are shown; they are evaluated from Lorentzian peak fittings. A black colored square indicates the absence of the hematite related two-magnon scattering line.

locally different. Not only a high temperature input, but also a mechanical deformation combined with a high pressure transfer in the GPa range may cause the formation of rhombohedral α-Fe_2O_3. Besides that, it is surprising that the surface positions without heavy wear history contain primarily magnetite as iron oxidation product. The line width is not broadened thus highlighting a homogeneous distribution of bond lengths and strengths. At those surface positions neither the temperature nor the pressure input was so high that the chemical bonding structure was considerably changed. The transition zone with its bright reflective character may hint at a white etching crack. At this domain the formation of iron oxides seems to be less probable, or the related Raman lines cannot be probed by the laser illumination due to a weak absorption. For future experiments, it would be interesting to identify the detailed chemical composition profiles of the surface domains and to solve the reason of their formation and their coupling with the substrate material. Since a phonon spectrum provides insight into the lattice vibrations and thermodynamic features, one may think about the description of the specific heat and magnetic contribution to the thermodynamic properties of the iron oxide compounds in further studies.

4. Conclusion

On the whole, we demonstrate that micrometer-sized areas of magnetite and hematite are detected at differently worn 18CrNiMo7-6 steel surfaces of screw rotors through confocal Raman scattering. A thermally activated transformation of magnetite to hematite during the liquid-free screw rotor operation is proposed to be the reason for the appearance of α-Fe_2O_3 at the heavily worn surface domain, where material was worn out. Magnetite-related Raman signals are found at the heavily worn and, in particular, worn surface domains. At the latter area their line widths do not change significantly thus indicating a weak distortion of the Fe-O bonding length and strength.

References

[1] Feldmann M, Dietzel D, Tekiel A, Topple J, Grütter P and Schirmeisen A 2016 *Phys. Rev. Lett.* **117** 025502
[2] Bhushan B (Ed.) 2017 *Nanotribology and Nanomechanics* (Berlin: Springer)
[3] Urbakh M and Meyer E 2010 *Nat. Mater.* **9** 8
[4] Kapoor A and Franklin F J 2000 *Wear* **245** 204
[5] Chung Y W 2011 *Micro- and Nanoscale Phenomena in Tribology* (Boca Raton: CRC Press)
[6] Toporski J, Dieing T and Hollricher O (Eds.) 2018 *Confocal Raman Microscopy* (Berlin: Springer)
[7] Weber W H and Merlin R 2000 *Raman Scattering in Materials Science* (Berlin: Springer)

[8] Gouadec G and Colomban P 2007 *Progr. Crystal Growth Charact. Mater.* **53** 1
[9] Rinder L 1979 *Schraubenverdichter* (Wien: Springer)
[10] Dreifert T and Müller R 2014 *9th Int. Conf. on Screw Machines* 29
[11] Debus J, Schindler J J, Waldkirch P, Goeke S, Brümmer A, Biermann D and Bayer M 2016 *Appl. Phys. Lett.* **109** 171601
[12] Hutchings I and Shipway P 2017 *Tribology: Friction and Wear of Engineering Materials* (Oxford: Butterworth-Heinemann)
[13] Parkinson G S 2016 *Surf. Sci. Rep.* **71** 272
[14] Yang X X, Li J W, Zhou Z F, Wang Y, Yang L W, Zheng W T and Sun C Q 2012 *Nanoscale* **4** 502
[15] de Faria D L A, Venancio Silva S and de Oliveira M T 1997 *J. Raman Spectros.* **28** 873
[16] Shebanova O N and Lazor P 2003 *J. Solid State Chem.* **174** 424
[17] Shebanova O N and Lazor P 2003 *J. Chem. Phys.* **119** 6100
[18] Jubb A M and Allen H C 2010 *ACS Appl. Mater. Interfaces* **2** 2804
[19] Grau-Crespo R, Al-Baitai A Y, Saadoune I and De Leeuw N H 2010 *J. Phys.: Condens. Matter* **22** 255401
[20] Martin T P, Merlin R, Huffman D R and Cardona M 1977 *Solid State Commun.* **22** 565
[21] Nadoll P, Angerer T, Mauk J L, French D and Walshe J (2014) *Ore Geology Reviews* **61** 1

IOP Conf. Series: Materials Science and Engineering **425** (2018) 012017 doi:10.1088/1757-899X/425/1/012017

Full 3D numerical analysis of a twin screw compressor by employing open-source software

Nicola Casari, Michele Pinelli, Alessio Suman
University of Ferrara, via Saragat 1, 44122 Ferrara, ITALY

Ahmed Kovacevic, Sham Rane
City, University of London, Northampton Square, EC1V 0HB, London

Davide Ziviani
Ray W. Herrick Laboratories, Purdue University, 177 S Russell Street, West Lafayette, IN, 47907-2099, USA

E-mail: nicola.casari@unife.it

Abstract. The push for having more reliable and efficient positive displacement machines (both compressors and expanders) for vapor compression and power generation (e.g., ORCs) applications has moved researchers to an always more spread employment of computational fluid dynamics (CFD). In particular, twin screw compressors, because of their high efficiency compared to other compressor types, have received interest over the last years. The numerical analysis of such machines is challenging: the deforming working chambers are very difficult to be correctly replicated. The relative motion of the rotors and the variation of the gaps during machine operation are few of the major difficulties in implementing reliable CFD models. A custom mesh generation algorithm is therefore often required for sumulating the machine operation.

In this work, SCORG-V5.2.2 was used to generate the meshes of the deforming domain around rotating parts of the machines. The open-source software OpenFOAM-v1606+ is then employed to compute the flow field associated with the operation of the twin screw. The coupling of the two tools has been carried out in this work, applying the methodology to a twin screw machine.

1. Introduction

Positive Displacement (PD) machines have a wide range of applications in modern engineering practice. Specifically, such machines have been employed as either compressor or expander in vapor compressor and power generation (e.g. ORCs). Several types of PD machines have been considered for these purposes, and an example of comparative analysis can be found in [1] for refrigeration and both [2] and [3] report a brief overview of the most used type of expanders in ORCs. A wider analysis collecting experimental data published in scientific journals, conference papers and PhD theses is presented in [4]. In that work, 102 unique ORC prototypes are analyzed and for each of them the type of expander is reported. A considerable part of these plants is equipped with a twin-screw machine. Such devices have received a constantly growing attention from the researchers, making screw compressors to comprise the majority of all positive displacement compressors in operation, as reported by [5, 6]. Either in compressor or expander configuration, the reliability and compactness they ensure are appreciated by designers and operators. Their cost-effectiveness is definitely the feature that makes this machine so spread, as suggested by [6] and [7].

The diffusion of such machines has pushed researchers to look into such devices, in order to improve their efficiency. A useful tool for the prediction of flow behaviour and performance is the Computational Fluid Dynamics (CFD) analysis, see [8, 9, 10, 11]. The experimental campaign may be very challenging, due to the geometrical complexity and the compatibility of the instrumentation. The numerical approach may be the only way to investigate the potential behaviour of the machine with new fluids without major changes to be carried out. Nonetheless, the complexity of the simulation has brought about the birth and the application of several numerical techniques. A brief yet comprehensive review of the possible methods for the simulation of PD machines that are usually available with common software is reported in [12]. Among the several issues that derives from the numerical approach applied to these machines, the high mesh deformation asks for extra care during the mesh generation and motion evolution phases. The deforming working chambers make the fluid domain to span from the volume of $\mathcal{O}(mm^3)$ to the $\mathcal{O}(\mu m^3)$ in the gap areas. The computational grid must follow this evolution, bearing such stretch without losing the quality, to correctly replicate the flow field evolution.

To tackle the mesh motion problem, different approaches have been developed over the years. Particularly, if referring to the review in [12], the Key Frame Remeshing algorithm and the Mesh Adaption - Dynamic Remeshing, are only some of the techniques that may be employed. The main idea underlying these two methods is that most of the general-purpose dynamic mesh solvers smooth the pre-defined boundary displacement inside the domain in such a way that the internal mesh is the solution of the boundary motion. The mesh is recomputed (globally or locally, according to the strategy chosen) if the cell quality falls below a certain threshold. A different approach is considered if one makes use of the overset grids, as reported for example by [13], or of the Immersed Boundary Method (IBM), see [14]. In the first case, the mesh deformation is not solved, and the domain motion is accounted for by defining different regions corresponding to different areas involved in the deformation. Such regions are rigidly displaced, passing from the actual time step to the next: the relative motion is accounted for by checking the overlapping cells. Fluxes between the two regions are exchanged only if a certain number of cells is overlapping. For what concerns the IBM, the motion is considered by modifying the source term in the flow equations. All the above-mentioned methods suffer from major drawbacks. Remeshing-based motion solvers are generally time-consuming and can easily lead to continuity errors, as suggested by [15]. Overset approach suffers from the same problem. IBM is poor in solving the boundary layer region, due to the non-aligned edges with respect to the wall.

An alternative approach is represented by the employment of body-fitted structured grids. Such meshes are easily stretched without losing quality, allowing for simulating the PD machine operation without need for remeshing, if the motion is properly solved. The mass conservation is not an issue any longer, since no remapping nor remeshing occurs. The alignment between wall and edge in the nearby of the wall is always respected and the resolution of the flow close to the boundary is generally well resolved. Unfortunately, the definition of a structured grid compatible with the rotors displacement is not straightforward. Besides, the mesh motion should be somehow pre-defined in order to ensure the overall quality of the grid to be maintained over the evolution of the motion. Cells can collapse or deform to an invalid shape if the internal-node displacement is calculated by the dynamic motion solver. A way to solve this is to use the custom predefined mesh generation: a set of meshes is prepared advanced, representing the control points through which mesh nodes pass. In such fashion, the mesh deformation is controlled. The "Custom Predefined Mesh Generation Algorithm" is used in this work [16]. SCORG-v5.2.2 was used for generating the set of meshes and OpenFOAM-v1606+ for the flow field solution. A new motion solver has been developed to force the mesh to pass through the control points, for modeling the rotor evolution.

2. Methodology

In this section the algorithm developed to make the flow solver to account for the mesh motion will be presented and applied on a simplified twin screw. The mesh generation will be discussed as well.

2.1. Geometry

The geometry of the machine has been generated with the SCORG software itself. A machine with a mild warp angle (i.e. 100°) has been considered. Also, the rotors are quite short (aspect ratio, i.e. length-to-diameter ratio, around 0.7) with respect to the more commonly employed machines (i.e. roughly 1.6). These simplifications have been introduced in order to have higher cell quality, due to the milder warpage of the thread, and a low number of grid elements due to the smaller domain. By employing these expedients, a simpler geometry yet representative of the screw operation has been generated. The so-obtained machine, together with a closer view of the rotors, is reported in Fig. 1.

The spread of the twin-screw type machine has been achieved thanks to the improvements in low-cost, high-precision manufacturing techniques. In light of this, a feature that is of paramount importance when dealing with this type of devices is the gap size. In this work the clearance between rotor and casing has been chosen to be equal to 60 μm like the interlobe gap. The axial gap has been chosen equal to 5μm.

2.2. Mesh Generation

In this paper, the deforming grid of the twin screw machine is generated using algebraic transfinite interpolation treated as an initial mesh upon which Elliptic Partial Differential

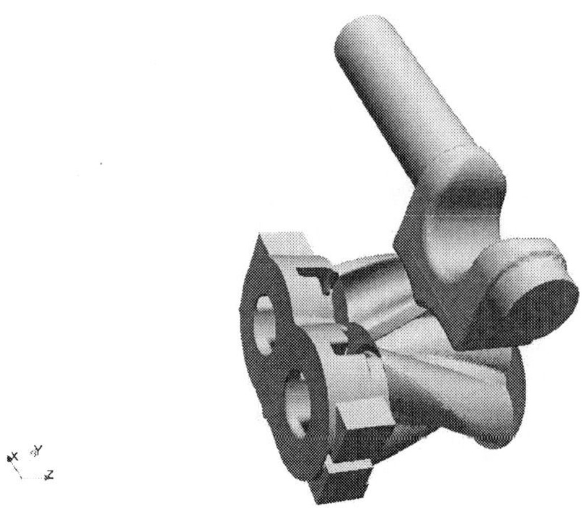

(a) Overall view of the machine investigated in this work

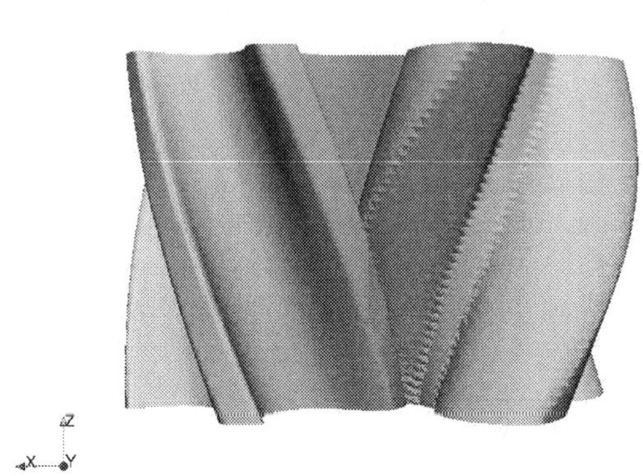

(b) Particular: upper view of the male and female rotors

Figure 1. Overall geometry and detail of the rotors of the twin screw investigated in this work

Equation (PDE) of the Poissons form is solved [17]. The resulting differential grid has highly improved cell quality and distribution. The initial algebraic grid obtained via transfinite interpolation is smoothed out using system of differential equation. The source function for such equations have been simplified to use only two input parameters based on the test function given by [18]. Particularly, the Radial Bias Factor and the Radial Bias Intensity have been added, for the inflation layer control. On the top of that, weighting coefficients have been introduced for an improvement of the orthogonality of the grid. The grid constraint (orthogonality, warpage, skewness,...) the generated mesh must satisfy have been defined in compliance with the threshold employed by the most used CFD tools. This does not automatically entails the compliance with the new tested CFD software (i.e. OpenFOAM) as described in section 3. In addition, a special procedure completely smooths the transition of the partitioning rack curve between the two rotors thus improving grid node movement and robustness of the CFD solver. A sample analysis of an oil injected twin screw compressor using the new Elliptic PDE grids has been presented [19] to compare the improvements in grid quality factors in the regions of importance such as interlobe space, radial tip and core of the rotors. A significant improvement in the grid quality and robustness of numerical solver with higher order schemes has been obtained with this differential implementation of the deforming mesh. The conforming of the two grids at the interface is such that the PDE mesher is used in a selective interlobe area of the mesh to convert the rack curve into a smooth transitioning curve across the specified number of angular positions of the rotor. This results into a gradually changing partition between the two O grids of the male and female rotors.

The rotor mesh must be coupled with the suction and the discharge ports. The mesh of such ports, has been generated with the OpenSource software cfMesh [20]: a cartesian cut-cell grid has been generated. The rotating domain is divided from the port static one by means of ACMI (arbitrary coupling mesh interface). Since the rotor axial patches do not completely overlap the ports, the interface has to automatically detect whether fluxes must be exchanged (when fluid-fluid cells are exposed) or not (e.g. when the interface is between fluid and solid cell).

Once the mesh is generated, it must be exported to OpenFOAM readable format. This feature is not yet available into SCORG. For the current analysis, the mesh translation was done via ANSYS-CFX and ANSYS-FLUENT. The rotors zone is exported as a whole, being recognized as a cellZone by OpenFOAM, thanks to the conformal interface. This expedient avoids conservation issues that may affect AMI interfaces [21]. This problem may be quite relevant in applications like the one presented in this work, since the velocity vector is highly tangential with respect to the interface, especially in the interlobe area. The resulting mesh is composed of roughly 1,700,000 cells. The rotor zone is composed of 200,000 hexaedral cells.

2.3. Numerical Analysis

As above mentioned, the twin-screw operation entails the rotors motion. Such rotation causes the deformation of the computational domain and thus the mesh should accomodate such displacement. The strategy adopted in this work is to generate a structured computational mesh for an initial position, as reported in the section 2.2. At this point a user-defined number of grids per gate-rotor pitch is generated and written in separate files. Each of the file contains the control points (coordinates) through which the computational nodes has to pass during the machine operation. This entire procedure is carried out within the SCORG-v5.2.2 framework. In the current work, a set of 50 files per groove are generated, meaning there are 50 points through which a moving mesh node has to pass to cover a pitch.

The dynamic motion solver of the CFD software therefore updates the mesh position for each of the time step, having the grid files as a reference. To calculate the position of a node for the $i-th$ time step, equation 1 is solved

$$x_{i,final} = \alpha x_{i,AG} + (1-\alpha)\, x_{i,NG} \tag{1}$$

where the subscript AG represents the actual grid file and NG is for the next grid file. AG and NG are the files written by SCORG containing the node postion at the two time instant we are interpolating among. x_i represents the position of the node at the corresponding grid file (when the subscript AG or NG is reported). This algorithm has been added to the mesh motion libraries available in OpenFOAM-v1606+ and OpenFOAM-v1712.

The case was simulated under the boundary conditions reported in Table 1. The high-Re k-ϵ model has been used for the performance evaluation with standard wall functions. It must be remarked here that the rotational speed of the two rotors is different due to the different number of threads. The algorithm proposed handles this difference automatically, asking the user to provide only the revolutions per minute of the male rotor. The rotational frequency of the female rotor can be easily derived from the number of threads ratio: in this work the female rotor has 4 threads and the male rotor has 3 threads. The diameter of the male rotor is 127.4 mm and the female one is equal to 120.4 mm.

Besides, the number of grooves has to be provided in order to compute the time it takes to cover one pitch for the male rotor. This allows the automatic scaling of the time step in order to satisfy the maximum CFL condition, to have a stable simulation. In this work the maximum CFL allowed to have stable simulation is bounded to 1. This limit is required to avoid the spikes of pressure/temperature that are outlined in section 3 . The solver employed in this work is *rhoPimpleDyMFoam*, which belongs to the compressible set of solver that comes with the software OpenFOAM-v1606+. This solver employs PISO algorightm [22] with inner integration performed with SIMPLE strategy [23]. The implementation in the OpenFOAM framework is described in [24] or [25]. This solver can automatically take into account any mesh motion input given by the user. In this work, the developed algorithm is enclosed in a library read by the

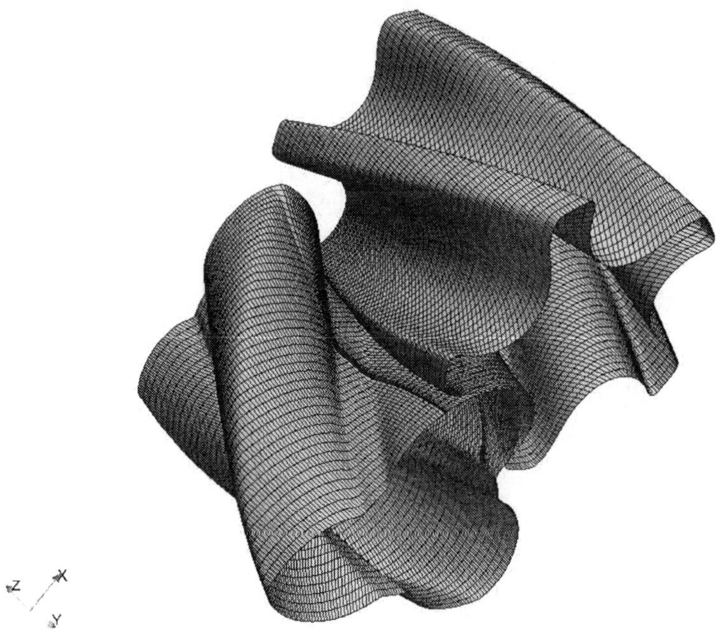

Figure 2. Computational mesh realized on the rotors

Table 1. Boundary conditions for the computation of the flow field

	Quantity	Value
Inlet	p_0	101,000 Pa
	T_0	300 K
	Turbulence intensity	1 %
	Turbulence mixing length	0.0021 m
Fixed Walls	T	adiabatic
Rotating Walls	ω_{male}	8000 rpm
	T	adiabatic
Outlet	p	131,000 Pa

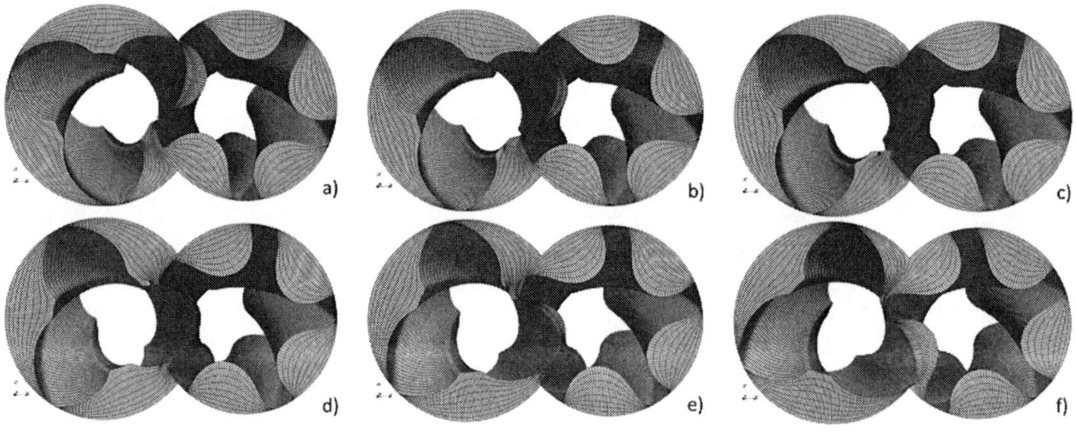

Figure 3. Evolution of the mesh during the CFD analysis, particular of one pitch.

dynamicMeshDict, that is the dictionary that asks for any mesh motion input. The library is dynamically linked during the compilation is such a way that is location insensitive and can be stored anywhere in the hard drive (i.e. it is not necessary to keep the files in the same directory of the case). The simulation run fully parallel thanks to the open MPI library. Extra-care should be taken when decomposing the domain, due to the presence of non-conformal interfaces at the inlet and outlet ports. To enforce mass conservation, both the sides of the ACMI interface are forced to stay on the same processor. This feature comes with the standard installation of OpenFOAM.

3. Results

The evolution of the grid position as provided by the motion solver algorithm is reported in Fig. 3. The high deformation of the conformal interface can be noticed.

The Rotor-To-Casing non-conformal approach cannot be used in this case. If the mesh was generated with this approach, the Rotor-Casing ACMI interface should communicate with two different neighbours: the discharge port and the interface among the two rotors. This is not supported by the ACMI implementation in OpenFOAM.

During the testing of the solver, higher Courant number was used, with respect to the value described in section 2.3. The simulations run with a CFL up to 50 (corresponding to a physical time step of roughly 10^{-6}) are stable, but the results where non-physical: the coupled PISO and SIMPLE loop makes the solver implicit, allowing for CFL higher than one. Pressure and temperature peaks in the gaps were found, especially in the suction area of the female rotor, as

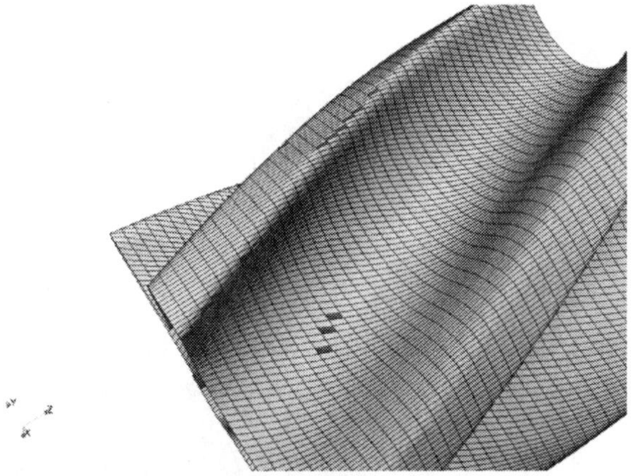

Figure 4. Faces having a wrong oriented normal

reported in Fig. 5. These spikes have occurred for all the tested numerical schemes both first and second order accurate. Limiting the gradients was not effective in this case, and the only solution was to lower the maximum time step of the simulation. This is probably related to the fact that few wrong oriented faces arise as a consequence of the rotation, as reported in Fig. 4. This seems not to be an issue when dealing with Courant number lower than one, but give problems when the time step exceeds such threshold. Further investigations are required to fix this problem.

The implementation presented in this work handles automatically this variable time step, as described in section 2.3. Future work is required to improve the robustness of the solver for higher time step. Spikes in temperature may be avoided by bounding the temperature in the rotor zone. This "clipping" is of help in stabilizing the simulation, but should be avoided since it acts as a source or sink of energy, causing a mismatch in the energy imbalance. Furthermore, it has been found that the artificial bounding is not effective if the time step is not reduced accordingly (as suggested above).

An example of the results that can be obtained if the CFL<1 condition holds is reported in Fig. 6, where the pressure fields on the male rotor is reported. The simulation has been carried out with the boundary conditions reported in Tab. 1. The pattern is in agreement with the expectations: the chamber in communication with the discharge port shows the highest pressure values. Immediately downstream the thread head, a slightly low pressure area can be detected as a result of the gap presence. The fluid expands through the gap, passing frome the high pressure groove to the lower one. Such pressure difference is then recovered inside the chamber.

With the mesh used in this work, the y^+ value required wall functions to be used. Particularly, a minimum y^+ of 50 has been found, whereas the average is around 120. The minimum value of the y^+ happens in the interlobe gap.

4. Conclusion

In this work, the simulation of a simplified twin screw compressor has been carried out. The simplified geometry was chosen in order to prove the capability of the algorithm developed in simulating the operation of a positive displacement machine. The simplifications reside in the low warp angle and the short axial dimension, whereas the gap size is very close to an actual machine.

The mesh motion routine responsible for the coupling of a pre-defined set of files that contains the control points for the mesh motion and the flow solver has been implemented. The set of

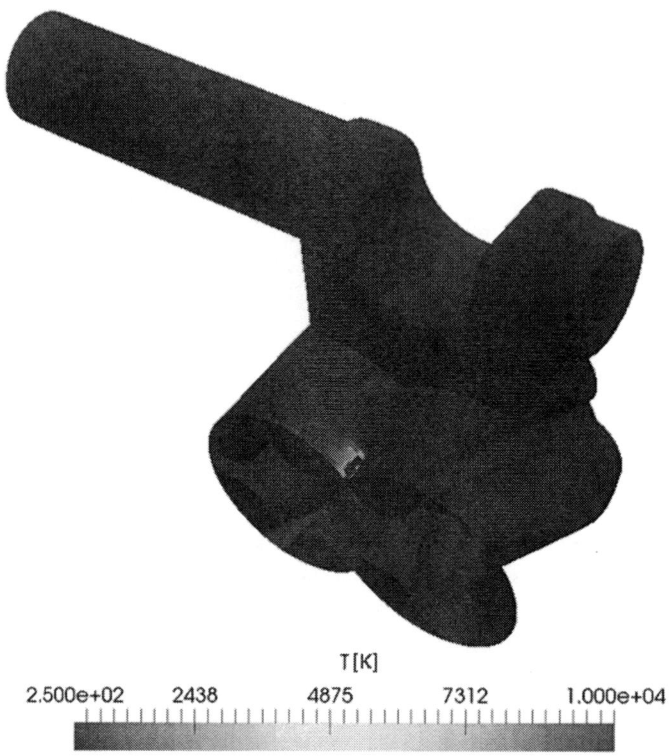

Figure 5. Temperature spike originated in the gap between thread head and casing on the female rotor

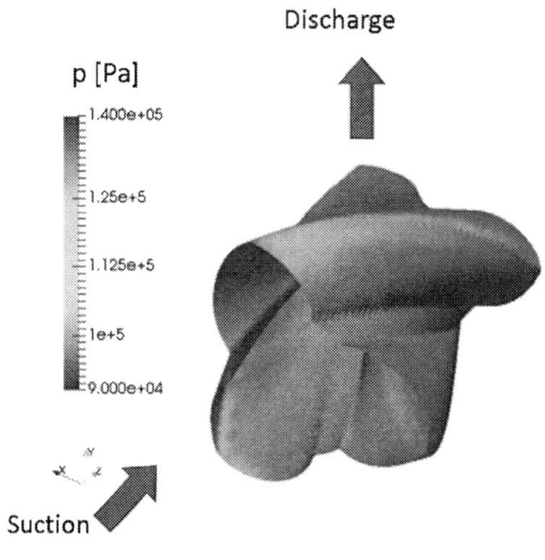

Figure 6. Pressure pattern on the male rotor: high pressure chamber and expansion downstream the gap

files has been generated by SCORG, a propretary software, and OpenFOAM, an open-source toolkit, was employed for solving the flow field. The coupling was carried out successfully in the case of single rotor domain, having a conformal interface in the interlobe area.

The dynamic mesh library here developed is able to make the mesh into OpenFOAM to move according to the output files generated by SCORG. The technique proposed is suitable for the simulation of positive displacement machines, not demanding for remeshing and thus not violating the space conservation law (entailing the non-conservativeness of the mass).

The application of this algorithm to a real twin-screw compressor, the comparison of the results with other CFD software and the validation is to be considered as a next step in this work.

5. Acknowledgement

The research was partially supported by the Italian Ministry of Economic Development within the framework of the Program Agreement MSE-CNR "Micro co/tri generazione di Bioenergia Efficiente e Stabile (Mi-Best)".

References

[1] Tassou S and Qureshi T 1998 *International Journal of Refrigeration* **21** 29–41
[2] Ziviani D, Gusev S, Lecompte S, Groll E, Braun J, Horton W T, van den Broek M and De Paepe M 2016 *Applied Energy* **181** 155–170
[3] Quoilin S, Declaye S, Tchanche B F and Lemort V 2011 *Applied thermal engineering* **31** 2885–2893
[4] Landelle A, Tauveron N, Haberschill P, Revellin R and Colasson S 2017 *Applied Energy* **204** 1172–1187
[5] Stosic N, Smith I K and Kovacevic A 2002 *A twin screw combined compressor and expander for CO2 refrigeration systems*
[6] Lemort V, Guillaume L, Legros A, Declaye S and Quoilin S 2013 *Proceedings of the 3rd International Conference on Microgeneration and Related Technologies*
[7] Tang H, Wu H, Wang X and Xing Z 2015 *Energy* **90** 631–642
[8] Kovacevic A, Stosic N and Smith I 2007 *Screw Compressors: Three Dimensional Computational Fluid Dynamics and Solid Fluid Interaction* vol 46 (Springer Science & Business Media)
[9] Suman A, Ziviani D, Gabrielloni J, Pinelli M, De Paepe M and Van Den Broek M 2016 *Energy Procedia* **101** 750–757
[10] Morini M, Pavan C, Pinelli M, Romito E and Suman A 2015 *Applied Thermal Engineering* **80** 132–140
[11] Chang J C, Chang C W, Hung T C, Lin J R and Huang K C 2014 *Applied Thermal Engineering* **73** 1444–1452
[12] Casari N, Suman A, Ziviani D, van den Broek M, De Paepe M and Pinelli M 2017 *Energy Procedia* **129** 411–418
[13] Suman A, Randi S, Casari N, Pinelli M and Nespoli L 2017 *Energy Procedia* **129** 403–410
[14] Mittal R and Iaccarino G 2005 *Annu. Rev. Fluid Mech.* **37** 239–261
[15] Rane S, Kovacevic A, Stosic N and Kethidi M 2013 *International Journal of Refrigeration* **36** 1883–1893
[16] Rane S 2015 *Grid generation and CFD analysis of variable geometry screw machines* Ph.D. thesis City University London
[17] Thompson J F, Soni B K and Weatherill N P 1998 *Handbook of grid generation* (CRC press)
[18] Knupp P and Steinberg S 2002 *Fundamentals of grid generation* (CRC press)
[19] Rane S and Kovacevic A 2017 *Advances in Engineering software* **107** 38–50
[20] Juretić F 2015 *Zagreb, Croatia*
[21] Buratto C, Casari N, Aldi N, Pinelli M and Suman A 2016 Cfd analysis of non-newtonian fluid processing pump 13th OpenFOAM Workshop, Guimaraes
[22] Issa R I 1986 *Journal of computational physics* **62** 40–65
[23] Patankar S 1980 *Numerical heat transfer and fluid flow* (CRC press)
[24] Jasak H 1996 *Error analysis and estimation for the finite volume method with applications to fluid flows.* (Imperial College London (University of London))
[25] Gonzalez E 2016 *46th AIAA Fluid Dynamics Conference* p 3960

IOP Conf. Series: Materials Science and Engineering **425** (2018) 012018 doi:10.1088/1757-899X/425/1/012018

Cfd simulation of a two stage twin screw compressor including leakage flows and comparison with experimental data

R Andres[1,3], J Hesse[1], F Hetze[1], D Low[2]

[1]CFX Berlin Software GmbH, Berlin, Germany

[2]Sullair, A Hitachi Group Company, Michigan City, United States,

[3]Corresponding Author

rainer.andres@cfx-berlin.de

Abstract. Computational Fluid Dynamics (CFD) is a common and validated simulation method in research and industry for the analysis of fluid systems. In the past years, it has proven to become more and more applicable for modeling the flow physics inside positive displacement (PD) machines. The working chamber and thus the discretized flow domain of PD machines are changing in time, characterized by complex thermodynamics. Compressible fluids, real-gas properties and leakage flows with trans- or supersonic characteristics are phenomena which have to be accounted for in order to properly model the behavior of the machine. As CFD methods evolve in general, but also for the application of PD machines in particular, the numerical model can replace a prototype during early stages of the product development. The desired simulation approach should be able to deliver sufficient accuracy at a feasible effort in terms of computational time and manpower to create the numerical model.

This paper presents the methodology of creating the numerical model for a sample screw compressor provided by Sullair for research purposes. It is a dry running two stage twin screw compressor running with air at a rated power range between 160 and 250 kW. The two stages are gear driven by the main shaft at rotational speeds between 1180 and 2100 rev/min. Each stage features different rotor profiles, where the first stage has a 4-6, the second stage a 5-7 lobe combination. The total pressure ratio of the two stages combined is up to 10:1. To enhance the performance of the compressor, discharged air from the first stage is cooled down before entering the second stage. A specific meshing method is used to model the size-changing working chambers between rotors and casing, where only hexahedral cells are used and mesh topology is constant. The model accounts for radial and axial clearances between rotors and stator, where rotors and stator are connected with interfaces. The transient simulation results are compared to experimental measurements for torque, and flow rate. Also discharge pressure and temperature after first and second stage are compared to the experimental results. In addition, the possibilities of the simulation are exemplified by the gathering of time- and space-resolved monitor points like temperature or pressure at distinct points within the compressor. Apart from direct comparison to the experiment, also a sensitivity study regarding the change of housing clearances is presented, as leakage flow has severe impact on the compressor performance. These clearances and the resulting leakages are often not exactly known whereas they also vary because of manufacturing tolerances or deformations due to the load on rotors and stator. Here, the numerical simulation can serve as a helpful tool to estimate the sensitivity and change of machine characteristics, which is hard to determine in the scope of experiments.

1. Introduction

While screw compressors are widely used in the industry, e. g. for pneumatic power tools or in the automotive industry oil free compressors are often used when purity of the air is critical. Within oil injected compressors, there is the risk of contaminated air which has an adverse effect on downstream processes and production. However, oil free compressors lack the sealing and cooling effect of the oil, thus in general the achieved pressure ratios are lower compared to oil flooded compressors. Compression ratio as well as the compression efficiency can be increased by using multi-stage compressors. The sample dry screw compressor from Sullair consists of two stages with pressure ratios up to 10:1 where each stage consists of two coaxial rotors. Furthermore, an intermediate cooler is located between the stages to cool the heated air discharged by the first stage in order to increase the compressor efficiency. For the manufacturer, reliable tools to evaluate overall machine performance such as compression ratio and power consumption as well as calculate losses and efficiency are of great desire. Also, the impact of design changes should be evaluated before actual prototypes are built so that developing costs can be reduced.

Screw compressors and their numerical simulation are quite complex, where for a multi-stage compressor, additional challenges are imposed. One approach is to treat both stages separately. However, this approach cannot describe direct interaction between the stages (e.g. pressure pulsation). Moreover, the requirement of setting a boundary condition at the discharge side of the first stage is apparent, since influence of the further downstream located second stage (and additional parts like an intermediate cooler) is a priori not known. Thus, an iterative simulation loop can be necessary in order to get a consistent solution.

The paper presents another strategy, where both compressor stages are coupled by including them in a single simulation run. The modeling methodology is analogue to single stage screw compressors. The deforming rotor chambers are separated and meshed independently from the stator parts and connected for the simulation run via fluid-fluid interfaces. During the transient run, rotor chamber meshes are deformed according to the current rotor position. For the modeled two stage screw compressor in this paper, both stages feature different rotor profiles with different rotating speeds, which the rotor chamber modeling has to account for. Also the intermediate cooler is modeled in a simplified way in the simulation setup, by decreasing the temperature between the two stages. The paper presents the modeling approach for a simplified setup (e. g. simplified cooler, adiabatic walls) to evaluate the feasibility of the CFD approach. For this purpose, achieved results are compared with measurements of the real compressor.

2. Rotor chamber modeling

The spatial discretization of the rotor chambers can be realized with different approaches which all have to account for the deformation of the rotor chambers over time. The immersed solid method, remeshing or

IOP Conf. Series: Materials Science and Engineering **425** (2018) 012018 doi:10.1088/1757-899X/425/1/012018

customized grid generation should be mentioned, where Spille-Kohoff points out the methodologies, advantages and disadvantages of the individual approaches more in detail [1]. The numerical grids for the present screw compressor analysis are generated with the mesh generator TwinMesh, which provides automatically generated hexahedral grids for PD machines with two coaxial rotors. For the performed simulations, 3D meshes of the chamber volumes for all rotor positions that are considered in the transient simulation are generated prior to simulation. An O-type grid topology is used, where inflation layers in the radial and axial directions are used to provide resolution in the boundary layers as well as a smooth volume change of grid cells in the regions of the axial clearance between rotors and housing. There are two O-type grids for each stage, which have a 1:1 connection. A detailed explanation of the utilized approach is given by Hesse [2]. A three dimensional view of the rotor chamber grids is shown in Figure 1.

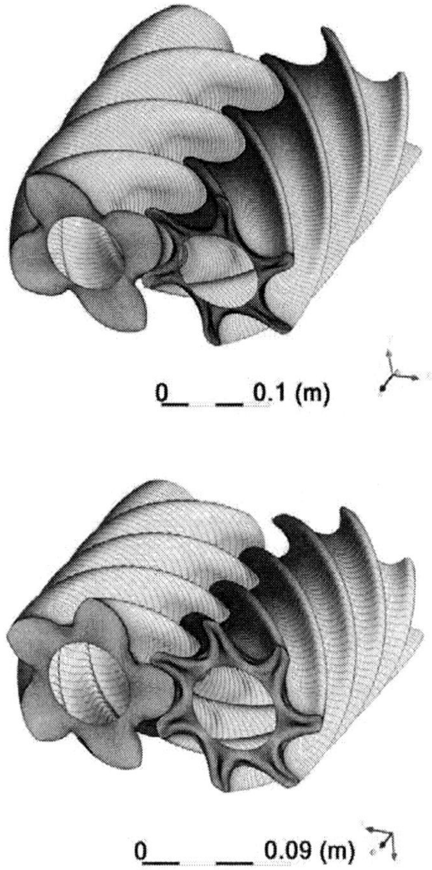

Figure 1: Grids for first stage (top) and second stage (bottom) of the compressor

Given that approach, it has to be considered that both compressor stages are modeled in one simulation setup, where the physical time step in the transient simulation is equal for the whole model. Thus, angle increments of the generated grids cannot be chosen arbitrarily for each stage. The ratio of angle increments for both stages has to represent the speed ratio of the two male rotors. The tooth pitch angles and angle increments for both stages are given in the following table:

Table 1: Angle increments according to the speed ratio of the modeled two stage screw compressor

	Stage 1	Stage 2
Speed ratio (Stage 2/Stage 1)	1.53	
Tooth pitch angle	90°	72°
Number of grids per pitch angle	90	47
Angle increment	1°	1.532°
Angle increment ratio (Stage2/Stage1)	1.532	

The angle increments are determined by a specified number of meshes per pitch angle, therefore the corresponding ratio slightly deviates from the exact ratio of 1.53. Hence the simulated revolution speed of the second stage will deviate from the reference value to the same small extend.

For both stages, rotational periodicity is taken into account in a way, that only the pitch angle for each stage (360° divided by the number of lobes of the male rotor) is meshed with a fixed angle increment. The resulting set of grids enables to run the simulation for as many revolutions as required in order to get a periodic solution over time (i. e. the time averaged value of monitored flow quantities is constant over time). It should be noted, that reuse of the same grid sets for all subsequent rotation angles higher than the meshed pitch angle range is not generally possible. It requires that grid nodes (with fixed indices) are at the same location within the fluid volume for initial and last rotor position within the pitch angle range. If this is not the case, the simulation has to be restarted, where flow quantities from the final rotor chamber grid are interpolated to the first grid. For the presented simulation, grid nodes attached to the stator curves (i. e. the stator walls around the rotors) are fixed in space, while all others can move according to the rotor motion. In that way, it is ensured that node positions of first and last generated grid match and no interpolation is required to reuse the same grid set during the simulation run.

The generated grids for the rotor chambers (including all clearances) consist of 1 873 680 elements for the first stage and 1 392 880 elements for the second stage. The radial and axial clearances are fully resolved with a mesh, where 20 elements in radial direction are present within the housing gaps and 40 within the intermesh clearance. In axial direction 6 elements are present over the axial gaps. No additional model is utilized to resolve the flow through clearances.

3. Simulation setup

All simulations for the sample Sullair screw compressor are performed in the commercial solver ANSYS CFX. Both stages are modeled within one case setup. The stator grids are connected to the rotor chamber grids with fluid-fluid interfaces, where the latter include the volume for the axial and radial clearances. During the simulation, the rotor positions and so the corresponding rotor chamber grids are updated automatically by a FORTRAN routine based on the generated grids prior to the simulation start.

The geometry of the intermediate cooler between the first and second stage is not discretized. Instead, the stages are connected with a free form duct, where the temperature loss caused by the cooler is modeled with an energy sink acting in the fluid volume of the duct. The location of the two stages relative to each other corresponds to the real compressor. Figure 2 shows the assembly of the case setup. The pressure loss due to the cooler was not measured in the experiment. It is expected to be in the magnitude of 0.10 to 0.15 bar. In the present setup, no pressure loss is defined in addition to the present pressure loss due to flow through the duct. However, it is an option in further investigations analogues to the modeling of temperature loss.

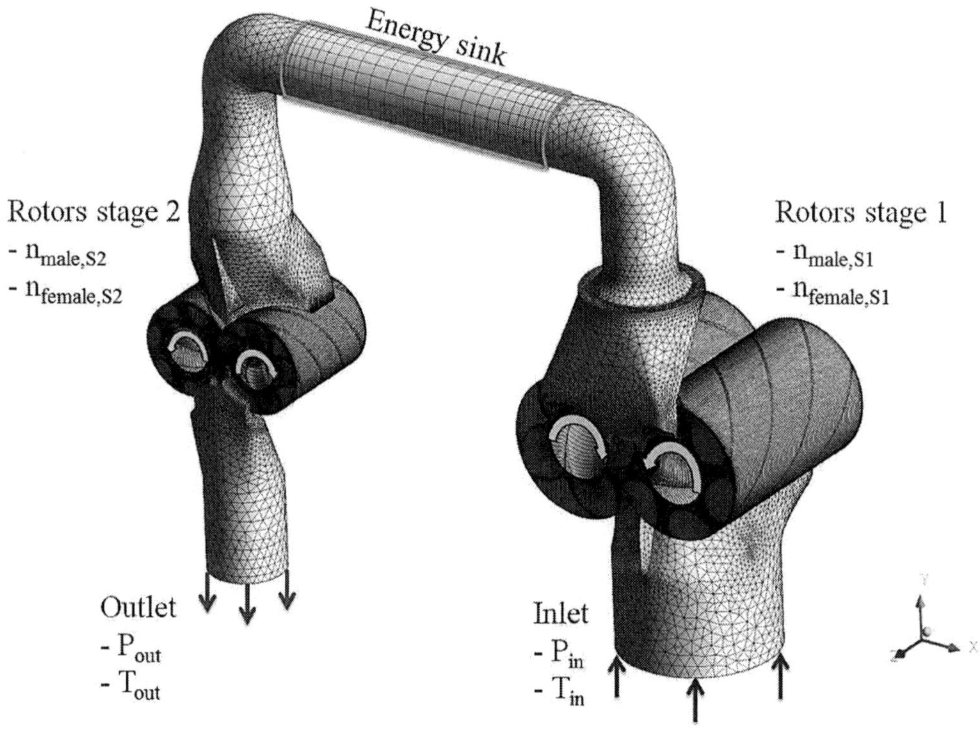

Figure 2: Numerical model of the screw compressor with boundary conditions and grid lines

The case is set up for three different operating points (OP), where the rotational speed of the stages is varied as well as inlet (first stage suction port) and outlet temperature (second stage discharge port, temperature is only used in case of backflow) according to the conditions indicated by experimental data. In addition, one case with smaller housing clearances (uniformly decreased by approximately 20%) and smaller intermesh clearance (closest distance between the rotors decreased by approximately 50%) for each stage is simulated, while axial clearances remain unchanged. Table 2 shows the four operating points taken into account:

Table 2: Boundary conditions for the performed simulations

Case	Main shaft speed [rev/min]	Inlet Pressure [bar(a)]	Outlet pressure [bar(a)]	Inlet temp. 1st stage [C]	Inlet temp. 2nd stage [C]	Outlet temp. 2nd stage [C]
OP1	1490	1.0	7.98	30.8	31.9	136.1
OP2	1790	1.0	7.98	28.6	34.1	143.0
OP3	2100	1.0	7.98	27.0	37.9	150.8
OP4 (Decreased radial clearances)	1790	1.0	7.89	28.6	34.1	143.0

IOP Conf. Series: Materials Science and Engineering **425** (2018) 012018 doi:10.1088/1757-899X/425/1/012018

The individual stages are driven by the main shaft over a friction gear. The female rotor speed of each stage is given by the lobe ratio between male and female rotor. The ratios are 2/3 for the first stage and 5/7 for the second stage.

All simulations are transient, where the angle increment was doubled by using every second of the generated rotor chamber grids. To model the accurate rotation speed of each rotor, the time step has to be doubled as well. The fluid is air as an ideal gas, with constant fluid properties except for the density. Turbulence is taken into account with the SST turbulence model. All walls are treated as adiabatic.

4. Simulation results

For both stages of the screw compressor, the compression inside the working chamber is visualized in Figure 3 by the instantaneous pressure distribution on the rotor walls. It is shown over one pitch angle of the corresponding stage. At the depicted rotor positions (0° to 60° for the first stage, 0° to 49° for the second stage), the pressure in the rotor chambers formed by male and female rotor lobes increases as the volume of the individual chambers is decreasing. Once the lobes reach the control edges, the connection of the chambers with the discharge port is established and compression process is finished.

Time averaged over one male rotor revolution, the discharge pressure of the first stage is 3.33 bar at 1790 rpm (average value on a plane, approx. 10 mm above the discharge port), where the suction pressure upstream of the second stage is 3.23 bar in average (average value on a plane, approx. 30 mm above the suction port). A time resolved depiction of discharge and suction pressure is given in Figure 4. This is in good comparison to the experiment, where for the same shaft speed, the discharge pressure of the first stage is about 3.5 bar. Inlet and outlet pressure (i. e. upstream of the first stage and downstream of the second stage), are at a fixed value as a boundary condition according to measurements (see Table 2).

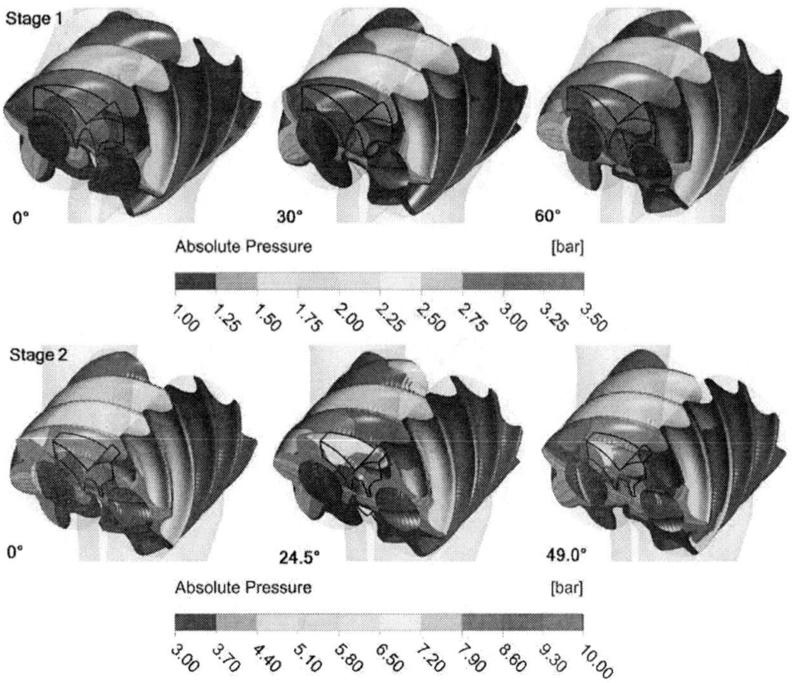

Figure 3: Pressure distribution at 1790 rpm for first (top) and second stage (bottom) over one pitch angle

IOP Conf. Series: Materials Science and Engineering **425** (2018) 012018 doi:10.1088/1757-899X/425/1/012018

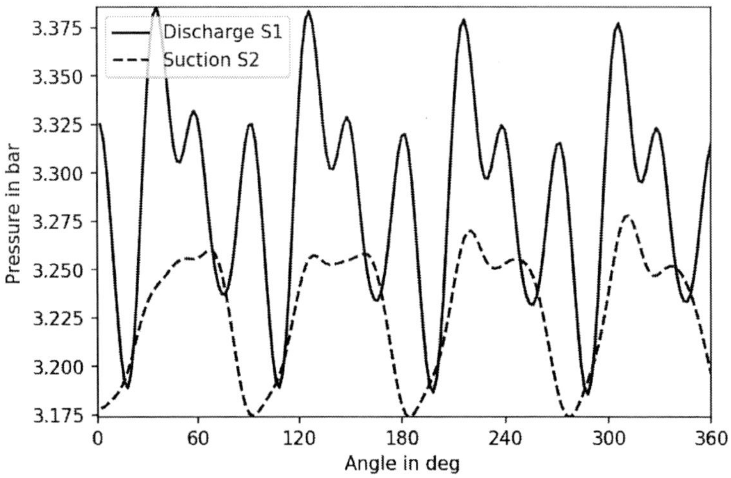

Figure 4: Discharge pressure of stage 1 (S1) and suction pressure of stage 2 (S2) over one male rotor revolution of the first stage at 1790 rpm

The velocity field as well as the temperature field on a plane within the duct is visualized in Figure 5. As specified by the applied energy sink within the duct, the temperature drops gradually in order to provide cooled air at the suction side of the second stage.

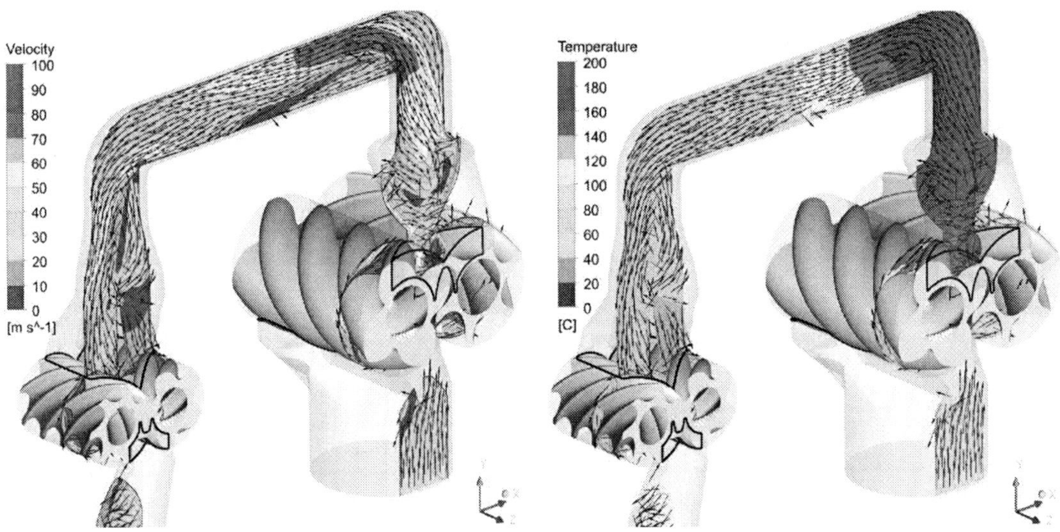

Figure 5: Instantaneous velocity and temperature field on a plane in the center of the duct between the two stages

The discharge temperatures at 1790 rpm of the first and second stage are quite constant with a maximum variation of about 8°C (first stage) and 0.5°C (second stage) respectively. The time averaged temperatures are approximately 163°C for the first stage and 128°C for the second stage. Measured in the experiment, the discharge temperature of the first stage is around 175°C and 158°C for the second stage.

The volumetric flow rates for all operating points at inlet and outlet of the compressor are shown in Figure 6. The given values correspond to the calculated mass flow rates divided by the reference density according to the inlet pressure and temperature. Due to numerical errors, the time averaged values of inlet and outlet flow rates are not exactly equal. However, the deviation is below 1% for all cases. The resulting volumetric flow rates are the mean values between inlet and outlet, each time averaged over 1 male rotor revolution of the first stage.

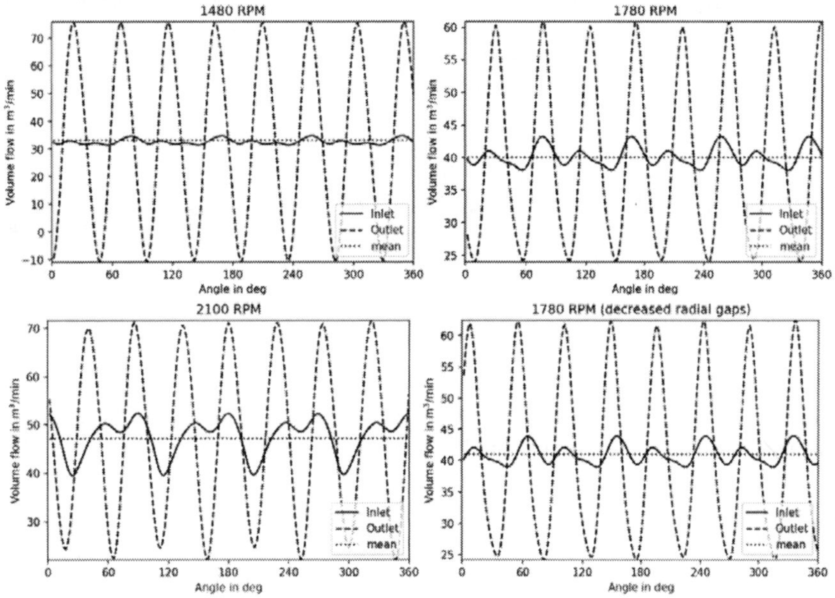

Figure 6: Volumetric flow rates at inlet and outlet of the compressor for the different operating points over one male rotor revolution of the first stage

An increase of the average flow rate with an increasing shaft speed is apparent (while inlet and outlet pressure are kept unchanged). For the lowest simulated shaft speed of 1480 rpm, also backflow at the outlet can be identified over a certain angle range. The flow rates are 32.6 m³/min at 1490 rpm, 40.3 m³/min at 1790 rpm and 47.5 m³/min at 2100 rpm shaft speed respectively. Qualitatively, it can be observed that the flow rate fluctuation at inlet and outlet is also more pronounced at higher shaft speeds.

Analogue to the flow rates, the indicated power is examined for all operating points. It is only calculated based on the torque which results from pressure forces acting on the rotors during the compression of the air inside the working chambers. Mechanical losses, like e. g. friction due to bearings, shaft sealing or lobe intermeshing are not included. Like the flow rates, power is increasing nearly linearly with rising shaft speed, what matches the expected behavior of the screw compressor. The calculated power for both stages combined is 188 kW at 1490 rpm, 230 kW at 1790 rpm and 276 kW for 2100 rpm respectively. For all shaft speeds, the first stage contributes about 55% of the total power. The time resolved power data are shown in Figure 7.

IOP Conf. Series: Materials Science and Engineering **425** (2018) 012018 doi:10.1088/1757-899X/425/1/012018

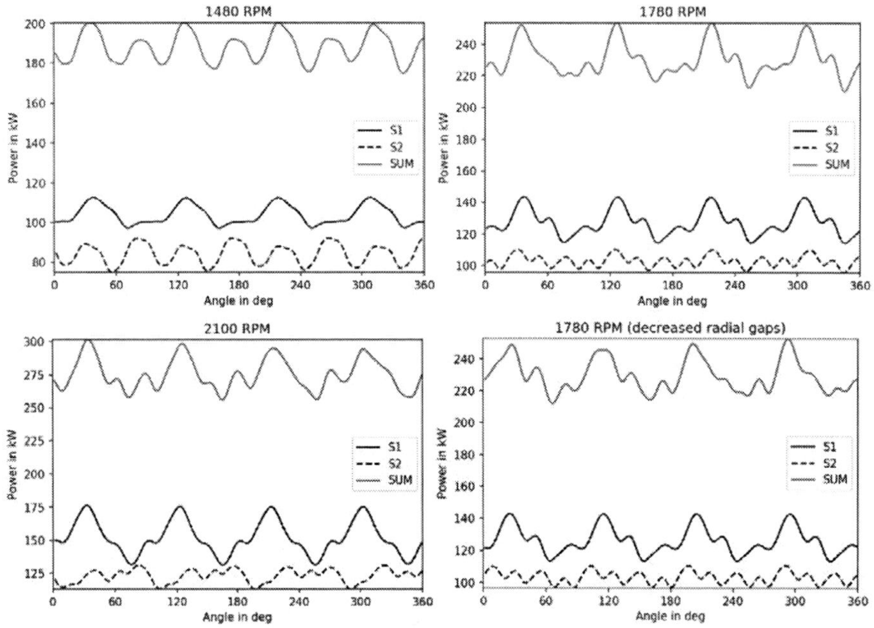

Figure 7: Power of the compressor for the different operating points over one male rotor revolution of the first stage

The simulation results are compared to measurements in Figure 8 and Table 3 of the real screw compressor in terms of measured volumetric flow, power and specific power.

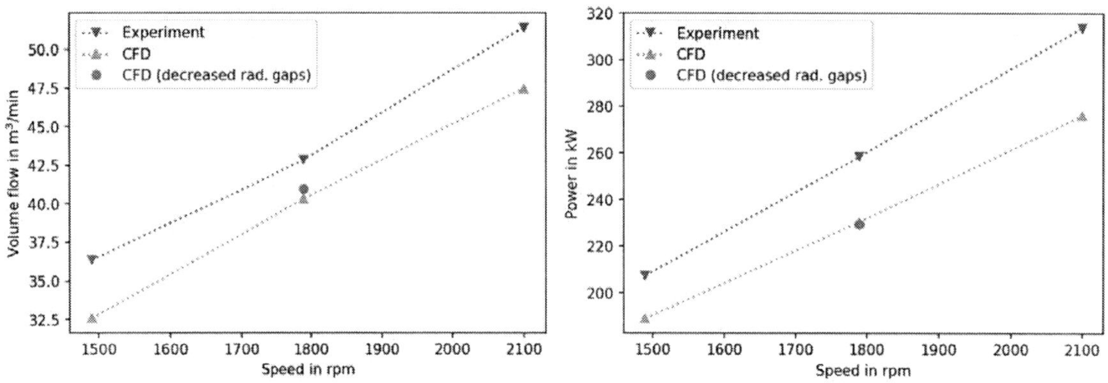

Figure 8: Comparison of experimental measurements and CFD results for all considered operating points

Table 3: Comparison of experimental measurements and CFD results for all considered operating points

	Measurements			CFD Results			Relative Deviation		
	Flow Rate [m³/min]	Power [kW]	Specific Power [kW/(m³/min)]	Flow Rate [m³/min]	Power [kW]	Specific Power [kW/(m³/min)]	Flow Rate [%]	Power [%]	Specific Power [%]
OP1	36.35	207.4	5.705	32.58	188.0	5.770	-10.4%	-9.4%	1.1%
OP2	42.85	258.4	6.030	40.33	230.4	5.712	-5.9%	-10.8%	-5.3%
OP3	51.43	313.8	6.102	47.45	276.0	5.817	-7.7%	-12%	-4.7%
OP4	-	-	-	40.96	229.4	5.601	-4.4%	-11.2%	-7.1%

It can be seen that volumetric flow rate as well as power are computed lower than measured for the real compressor. For the flow rates, CFD results are about 8% to 10% lower, whereas the power is about 9% to 12% lower. Given the lower flow rate, a lower power is evident. Hence, also the specific power is compared. Due to the lack of mechanical losses present in the CFD simulation, the specific power is expected to be lower than of the real compressor. This is the Case for OP2 to OP4, where the specific power is about 5% to 7% below the measured value. However, for OP1 the calculated value exceeds the experimental value by 1%. At this point it is not clear why this is the case and it needs further investigation. After all, it is noteworthy that the specific power for OP1 also deviates from OP2 and OP3 in the experiment.

A significant influence of the decreased clearances for 1790 rpm (i. e. the same as for OP2) cannot be observed, as flow rate and power remain approximately the same. Concerning the power, the averaged value is even slightly lower, whereas a higher value would be expected. One reason for the miniscule influence of the decreased clearances could be the utilized meshing strategy, where nodes are fixed on the stator curves and slide over the rotor walls. The 3D mesh including the twist angle is generated by 2D mesh slices where linear interpolation is present between the individual slices depending on the number of grid nodes in axial direction. Within the 3D mesh, this can lead to higher clearances than present in the 2D mesh slices. A different meshing strategy (e. g. fixing the grid nodes on the rotor curves instead of on the stator curves) could achieve better control over the gap sizes, however this approach was not realized because of the reasons stated in chapter 2 (i.e. the ability to combine both stages in one setup and using a small set of meshes to perform an arbitrary amount of time steps without interpolation).

In addition to direct comparison to experimental data of the real compressor, a simple estimation of the theoretical compressor power is done according to [3], assuming an isentropic compression:

$$P=\frac{Q_I p_I}{64.85}\left[\left(\frac{p_2}{p_I}\right)^{(k-1)/k}-1\right] \tag{1}$$

This approach is valid for single stage compressors but still utilized for convenience. A more reliable estimation of the compressor power would necessitate a split into first and second stage. In this estimation, Q_1 is the volumetric flow rate at inlet conditions, p_1 and p_2 the inlet and outlet pressure and k the isentropic exponent, which is about 1.4 for air. It should be noted that for this formula, pressure units are in psi and flow rates in cfm, where the theoretical power is in hp. For convenience, the results are presented in SI units hereinafter. Since this theoretical power assumes isentropic compression without any mechanical losses, the efficiency of the screw compressor can be taken into account. The total efficiency η_{tot} is given by Sullair for the individual operating points and includes all losses, i. e. thermodynamic and mechanical losses. Hence, the additional power due to the losses can be added to the theoretically determined power:

$$P_t = \frac{P}{\eta_{tot}} \tag{2}$$

The power estimation and comparison to both, measurements and CFD results is shown in Table 4. The theoretical assumption combined with the stated total efficiency by Sullair is very close to the measured power. Compared to the CFD results, it is also in good agreement. For OP1, the power is slightly overestimated by 0.5%, where it is about 6% to 8% lower for OP2 and OP4. This corresponds to the deviations regarding the specific power shown in Table 3. Analogue to the results for the specific power, it should also be noted here that a lower compressor power value would be expected for OP1 (i.e. it should be lower than the theoretical combined with the total efficiency). It remains uncertain if there are inaccuracies in the CFD calculation or the measurements as well as the determined total efficiency which is the basis of the theoretical power calculation.

Table 4: Theoretical power estimation in comparison with experimental measurements and CFD results

		Measurements			Relative Deviation	CFD Results			Relative Deviation
	η_{tot} [%]	P [kW]	P_t [kW]	Power [kW]		P [kW]	P_t [kW]	Power [kW]	
OP1	82.01	171.1	208.6	207.4	-0.6%	153.3	187.0	188.0	0.5%
OP2	77.59	201.7	260.0	258.4	-0.6%	189.8	244.7	230.4	-6.2%
OP3	76.60	242.1	316.0	313.8	-0.7%	223.3	291.5	276.0	-5.6%
OP4	(77.59)	-	-	-	-	192.8	248.5	229.4	-8.3%

5. Conclusion and outlook

This paper shows a CFD approach to simulate the sample dry screw compressor provided by Sullair. This compressor consists of two compressor stages with an intermediate cooler. In the presented approach, both stages as well as the interstage cooling are modeled within one setup, allowing a direct coupling of the stages without the need to specify boundary conditions at the outlet of the first stage or at the inlet of the second stage respectively. The cooler is replaced with a simple duct, where the temperature drop (i. e. a specified suction temperature for the second stage is reached) is realized with an energy sink.

The simulations are performed for a total pressure ratio (both stages combined) of 7.98 for three different rotational speeds. In addition, for one shaft speed, radial clearances (i.e. housing and intermesh gaps) are reduced. The CFD results are compared with measurements of the real compressor. The values for the flow rate and compressor power are in good match with the experiment, so it is indicated that the overall working mechanism of the two stage compressor is appropriately captured. In general the simulations with reference clearances underestimate the flow rates by about 6% to 10%. The uncertainties regarding clearance sizes in the 3D mesh mentioned in chapter 4 are expected to contribute to these findings. Moreover it has to be noted that deformations due to pressure and temperature loads are not taken into account in the simulation. These deformations present in real compressors after heating up can also lead to a change in clearances sizes, affecting the flow rate. Being a highly complex machine, other influences like e. g. the flow conditions upstream, downstream and between the stages as well as the wall heat transfer, which are all neglected in the simulation approach, are expected to have a significant influence on the compressor performance.

However, the presented model serves for a solid basis to further investigations and allows including or enhancing neglected or simplified parts of the model (e. g. the intermediate cooler, the assumption of adiabatic walls or a better representation of the cooler regarding total pressure loss). Future work on the model will focus on a more detailed investigation of clearances and leakage flows and the associated meshing strategies. Also the heat transfer is of interest, thus rotor and stator solids as well as active cooling of stator components should be included to the existing model to validate and improve the simulation results. As a goal for the industry, a precise distinction and quantification of thermal and mechanical losses is desired in order to optimize their compressors.

Nomenclature

P	Theoretical compressor power	[W]	k	Isentropic exponent	[-]
P_t	Theoretical compressor power including total compressor efficiency	[W]	Q_1	Volumetric flow rate at inlet conditions	[cfm]
η_{tot}	Total compressor efficiency	[%]	p_1, p_2	Suction and discharge pressure	[psi]

References

[1] Spille-Kohoff A, Hesse J and El Shorbagy A 2015 Cfd simulation of a screw compressor including leakage flows and rotor heating *9 th International conference on compressors and their systems* (London)

[2] Hesse J, Spille-Kohoff A, Hauser J and Schulze-Beckinghausen P 2014 Structured meshes and reliable CFD simulations: TwinMesh for positive displacement machines *9th International Conference on Screw Machines (*Dortmund)

[3] Compressed Air & Gas Institute 2016 *Compressed Air and Gas Handbook (Seventh Edition)* (Cleveland)

IOP Conf. Series: Materials Science and Engineering **425** (2018) 012019 doi:10.1088/1757-899X/425/1/012019

Modeling a dry running twin-screw expander using a coupled thermal-fluid solver with automatic mesh generation

David Rowinski[1*], Alexander Nikolov[2], Andreas Brümmer[2]

1. Convergent Science, Inc., Madison, WI, USA
2. Chair of Fluidics, TU-Dortmund University, Dortmund, Germany

* Corresponding Author. Email address: david.rowinski@convergecfd.com

Abstract. Understanding the details of the internal flow processes in screw compressors and expanders is very important for their efficient and robust design. Computational fluid dynamics (CFD) provides full access to a modeled three dimensional flow field and its variation in time. However, the application of CFD to screw compressors and expanders can be difficult because of the complicated geometries involved and the need to supply the computational grid on which the modeled equations are solved over a large number of time steps. While the majority of previous research on CFD applications to screw machines features inventive techniques for generating meshes that adequately resolve the flows in the small clearances, an alternate approach is demonstrated in this work. The screw expander SE 51.2 from TU Dortmund University is analyzed here through a CFD model which generates the grid automatically based on a modified Cartesian cut-cell approach. The grid is then adaptively refined based on local gradients of velocity and temperature. At each time step, the grid is regenerated based on the geometry motion. As opposed to resolving the flow in the clearances, a model is applied so that the cells in the clearance can remain relatively large. The detailed measurements of the screw expander are used to validate the model. The operating conditions investigated include the expansion of dry air at a four-to-one pressure ratio for four different rotational speeds. The measured internal chamber pressures are compared to the results from the model, as are the average mass flow rate, indicated power, and outlet temperature. A coupled thermal-fluid approach is used to model the rotor temperatures and corresponding thermal deformation of the rotors and housing. In this approach, the fluid and solid temperatures are solved together; to deal with the problem of the disparate time scales between the fluid and solid heat transfer, the solids are periodically solved to steady state using heat transfer coefficients and near-wall temperatures computed from an energy conserving averaging over several cycles. The effects of the various leakage flow paths in the model, including the rotor-to-rotor, rotor-tip-to-housing, and bearing leakage are demonstrated and quantified. Finally, simulation and experimental results are compared in terms of different rotor-tip-to-housing clearance heights. The model considering the appropriate thermal deformation of the rotors is shown to yield the best agreement with the measurements, however there is work remaining to reduce the model calculation time, especially at low rotational speeds.

1. Introduction

Screw expanders are mechanical devices used for extracting energy from waste heat in organic Rankine cycle and trilateral flash cycle systems. As energy demands around the world grow and regulations continue to drive energy generating devices to higher efficiencies, the importance of devices such as screw expanders for energy recovery continues to grow [1]. In order to design these machines for optimal efficiency and extract as much energy from the flow as possible, it is important to understand the nature of the flows and how they interact with the machine geometry. Mathematical models of these processes are invaluable in both optimization and analysis.

There are several different levels of fidelity at which the flow and thermodynamics in these devices can be modeled, and as the fidelity of the models increase so does the cost of each method. Chamber models [2], [3], [4], [5], also developed for the screw compressor and vacuum pump counterparts to screw expanders, are the most compact of these, and offer a very low computational cost while providing a modeled solution for the flow and thermodynamic state variables for individual homogeneous chambers. Advancements in modeling efforts have evolved some of these approaches to become even more sophisticated, including multi-phase effects [6], thermal and mechanical deformation [7], and advanced analytical models for the leakage flows [8]. Overall, though the chamber models are simplified, the cost is low enough to allow many design iterations for practical optimization use.

On the other hand, the most detailed of the modeling approaches for screw expanders is the use of computational fluid dynamics (CFD) to solve for the complete flow path in space and time using a set of modeled equations. While this approach is inherently more computationally expensive than the chamber models, providing access to a modeled three-dimensional transient flow path gives the designers and analysts the most detailed information about the flow. The primary challenge in applying CFD to screw machines is the construction of the numerical grid on which the model equations are solved. The complex rotor geometries, complex sealing lines and clearances, and the impact these have on the device performance make the direct application of CFD difficult.

The majority of CFD modeling studies on screw compressors, vacuum pumps, and expanders has been based on analytical grid generation methods, including the works in [9], [10], and [11] for screw compressors and in [12], [13], and [14] for screw expanders. These methods have been successful in computing the solutions to the modeled equations and have been validated in a wide variety of screw machines. However, these methods are not without their own limitations, some of which include a restriction on the geometry of the rotors and housing, relatively large computational cost, and numerical diffusion from moving meshes. In this study, an alternate approach for generating the grid and solving the modeled equations is used.

This study uses a body-fitted Cartesian cut-cell grid to enclose the fluid volume and provide the finite volume cells on which the modeled equations are solved as demonstrated in [15]. Adaptive mesh refinement (AMR) is used to increase grid refinement at points in time and space where the gradients in velocity and temperature are large. As opposed to resolving the clearance flow, as is done in many of the aforementioned CFD studies, the clearance flow is modeled here. This permits the cells in the clearances to be relatively large. All together, these features result in a CFD model which is easily applied to arbitrary moving geometries, constructs the mesh automatically at every time step with little user effort, and computes the solutions with high stability and accuracy due to the interior grid being composed of stationary orthogonal hexahedral elements.

2. Modeling methodology

The modeling process begins with the creation of the discretized mesh where the modeled equations are solved, the description of the computational domain and boundary conditions, and the specification of the modeled equations. In this section, these processes are described in more detail.

2.1. Grid discretization

The primary feature which makes this work different from the majority of previous modeling studies of screw compressors or expanders is the way in which the numerical grid is created [16]. The procedure begins from computer-aided design (CAD) software, where a geometry of the fluid domain is created. The surface surrounding this fluid volume is discretized into a triangulated surface which is a direct import for the modeled geometry. Each of the discretized triangles is then associated to a boundary condition for the flow model.

The actual volume mesh is created at run time by the solver and is illustrated in figure 1. At every time step, an orthogonal hexahedral mesh is overlaid on the triangulated surface. The cells that lie entirely inside the surface are identified as interior cells. The cells which are intersected by the bounding surface are cut into arbitrary-sided polyhedral cells. These cells are identified as boundary cells. Every boundary cell retains exactly the geometry it has when cut by the bounding surface regardless of the cell size. It is important to point out that while the triangulated surface will determine the shape of the computational domain, it is the grid cells and not the triangles where the equations are being solved. It is also important to point out that the geometrical features of the computational domain are determined solely by the triangulated surface and not the grid; since the boundary cells are pure cuts, they always retain the same volume regardless of their size. In figure 1, the grid of the screw expander model is shown from several directions.

Figure 1. Demonstration of the grid methodology used in the CFD model. On the top left and right are cut-planes through the rotors colored by velocity and displaying the mesh lines. On the bottom left is another cut-plane, with a detail of the boxed region showing the adaptive mesh refinement resolving the velocity jet at the inlet port on the bottom right.

Any boundary in the surface may move arbitrarily so long as it does intersect another boundary. At every time step, the triangles that move are swept through the boundary cells and that swept volume along with the boundary flux is calculated and integrated into the modeled equation solution. Thus, the solution of the equations and grid generation are tightly coupled. Cells may be refined based on predefined volumes or adaptively based on local gradients as in [17] down to a user-specified size. In the close-up in figure 1, the additional refinement at the high-velocity jet in the inlet port is shown. At every cell, the gradients are evaluated and if they exceed a user-specified threshold, the cells are refined in all three directions.

2.2. Modeled equations

The flow of a compressible Newtonian fluid is considered in this study. Conservation of mass, momentum, and total energy along with an equation of state relate the six flow variables of pressure, density, temperature, and velocity vector field. An unsteady Reynolds-Averaged Navier Stokes (RANS) turbulence model is used to account for the effects of turbulence on the velocity field; in this case, the Renormalization Group (RNG) k-epsilon turbulence model [18] is employed, which requires the solution of two extra fields, the modeled turbulent kinetic energy and turbulent dissipation rate.

The equations are solved with Pressure-Implicit Splitting of Operators (PISO) pressure-velocity coupling [19], along with a bi-conditioned conjugate gradient solution for pressure and successive over relaxation (SOR) iterative solution procedure for the remaining equations. The method is formulated in the finite volume form, where each cell (including both the interior cells and boundary cut cells) is treated as a finite volume, and all quantities are computed and stored at the cell centers. The scheme of Rhie-Chow [20] is used to prevent checker-boarding on this collocated grid.

2.3. Conjugate heat transfer

Many previous studies on similar cases have shown that the wall temperatures, both on the fluid and the solid, have important consequences [10], [11], [12], [14] in terms of the rotor mechanics, the fluid heat transfer to the wall, and the clearance flow characteristics. In this study, the solid domain is considered along with the fluid domain for the purposes of calculating a modeled temperature for the solid walls. The combined domain is shown in figure 2 and includes the fluid, the solid housing, and the solid rotors. In the solid, a single heat conduction equation is solved.

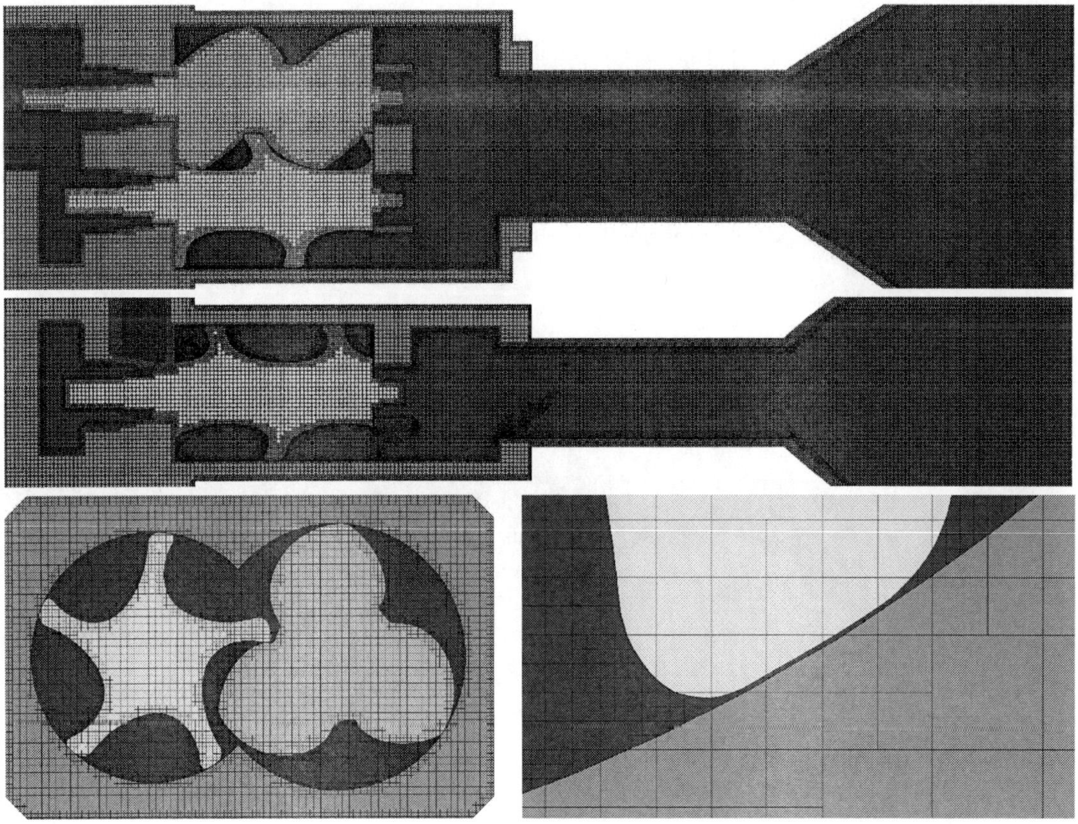

Figure 2. Views of a grid used in the coupled fluid-solid model. On the top and middle rows are views of the mesh through the rotor axes. On the bottom left is a view in the transverse plane across the rotors

and on the bottom right is a zoomed in view of one of the clearances showing a location where the clearance flow is modeled. All planes are colored by region: blue (dark) is the fluid region, green (medium dark) is the joined housing elements, yellow (medium light) is the male rotor, and pink (light) is the female rotor.

At the interfaces between the solid and the fluid, both the heat flux and the temperature are coupled. At every time step, both the fluid and the solid are solved in a time-accurate fashion. An inherent difficulty with this type of problem is the disparate time scales between the fluid and the solid; while the heat transfer processes in the fluid occur on very small time scales, the time scales associated with the conduction in the solid are generally much longer. Furthermore, since the temperature fields of the fluid and solid are coupled and the flow is linked through the density, both the solid and the fluid evolution will affect one another.

To remedy this problem, a technique referred to as super-cycling is used. With super-cycling, the near-wall temperatures and heat transfer coefficients are averaged over several cycles. Once every lobe passing time, the solid is solved completely to steady state using the thermal information from the fluid over several cycles. This process is completely automated in the solver, but it does require running for several cycles since the solution of both solid and fluid affect one another. One component missing from the model is directly including the effect of the thermal deformation in the same calculation. At the conclusion of the super-cycling simulation, the rotor profiles are scaled from their axes of rotation using the computed solid temperatures, and an additional calculation proceeds using this altered rotor geometry.

2.4. Modeling of clearance flows

The mesh shown in figure 2, in particular the detail in the lower right, demonstrates that the clearance flow is not fully resolved on the grid sizes considered. In order to account of the effects of the mesh being too coarse to fully resolve the clearance flow, a simplified calibrated model [15] is used. This model relates a clearance Reynolds number to an additional dissipative term in the momentum equation to correctly capture the mass flow through the clearance. The total mass flow rate through each clearance depends on the local conditions, including the actual clearance and the grid resolution, but generally this simplified model reduces the flow between 0-20% from the under-resolved model.

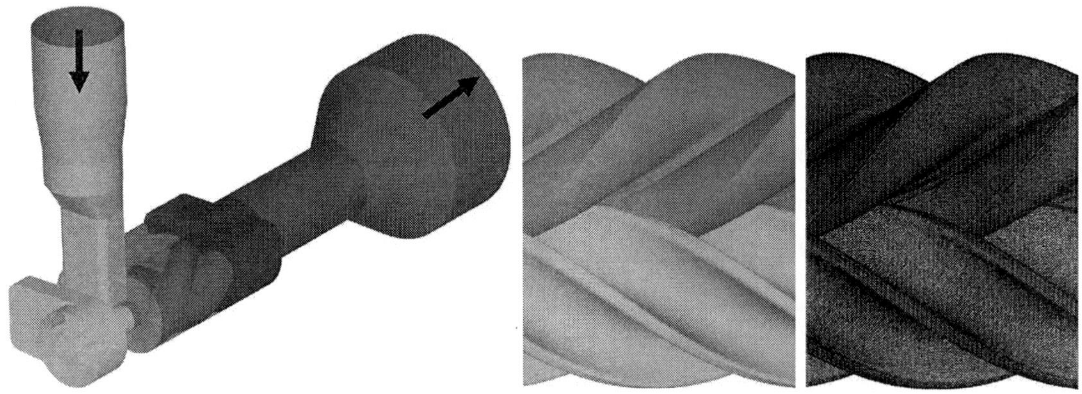

Figure 3. To the left is the fluid region from the SE 51.2 screw expander CAD model. On the right are the male and female rotors; to the far right, the triangles representing the discretized surface of the rotors are overlaid.

3. Test case description

The screw expander studied in this work is SE 51.2 designed at TU Dortmund University. SE 51.2, originally based on a similar twin screw supercharger, features a relatively small size including a displacement volume of 285 cc per male rotor revolution, a rotor center distance of 51 mm, and a rotor length of 101 mm. The built-in volume ratio is 2.5, though it is designed for convenient modification through the use of interchangeable control edge plates. The expander can handle pressure ratios between the inlet and outlet as large as 6:1 and speeds as high as 20,000 based on the male rotor. The performance of the expander has been extensively characterized experimentally in [21] over a wide variety of operating conditions, and more detailed experimental measurements have been performed in [22] and [23], including both dry running and water injected conditions.

The flow path of the twin screw expander and detail of the rotors is shown in figure 3. High pressure air enters from the top left, is drawn down into the expanding rotor chambers through the inlet port as the rotors counter-rotate. Once the working chamber is disconnected from the inlet port after the rotor flanks have rotated beyond the inlet control edges, the chamber is sealed and the internal expansion processes. Around the point where the maximum chamber volume is reached, the rotors continue to rotate exposing the control edges of the outlet port, at which point the outflow from the chambers begins. The rotation of the rotors reduces the volume of the expanded fluid and it is pushed out of the outlet port. The expander SE 51.2 is a gearless expander, so the counter-rotation of the rotors occurs with rotor contact.

In this investigation, only dry-running operating conditions are considered as have been the focus of previous CFD studies of this device in [12] and [14]. For the conditions investigated in this work, air enters at a pressure of 4 bar absolute and 90 °C and exits at a pressure of 1 bar absolute. These are incorporated into the model through setting a constant total pressure and temperature at the inlet and a constant static pressure at the outlet. Measurements of mass flow rate, indicated power, outflow temperature, and internal chamber pressure have been previously conducted for male rotor speeds between 500 and 18000 rpm, and the model considers four of these speeds (1000, 4000, 10000, and 16000 rpm) to compare calculated and measured indicator diagrams. The nominal rotor-tip-to-housing clearance in the base cases is 0.08 mm. Additional measurements have been conducted in modified housing geometries where the bore is increased to yield rotor-to-housing clearances of 0.16 mm and 0.24 mm, and these conditions are the focus of the second part of the CFD modeling study.

4. Results and discussion

The model described in section 2 is applied to the test case described in section 3, and the results are presented and discussed here. First presented is the steady state thermal predictions, followed by grid and leakage treatment, and lastly the comparison of different rotor-tip-to-housing clearances.

4.1. Base case temperature comparison

Since there is a close coupling between the solid and fluid solutions and a large discrepancy in their time scales, it is very important to show that the model has reached an appropriately temporally converged steady state (or a statistically stationary state). Figure 4 shows the inflow, outflow, and rotor temperatures for a case to demonstrate how this is achieved and how the results compare to the measured fluid temperatures.

As shown in the temperatures plotted on the left side of figure 4, the model is first run for a number of cycles. After five revolutions of the male rotor, the solid is solved to steady state to yield new wall temperatures seen by the fluid (and evident by the jump in rotor temperature at 5 revolutions). From there, the fluid and solid are continued to be solved together at each time step in a time-accurate fashion, and at every lobe passing the solid is again solved to steady state using averaged fluid temperatures for the previous five lobe passing times. It takes about an additional five revolutions to arrive at steady state for most of the solid temperatures and fluid temperatures around the rotors. However, because there is some additional time required to flow through to the outlet, it takes longer (about twenty revolutions of the male rotor in total) to reach a steady state for the average temperature at the outflow.

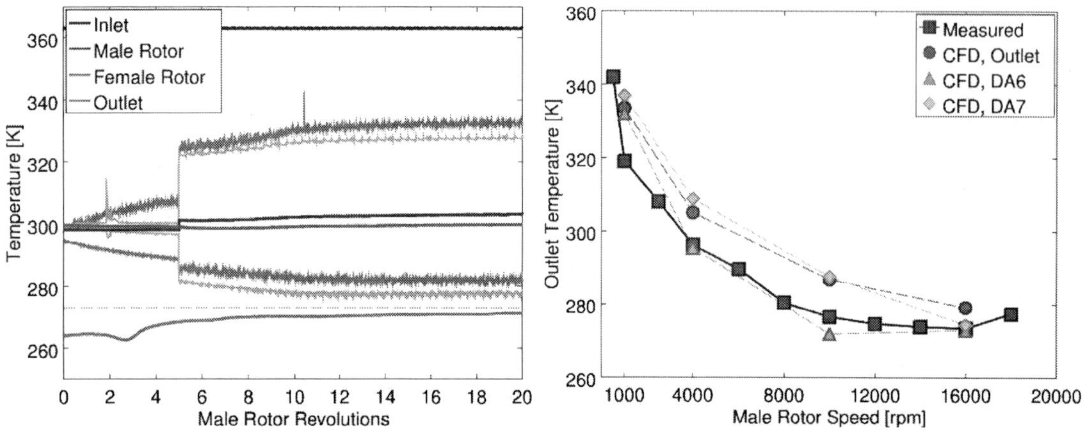

Figure 4. On the left is the evolution in time of temperatures (outlet fluid measurement shown in a brown dashed line, calculated outlet fluid spatial mean in the solid brown line, female rotor spatial mean temperature in a solid dark green line, male rotor spatial mean in a dark blue line, and inlet temperature in a dark purple line) in the coupled fluid-solid model as they reach steady state through the super-cycling procedure. On the right are the outlet temperatures at steady state comparison between measurements (black squares) and CFD model at various locations (outlet in blue circles, monitor point DA6 on the housing side in green triangles, and monitor point DA7 on the outlet housing underside in orange diamonds) as functions of male rotor speed.

The right side of figure 4 shows the modeled temperatures compared to the predicted temperatures at the outlet. In general, at high rotational speeds the leakage is proportionally less, so the outlet temperature is controlled primarily by the balance between the isentropic outlet temperature and the solid wall temperatures. As the speed decreases, the internal leakage effects become more important, and more high temperature gas from the inlet makes its way to the outlet, resulting in both higher outlet temperatures and higher solid temperatures. In general, this trend is well-produced by the model. Figure 4 shows the modeled temperature at three different locations: the actual domain outlet, monitor point DA6 near the outlet control plate, and monitor point DA7 slightly downstream of the outlet plate (where the measured temperature is actually reported).

At 16000 rpm, the agreement at DA7 is very good, but for 10000 and 4000 rpm the temperature is over-predicted by about 10 K. At 1000 rpm, the temperature is over-predicted slightly more, by nearly 20 K. However, at this point in the curve there is a large sensitivity of outlet temperature to rotor speed. As seen by the other modeled temperatures, there is some sensitivity to the spatial location of the monitor point and the solid wall temperature there. More work could also be done to characterize the convective boundary condition applied to the outer housing. An additional consideration which is also very important is the variation in time to steady state with rotor speed. While the model time for the low rotational speeds represented a greater physical time, it also corresponded to running fewer lobe passings, so more work should also be done to determine the quality of the transient solution reaching steady state for these lower speed conditions.

Figure 5 shows the rotors and a cut-plane over the fluid and housing colored by the temperature. The difference in outlet temperature shown in figure 4 is shown more qualitatively here. The higher fluid outlet temperatures, as well as the higher solid wall temperatures, can easily be seen here as the rpm decreases from 16000 to 1000.

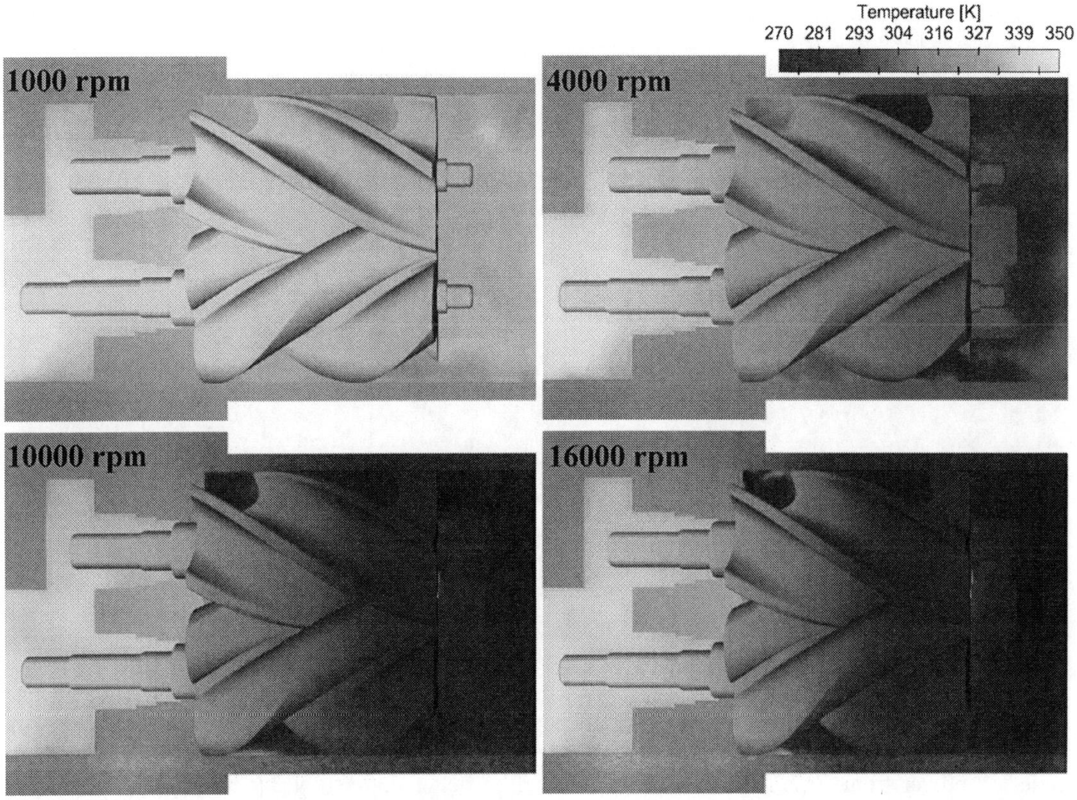

Figure 5. Rotors and a cut-plane across the rotor axes colored by temperature near steady state for male rotor speeds of (from top-left) 1000, 4000, 10000, and 16000 rpm.

4.2. Effects of the grid resolution, leakage paths, and clearance treatment

Once the procedure for obtaining the steady-state thermal solution is validated, the behaviour of the mass flow rate, indicated power, and internal chamber pressure is examined for different setup configurations of the base case (4:1 pressure ratio, 90 °C inlet temperature, and 0.08 mm cold rotor-to-housing clearances). Three different configurations are examined: Case 1 which has cold rotor clearances and under-resolved clearance flow, Case 2 which has hot rotor clearances (determined from the steady state thermal load) and a clearance model described in reference [15] for modeling the clearance flow, and Case 3 which is identical to Case 2 but also includes an additional bearing leakage path on the inlet side of the rotors and the bearing casing which is visible in the top-left of figures 2 and 3. The pressure in this additional region is supplied based on measurements and is referred to here as the cushion pressure.

First, the mass flow rate and indicated power for these three cases are compared in figure 6 for varying rotational speeds. The simplest and least expensive configuration, Case 1, results in the mass flow rate and indicated power in closest agreement with the measurements. However, it is clear when looking in deeper detail at the internal chamber pressures in figure 7, that those models do not yield the correct mass flow rate and indicated power for the right reason; likely offsetting errors resulting from over-predicting the leakage rates out of the chamber during the initial part of the expansion process and then over-predicting the leakage rates into the chamber during the end of expansion. So much can be learned from Case 1: even though the mass flow rate and indicated power may yield good agreement with the measurements, the offsetting errors are deceiving in predicting the right answers for the wrong reason. It is very valuable to have more complete description of the flow, particularly the internal chamber pressures for a thorough analysis and making sure the model results are correct for the right reasons.

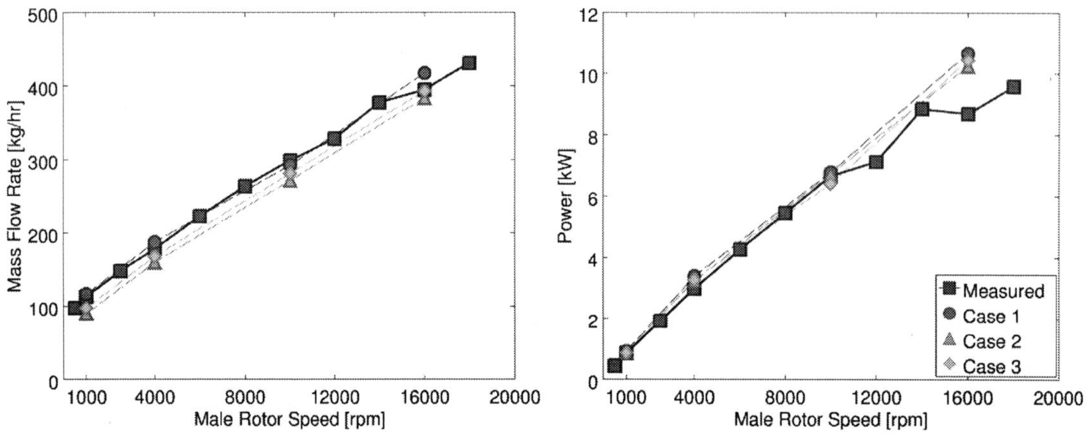

Figure 6. Mass flow and indicated power for measurements (black squares), and CFD models for cases 1 (blue circles), 2 (green triangles), and 3 (orange diamonds) as functions of male rotor speed.

Progressing from Case 1 to Case 2, one clear difference is that the mass flow rate drops consistently from 1 to 2. This occurs because of a generally greater restriction, both in the clearance flow model and the generally smaller clearances in Case 2. Since the flow between chambers is nearly choked for a large part of the expansion process, the clearance model mainly reduces the flow rate from that on the under-resolved grid, which is at a point around plug flow at the sonic velocity. There is a slight drop in indicated power from Case 1 to Case 2, which can be better understood by examining the internal chamber pressure traces in figure 7. From an angle around 240°, there is less refilling of the chamber (from the higher pressure chamber on the inlet side) due to the restricted leakage in Case 2, so the pressure is generally lower around this point.

From Case 2 to Case 3, where the bearing leakage is added into the model, the mass flow rate increases slightly, closer to the measured values, but still less than those occurring in Case 1. This is mainly contributed at the point in the cycle where the internal chamber pressure decreases beyond the cushion pressure but the chamber is still in communication with the bearing leakage passage. The effects on the internal chamber pressure are strongest in Case 1, but the effect on the mass flow is present in all cases. The average error in the mass flow rate for Case 3 is about 4% too low, however considering the overall agreement in all compared metrics, there is significant improvement overall from other less sophisticated configurations, Cases 1 and 2.

The model does a good job at predicting that, in general, as the rotational speed increases so do the pressure wave dynamics in the inlet port. However, there is considerable deviation in the actual pressure signal for the points with higher rotational speeds, specifically at 10000 and 16000 rpm. This is perhaps a result of the inlet boundary condition, which allows pressure reflections. Further investigations with non-reflective boundary conditions at the inlet and outlet, as used in [14], are currently being examined.

In general, the prediction of indicated power (which in the CFD model is derived from the pressure and viscous torque on the rotors and the rotational speed) follows that of the mass flow rate; the models which predict the higher flow rate also yield higher power. This is explainable through the comparison of the internal pressure, where the cases with higher mass flow rate also generally had higher internal chamber pressures, which translates to higher power consumption when computing the indicated work from the pressure and volume integral.

The refilling of the chamber from the inlet side higher pressure chamber that occurs around 250° and is most evident at the low rotational speeds is in general well-predicted by the model. At some of the conditions (specifically at 4000 and 16000 rpm), the decrease in the pressure is offset from the measurements by about 10°. This occurs from the start of expansion, so it is possibly a result of incorrect phasing of the pressure fluctuations at the inlet and not as much in the expansion itself.

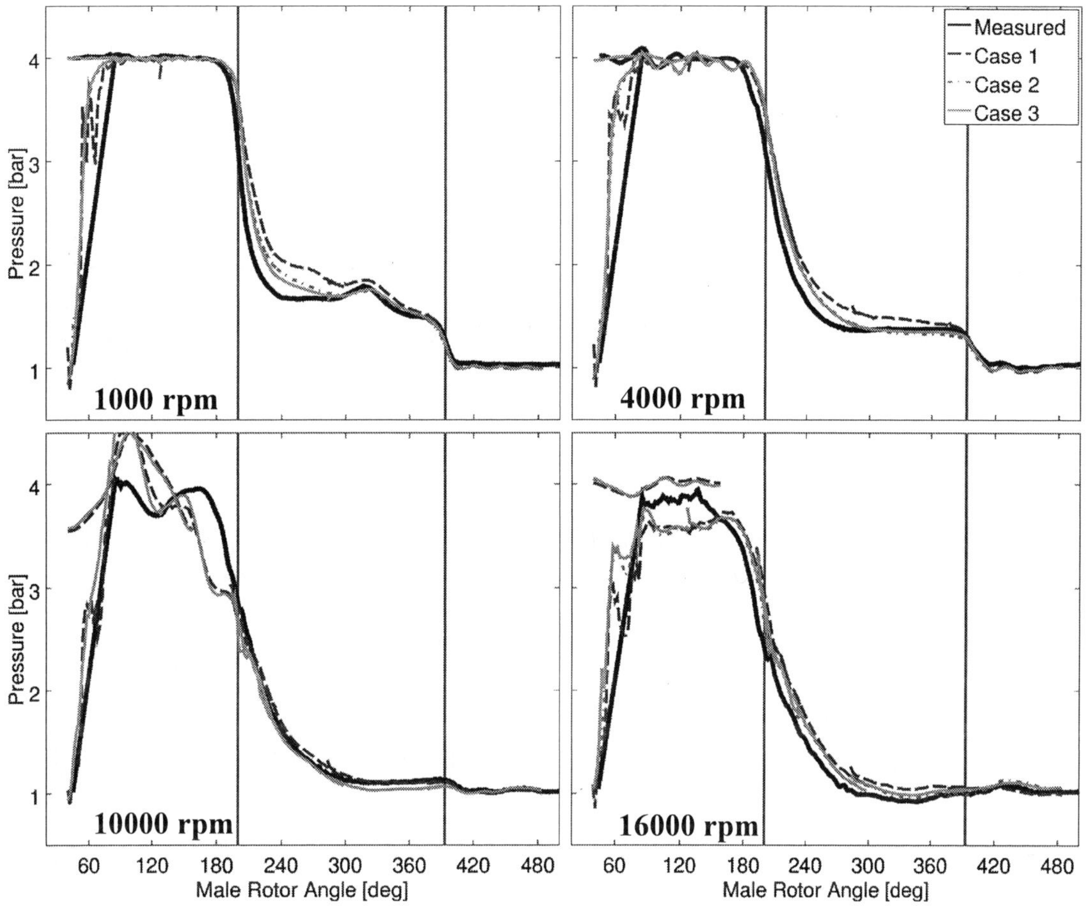

Figure 7. Internal chamber pressures for measurements (black solid lines), and CFD models for cases 1 (dark blue dashed line), 2 (medium green dashed-dotted lines), and 3 (thin light orange lines) for male rotor speeds of (from top-left) 1000, 4000, 10000, and 16000 rpm.

4.3. Effects of rotor-tip-to-housing clearances

In the final investigation, rotor-to-housing clearances of the base case geometry at 0.08 mm are compared to clearances of 0.16 mm and 0.24 mm using newly acquired experimental measurements. In the experiments, the rotor bores are increased so that the rotor-to-housing clearances are increased equally around the bore. For the CFD model, the configuration equivalent to Case 3 (with both higher grid resolution, hot clearances, modeled clearance flow, and bearing leakage paths) is assessed for the two new clearances.

The mass flow rate and indicated power for the measurements and CFD models are shown in figure 8. As the clearances increase, the measured flow rate increases nearly proportionally to the clearance for most of the points, especially at lower rotor speeds. For the higher speeds, there is slightly more variation possibly due to the varying throttling and acoustic effects in the inlet port. The CFD models yield flow rates with very similar trends, that is, increase in flow rate proportionally to the clearance. At low rotor speeds of 1000 and 4000 the results coincide very well for all clearances. At 10000 rpm, the CFD model prediction is significantly lower than the measured values, which will be evaluated and explained after examining the internal chamber pressure. At 16000 rpm, there is a little more deviation in the opposite direction, with the CFD model over-predicting the flow rate.

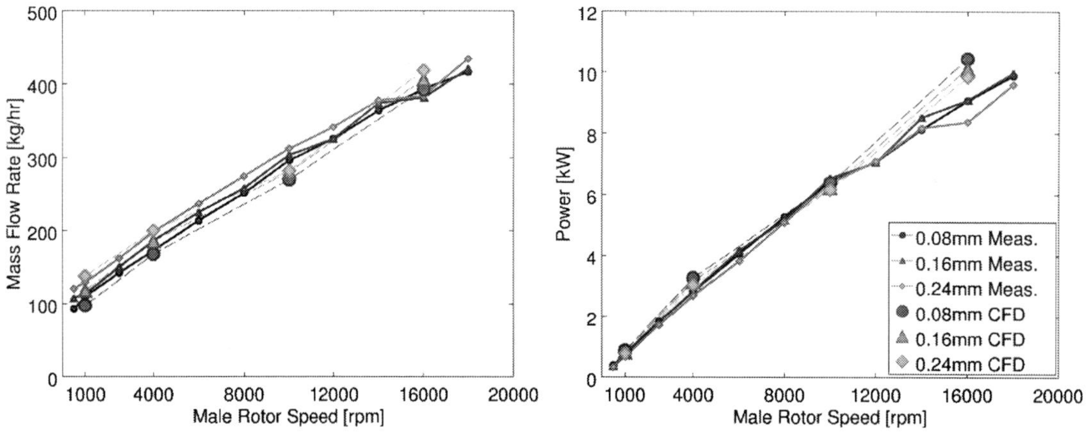

Figure 8. Effect of rotor-tip-to-housing clearances on global measurements and model results. To the left is the mass flow rate and to the right is the indicated power, both as functions of male rotor speed. The measurements are shown in the small gray symbols for clearances of 0.08 mm (dark circles), 0.16 mm (triangles), and 0.24 mm (light diamonds). The model results are shown in larger colored symbols of 0.08 mm (blue circles), 0.16 mm (green triangles), and 0.24 mm (orange diamonds).

The comparison between the measured and modeled indicated power is also shown in figure 8. In the measurements, there is a very slight decrease in the indicated power as the clearances increase. This physically represents the power lost from the excess leakage mass. In the CFD models, the difference is slightly exaggerated. At all the different speeds there is a decrease in indicated power as the clearance increases. For all points except 10000 rpm, the models yield a slightly higher indicated power. More insight on the power generation will be extracted from the internal chamber pressure comparison in figure 9.

Figure 9 shows the measured and computed chamber pressure for all four rotor speeds. At the lowest speeds, 1000 and 4000 rpm, the effects of increasing the clearance are strongest. As the expansion process begins (when the inlet port control edges are closed around 200°), the pressure inside the chamber decreases more rapidly with larger clearances. This flow is presumably directed mainly at the adjacent low pressure chamber, so there is some of the decrease in pressure which is not due to expansion itself but instead inter-chamber leakage. Very similarly, around 270° the pressure in the chamber can be seen to increase more (though it is still lower) when the clearance is large due to the chamber receiving more mass from the adjacent high pressure chamber.

At the higher speeds of 10000 and 16000 rpm, there is considerably less difference in the internal pressure due to the leakage being less important at high speeds. There is, however, considerably more effect of the acoustic waves and the throttling. For the 10000 rpm case, which had an unexpectedly low mass flow rate and indicated power across all clearances, the reason is more obvious when examining the chamber pressure. A large difference in the pressure behavior in the inlet port is the reason, as was observed in the previous cases at 10000 rpm. At 16000 rpm, the differences in inlet port pressure are not so extreme, however there is still a large decrease in pressure observed between the inlet port and the rest of chamber during filling which is due to the throttled flow.

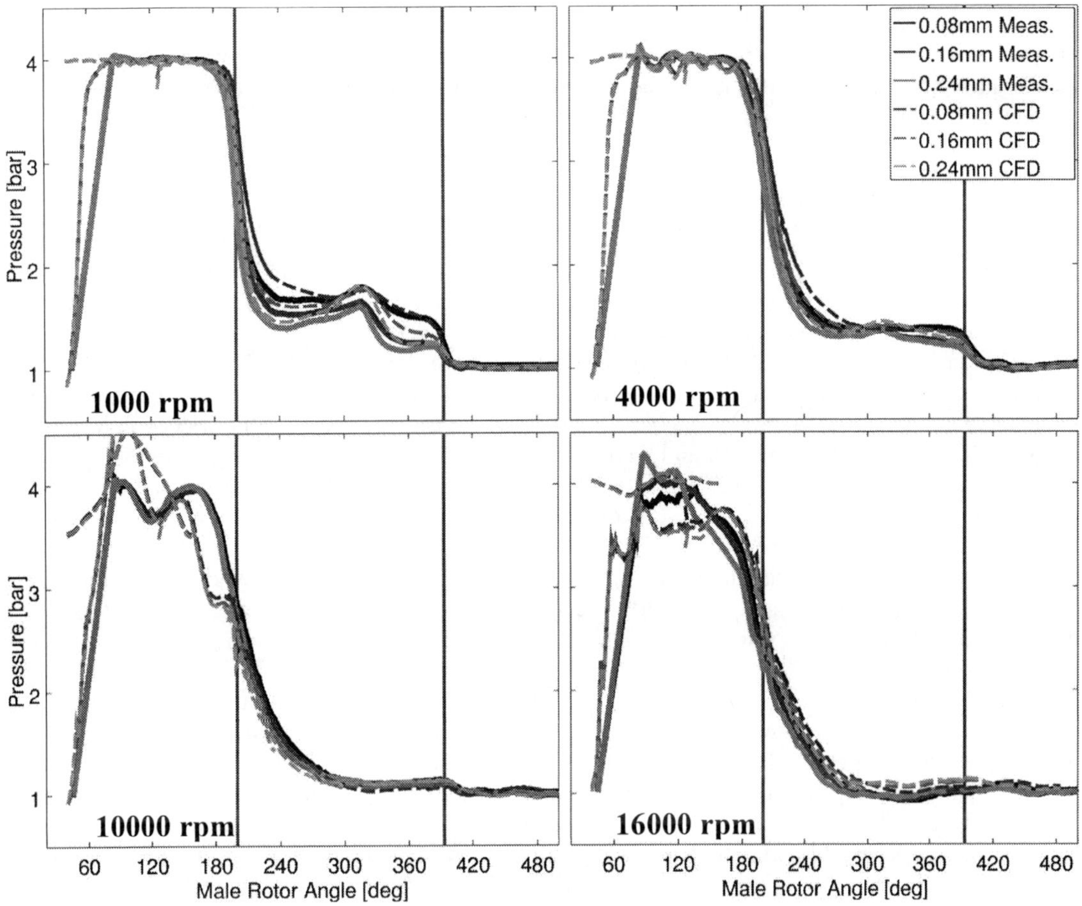

Figure 9. Effect of rotor-to-housing clearances on measured and model internal chamber pressures. Solid thick gray lines are from measurements at clearances of 0.08 mm (dark), 0.16 mm (medium), and 0.24 mm (light). Dashed colored lines are from the model at clearances of 0.08 mm (dark blue), 0.16 mm (medium green), and 0.24 (light orange). Male rotor speed differs between each figure at (from top left) 1000, 4000, 10000, and 16000 rpm.

5. Conclusions and future work

This study has several important findings relevant to the topic of modeling flows in screw machines. The first of these is the validity of the cut-cell based method for CFD modeling. Key advantages of this approach include general applicability regardless of geometry or motion types, and relatively easy process of beginning a simulation from CAD without the need to create a volume mesh. The method conserves volume and mass, and it utilizes stationary orthogonal hexahedral grid for low numerical diffusion, high accuracy, and high stability. A key limitation to this approach relevant to this application is the need to model the clearance flow in order to keep the computational cost manageable. This issue was addressed in comparing the under-resolved flow to cases in which a simplified model developed in another work was utilized.

The use of a semi-transient approach called super-cycling is shown to solve a problem of the disparate time scales between the solid and the fluid without the need to artificially change the solid or fluid thermal properties or conduct multiple simulations. Approximately 10 revolutions of the male rotor were required to arrive at a steady state solution in the solid, after which the flow and solid could be continued either with super-cycling or in a purely transient, time-accurate way. The thermal solutions were validated against measurements of the temperature and found to be in very good agreement for the highest speed, but the agreement deviated more as the speed was reduced. Current work is ongoing to determine if this is a result of differences in the flow rate predictions, interaction with the leakage flow, heat transfer model applied through thermal boundary layer profiles, or an artifact of not actually reaching a steady state condition.

The model behavior is compared to experimental results over a wide range of rotor speeds and clearances. The model results which considered the thermal deformation, clearance flow model, and bearing leakage demonstrated the best overall agreement with measurements. As was observed early on in Case 1, one must be careful to determine the model validity from mass flow rate and indicated power alone. The internal chamber pressure measurements conducted at TU Dortmund University are very valuable in determining whether or not the model is correct for the right reasons. There is still more improvement to be made in the prediction of the internal chamber pressures for two main reason: (1) the clearance flow model must be generalized and evaluated specifically for the operating conditions evaluated here, and (2) the pressure reflections at the inflow boundary must be evaluated and alternative boundary conditions, including a Navier-Stokes characteristic boundary condition with controlled degree of impedance at the inlet, should be investigated.

The measurements with varying rotor-to-housing clearances are very helpful in understanding the model behavior. It is clear from the results of the low speed cases that there is still more work required to predict accurately the internal chamber pressures and leakage flows, though the current results show levels of agreement which are roughly similar or less error than other reported works on the same case. More work is currently required to quantify precisely the leakage flow rates into distinct categories of blowhole leakage, axial rotor-to-housing leakage, radial rotor-to-housing leakage, and rotor-to-rotor leakage.

One additional improvement necessary for the solver in these cases is to improve the computationally efficiency. While these runs were not necessarily optimized for fastest turn-around time, there is much room for improvement here. Especially at the low rotational speeds (the run time scales almost inversely proportionally to rotor speed), the run times could become prohibitively long for application in an engineering context (run times at 1000 rpm were generally around 2 weeks total turn-around time running in parallel on 32 cores; the goal is typically to have a simulation result in under 15 hours, a target which is currently achieved only in the 16000 rpm cases). Current work is ongoing in several areas, including tighter coupling to allow running with larger time steps and higher parallel efficiency related to moving geometries and surface storage.

References

[1] Smith I K, Stosic N and Kovacevic A 2014 *Power recovery from low-grade heat by means of screw expanders* (Chandos Publishing)

[2] Stosic N, Kovacevic A, Hanjalic K and Milutinovic L 1988 Mathematical Modelling of the Oil Influence Upon the Working Cycle of Screw Compressors *Proc. of the International Compressor Engineering Conference at Purdue* (Paper 645) West Lafayette, USA: Purdue University

[3] Kauder K and Rau B 1994 *Auslegungsverfahren fur Schraubenkompressoren, Design Procedure for Screw Compressors* (Düsseldorf: VDI-Verlag)

[4] Hanjalic K and Stosic N 1997 Development and Optimization of Screw Machines with a Simulation Model, Part II: Thermodynamic Performance Simulation and Design Optimization. *Transactions of the ASME* 664(119)

[5] Janicki M 2007 *Modellierung. und Simulation von Rotationsverdrängermaschinen.* Dissertation, TU Dortmund University, Dortmund, Germany

[6] Feng C, Yueyuan P, Ziwen X and Pengcheng S 2001 Thermodynamic Performance Simulation of a Twin-Screw Multiphase Pump *Proc. of the Institution of Mechanical Engineers, Part E: Journal of Process Mechanical Engineering* 215(2) 157-163

[7] Nikolov A, Huck C and Brümmer A 2012 Influence of Thermal Deformation on the Characteristic Diagram of a Screw Expander in Automotive Application of Exhaust Heat Recovery *Proc. of*

the 22rd International Compressor Engineering Conference at Purdue (Paper 2128) West Lafayette, USA: Purdue University

[8] Gräßer M and Brümmer A 2014 An Analytical Model of the Incompressible One-Phase Clearance Flow in Liquid Injected Screw Expanders *Proc. of the International Conference on Screw Machines* Dortmund, Germany: TU Dortmund

[9] Kovacevic A 2002 *Three-Dimensional Numerical Analysis for Flow Prediction in Positive Displacement Screw Machines* Ph.D. Thesis, School of Engineering and Mathematical Sciences, City University London, UK

[10] Voorde Vande J, Vierendeels J, Dick E 2005 Development of a Laplacian-Based Mesh Generator for ALE Calculations in Rotary Volumetric Pumps and Compressors *Comput. Methods Appl. Mech. Eng.*, 193, 4401–4415

[11] Kovacevic A, Rane S, Stosic N, Jiang Y, Furmanczyk M and Lowry S 2014 Influence of Approaches in CFD Solvers on Performance Prediction in Screw Compressors *Proc. of the 22rd International Compressor Engineering Conference at Purdue* (Paper 2252) West Lafayette, USA: Purdue University

[12] Kovacevic A and Rane S 2013 3D CFD Analysis of a Twin Screw Expander *Proc. of the 8th International Conference on Compressors and Their Systems* London, UK: City University London

[13] Papes I, Degroote J and Vierendeels J 2015 Numerical Simulation of a Twin Screw Expander for Performance Prediction *Proc. of the 9th International Conference on Compressors and their Systems* London, UK: City University London

[14] Rainer A, Hesse J, Babic H, Salecker U, Spille-Kohoff A, Nikolov A and Brümmer A 2016 CFD Simulation of a Twin Screw Expander including Leakage Flows *Proc. of the 23rd International Compressor Engineering Conference at Purdue* (Paper 2497) West Lafayette, USA: Purdue University

[15] Rowinski D H, Li Y and Bansal K 2018 Investigations of Automatic Meshing in Modeling a Dry Twin Screw Compressor *Proc. of the 24th International Compressor Engineering Conference at Purdue* West Lafayette, USA: Purdue University

[16] Senecal P K, Pomraning E, Richards K J, Briggs T E, Choi C Y, McDavid R M, Patterson M A, Hou S and Shethaji T 2007 A New Parallel Cut-Cell Cartesian CFD Code for Rapid Grid Generation Applied to In-Cylinder Diesel Engine Simulations *SAE Technical Paper* #2007-01-0159

[17] Pomraning E, Richards K J and Senecal P K 2014 Modeling Turbulent Combustion Using a RANS Model, Detailed Chemistry, and Adaptive Mesh Refinement *SAE Technical Paper* 2014-01-1116

[18] Yakhot V and Orszag S A 1986 Renormalization Group Analysis of Turbulence. I. Basic Theory. *J. Sci. Comput.*, 1, 3

[19] Issa R I 1985 Solution of the Implicitly Discretised Fluid Flow Equations by Operator-Splitting *Journal of Computational Physics*, 62

[20] Rhie C M and Chow W L 1983 Numerical Study of the Turbulent Flow Past an Airfoil with Trailing Edge Separation *AIAA J.*, 21, 1525-1532

[21] Hütker J and Brümmer A 2013 Physics of a Dry Running Unsynchronized Twin Screw Expander *Proc. of the 9th International Conference on Compressors and their Systems* London, UK: City University London

[22] Nikolov A and Brümmer A 2014 Influence of Water Injection on the Operating Behaviour of Screw Expanders: Experimental Investigation *Proc. of the International Conference on Screw Machines* Dortmund, Germany: TU Dortmund

[23] Nikolov A and Brümmer A 2016 Analysis of Indicator Diagrams of a Water Injected Twin-shaft Screw-type Expander *Proc. of the 23rd International Compressor Engineering Conference at Purdue* (Paper 2492) West Lafayette, USA: Purdue University

IOP Conf. Series: Materials Science and Engineering **425** (2018) 012020 doi:10.1088/1757-899X/425/1/012020

A Similarity based Efficiency Model of Spindle Screw Pumps

C Schänzle, T Corneli, P F Pelz

Technische Universität Darmstadt, Department of Mechanical Engineering, Chair of Fluid Systems, Otto-Berndt-Str. 2, 64287, Darmstadt, Germany

peter.pelz@fst.tu-darmstadt.de

Abstract. Spindle screw pumps are used in numerous different applications and industrial sectors. When selecting a suitable spindle screw pump for a specified application, manufacturers are often confronted with a lack of comparable measurement data for the desired combination of operating conditions and pumping fluid. Consequently, the volumetric and mechanical-hydraulic efficiency of a pump under the operating conditions of the application need to be estimated. In this context, this paper discusses the application of similarity for three-spindle screw pumps and introduces a similarity based efficiency model. The model is validated by means of a measured pump characteristic at different operating conditions varying pressure, rotational speed and the viscosity of the pumping fluid. The validation results prove that a pump characteristic can be represented over a wide operating range based on similarity. An estimation of the volumetric and the mechanical-hydraulic efficiency at a changed viscosity is achieved with high accuracy. Furthermore, a new approach to monitor and to evaluate manufacturing uncertainty based on the model parameter relative gap is presented. Finally, the paper gives an outlook on future experimental investigations at TU Darmstadt on model series of three-spindle screw pumps containing pumps of different displacement volume.

1. Introduction

Spindle screw pumps are characterized by their wide operating range, low-pulsation performance and high volume specific delivery capacity. This motivates their utilization in numerous different applications and industrial sectors, e.g. coolant supply in machine tools, fuel injection in marine diesel engines or lubricant supply in technical facilities.

When selecting a spindle screw pump for a specified application of a customer, manufacturers are often confronted with a lack of comparable measurement data for the desired combination of operating conditions and pumping fluid. Especially varying viscosities of the pumping fluid immensely affects the efficiency of spindle screw pumps. However, highly precise calculations based on numerical simulations or experimental tests usually are too costly and time consuming and are therefore neglected. This makes it necessary to estimate the volumetric efficiency and mechanical-hydraulic efficiency of a pump under the operating conditions of the application. A useful estimation needs to be easy to apply, physically based and sufficiently accurate. Furthermore a reduction of the complexity is advantageous and requires to focus on the major influence parameters.

Against this background, similarity of spindle screw pumps promises to be a simple and effective approach to estimate the volumetric and mechanical-hydraulic efficiency. On this basis, the authors introduced a dimensionless and type-independent efficiency model of positive displacement pumps that meet the above conditions [1]. Furthermore they discuss the application of this model in the context of legislative efficiency regulations [2] and they apply similarity on external gear pumps with promising results [3]. The model considers the operating conditions, i.e. pressure difference Δp and rotational speed n, the fluid properties, i.e. kinematic viscosity ν, density ϱ and compressibility κ, and the machine parameters, i.e. displacement volume V and average gap height $\bar{s}$.

As a result, four dimensionless variables are identified that describe the efficiency behavior: (i) specific pressure Δp^+, (ii) Reynolds number Re, (iii) specific compressibility κ^+, and (iv) relative gap ψ:

$$\Delta p^+ := \frac{\Delta p}{\nu^2 \varrho V^{-2/3}}, \qquad Re := \frac{nV^{2/3}}{\nu}, \qquad \kappa^+ := \kappa \Delta p, \qquad \psi := \frac{\bar{s}}{V^{1/3}}. \tag{1}$$

A fundamental basis of the previous work [1,2,3] were experimental investigations on spindle screw pumps conducted by Corneli et al. [4] that have not been published until now. These experimental investigations on identical screw pumps with modified gaps are used to validate the efficiency model of Pelz et al. [1] in terms of spindle screw pumps and the results are presented in this paper. Furthermore, the approach of another study that will be carried out in the second half of 2018 is introduced. This future study will focus on model series of spindle screw pumps and will examine the scalability of spindle screw pump characteristics based on the displacement volume.

This paper is structured as follows: The paper begins with a brief literature overview on efficiency models of positive displacement pumps. In the second section, similarity of spindle screw pumps is discussed. Subsequently, the experimental setup of Corneli et al. [4] that was used for the experimental investigations on identical spindle screw pumps with modified gaps is described. In the 'result' section, three research questions are discussed based on the experimental investigations of Corneli et al. [4]: (i) How accurate can a spindle screw pump characteristic be represented based on similarity? (ii) How accurate is an estimation of the volumetric and mechanical-hydraulic efficiency based on similarity? (iii) Is a representation of manufacturing tolerances based on the model parameter relative gap ψ useful? The third research question is discussed on the dimensionless pump characteristics of identical screw pumps with modified gaps. The paper closes with conclusions and an outlook on future work.

2. Literature

The literature gives numerous studies on the efficiency of positive displacement pumps. The most comprehensive overviews give Ivantsyn and Ivantysynova [5] and Kohmäscher et. al [6]. The authors of this paper already summarized these overviews in [1,33]. Nonetheless, for reasons of completeness and readability, this paper also gives this brief summary of the most important studies on efficiency of positive displacement pumps.

Ivantsyn and Ivantysynova [5] and Kohmäscher et. al [6] differentiate efficiency models into physical, empirical and data-driven models.

The physical models describe the volumetric and mechanical-hydraulic losses in positive displacement pumps. On this basis, Wilson [7] developed an efficiency model in the 1940s that assumes the leakage to be a laminar flow and describes the mechanical-hydraulic losses as dependent on viscous friction only. Schlösser and Hilbrands [8,9,10,11] extend the leakage model with a turbulent flow component. The mechanical-hydraulic losses are complemented with a pressure-related and an inertia-related loss. Thoma [12] und Bravendik [13] further develop these models focusing on pumps with adjustable displacement volumes.

All these physical models rely on dimensionless loss coefficients, comparable with pressure loss coefficients. These loss coefficients need to be determined empirically and are assumed to be constant. However, Zarotti and Nervegna [14], Rydberg [15] as well as McCandlish and Dorey [16] conclude from their studies that this assumption can be violated, e.g. by changing gap height due to varying operating conditions. Consequently, this may lead to inaccuracy and model uncertainty. That is why they complete the physical description with empirical formulas based on experimental findings and increase the complexity of the models according to Ivantsyn und Ivantysynova [5].

Following Kohmäscher et. al. [6], other model based approaches are the numerical or data-driven models of Ivantsyn and Ivantysynova [5], Huhtala [17] and Baum [18]. These models require a large amount of measurement data that is approximated by different numerical methods, e.g. non-linear polynomial approximation or neural networks.

In summary, a lot of research on the efficiency of positive displacement pumps was carried out in the past decades, leading to more complex and detailed models.

However, regarding the requirements for an estimation, i.e. being easy to apply, physically based and sufficiently accurate, physical models are still of high value. In this context, Pelz et. al. [1] introduces a semi-analytical efficiency model based on dimensional analysis [19] that reduces the complexity due to a variety of different gap geometries. In the following sections the similarity of spindle screw pumps is discussed and the experimental validation is presented.

3. Similarity of three-spindle screw pumps

This section introduces the physical model approach and the application of similarity on spindle screw pumps. The authors validate this approach for external gear pumps based on experimental data and prove its beneficial use [3].

Of major importance is the isentropic efficiency η that represents a measure of the energetic quality of a pump. Based on the first law of thermodynamics, for a time averaged stationary and thermally isolated machine the isentropic efficiency is defined as the hydraulic power divided by the shaft power P_S. The hydraulic power is obtained as the product of the volume flow Q_1 at the inlet of the pump, the discharge pressure Δp and a correction factor that depends on the compressibility κ. Considering the shaft power is the product of the shaft torque M_S and the rotational speed n, one obtains the well-known definition of the efficiency η as (cf. [1])

$$\eta := \frac{Q_1 \Delta p}{2\pi M_S n}\left(1 - \frac{\kappa \Delta p}{2}\right). \tag{2}$$

In the following, the influence of the compressibility is neglected due to low discharge pressures, i.e. $\kappa\Delta p \ll 1$. Consequently, the volume flow rate $Q_1 = Q$ is assumed to be constant for $\kappa\Delta p \ll 1$. Extending equation (2) with the displacement volume V, the efficiency can be written as the product of the volumetric efficiency η_{vol} and the mechanical-hydraulic efficiency η_{mh}

$$\eta = \eta_{vol}\eta_{\mathrm{mh}}, \qquad \eta_{vol} := \frac{Q}{nV}, \qquad \eta_{\mathrm{mh}} := \frac{\Delta p V}{2\pi M_S}. \tag{3}$$

Both partial efficiencies can be represented as a function of the respectively responsible loss: for the volumetric efficiency, this is the leakage Q_L. Taking the theoretical volume flow rate $Q_{th} = nV = Q + Q_L$ into account, the volumetric efficiency η_{vol} can be written as

$$\eta_{vol} := \frac{Q}{nV} = 1 - \frac{Q_L}{nV}. \tag{4}$$

The friction torque M_{mh} represents the mechanical-hydraulic losses. Considering the shaft torque is the sum of the hydraulic torque $M_{hyd} = \Delta p V / 2\pi$ and the friction torque M_{mh}, gives the mechanical-hydraulic efficiency η_{mh}

$$\eta_{mh} := \frac{\Delta p V}{2\pi M_S} = \frac{1}{1 + 2\pi \frac{M_{mh}}{\Delta p V}}. \tag{5}$$

Following the above approach, a description of the losses, i.e. the leakage Q_L and the friction torque M_{mh}, results in a description of the volumetric and mechanical-hydraulic as well as total efficiency.

On this basis, we continue performing a dimensional analysis, the basis of our modeling and of similarity. The procedure is as follows:

firstly, all major influencing variables on the losses are determined. The following six influencing variables are considered: the operational parameters discharge pressure Δp and rotational speed n, the properties of the pumping medium density ϱ and kinematic viscosity ν, and the geometric parameters displacement volume V and average gap height $\bar{s}$ of the pump. The average gap height $\bar{s}$ is a newly introduced size representing an average height all gap heights of the several different gaps of a screw pump. It is motivated by the analogy of a hydrodynamic journal bearing with the average height of the lubrication gap $\bar{h}$, i.e. $\bar{s}$ is interpreted as $\bar{h}$. The characteristic length of the pump is $V^{1/3}$.

Secondly, performing a dimensional analysis reduces the number of model variables and, thus, simplifies the model while maintaining the physical significance [19]. This yields five dimensionless variables characterizing the operating state of a pump: The specific pressure Δp^+, Reynolds number Re, and relative gap size ψ are the independent dimensionless variables and are defined as

$$\Delta p^+ := \frac{\Delta p}{\nu^2 \varrho V^{-2/3}}, \quad Re := \frac{nV^{2/3}}{\nu}, \quad \psi := \frac{\bar{s}}{V^{1/3}}. \tag{6}$$

Furthermore, both the leakage Q_L and the friction torque M_{mh} are also represented by dimensionless variables, the specific leakage Q_L^+ and the specific friction torque M_{mh}^+. These dependent dimensionless variables are the defined as

$$Q_L^+ := \frac{Q_L}{\nu V^{1/3}}, \qquad M_{mh}^+ := \frac{M_{mh}}{\Delta p V}. \tag{7}$$

Finally, the specific leakage $Q_L^+ = Q_L^+(\Delta p^+, Re, \psi)$ and the specific friction torque $M_{mh}^+(\Delta p^+, Re, \psi)$ are functions of the specific pressure, Reynolds number and relative gap, which need to be determined. This leads directly to the descriptions of the volumetric, of the hydraulic-mechanical, and of the total efficiency:

$$\begin{gathered}
\eta_{vol} = 1 - \frac{1}{Re} Q_L^+(\Delta p^+, \psi), \\
\eta_{mh} = \frac{1}{1 + \frac{2\pi}{1 - \kappa \Delta p/2} M_{mh}^+(\Delta p^+, Re, \psi)}, \\
\eta = \frac{1 - \frac{1}{Re} Q_L^+(\Delta p^+, \psi)}{1 + \frac{2\pi}{1 - \kappa \Delta p/2} M_{mh}^+(\Delta p^+, Re, \psi)}.
\end{gathered} \tag{8}$$

On the basis of measurement data provided by pump manufacturers, Pelz et. al. [1] illustrate that the specific leakage Q_L^+ can be approximated by a semi empirical model in terms of a power law.

$$Q_L^+ = L * (\Delta p^+ \psi^3)^m. \tag{9}$$

The dimensionless leakage coefficient L and the exponent m are the only model parameters. The leakage coefficient includes the ratios of leakage specific geometric parameters of the pump, e.g. gap length and gap width, to the characteristic length of the pump $V^{1/3}$. At the same time, it was shown, that the influence of the Reynolds number is negligible for screw pumps. However, the investigations of Schänzle et al. [3] on external gear pumps at low operating pressures show a clear dependency of the specific leakage on the Reynolds number. The influence of the Reynolds number represents the drag flow of the leakage caused by the rotation of gears. In screw pumps the drag flow is negligible in comparison to the pressure driven flow which is represented by the specific pressure.

On the other hand, Schlösser and Hilbrands [10] introduced a physically based approach for the estimation of the friction torque M_{mh} that represents a linear combination of a pressure-related loss, a viscous friction-related loss and inertia-related loss:

$$M_{mh} = C\Delta pV + R_{\mu}\frac{\mu nV}{s} + R_{\varrho}\varrho n^2 V^{5/3}. \tag{10}$$

Applying the dimensionless variables given by the equations (6) and (7) to this approach yields a description of the specific friction torque

$$M_{mh}^{+}(\Delta p^{+}, Re, \psi) = C + R_{\mu}\frac{Re}{\Delta p^{+}\psi} + R_{\varrho}\frac{Re^2}{\Delta p^{+}}. \tag{11}$$

C, R_{μ} and R_{ϱ} are the dimensionless loss coefficients of the different loss terms. These loss coefficient also include the ratios of loss specific geometric parameters to the characteristic length of the pump $V^{1/3}$ (see [8,9,10,11]).

Altogether, there are five model parameters, namely L, m, C, R_{μ} and R_{ϱ}, that need to be identified, i. e. calibrated, using measurement data. The model parameter identification is based on robust linear regression. In summary, the semi-analytical equations (9) and (11) together with equation (8) give the mathematical formulas which are applied on spindle screw pumps in this paper.

Before that, we take a closer look on the prerequisite of a useful application of the concept of similarity and dimensional analysis. This is the presence of geometric similarity being a part of full similarity. For this study and the future study, this is crucial for two reasons:

firstly, the average gap height $\bar{s}$ of a series of identical pumps will always vary from the nominal value due to manufacturing (including assembly) uncertainty. Hence, full geometric similarity in spindle screw pumps is never present. Furthermore, from the practical point of view, it is advantageous to set a detailed geometrical modelling of the several different gaps of a screw pump aside in favor of determining the average gap height by means of the measured pump characteristic. For this purpose, Pelz et al. [1] introduce the relative gap $\psi := \bar{s}/V^{1/3}$ and the class of gap ψ/ψ_{ref} to provide an approach for relative considerations. In this paper we follow this approach and validate it based on experimental investigations of identical screw pumps with modified gaps. On the one hand, the approach allows a characterization of the manufacturing uncertainty and, on the other hand, a comparison of pumps based on one single characteristic quantity.

Secondly, spindle screw pumps need to be investigated in regard to their geometric similarity when aiming to scale the pumps characteristic based on the displacement volume. Typically, within a model series of spindle screw pumps, changes of the displacement volume are realized by varying the main geometry parameters of a spindle. Those are the head circle diameters d_A, the pitch H and the length l. In this context, Thurner [20] presents a pure analytical description of the displacement volume of a three-spindle screw pump as a function of the pitch and the square of the head circle diameter of a spindle:

$$V \propto Hd_A^2. \tag{12}$$

Furthermore, within a model series of spindle screw pumps, the ratios of these above main geometry parameters remain equal and, thus, screw pumps of the same model series can be considered geometrically similar in terms of the main geometry parameters. Regarding the leakage and the viscous-driven losses, a closer look on the gap geometries is also necessary. In this regard, Rossow [21] identifies the length and wide of the gaps with major influence on the above losses which are the tooth sided gaps and the contour gaps [4]. The length and the width of the gaps can also be represented as a function of the pitch and the head circle diameter. From their actual mathematical description follows that the ratios of these geometric parameters remain equal if geometric similarity of the above main geometric parameters is fulfilled. From this point of view, only the ratio of the gap heights is expected to change when varying the displacement volume. As mentioned above, this is due to manufacturing uncertainty. Consequently, within a model series the relative gap ψ is expected to change as well and, hence, the relative gap is identified as the one measure that represents the incomplete geometric similarity of spindle screw pumps. Following this reasoning, the five model parameters, L, m, C, R_μ and R_ϱ are expected to remain equal for screw pumps of one model series that fulfills the above requirements. The validation of this approach will be core of the future study on model series of spindle screw pumps.

4. Experimental setup

The following section presents the experimental setup of Corneli et al [4] of the investigations on identical spindle screw pumps with modified gaps. The validation results are presented in the following section.

The hydraulic circuit diagram of the experimental setup is shown in figure 1. Pressure and temperature are measured at the inlet and outlet of the pump. A piezoresistive sensor and a Pt-100 resistance thermometer are utilized, respectively. A torque meter with built-in speed sensor operating on the strain gage principle measures the shaft torque and rotational speed of the pump. Furthermore, a screw type flow meter is used to measure the volume flow rate at the outlet of the pump. The pressure at the outlet of the pump is adjusted with a throttle valve. The cooling of the pumping fluid is realized using a pump-transfer cooler filtration unit. Table 1 summarizes the measurement range and accuracy of the used measurement equipment.

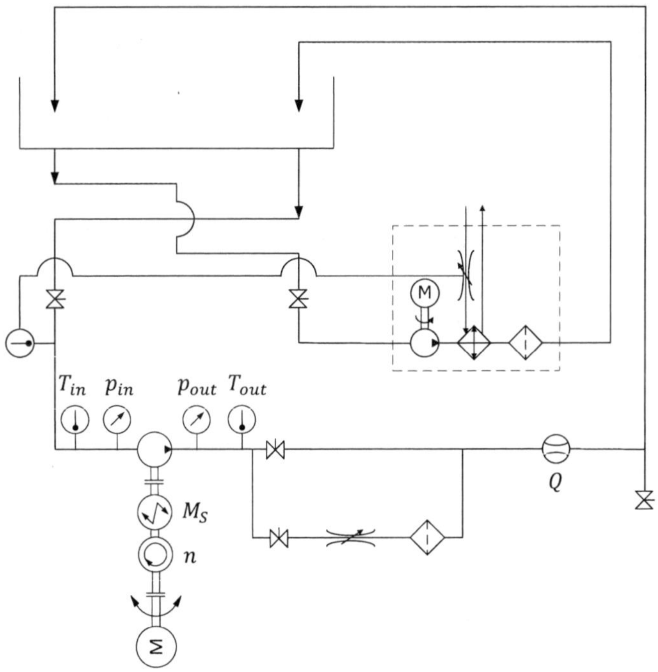

Figure 1: Hydraulic circuit diagram of the test bench.

The pumping fluids utilized in this study were standard hydraulic oils of the viscosity classes 2, 7, 22 and 68. The viscosity-temperature curves and the density-temperature curves of all hydraulic oils were measured with a highly accurate glass capillary viscometer and a density meter.

In total, the investigations of Corneli et al. include pump characteristic measurements of eight three-spindle screw pumps: one original pump without modified gaps and seven pumps with modified gaps. A detailed description of the gaps and their location in the screw pump can be found in [4] (see fig. 4 and 5). However, for reasons of confidentiality a detailed description of the modifications is not given.

All measurements were carried out at rotating speeds between 650 and 1450 rpm, a fluid temperature of 40 °C and discharge pressure between 2 and 28 bar. Table 2 summarizes the modifications of all three-spindle screw pumps.

Table 1: Measurement ranges and uncertainty of measured operation variables (MW= measured value, FS=full scale).

MEASURED VARIABLES	MEASUREMENT RANGE	MEASUREMENT ACCURACY
p_{in}	0 ... 5 bar	0.15 % FS
p_{out}	0 ... 30 bar	0.15 % FS
Q	120 l/min	0.5 % MW
M_S	40/200 Nm	0.2/0.1 % FS
n	12000 rpm	0.1 % MW
T_{in}, T_{out}	0 ... 100 °C	± 0.6 °C

Table 2: Eight three-spindle screw pumps with modified gaps investigated by Corneli et al. [4].

STATE OF MODIFICATION	MODIFICATION	AFFACTED GAPS
original	-	-
1	reduced tip diameter of the choke plunger	Throttle gap - dsc
2	reduced tip diameter of the choke plunger	Throttle gap - isc
3	regrinded tooth side of the drive screw	thooth sided gap - dsc
4	reduced inside diameter of the drive screw	thooth sided gap - dsc, radial gap - dsc
5	reduced tip diameter of the idle screw	radial gap - dsc, contour gap - isc
6	reduced inside diameter of the idle screw	thooth sided gap - dsc, radial gap - isc
7	reduced tip diameter of the drive screw	radial gap - isc, contour gap - dsc

5. Results

As the introduction states, this paper raises three research questions: firstly, how accurate can a spindle screw pump characteristic be represented based on similarity? Secondly, how accurate is an estimation of the volumetric and mechanical-hydraulic efficiency based on similarity? Thirdly, is a representation of manufacturing tolerances based on the model parameter relative gap ψ useful? The following section discusses the first two research questions in detail based on one spindle screw pump measured by Corneli et al. This is a three-spindle screw pump without modified gaps (cf. tab 2.: 'Original'). However, the approach and the results are transferable to the screw pumps with modified gaps as well. The third research question is discussed on the dimensionless pump characteristics of all identical screw pumps with modified gaps. The operating range of all measurements carried out in this study was limited by the maximum pressure of 28 bar and the maximum rotating speed of 1450 rpm. As section 4.1 states, four different hydraulic oils were utilized.

First of all, the model of the leakage flow rate is validated. Equation (9) gives the mathematical relationship between specific leakage and specific pressure by means of a power law. Figure 2 shows the specific leakage versus specific pressure for all measured operating points, i.e. varying the pressure, rotating speed and viscosity, in one double logarithmic diagram. The power law in equation 9 describes the connection between the specific leakage and the specific pressure with good accuracy. The exponent m has a value of 0.72 and the leakage constant L has a value of $10^{-4.7}$. Furthermore, it is evident that the Reynolds number does not affect the specific leakage.

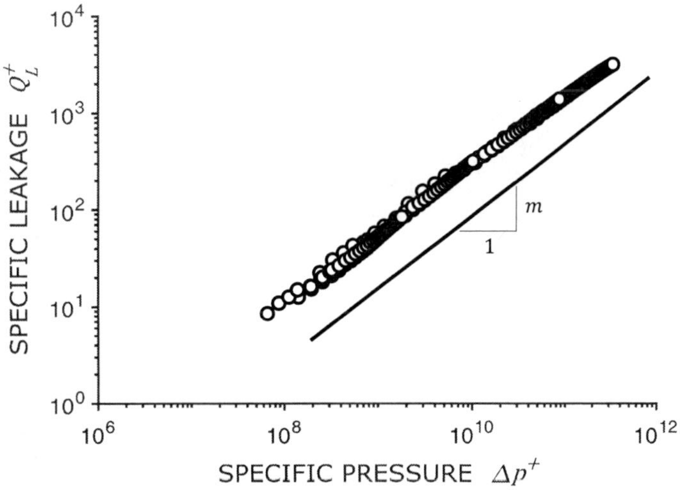

Figure 2: Specific leakage versus specific pressure in a double logarithmic diagram.

In the next step, the accuracy of the leakage model is determined. For this purpose, the relative deviation of the leakage model $\delta(Q_L)$ and the relative deviation of the volumetric efficiency model $\delta(\eta_{vol})$ (see eqn. 8 and 9) are considered. Those relative deviations are defined as follows

$$\delta(Q_L) := \frac{|Q_{L,model} - Q_{L,measurement}|}{\mathrm{m}},$$
$$\delta(\eta_{vol}) := \frac{|\eta_{vol,model} - \eta_{vol,measurement}|}{\eta_{vol,measurement}}. \tag{13}$$

The model parameters L and m are identified using all available measurement points. Hence, the relative deviation is a measure for the accuracy of the leakage model to represent the screw pump characteristic. Figure 3 shows both relative deviations between the measurement points and the model.

The relative deviation of the leakage model is below 10 %, predominantly. Solely for low pressures, the deviation increases. Consequently, the relative deviation of the volumetric efficiency model is very low and mostly below 2 %.

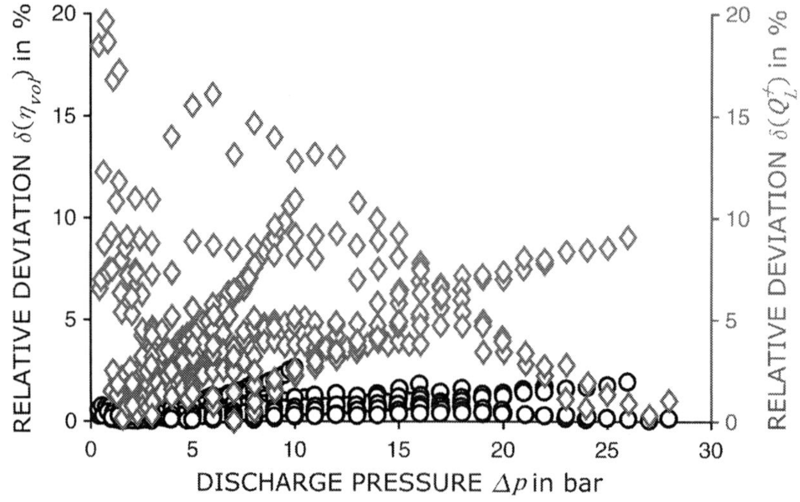

Figure 3: Relative deviation of the leakage model (grey diamond marker) and the volumetric efficiency model (black circle marker) versus the discharge pressure.

The following section presents the validation of the dimensionless friction torque model. The mathematical description is given in equation 11. Similar to the leakage, the model parameters of friction torque model, i.e. the loss coefficients C, R_μ and R_ϱ, are identified using all available measurement points: C has a value of $6.08\text{x}10^{-4}$, R_μ has a value of $2.87\text{x}10^{4}$ and R_ϱ has a value of 6.35

Figure 4 shows a 3-dimenional logarithmic diagram with all measured operation point and the friction torque model surface as a function of the Reynolds number and the specific pressure.

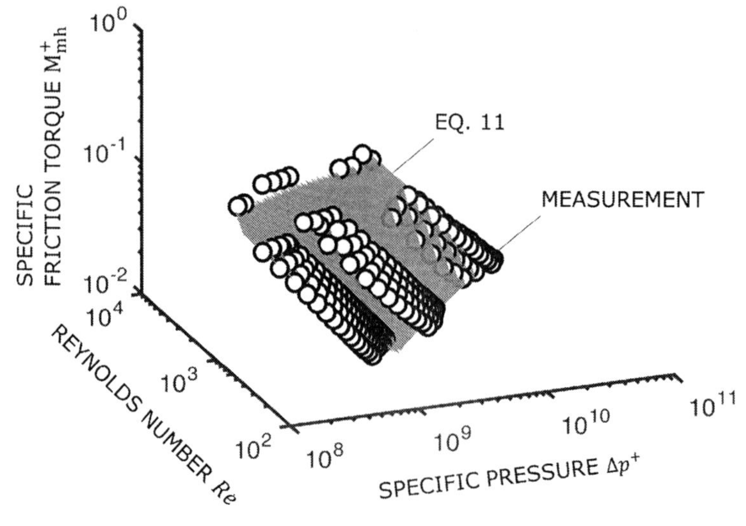

Figure 4: 3-dimensional logarithmic diagram of specific friction torque vs. Reynolds number and specific pressure.

The accuracy of this model description can be determined in the same way as the accuracy of the leakage model. Hence, the relative deviation of the friction torque model $\delta(M_{mh})$ and the relative deviation of the mechanical-hydraulic efficiency model $\delta(\eta_{vol})$ (see eqn. 8 and 11) are defined as

$$\delta(M_{mh}) := \frac{\left|M_{mh,model} - M_{mh,measurement}\right|}{M_{mh,measurement}},$$
$$\delta(\eta_{mh}) := \frac{\left|\eta_{mh,model} - \eta_{mh,measurement}\right|}{\eta_{mh,measurement}}. \quad (14)$$

All model parameters of the friction torque model, namely C, R_μ and R_ϱ are identified using all available operating point excluding the hydraulic oil of viscosity class 2. For operating points with a very low viscosity, the mechanical-hydraulic efficiency is mostly above 96 %, and therefore the friction torque is low and the measurement uncertainty increases. For this reason, these measurement points are neglected in the model parameter identification.

The relative deviation of the friction model is below 15%, predominantly. The relative deviation of the mechanical-hydraulic efficiency is also very low and mostly below 2%. Due to the high mechanical-hydraulic efficiency even high deviation of the loss model have only little influence. In comparison with the leakage model, the friction torque model is less accurate. Nonetheless, the loss models and the corresponding partial efficiency models are able to represent the three-spindle screw pump characteristic in the given operating range.

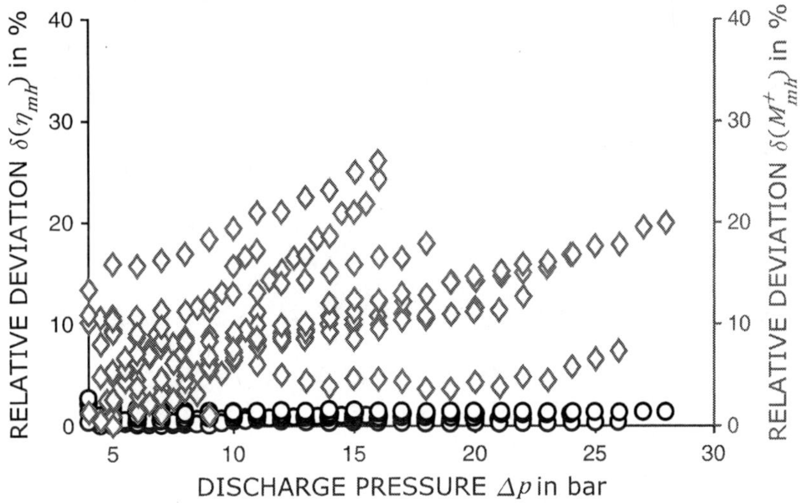

Figure 5: Relative deviation of the friction torque model (grey diamond marker) and the mechanical-hydraulic efficiency model (black circle marker) versus the discharge pressure.

The second research question is relevant for the practical use of similarity. Measuring pump characteristics for more than one or two commonly used pumping fluids is costly for the manufacturers. Therefore, when selecting a suitable spindle screw pump for a specified application, they are often confronted with a lack of comparable measurement data for the desired combination of operating conditions and pumping fluid viscosity. For this reason, the practical benefit of the volumetric efficiency model and the mechanical-hydraulic efficiency model is examined on the basis of the following case study: The pump characteristic measured with a hydraulic oil of viscosity class 7 are used to identify the five model parameters L, m, C, R_μ and R_ϱ. Subsequently, the calibrated models are used to estimate the pump characteristic at a hydraulic oil of viscosity class 22.

Figure 6 shows the estimation of the volumetric efficiency together with the measurements at 22 cSt at four different Reynolds numbers. As can be seen, the estimations match the measurements with high accuracy.

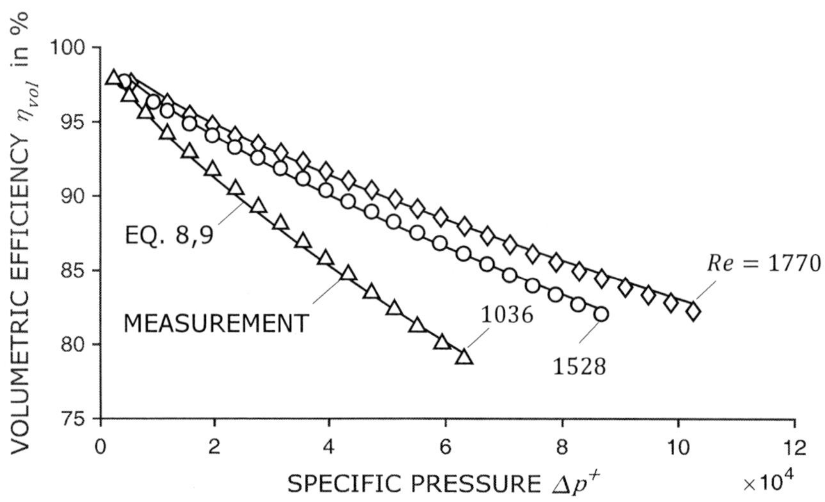

Figure 6: Efficiency model and measurements at 22 cSt.

Figure 7 shows the estimations of the mechanical-hydraulic efficiency together with the measurements at 22 cSt at four different Reynolds numbers. Similar to Figure 6, the estimation of the mechanical-hydraulic efficiency matches the measurements with high accuracy, too.

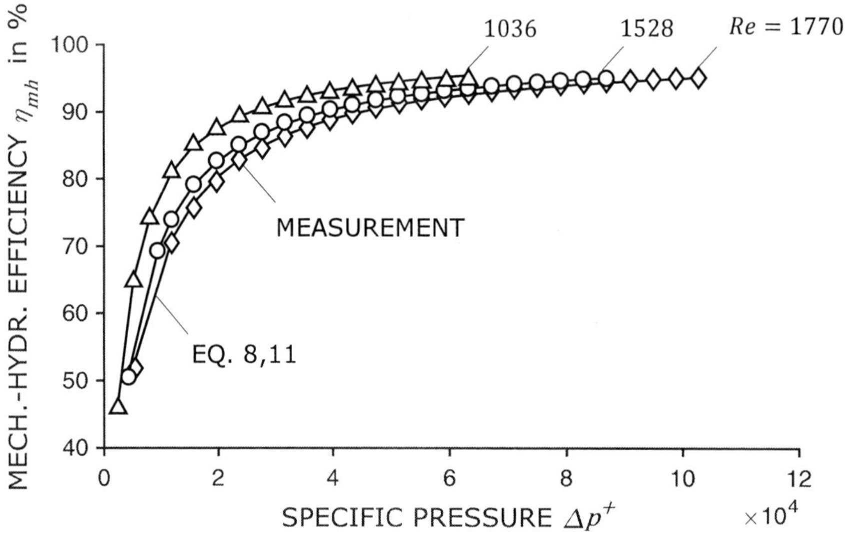

Figure 7: Friction torque model and measurements at 44 cSt.

In summary, the above results prove that the application of similarity and of the presented model is useful and beneficial to calculate the pump characteristics at different viscosities. Following the approach of this paper, the prevalent lack of comparable measurement data for desired combinations of operating conditions and pumping fluid viscosity can be compensated.

The following section discusses the third research question regarding the representation of manufacturing tolerances based on the relative gap ψ. For this reason, the measurements of all identical three-spindle screw pumps with and without modified gaps (see. tab. 2) are evaluated under the following considerations: The pump characteristic of the original three-spindle screw pump serves as a reference with an assigned relative gap $\psi := 1$. The other three-spindle screw pumps with modified gaps represent deviations from this reference due to manufacturing tolerances. For the spindle screw pumps with modified gaps the relative gap is determined on the basis of a relative comparison with the reference pump. This approach was already applied by the authors (cf. [1]) comparing different positive displacement pump types.

However, at this point it is to be emphasized that the study of this paper is a hypothetical one, as the real manufacturing tolerances are actually lower than the gap modification of the pumps [4]. Nonetheless, the approach is transferable to pumps of a real production process.

Against this background, figure 8 shows the double logarithmic diagram of specific leakage versus specific pressure for all measured three-spindle screw pumps. For reasons of clarity, all spindle screw pumps are represented by their model curve (Eq. 9). All model curves of the spindle screw pumps with modified gaps, modification 1 7 (see tab. 2) lie above the original spindle screw pump. The model parameter m is nearly constant for all spindle screw pumps varying between 0.71 and 0.74. Assuming the model parameter m is equal for all spindle screw pumps with $m = 0.72$ (original spindle screw pump), the relative gap ψ for all spindle screw pumps with modified gap is determined on the basis of a relative comparison with the original spindle screw pump. Hence, the relative gap is a measure for the relevance of the corresponding modification. Table 3 lists the relative gap for all three-spindle screw pumps.

For manufacturers, the above approach can be used as follows: In the monitoring process of manufacturing quality newly manufactured pumps can be rated against a referential pump characteristic. Usually a manufacturer proves that each pump delivers the required volume flow rate at the customer's operating point, i.e. discharge pressure and rotating speed, by means of a single measurement. Is a pump's performance significantly worse than expected, the deviation results from the sum of all manufacturing uncertainties of those geometric sizes that influence the gap heights of a pump. A back tracking of a possible exceedance of a manufacturing tolerance of a single geometric size is time consuming and costly and, furthermore, not always effective.

On the other side, measuring the whole characteristic curve of newly manufactured pumps on a random basis additionally, the manufacturing process can be monitored reasonably by only one single parameter: the relative gap ψ of each random sample. The relative gap represents the accumulated manufacturing (including assembly) uncertainty that influence the gap heights in a pump. Consequently, an evaluation of the manufacturing process is possible based on the distribution of the relative gap of all random samples.

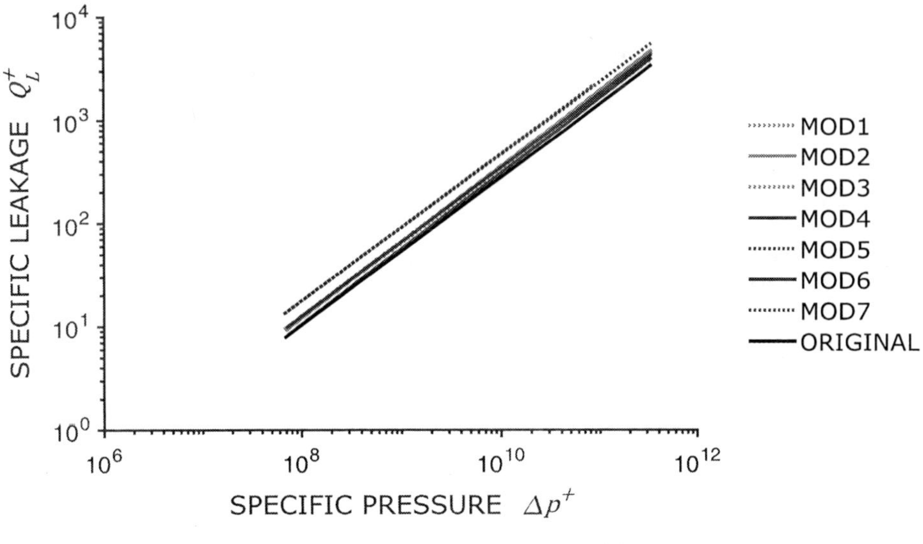

Figure 8: Specific Leakage versus specific pressure in a double logarithmic diagram.

Table 3: Relative gap ψ for all spindle screw pumps.

MODIFICATION	RELATIVE GAP ψ
original	1 (by definition)
1	1.12
2	1.13
3	1.08
4	1.05
5	1.28
6	1.11
7	1.27

6. Conclusion and Outlook

The validation results prove that the presented model approach based on similarity and dimensional analysis are useful to represent a pump's characteristic over a wide operating range. The estimation of the volume flow rate and power consumption at a changed viscosity show high accuracy and are of high value for the manufacturer and operators. Furthermore, this paper presents a new approach to monitor and to evaluate the manufacturing uncertainty on the basis of the model parameter relative gap ψ.

An experimental study on model series of three spindle screw pumps will be carried out in the second half of 2018 at TU Darmstadt. The focus will be on similarity based scaling of spindle screw pump characteristics when varying the displacement volume. These investigations will be carried out within the research project No. 10887/16, funded by budget resources of the Bundesministerium für Wirtschaft und Technologie (BMWi) approved by the Arbeitsgemeinschaft industrieller Forschungsvereinigungen "Otto von Guericke" e.V. (AiF).

Acknowledgments

We would like to thank the Leistritz Pumpen GmbH for funding and their substantial technical contributions to carry out the investigations presented in this paper.

Nomenclature

d_A	head circle diameter of spindle
dsc	drive screw
$\bar{h}$	average height of the lubrication gap
H	pitch of spindle
isc	idle screw
l	length of spindle
L	dimensionless leakage coefficient
m	exponent for power law of specific leakage
M_{hyd}	hydraulic torque
M_{mh}	friction torque
M_{mh}^{+}	specific friction torque
M_S	shaft torque
n	rotational speed
Δp	discharge pressure
Δp^{+}	specific pressure
p_{in}	pressure at inlet of pump
p_{out}	pressure at outlet of pump
P_S	shaft power
Q	volume flow rate
Q_1	volume flow rate at inlet of pump

Q_L	Leakage
Q_L^+	specific leakage
Q_{th}	theoretical volume flow rate
C	pressure-related loss coefficient
R_μ	viscous friction-related loss coefficient
R_ϱ	pressure-related loss coefficient
Re	Reynolds number
$\bar{s}$	Average gap height
V	displacement volume
$\delta()$	relative deviation of a quantity
η	total efficiency
η_{mess}	measured efficiency
η_{mh}	mechanical-hydraulic efficiency
η_{vol}	volumetric efficiency
κ	compressibility
μ	dynamic viscosity
ν	kinematic viscosity
ϱ	Density
ψ	relative gap
ψ/ψ_{ref}	class of gap

References

[1] Pelz P F, Schänzle C and Corneli T 2016 Ähnlichkeits-beziehungen bei Verdrängermaschinen-eine einheitliche Wirkungsgradmodellierung *O+P – Ölhydraulik und Pneumatik 1-2* pp 104-113

[2] Schänzle C, Ludwig G and Pelz P F 2016 ERP Positive Displacement Pumps – Physically Based Approach towards an Application-Related Efficiency Guideline *Proceedings of 3rd International Rotating Equipment Conference* Düsseldorf Germany

[3] Schänzle C, Störmer N. and Pelz P F 2018 Modeling The Efficiency of External Gear Pumps based on Similarity Considerations *Proceedings of ASME Symposium on Fluid Power and Motion Control 2018* University of Bath United Kingdom

[4] Corneli T, Preuß N, Trossmann O and Pelz P F 2014 Experimental studies on the volumetric efficiency of triple screw pumps *International VDI Conference Screw Machines 2014* Dortmund Germany

[5] Ivantysyn J and Ivantysynova M, 1993 Hydrostatische Pumpen und Motoren Vogel Verlag Würzburg Germany

[6] Kohmäscher T, Rahmfeld R, Skirde E and Murrenhoff H 2007 Improved loss modeling of hydrostatic units- requirements for precise simulation of mobile working machine drivelines *Proceedings of International Mechanical Engineering Congress and Exposition* Seattle USA

[7] Wilson W E 1950 Positive displacement pumps and fluid motor Publication Corporation New York USA

[8] Schlösser W M J 1961 Ein mathematisches Modell für Verdrängerpumpen und –motoren *O+P-Zeitschrift 5* pp 122-129

[9] Schlösser W M J and Hilbrands J 1963 Der volumetrische Wirkungsgrad von Verdrängerpumpen *O+P-Zeitschrift 12* pp 469-476

[10] Schlösser W M J and Hilbrands J 1965 Über den hydraulisch-mechanischen Wirkungsgrad von Verdrängerpumpen *O+P-Zeitschrift 9* pp 333-338 1965

[11] Schlösser W M J Über den Gesamtwirkungsgrad von Verdrängerpumpen *O+P-Zeitschrift 12* pp 415-420

[12] Thoma J 1970 Mathematische Modelle und die effektive Leistung hydrostatischer Maschinen und Getriebe *O+P Zeitschrift 6* pp 233-237

[13] Bravendik R 1987 Neue Methode zur Bestimmung der Verluste (Wirkungsgrade) an hydrostatischen Maschinen *O+P Zeitschrift 11* pp 861-866

[14] Zarotti L and Nervegna N 1981 Pump efficiences approximation and modeling *6th International Fluid Power Symposium* Cambridge UK

[15] Rydberg K-E 1983 On performance optimization and digital control of hydrostatic drives for vehicle applications Linköping Studies in Science and Technology Ph.D Thesis No. 99 Linköping Sweden

[16] McCandlish D and Dorey R 1988 The mathematical modelling of hydrostatic pumps and motors *Proceedings of the Institution of Mechanical Engineers Vol 198 Both International Fluid Workshop* University of Bath UK

[17] Huhtala K 1996 Modelling of Hydrostatic Transmission – Steady State, Linear and Non-Linear Models *Mechanical Engineering Series No. 123* Tampere Finland

[18] Baum H 2001 Einsatzpotentiale neuronaler Netze bei der CAE-Tool unterstützten Projektierung fluidtechnischer Antriebe Ph.D Thesis Insitut für fluidtechnische Antriebe und Steuerungen IFAS RWTH Aachen Germany

[19] Simon V, Weigand B and Gomaa, H 2017 Dimensional Analysis for Engineers Springer International Publishing

[20] Thurner J 2013 Last und Lastausgleich zykloidenverzahnter Schraubenpumpen PhD-Thesis Shaker-Verlag Aachen Germany

[21] Rossow P 2014 Development of analytical models to determine the driving torques and leakage flows of triple screw pumps Master Thesis Darmstadt Germany

[22] ISO 4409:2007 2007 Hydraulic fluid power – Positive displacement pumps, motors and integral transmissions – Methods of testing and presenting basic steady state performance Beuth Verlag

Model tests on the control behaviour of a test air supply system in open or closed-loop operation

L de Buhr[1,2], H Fleige[2] and J Seume[1]

[1] Institute of Turbomachinery and Fluid Dynamics, Leibniz Universität Hannover
[2] Aerzener Maschinenfabrik GmbH

hans.fleige@aerzener.de

Abstract. The Leibniz Universität Hannover is currently establishing a new mechanical engineering campus which includes a new research building "Dynamics of Energy Conversion" (DEW). This building provides a large compressor station for either steady or dynamic (transient) operation of turbomachinery and power plant test rigs (e.g. air turbine, axial compressor, combustion chamber, planar cascade, acoustic wind tunnel). The test air supply system is designed to enable investigations under high load gradients over wide operating ranges with Reynolds and Mach numbers controlled independently in order to fulfil aerodynamic similarity conditions between reality and model experiments. This is achieved by closed loop operation of the test air supply system which allows independent adjustment of pressure, temperature and volume flow rate as well as independence from environmental influences such as temperature or humidity. The compressor station utilizes as first stage two parallel Roots-type PD compressors and as second stage two parallel screw compressors. The test rigs operate at expansion ratios between 1 and 6. Test rig inlet pressures range from 1 bar(abs) to 8 bar(abs) with a maximum mass flow rate of 25 kg/s. At all conditions temperatures can be regulated between 60°C and 200°C. The test air supply system has a maximum electric power input of approximately 6 MW. As stringent demands on stability and reproducibility have to be met and automatic operation was requested, a scaled and simplified but fully functional model of the test air supply system was built, mainly to enable testing of control methods and devices prior to their final implementation on site. The functional model uses DN150 piping and consists of one Roots-type PD compressor as first stage and one screw compressors as second stage. Both compressors are driven by electric motors regulated by frequency converters. A turbine test rig is represented in the model by an adjustable throttle valve. Precise control of the mass flow rate is provided by a cascaded adjustable bypass around the test rig. The paper describes the test air supply system and the scaled model and presents experimental results on the achievable stability of pressure, temperature and mass flow rate at the test rig inlet in steady operation at several operating conditions of the model.

1. Introduction

The erection of a new mechanical engineering campus (Campus Maschinenbau Garbsen – CMG) for the Leibniz Universität Hannover situated in Garbsen (north of Hanover) started December 2015. Planned for 2019, the campus will provide capacity for about 4500 students and 800 faculty staff members and will concentrate all 18 institutes of the Faculty of Mechanical Engineering in one location.

A large part of the total investment costs of approximately 145 million Euro is earmarked for the research building “Dynamics of Energy Conversion” (DEW). The building, costing around 42 million Euro, includes a number of test benches for experiments on turbomachinery and power plant components such as engines, generators, turbines, diffusers and compressors.

Research will focus on aerodynamics, aero elasticity and combustion processes in order to better understand and improve the energy conversion processes with the aim to increase efficiency and reduce emissions. Typical investigations are carried out using e.g. a seven stage model turbine [1], a linear cascade wind tunnel [2] or a high-speed axial compressor [3].

Due to the constantly increasing possibilities of simulation methods and the decreasing remaining improvement potential of the components under test, the requirements for test benches are increasing likewise. Today, simulations are used for test pre-prediction which allows for improved positioning and adequate selection of measurement sensors and equipment as well as for optimized overall planning of the experiments. By this, costs are decreased while the quality of the results is enhanced but the requirements for reproducibility, stability and similarity of the experiments increase as well.

As a result, there are stringent requirements for the test air supply system as test air is needed for operation of most of the test benches. Other test benches like the high speed axial compressor that produce their own air flow need to use components of the supply system like e.g. piping, mass flow rate measurement and throttling devices.

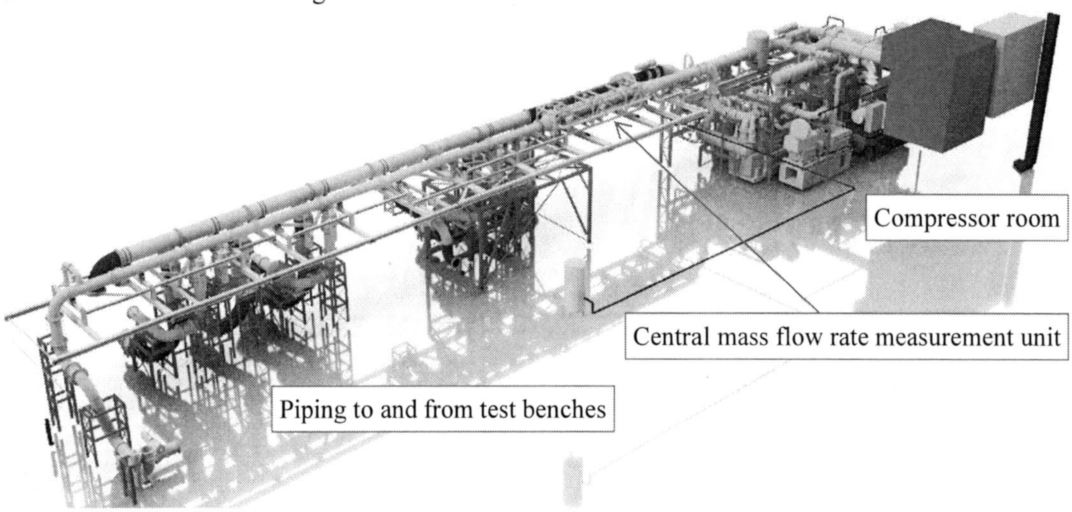

Figure 1. Complete view of the test air supply system (from south).

2. The full-scale test air supply system

Aerzener Maschinenfabrik acquired the order to develop and deliver such a test air supply system. The overall concept foresees a compressor station including temperature control of the test air, a mass flow rate control bypass, a central mass flow rate measurement device, an air distribution system to seven selectable test benches including piping, valves, silencers, coolers etc. and a control system for the selection of different operation modes, -types, -configurations and test bench inlet conditions. The system can be operated in open loop or closed loop mode. The latter allows changing the overall air mass content of the system by either pressurizing or evacuating. Figure 1 depicts a complete view of the test air supply system. The eastern part on the right of the picture shows the compressor room, the green box represents the inlet housing, the red box the outlet housing. In the center, the mass flow rate measurement is situated and in the western part the test air is distributed to the test benches. The overall dimensions are approximately 82 by 15 by 9 m.

The layout of the compressor station shows a symmetrical, modular structure. Two parallel identical lines each comprise a large Roots-type positive displacement compressor (PD blower) as a first and a dry screw compressor as a second stage. All four compressors are driven by separate 690 V electrical motors with separate frequency converters. Due to the modular structure the compressor station is highly flexible, has a large control range (see figure 6) and runs at high efficiencies even at part load conditions. In addition, a certain redundancy is implemented as the second stage of one line can be connected to the first stage of the other line as well. Operation using only screw compressors or only PD blowers is also possible. In these cases the stage which is not needed is bypassed. The compressor room is visible in figure 2. Differently from figure 1, inlet and outlet housings are not shown. The PD blowers are situated on the right and the screw compressors on the left.

Figure 2. Compressor room.

In order to fulfil aerodynamic similarity conditions between reality and model experiments as good as possible, the ability to independently control Reynolds and Mach numbers is necessary. Considering those values that can be influenced by the test air supply system, the Mach number Ma is proportional to the volume flow rate $\dot{V}$ and inversely proportional to the square root of the absolute test air temperature T

$$Ma = \frac{c}{a} = \frac{\dot{V}}{A}(\kappa RT)^{-0.5} \Rightarrow Ma \sim \frac{\dot{V}}{T^{0.5}} . \tag{1}$$

As air always being the test fluid, the material properties κ and R cannot be significantly altered, but ambient air can still slightly change composition due to variations in absolute humidity. This is the reason why for closed loop operation, a special operating mode to dry the air content is foreseen prior to critical test runs (see section 2.2.).

The Reynolds number Re is proportional to volume flow rate $\dot{V}$, pressure p and characteristical length L and inversely proportional to the dynamic viscosity η and the absolute test air temperature T

$$Re = c\frac{L}{\nu} = \left(\frac{\dot{V}}{A}\right) L \left(\frac{\rho}{\eta}\right) = \dot{V}\, L\, p\, (A\, \eta\, R\, T)^{-1} . \tag{2}$$

For moderate pressures (the compressor station can deliver an absolute pressure up to 8.0°bar) the dynamic viscosity of air is only slightly influenced by pressure and proportional to approx. the square root of the absolute temperature [4]. Therefore, from equation (*2*) follows that the Reynolds number is proportional to the volume flow rate, the pressure and the characteristical length and inversely proportional to the absolute test air temperature to the power of 1.5.

$$Re \sim \frac{\dot{V}}{T^{1.5}} p \ . \tag{3}$$

Comparing equations (*1*) and (*3*) shows that the influence of the temperature of the test air delivered by the supply system on the Reynolds number is higher than on the Mach number. Furthermore, pressure can be used to adjust the Reynolds number without a significant change of the Mach number.

2.1. Main components

2.1.1. Positive displacement blowers. The compressor station utilizes as first stage two parallel Roots-type PD compressors GM20.20, each with an intake volume flow rate between 9600 and 48600 m^3/h. They can operate with inlet pressures between 0.2 and 3.5 bar (abs). The maximum outlet pressure is 4.3 bar (abs) and the maximum pressure difference is 0.8 bar. The drive shaft sealing system uses oil lubricated mechanical face seals. The oil system is external. The PD blowers are driven by separate variable speed 690 V electrical motors with a nominal power of 1250 kW each.

2.1.2. Screw compressors. As second stage two parallel screw compressors VRa736S, each with an intake volume flow rate between 6900 and 21600 m^3/h are installed. They can operate with inlet pressures between 0.25 and 3.5 bar (abs). The maximum outlet pressure is 9 bar (abs) and the maximum pressure difference is 10 bar. The rotors are supported by hydrodynamic sleeve bearings. The screw compressors are driven by separate variable speed 690 V electrical motors with a nominal power of 2700 kW each.

2.1.3. Silencers. All four PD blowers and screw compressors are equipped on their high pressure sides with separate silencers in order to deliver pulsation free air to the test benches and to allow for low sound emission levels of the extended pipework outside the compressor room. Special care had to be taken of the blower silencers to be able to suppress the extremely wide frequency spectrum due to their large operating range. Simulation showed that it is necessary to use a combination of two reactive-type silencers and additionally two quarter-wave tubes to attenuate the very low frequencies created by the PD blowers running at minimum speed.

2.1.4. Coolers and temperature control. Between the PD blowers and the screw compressors a common intercooler is situated. All coolers are of the air – water type. Cooling water is centrally provided and re-cooled in a special building on campus.

A common aftercooler including temperature control of the test air is placed downstream of the screw compressors. At all conditions the inlet temperatures to test benches can be regulated between 60°C and 200°C. This is achieved by cooling the main part of the air flow to a temperature slightly lower than required and mixing this portion with a controlled amount of hot air that is bypassing the aftercooler. In those cases when the compressor outlet temperature is lower than the required inlet temperature of the test bench (because the pressure required at the test bench corresponds to a lower temperature) the air is compressed to a higher pressure (and therefore temperature) by using an additional throttling valve upstream of the aftercooler. Due to the maximum absolute pressure of the compressor station of 8 bar the test air can be treated as an ideal gas and therefore throttling is isothermal.

For closed loop operation it is necessary to keep the inlet temperatures to the PD blowers within their allowable limits wherefore a re-cooler is installed in the low pressure line coming from the test benches. The re-cooler is followed by a cyclone separator in order to protect the compressor station and the whole system from oil drops and solid particles, which may come from the test benches in the case of malfunction.

2.1.5. Mass flow control bypass. In order to precisely control the mass flow rate through a test bench, a combination of two measures is used. At first, the rotational speed of one or more compressors or PD blowers is set in such a way that the delivered mass flow rate is slightly higher than required. Second, the bypass upstream of the central mass flow rate measuring device is opened to achieve fine-tuning.

As the PD blowers and especially the screw compressors (equipped with hydrodynamic sleeve bearings) have a minimum allowable rotational speed, the mass flow rates corresponding to these speeds need to be bypassed almost completely for operation points with very low mass flow rates. Therefore, the bypass is cascaded and consists of four parallel valves ranging from DN 150 to DN 10. An additional DN 700 bypass using a fast-opening relief valve is provided for emergency situations to open when the safety shut-off valves up- and downstream of a test bench are closing.

2.1.6. Central mass flow measurement. The central highly precise mass flow rate measurement unit consists of a diffuser that distributes the

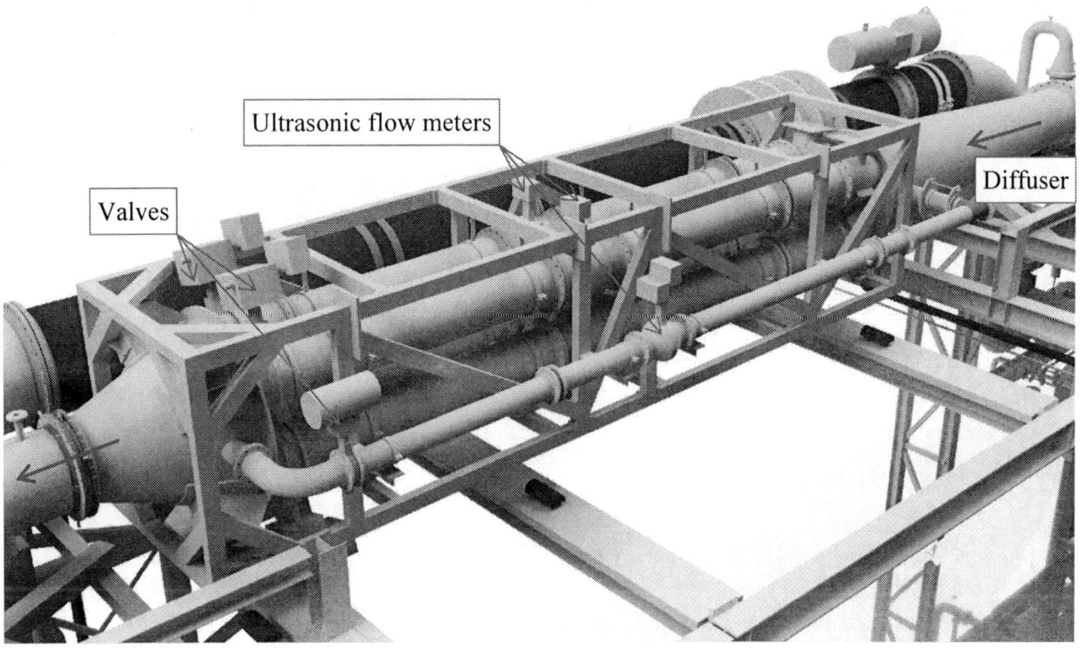

Figure 3. Central mass flow rate measurement unit. Flow from right to left.

flow to five parallel ultrasonic gas meters, four in DN 500 (each with a maximum volume flow rate of 25800 m³/h) and the fifth in DN 200 with a maximum volume flow rate of 4500 m³/h (see figure 3). The number of active flow lines is selected automatically by a control system (using valves) in such a way that all gas meters in use are operated in the turn down regime of lowest uncertainty (10…100 % of maximum volume flow rate). Each line is equipped with a flow straightener of the Etoile-type followed by a fine wire mesh. All ultrasonic gas meters use four measurement paths.

Inside the gas meters the static pressure measurements use high precision sensors with 0.01% IS-50-accuracy. This means that measurement uncertainty is 0.01 % of half of full scale when the measurement value is in the range between 0...50 % of the full scale, and 0.01 % of reading when the measurement value is in the range between 50...100 % of full scale. Temperature is measured using the average value of three precision-type PT100 sensors downstream of the nozzle where the separate lines merge again. The maximum combined uncertainty of the sensors and the 24 bit transmitter occurs at the lowest temperature (60°C) with 0.183 %. A total combined uncertainty of the mass flow rate measurement of 0.55 % can be anticipated including all influencing measurements (averaged squared uncertainty).

In order to achieve an even distribution of the flow into the four DN 500 branches and even velocity profiles at the gas meters, the diffuser including the upstream bends was CFD-simulated. As visible in figure 4 (flow from left to right), the bends can produce significant swirl and hence wall mounted vortex generators at the diffuser inlet and built-in components to reduce the diffuser outlet area were found to be necessary.

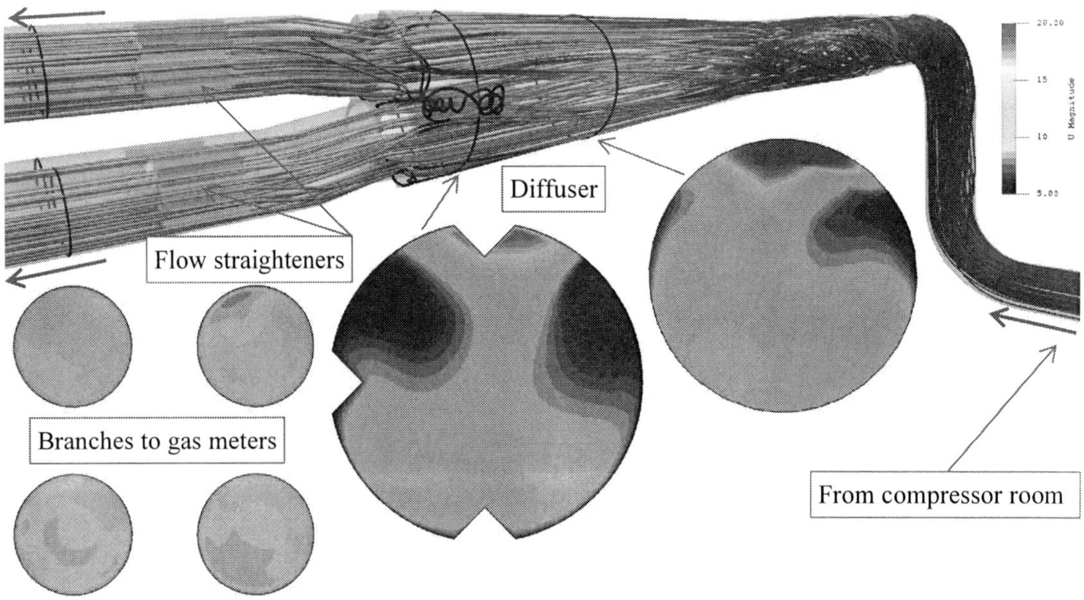

Figure 4. CFD-simulation of the distribution diffuser. Streamlines coloured by velocity magnitude.

For plausibility checks and detection of substantial leakages at the test benches, a second simple mass flow rate measurement device is positioned downstream of the test benches. The thermal flow sensor measures directly a flow velocity at reference conditions. The cross-sectional area and a velocity profile factor allow calculation of the mass flow rate. The profile factor can be derived by comparing with the readings from the central mass flow measurement.

2.1.7. Pipework and settling chambers. Piping between compressor station and test benches are sized in DN 700 on the high pressure side and DN 1000 downstream of the test benches. All pipework is designed for a nominal pressure of 10 bar(g) and implemented with combined sound and heat insulation, which allows reaching shorter times until stable operation with constant temperatures.

For sensitive test benches additional settling chambers are provided. CFD simulations were used to develop the design of the chambers (see figure 5) which incorporate multiple guide vanes in the upstream bends, a flow straightener followed by a wire mesh and an outlet nozzle. To allow future modifications, the settling chambers can be opened and built-in components can be changed.

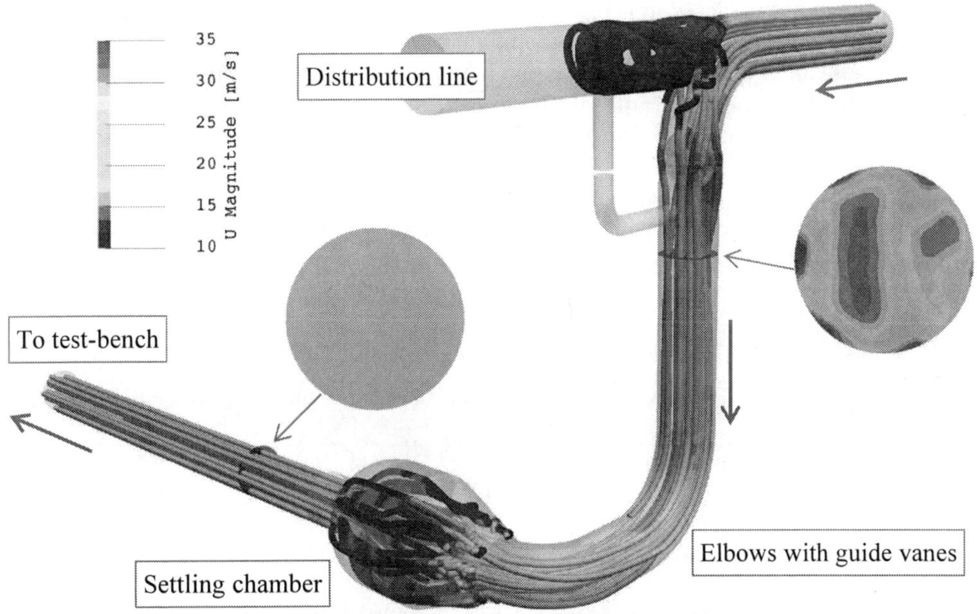

Figure 5. CFD-simulation of a settling chamber. Colours by velocity magnitude.

2.2. *Operation*

In open loop mode the test rigs operate at expansion ratios between 1 and 6. Test rig inlet pressures range from 1 bar (abs) to 6 bar (abs) with a maximum mass flow rate of 22 kg/s (79200 kg/h). In closed loop mode the test rigs operate at expansion ratios between 1 and 4.

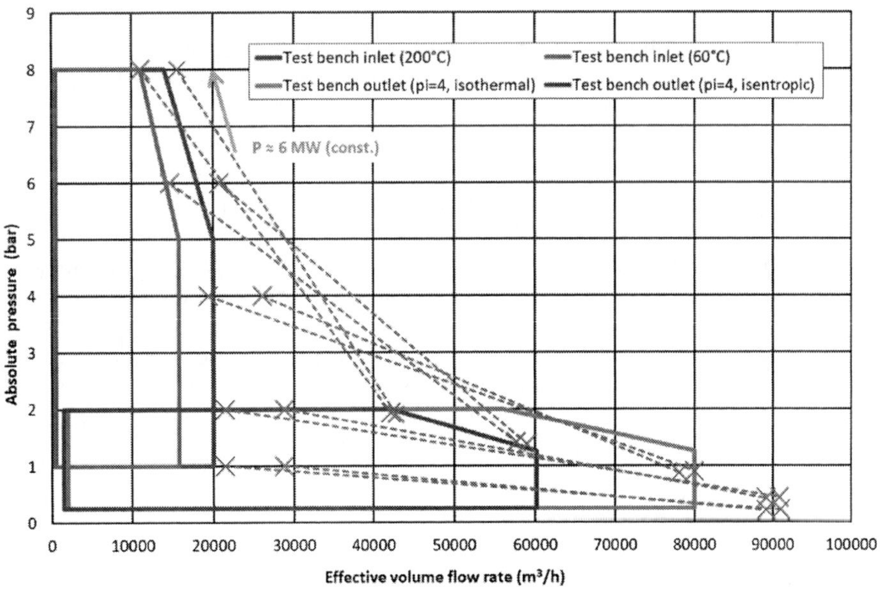

Figure 6. Operational area of the test air supply system.

The inlet pressures in closed loop mode range from 1 bar (abs) to 8 bar (abs) with a maximum mass flow rate of 25 kg/s (90000 kg/h). At all conditions temperatures can be regulated between 60°C and 200°C. The test air supply system has a maximum electric power input of approximately 6 MW. Figure 6 depicts the required operational area of the test air supply system in a p over $\dot{V}$ diagram. The dotted lines represent examples for the change of state over test benches with an expansion ratio of 4.

For the description of the different ways to operate the test air supply system it is helpful to distinguish between different operation modes (open loop or closed loop), operation types (steady or transient) and operation configurations (which of the four compressors are involved). In addition, the operation can be either pressure controlled or mass flow rate controlled. Table 1 lists the aerodynamic inlet conditions that different test benches require.

Table 1. Aerodynamic inlet conditions at different test benches.

Test bench	Pressure [bar]	Temperature [°C]	Max. mass flow rate [kg/h]
Wind tunnel + planar cascade	1…2	60…100	80 000
Combustion chambers	2…6	60	28 000
Turbines open loop (OL)	1…4	60…200	80 000
Turbines closed loop (CL)	1…8	60…200	92 000

If tests with elevated temperatures are necessary directly after starting of the air supply system, the special operating mode “warming up” helps to accelerate the heating-up of pipes and other components. An additional bypass downstream of the test bench with the largest distance from the compressor room is then used to run the system at high pressure and therefore high temperature before a certain test bench is connected.

In closed loop operation, it is reasonable to extend the “warming up” mode into a “drying” mode. In this mode, the system runs at highest outlet pressure (8 bar (abs)) and the aftercooler is adjusted to maximum capacity, so that most of the water content is extracted by condensation.

3. The scaled functional model

As stringent demands on stability and reproducibility have to be met and automatic operation has been requested, it was decided to build a scaled and simplified but fully functional model of the test air supply system. The model enables testing of control methods and devices prior to their final implementation on site which saves commissioning time and provides useful insights into the operational behavior of such a closed system where mass flow rate (or pressure) and temperature have to be precisely controlled over a very wide operational range. All investigations presented in this paper are carried out on the scaled functional model.

Design and construction of the model are described in [5] and depicted in the perspective sketch of figure 7. The main simplification of the model is that only one flow line is implemented which consists of one Roots-type PD compressor GM13.6 (enforced for elevated inlet pressures) as first stage and one screw compressor VM10 as second stage. The GM13.6 provides an intake volume flow rate between 150 and 1400 m³/h. It can operate with inlet pressures between 0.25 and 2.2 bar (abs). The maximum outlet pressure is 3 bar (abs) and the maximum pressure difference is 0.8 bar. The VM10 provides an intake volume flow rate between 122 and 600 m³/h. It can operate with inlet pressures between 0.25 and 3 bar (abs). The maximum outlet pressure is 8.5 bar (abs) and the maximum pressure difference is 6.5 bar. Both compressors are driven by electric motors regulated by frequency converters. Their respective turn-down ratios are limited to the same values as those of the full scale compressors and the ratios of the maximum inlet volume flows of the PD-blowers and the screw compressors are approximately the same in the full scale system (48600/21600 = 2.25) and in the model (1400/600 = 2.33). DN150 piping is mainly used on the low pressure side and DN100 on the high pressure side. The DN200 ultrasonic gas meter requires inlet and outlet tubing of the same

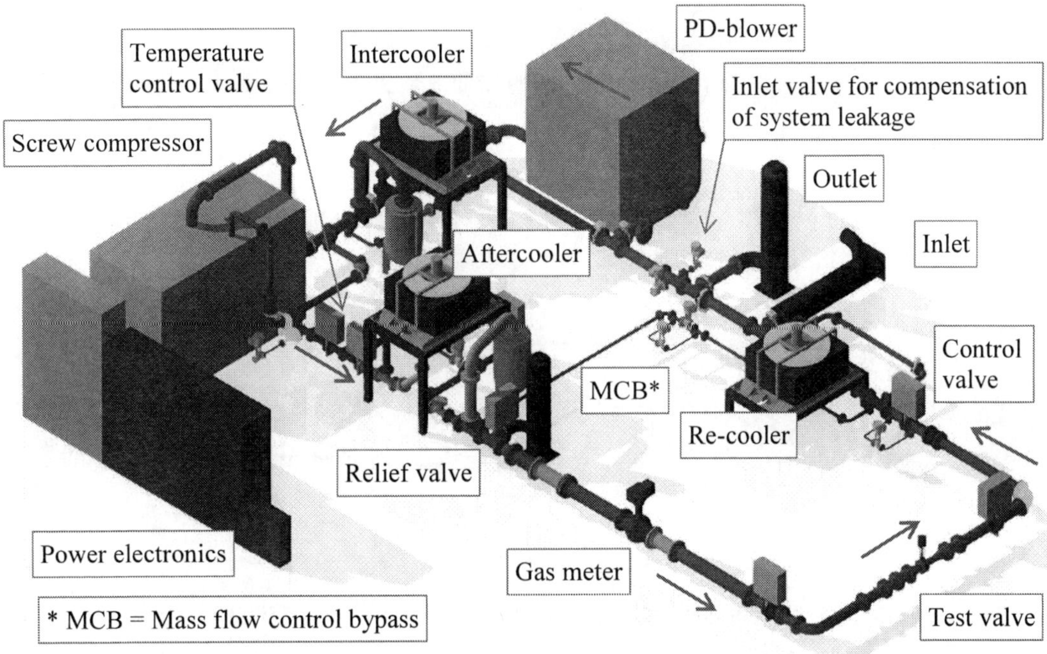

Figure 7. Outline of the functional model.

diameter. As these instruments are quite expensive, it was decided to use only one instead of five parallel meters and the DN200 meter will later on be re-used as the fifth meter in the full scale mass flow rate measurement unit. Only one test rig is present in the model which consists of an adjustable throttle valve representing a turbine. A settling chamber upstream of the test rig is not contained in the model. Precise control of the mass flow rate is provided by a cascaded adjustable bypass around the test rig consisting of only two instead of four parallel valves. In figure 7 the main components of the model are addressed. All coolers are of the air – air type with the possibility to control the cooling capacity via VSD-driven fans and bypass valves.

3.1. Control system

The operational limits of blower and screw compressor are monitored separately by standard compressor controls incorporated in each of the units. Alarms and trips of the compressors and all measurement values from the functional model are reported to a central Programmable Logic Controller (PLC). The PLC is connected with a Human-Machine Interface (HMI) panel to enable manual control and with a process control computer for automatic control. This computer is operated remotely by a separate PC which is using the specialized test stand automation software PAtools. Commands from the HMI or the PAtools software are transferred to the different actuators of the functional model via the PLC.

3.2. Experiments and results

First experiments were carried out manually using the HMI panel and have been successively replaced by programmed sequences and controls. As a starting point, simple proportional–integral–derivative controllers (PID-controllers) as provided by the PAtools software have been implemented. At the present state, only experimental data of steady operation conditions is available but transient tests are scheduled as well. All experiments described below are carried out using both stages. All figures that visualize stability of certain values represent stable operation points that develop some time after a target value of a control process was set.

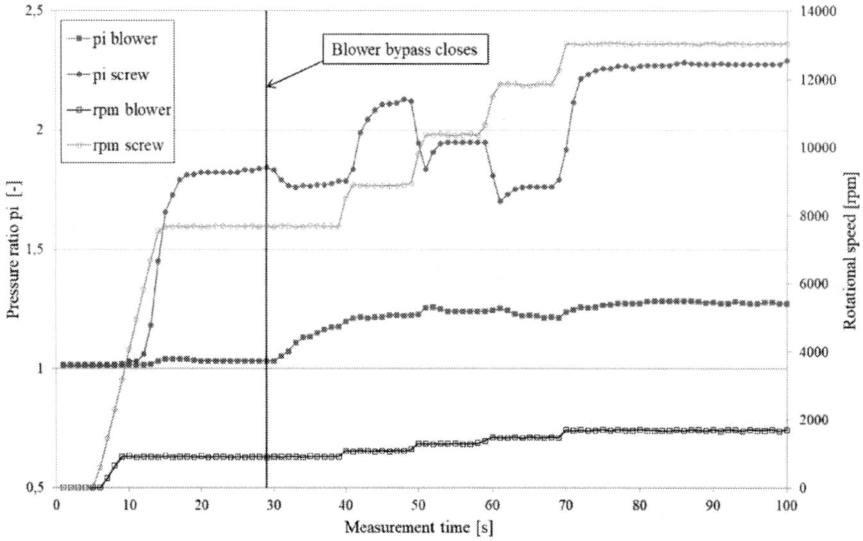

Figure 8. Start-up sequence - closed loop (CL), two stages.

3.2.1. Start-up procedure. The start-up procedure depicted in figure 8 is created in analogy to the full scale system where the rotors of the screw compressor are supported by hydrodynamic sleeve bearings. These bearings need a minimum rotational speed in order to develop their load capacity. It is therefore important to start with low load (low pressure difference) and accelerate quickly to the minimum rotational speed. Figure 8 shows that this is done within the first 10 s. After that, at 29 s, the bypass around the blower is closed and both machines accelerate (here in steps) until another stable operation point is reached at approximately 80 s with a screw compressor outlet pressure at 2.8 bar (abs) and a mass flow rate at 760 kg/h.

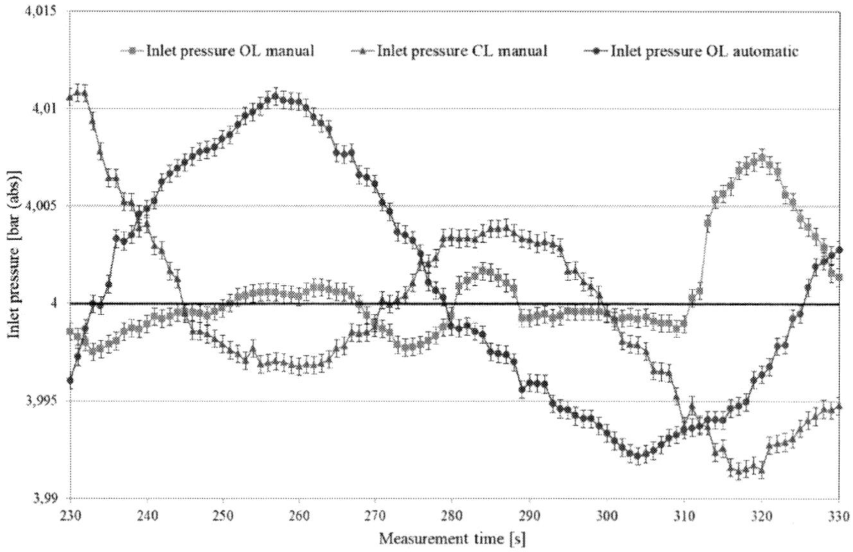

Figure 9. Stability of test rig inlet pressure.

3.2.2. Pressure control and back-pressure control. An example how precisely a constant pressure of 4.0 bar (abs) upstream of the test valve (representing e.g. an air turbine) can be held while the valve settings are unchanged is displayed in figure 9. The expansion ratio over the test valve is 2.0 while the mass flow rate has been adjusted in order to reach and hold the target pressure upstream of the test valve. The system is operated with both stages at a mass flow rate of approximately 350 kg/h. In the diagram, vertical bars represent the measurement uncertainty of the pressure measurement.

Comparison of the three graphs reveals that manual control gives superior results and that therefore the control settings still have to be improved. Best results are achieved in open loop (OL) operation mode because then no compensation of system leakages is needed in order to keep the PD blower inlet pressure (see figure 7) constant which simplifies the overall control task.

During the same tests as described above, the outlet pressure of the test valve had to be controlled as well in order to keep the expansion ration of 2.0 constant using the control valve downstream of the test valve as depicted in figure 7. Measurement of this pressure is done with a sensor having a significantly higher uncertainty than on the pressure side. In addition, an AD-converter with only 12 bit resolution is used, which results in the stepped curves as visible in figure 10. The target pressure of 2.0 bar (abs) is depicted together with the upper and lower values resulting from the measurement uncertainty as given by the sensor manufacturer. As visible in figure 9, at this stage of the control system development, manual control still gives better results than automatic control.

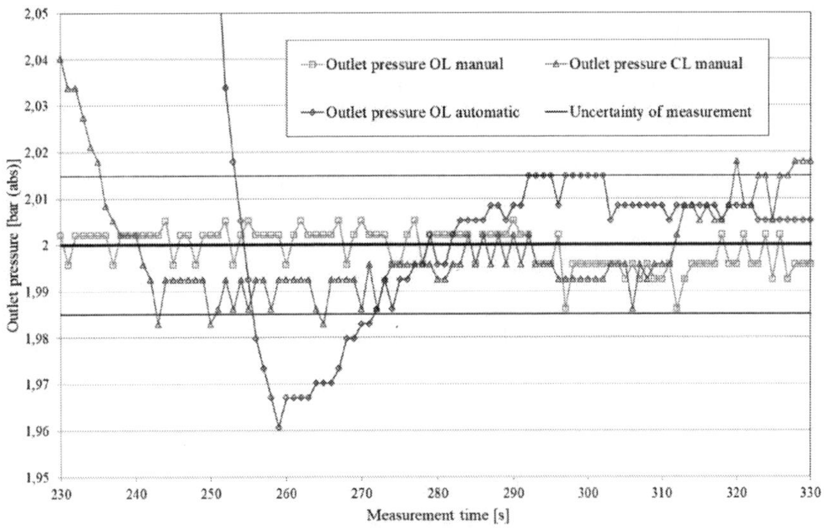

Figure 10. Stability of test rig outlet pressure.

3.2.3. Mass flow rate control. As the test valve setting was not changed during the pressure control tests of the section above, the inlet pressure variation (see figure 9) is a measure for the mass flow rate variation as well. As for certain applications it is important to run the test air supply system in mass flow rate controlled mode, the stability of the directly measured mass flow rate is investigated in this section. Figure 11 shows mass flow rate variations and deviations from the target value 400 kg/h in different control and operation modes. The system is operated with both stages and an expansion ratio of 2 over the test valve and a test valve inlet pressure of 2.5 bar(abs). In automatically controlled CL mode, the standard deviation from the target value is the highest (1.18 %) of the three curves. This is to be expected as in this case the highest number of control circuits is simultaneously active.

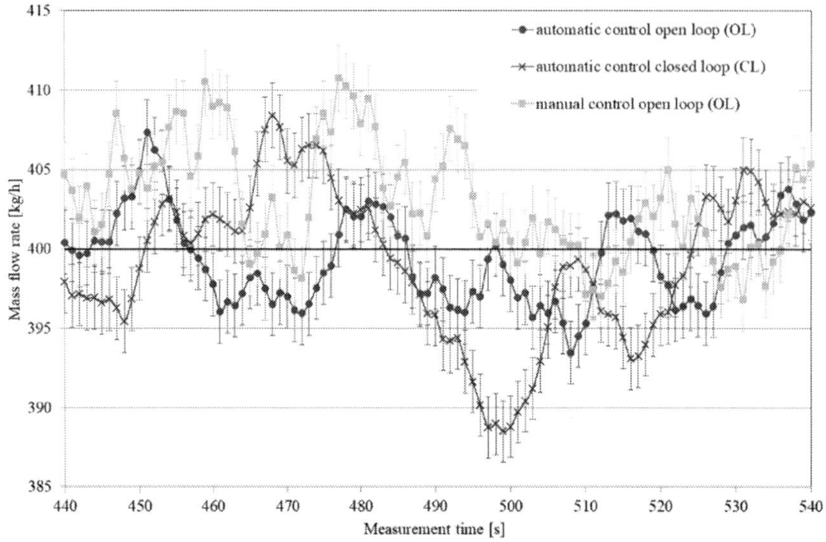

Figure 11. Mass flow rate control.

3.2.4. Inlet pressure repeatability. Figure 12 proves how precisely the same test valve inlet pressure 4 bar(abs) can by adjusted in open loop operation with both stages and the same settings for the test valve on two different test days at 365 kg/h. PD blower speed and screw compressor speed are set to the same values and manual fine tuning using the small mass flow bypass valve is applied. The pressure variation achieved in this example is on both days within +/- 0.11 % of the target value with a standard deviation of less than 0.06%. The repeatability of the average values is within 0.006%.

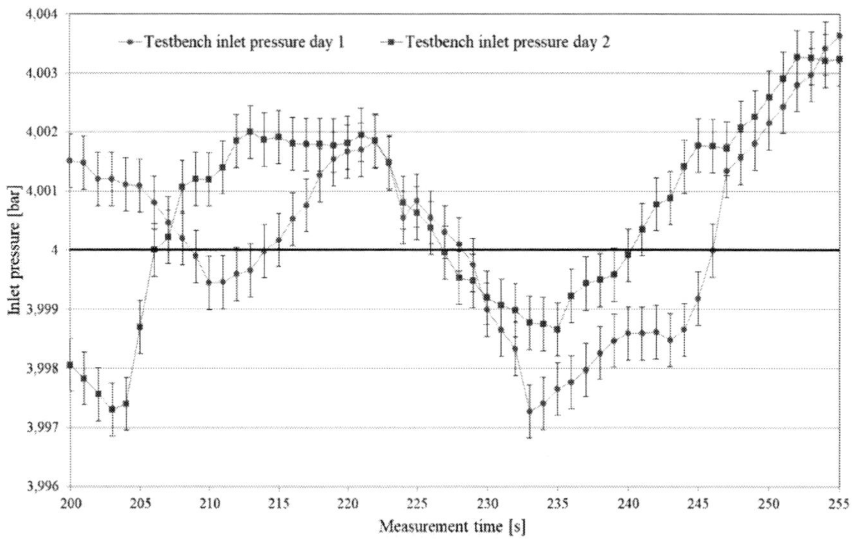

Figure 12. Inlet pressure repeatability in open loop manual control operation.

3.2.5. Temperature control. The temperature at the test valve is influenced by the compressor outlet temperature (and therefore by the compressor pressure ratio) but has to be controllable independently. If the compressor outlet temperature is significantly higher than the temperature which is desired at the

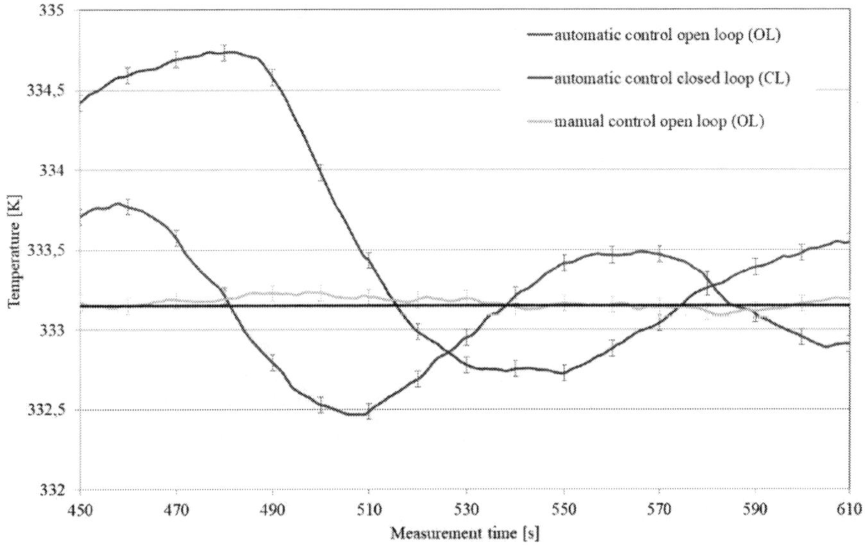

Figure 13. Stability of temperature.

test valve, control is possible by using the aftercooler. Its cooling performance can be controlled by the cooler fan speed and the settings of the bypass valves parallel to the cooler. If the compressor outlet temperature is too low (due to a low pressure desired at the test valve) than the screw compressor outlet has to be throttled additionally using the temperature control valve (see figure 7). Both cases are shown in the following figures for a test valve pressure of 2.8 bar (abs). Temperature stability at a target value of 333.15 K without the need to use the temperature control valve is depicted in figure 13 and for a target value of 342.15 K and the use of the temperature control valve in figure 14. In both cases the required stability (max. +/- 3K deviation from the target value) is reached.

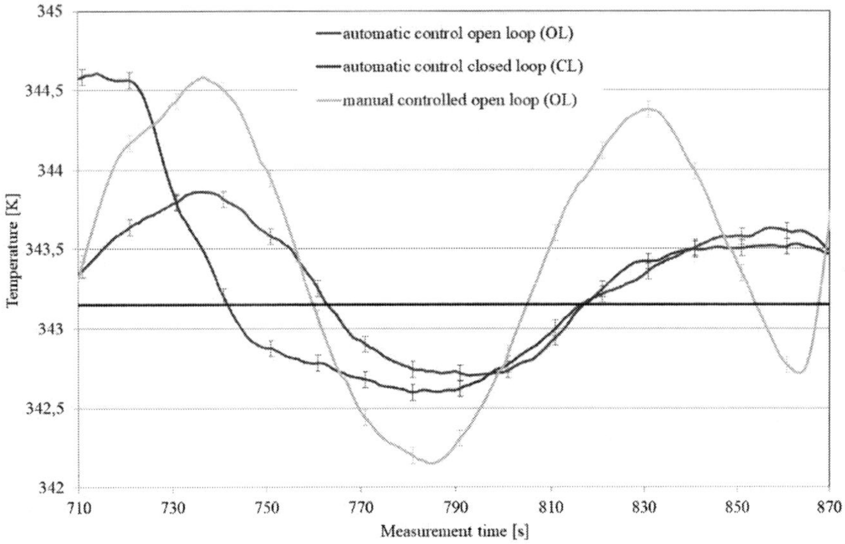

Figure 14. Stability of temperature using temperature control valve

Figure 14 reveals a maximum deviation of 1.6 K (standard deviation of 0.72 K) in automatically controlled closed loop operation. After a certain settling time, automatic control in closed loop (CL) as well as in open loop (OL) operation mode show similar deviations from the target value. Obviously, the automatic compensation of system leakages which is needed in CL in order to keep the PD blower inlet pressure constant has no significant influence. Again the performance of manual control is superior, which shows the potential for fine tuning of the control parameters.

Figure 14 shows that the additional use of the temperature control throttle does not significantly change the performance in automatically controlled closed loop operation. But in open loop operation this time manual control shows oscillations that are much less pronounced in automatic control.

4. Conclusion and outlook

The control system for a large scale test air supply unit was developed and successfully tested using a scaled functional model. At all operational modes that were tested, the required stability and precision criteria were fulfilled, although a significant improvement potential was shown as well. In many cases manual control gave superior results over automatic control which reveals that fine tuning of the controller parameters has a potential to decrease deviations from target values. Additional performance can be expected from using PID-controllers with adaptive values of the controller parameters. In a next step, further operating points distributed over the large operational area and transient behavior should be investigated. For the experiments in this paper, pre-selection of compressor operating conditions was done manually (see figure 9). The implementation of the two-stage performance calculation into the software is planned.

Acknowledgments

The authors wish to express their gratitude to Stefan Gärling for performing the CFD simulations and component design, Christian Hage for silencer simulation and design, Henning Rätz for providing his expertise on the PAtools software and all colleagues who contributed to the construction of the scaled functional model.

References

[1] Binner M and Seume J 2014 Flow patterns in high pressure steam turbines during low-load operation, *J. Turbomachinery* **136**, Issue 6, 061010

[2] Mulleners K, Gilge P and Hohenstein S 2014 Impact of surface roughness on the turbulent wake flow of a turbine blade, *J. Aerodynamics*. 2014

[3] Siemann J, Krenz I and Seume J 2016 Experimental investigation of aspiration in a multi-stage high-speed axial-compressor, *Proc. ASME Turbo Expo* Seoul, GT2016-56440

[4] Sutherland W 1893 The viscosity of gases and molecular force, *Phil. Mag. Ser. 5*, **36**., pp 507-531.

[5] Wald F 2017 *Funktionsmodell für eine zweistufige Verdichteranlage mit closed-loop-Betrieb und hochgenauer Massenstromregelung*. Bachlor-Thesis, Jade Hochschule Wilhelmshaven and Aerzener Maschinenfabrik GmbH.

[6] De Buhr L 2018 *Entwicklung einer hochgenauen Massenstromregelung für eine zweistufige Verdichteranlage mit Closed-Loop-Betrieb*. Master-Thesis, Institut für Turbomaschinen und Fluid-Dynamik, Leibniz Universität Hannover and Aerzener Maschinenfabrik GmbH.

IOP Conf. Series: Materials Science and Engineering **425** (2018) 012022 doi:10.1088/1757-899X/425/1/012022

A model for the transient pulsation generation at the discharge of a screw compressor by a shock tube analogy

Paul Xiubao Huang

Hi-Bar MC Technologies, LLC., Fayetteville, GA, USA

E-mail: paulxbh@yahoo.com

Abstract. Although screw compressors (dry and wet) have been widely used in many applications (such as HVAC&R, petrochemical and gas transmission, plant air, etc..) for more than half a century, limited information is available regarding the genesis and mechanism of gas pulsations at pocket passing frequency (PPF) dominating at discharge side of these compressors [1]. The understanding of its physical nature, magnitude and affecting parameters, the location and moment of generation, and the velocity it propagates are all of fundamental, and will help to determine how it will interact with downstream components such as silencers and piping system, especially for accurately predicting downstream pulsation levels and excited vibration and noise at the higher multiples of the PPF.

The primary goal of this paper is to apply the shock tube theory [2] to screw compressors during the transient generation phase of gas pulsations at discharge. It will reveal that the nature of gas pulsation is a combination of large amplitude compression waves (CW) and expansion waves (EW) accompanied with an induced fluid flow (IFF). The most dominant gas pulsation at discharge is directly caused by either an over compression (OC) or a under compression (UC) suddenly discharging to compressor outlet. Therefore its exact location and moment of generation, magnitude, travelling directions and velocity can all be predicted based on design parameters and operating conditions of those machines. With this knowledge, its interactions with the compressor cavity and downstream piping and silencer are conducted for a typical UC and case, and results are in good agreement with the test data of previous researchers. The analysis suggests that the shock tube model can provide a pedagogic tool in understanding the physics of the transient phase of pulsation generation and subsequent formation. As such it can provide valuable insight to developers of more precise CFD calculations and a comprehensive algorithm of gas pulsations in the future.

1. Introduction

1.1 The myth of gas pulsations with screw compressors

During the past 60 years and especially during the last 30 years, screw compressors have become very popular because they combine the advantages of various PD compressors and can provide a wide range of operating conditions. In 1975, screw compressors gained more acceptance after the publication of "API Standard 619 (1st Edition)". Today, screw compressors are widely used to compress a wide range of gases in many different industries including chemical and petrochemical,

IOP Conf. Series: Materials Science and Engineering **425** (2018) 012022 doi:10.1088/1757-899X/425/1/012022

food processing, pulp and paper, power generation, natural and process gas applications, refrigeration and in vapour recovery services [1]. A typical dry screw compressor is shown in Figure 1.

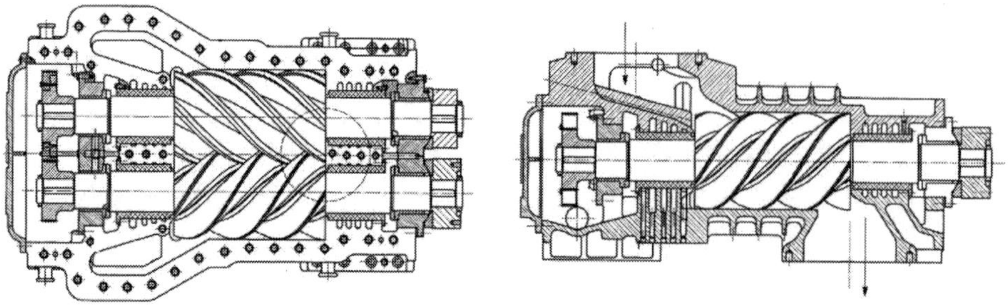

Figure 1. A Typical Dry Twin Screw Compressor per API 619 [1].

Screw compressors (dry and wet) do not have suction or discharge valves like reciprocating compressors. There is a common misconception that the gas is continuously "extruded" from the compressor, like a sausage grinder, without any significant pulsations [1]. Many sales brochures and instruction manuals try to create a myth that suggests that screw compressors do not generate any pulsations and, if they do, the pulsation levels are expected to be very low. Statements that screw compressors do not produce pulsation are in stark contrast to the technical literature. For example, API Standard 619 5th Edition [3] states that "In screw compressor systems, the flow is not steady, but moves through the piping in a series of flow pulses that are superimposed upon the steady (average) flow". As observed and reported recently [1,4,5,6], the pulsation levels are minimized when the compressors are operating near the design conditions, but can be significantly increased at off design conditions, as much as 10 – 20 percent of the absolute discharge pressure. Moreover, the waveform of the pressure pulsations are first revealed by Fujiwara and Sakurai [7] and later confirmed by Mujic et al [8] as of a general sawtooth/triangular shape as shown in Figure 2.

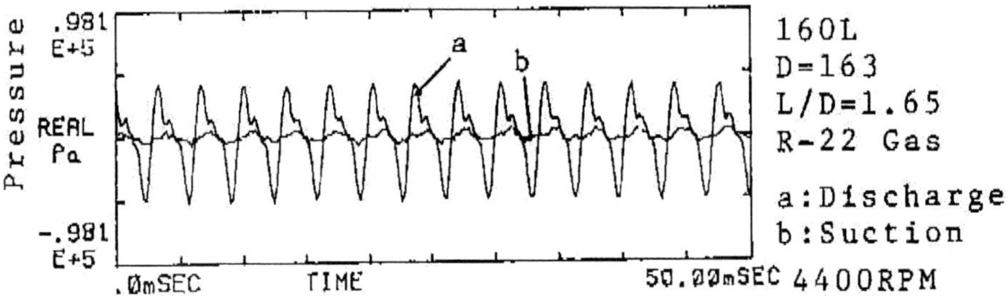

Figure 2a. Measured pressure pulsation waveform at the screw discharge from Fujiwara et al [7].

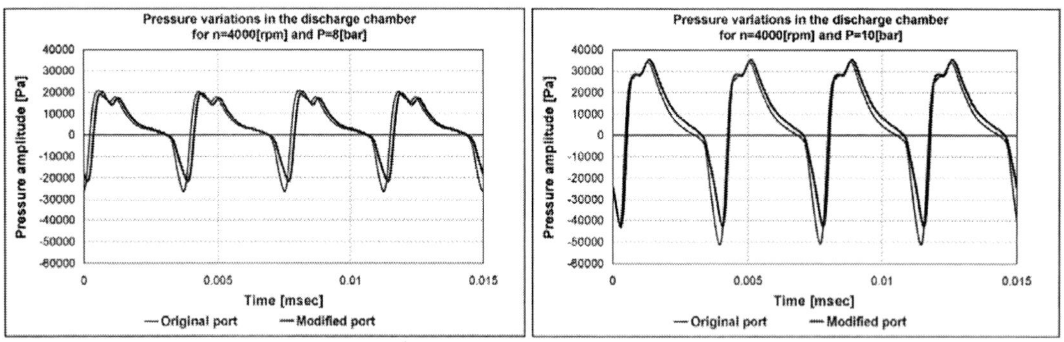

Figure 2b. Measured pressure pulsation waveform at the screw discharge from Mujic et al [8].

The high levels of gas pulsations are harmful to the downstream piping system if left uncontrolled and could potentially damage pipe line, in-line equipments, and excite severe vibrations and noises. So API 619 requires pulsations dampeners or silencers be installed at inlet and outlet of compressor. The knowledge of their physical nature, magnitude and frequency under various operating conditions is critical to the design of silencers and downstream system.

1.2 Screw compression cycle and causes of gas pulsations

Both wet and dry screw compressors generate pulsations at multiples of the pocket-passing frequency (PPF), which is defined as the number of lobes on the male rotor multiplied by the compressor running speed in Hz. The maximum generated pulsation levels normally occur at 1x PPF and are generally reduced at the higher harmonics.

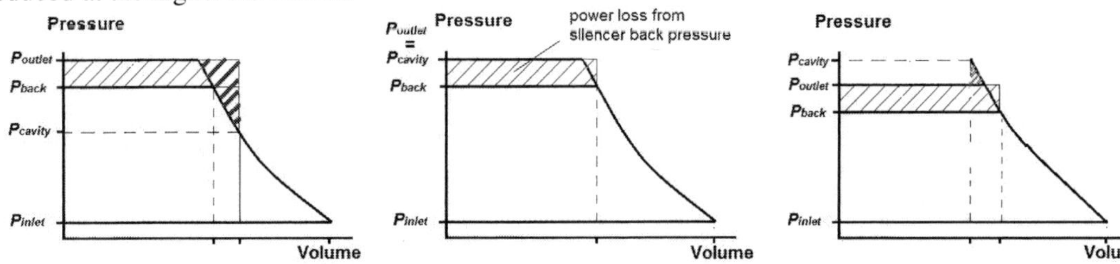

Figure 3. A P-V Diagram showing different modes of operation and power losses.

The pulsation amplitudes generated by screw compressors are affected by many variables as observed by various researchers [1,4,5,6,7,8], such as the shape of discharge port, screw profile, internal clearances and gas type. But the most significant contribution comes from operating under high pressures at off-design conditions of either an under-compression (UC) or over-compression (OC) as shown in Figures 3a and 3c. An UC happens when the pressure at the discharge line (p_{outlet}) is greater than the pressure of the compressed gas within the cavity (p_{cavity}) just before the opening. This would result in a rapid backflow into the cavity, a pulsed gas flow, according to the conventional theory. On the other hand, an OC takes place when the pressure at discharge line (p_{outlet}) is less than the pressure inside the cavity (p_{cavity}), causing a rapid forward flow of the gas into the discharge system. Screw compressors suffer from UC or OC due to the inherent mis-match of the fixed design pressure and ever-changing system pressures and operating conditions. The black and blue shaded areas in Figure 3 are the power losses due to discharge silencer (back pressure loss = $p_{outlet} - p_{back}$) and from UC or OC mode respectively.

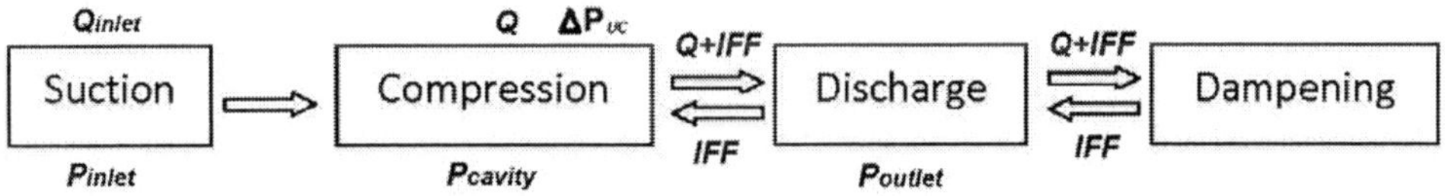

Figure 4. Phases of a traditional screw compression cycle under UC mode.

Figure 4 shows a flow chart for phases of a traditional compression cycle of a screw compressor system in under compression mode (UC). The inlet volume flow Q_{inlet} goes from suction to compression, and to discharge phase when pulsation is generated as the induced fluid flow (IFF) or backflow in case of UC (This will be explained in detail by Figure 6 in paragraph 2.2), which is then dampened by a dampening device typically connected downstream of the compressor discharge.

1.3 The state of gas pulsation modelling and the goal of this paper

Naturally, as important matters as gas pulsations, there have been tremendous efforts from both academia and industry made to model the phenomena, especially since the late 1980s by researchers [4,5,6,8]. The approaches are mostly CFD based having different schemes solving non-linear differential equations (DE) and the results can explain many observations for various applications. However, a comprehensive algorithm is still lacking and needs to be developed for accurately predicting the pressure and flow modulation (pulsation) generated by screw compressors at different operating conditions, according to [1]. For example, it is difficult to accurately predict the pulsation levels, especially at the higher multiples of the PPF, in the silencer and downstream piping when the modulation produced by the compressor is not accurately known. Many compressor manufacturers do not know the exact pressure and flow modulation (pulsation) generated by the compressor and often use an approximate value, such as 10 – 20 percent of the discharge pressure, when evaluating silencer designs.

Instead of directly tackling DEs, a diversionary approach has been attempted by Huang [2,9] using a shock tube analogy to simulate the transient generation of gas pulsations at the PD compressor discharge under UC or OC at off-design conditions, hence to gain a quantitative understanding of the affecting parameters and its physical nature. This paper will explore more deeply into the shock tube model. The aim is to apply the shock tube model to the screw compressor discharge in order to get a clear physical picture of the transient generation of gas pulsations, such as the exact location and moment of its generation, magnitude, travelling directions and velocity based on the design parameters and operating conditions of screw compressors. Then this information is used to determine or derive how it will interact with a quasi 1-D compressor system consisting of the cavity, silencer and downstream piping system. In other words, this paper intends to correlate and explain how the operating conditions of UC and OC shown in Figure 3 could produce the shape of the waveforms and magnitude of pressure and flow pulsations as measured in Figures 2a and 2b. As such it is expected to provide valuable insight to developers of more comprehensive and precise CFD calculations of gas pulsations in the future.

2. Shock tube mechanism of the transient pulsation generation at the discharge of a screw

Refer to [2,9,10] for more detailed descriptions of the shock tube analogy. However, some main points and figures are taken from these references and used below in order to keep a smooth context flow. According to the transient analysis of a screw compressor discharge by the shock tube analogy, the sudden communication between the cavity and the compressor outlet at different pressures is compared with (analogous to) the diaphragm bursting of the shock tube and would trigger gas pulsation generations in the form of CW-IFF-EW, as illustrated in Figure 5 for the under-compression (UC) and over-compression (OC) modes. The main results of the shock tube analysis can be summarized by the three Transient Gas Pulsation Generation Rules below.

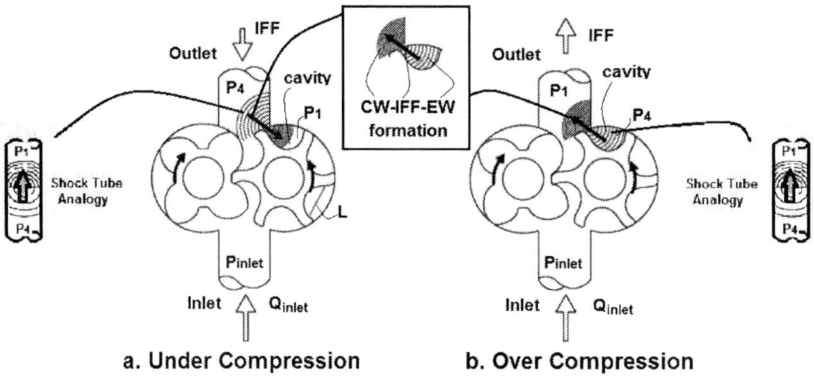

Figure 5. Transient phase of screw compressor discharge by the Shock Tube Analogy [2].

2.1 Rules of the transient gas pulsation generation based on the shock tube mechanism

1. Rule I: For two divided compartments (either moving or stationery) with different gas pressures p_4 and p_1, there will be no or little gas pulsations generated if the two compartments stay divided or isolated;
2. Rule II: If the divider between the high pressure gas p_4 and the low pressure gas p_1 is suddenly removed, gas pulsations are generated at the location and moment of the opening as a composition of a fan of Compression Waves (CW) or a quasi-shock wave, a fan of Expansion Waves (EW) and an Induced Fluid Flow (IFF) with magnitudes as follows:

$$CW = p_2 - p_1 \tag{1}$$

$$EW = p_4 - p_2 \tag{2}$$

$$\Delta U = (p_2 - p_1)/(\rho_1 \times W) \tag{3}$$

where $(p_2 - p_1)$ is the over-pressure from CW, $(p_4 - p_2)$ is the under-pressure from EW, ρ_1 is the gas density at low pressure region, W the speed of the shock wave, ΔU the velocity of Induced Fluid Flow (IFF);

3. Rule III: Pulsation component CW is the action by the high pressure gas p_4 to the low pressure gas p_1 while pulsation component EW is the reaction by the low pressure gas p_1 to the high pressure gas p_4 in the opposite direction, and their magnitudes are such that they approximately divide the initial pressure ratio p_4/p_1 equally into CW and EW [equation (4)]. At the same time, CW and EW induce a unidirectional fluid flow pulsation IFF in the same direction as the CW.

$$p_2/p_1 = p_4/p_2 = (p_4/p_1)^{1/2} \tag{4}$$

Rule I implies that there would be no or little pulsations during suction (if there is no inlet valve), transfer and compression (expansion) phases of a screw compression cycle because of the absence of either a pressure difference or sudden opening. Focus instead should be placed upon the discharge phase, especially at the moment of the discharge when the cavity is suddenly opened to outlet line during off-design conditions such as UC or OC.

Rule II indicates specifically the ***moment*** of gas pulsation generation as the instant of the communication between gas p_4 and gas p_1, and the ***location*** as the discharge port opening. Moreover, it defines ***two sufficient conditions*** for gas pulsation generation:

a) The existence of a pressure difference Δp_{41} which is a measure of how severe the UC or OC is;
b) The sudden opening of the divider separating the pressure difference, a measure of the effect of rotor speed.

Rule II also reveals the composition and magnitudes of gas pulsations at onset as a combination of large amplitude Compression Waves (CW) or a quasi-shockwave, a fan of Expansion Waves (EW) and an Induced Fluid Flow (IFF). These waves are non-linear with changing wave form during propagation. This is in direct contrast to the acoustic waves that are linear and do not induce a mean through flow. It is important to note the wholeness of three pulsation components (CW, EW and IFF) that are generated simultaneously and one cannot be produced without the others.

Rule III shows further the interactions between two gases of different pressures are mutual so that for every CW pulsation component, there is always an equal but opposite EW pulsation component in terms of pressure ratio [equation (4)]. Working together, they induce a unidirectional fluid flow pulsation (IFF) component in the same direction as the compression waves (CW). Moreover, Rule III quantitatively correlates, by equation (4), the post-discharge opening induced gas pulsations (Δp_{21}, Δp_{42}) with the pre-opening pressure difference Δp_{41} between the compressor cavity and discharge line.

2.2 Application of the transient gas pulsation rules to screw compressor discharge

Now let's take a new look at the traditional compression cycle of a screw compressor operating at under compression mode (UC) in light of the Shock Tube Theory. Figure 6 illustrates the same process shown in Figure 4 again in time domain, emphasizing the events during the transient backflow phase of discharge.

The transient phase starts when the compressor cavity suddenly opens to the outlet line to expose the pressure difference between the design pressure in the compressor cavity P_{cavity} and the outlet pressure P_{outlet}, i.e. ΔP_{UC} (= ΔP_{41}). According to the transient analysis [2,10] of the shock tube analogy, the sudden communication between the cavity and the outlet would trigger the generation of gas pulsations in the form of CW-IFF-EW, which originate at the discharge port as shown in Figure 5a. CW would go into the cavity to wave compress the cavity gas Q (traditionally called backflow compression) and induce IFF, while EW comes out to the outlet pulling IFF and causes gas pulsating downstream. The magnitude of CW and EW is approximately equal by dividing the ΔP_{UC} (= ΔP_{41}) in terms of pressure ratio. Since the waves (both CW and EW) travel much faster (10-20 times) than the screw rotor, the time it takes to wave compress and induce IFF is much shorter than the traditional rotor driven compression and discharge process (which starts as the cavity is closed to the inlet and ends until all gases are emptied into the outlet) as shown by Δt_{UC} and the sharp rise of pressure ΔP_{UC} in Figure 6b. It should be emphasized that it is in the transient phase that the pressure in the cavity gets equalized with the outlet by incoming CW and volume flow IFF while simultaneously the induced downstream pulsating gas flow IFF and EW are dampened by the silencer if there is one. Though the transient time is short, it is finite and can be calculated approximately by

$$\Delta t_{UC} = 2Lc / W_{CW} \tag{5}$$

$$\Delta t_{OC} = 2Lc / W_{EW} \tag{6}$$

where Lc is a characteristic length of the cavity at the moment of the discharge opening that CW or EW has to reach to be reflected and W_{CW} or W_{EW} is the wave velocity corresponding to CW and EW. Factor 2 means a round trip (incident and reflective waves) is needed to wave compress the cavity gas Q to diminish the initial ΔP_{UC} or ΔP_{OC} to less than 5% [2,10]. For an industrial size air compressor, the magnitude of Δt_{UC} is typically less than a fraction of one 1/1000 s as shown in Table 2 below.

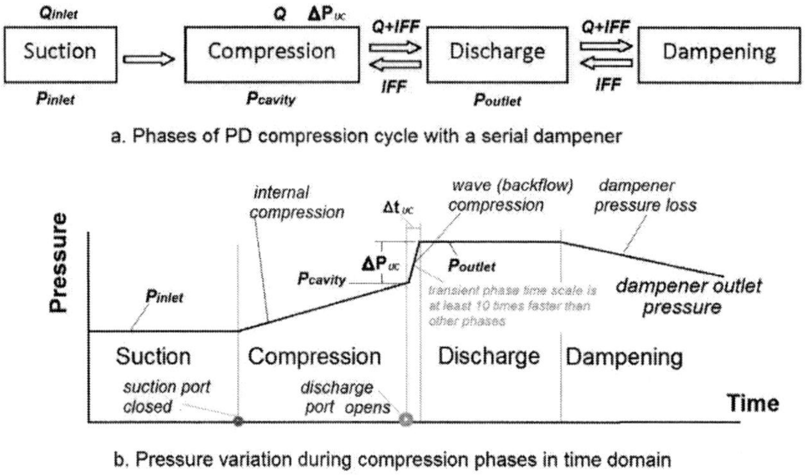

Figure 6. Phases of a screw compression cycle for traditional serial dampening under UC mode.

After the transient backflow phase, the combined cavity volume flow and back flow, Q+IFF, is discharged into the discharge line by the positive movement of the rotors that takes much longer time as marked as discharge phase in Figure 6b. Q+IFF would go into a downstream silencer (second time for IFF) to be further dampened. The silencer type used is mostly a reactive or combo type sometimes with an orifice (called Kotter plate) plate at the silencer inlet as shown in Figure 7.

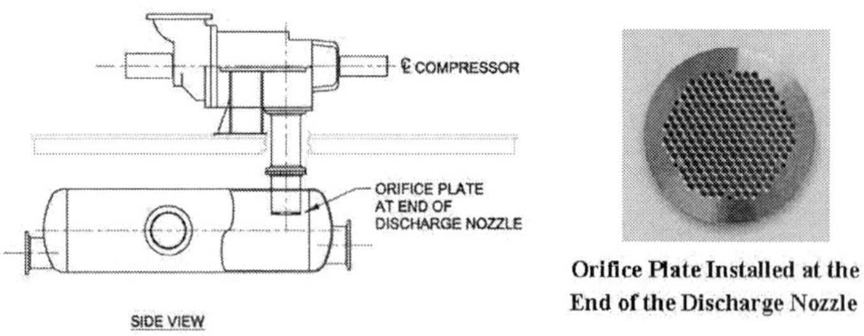

Figure 7: Example of a discharge system with silencer and Kotter plate (orifice plate)

Figure 8 shows the phases of the screw compression cycle under an OC mode. Note that for OC mode, the IFF goes forward through the dampener during the transient phase before the cavity flow Q, while Q flows through the same silencer after IFF passes. It is different from the UC mode in that IFF only goes through the silencer once instead of twice for UC mode.

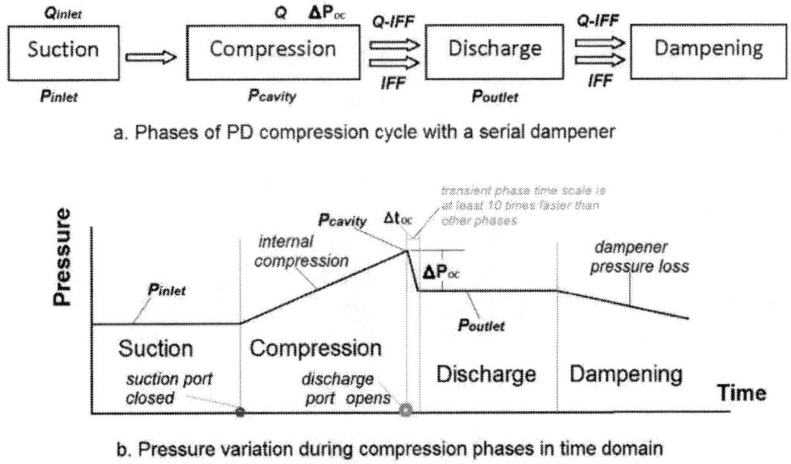

Figure 8. Phases of a screw compression cycle for traditional serial dampening under OC mode.

2.3 The non-linearity and nature of gas pulsations of screw compressors

To get a feel for the magnitudes of gas pulsations as non-linear waves that induces strong fluid flow in stark contrast to the linear acoustic waves, Table 1 lists a practical range of fluctuating pressures, from 0.00002 – 2 bar (0.0003 - 30 psi) or equivalent to 100-200 dB, for comparison. The high end (2 bar or 30 psi or 200 dB) represent a typical pressure pulsations value for most of the industrial screw compressors while the low end (00002 bar or 0.0003 psi or 100 dB) represent a typical pressure fluctuation for high sound. It will be shown that, because of the orders of magnitude differences, gas pulsations and sound (linear acoustic waves) behave very differently, hence the way to model them should differ too.

Table 1. Magnitude of acoustic waves in comparison with industrial gas pulsations [2].

	Sound			Gas Pulsations		
Pressure Fluctuation, bar	0.00002	0.0002	0.002	0.02	0.2	2
Pressure Fluctuation, psi	0.0003	0.003	0.03	0.3	3	30
Sound Pressure level, dB	100	120	140	160	180	200

From Pulsation Rules, the pulsation component of the compression waves (CW= $p_2 - p_1$) can be expressed by Equation (3), also known as the Shock Wave Equation, relates the abruptly arisen pressure of a shock wave with strength, ΔP_{21}, propagating in air of density, ρ, in shockwave speed of W, and inducing a flow velocity of ΔU. When the magnitude of ΔP becomes smaller and smaller, Equation (3) is reduced to Equation (7) as follows,

$$dp = \rho \; a \; du \tag{7}$$

which is the classical acoustic equation, representing small amplitude pressure wave front, dp, propagating in air of density, ρ, at the speed of sound, a, and inducing an air velocity of du, (Capital P and U are used to indicate their larger magnitude). To get a sense of their quantitative differences, Table 2 lists the induced fluid velocity and other variables by the same range of pressure fluctuations listed in Table 1 for air propagating one-dimensionally in a shock tube. An on-line WiSTL Shock Tube Calculator [11] is used, from known W, to calculate the values of Δp_{41} (=p_4-p_1), Δp_{21} (=p_2-p_1), speed of sound, speed of CW and EW, ΔU (IFF) under the following initial conditions: for air with the same initial temperature of 300 K on both sides and low pressure of 1.0 bar for a diaphragm-triggered shock tube. The transient time is calculated from wave velocity of CW & EW by Equation (5) assuming Lc=0.05 m. The bottom two rows of Table 2 list the calculated pressure ratios for both CW and EW gas pulsations to show the non-linearity and for validating Equation (4) in the pressure range for typical engineering applications.

Table 2. Comparison of behaviours between linear sound and non-linear gas pulsations.

	Acoustic Waves			Gas Pulsations		
CW Strength, Δp_{21}, bar	0.00002	0.0002	0.002	0.02	0.2	2
EW Strength, Δp_{42}, bar	0.00002	0.0002	0.002	0.02	0.3	8.5
UC or OC, Δp_{41}, bar	0.00004	0.0004	0.004	0.04	0.5	10.5
CW speed in Mach Number	1.000009	1.00009	1.0009	1.009	1.09	1.65
IFF, ΔU, m/s	0.0052	0.052	0.52	5.2	47	302
Speed of sound (a_1), m/s	347	347	347	347	347	347
Speed of CW, m/s	347.4	347.5	348	351	379	573
Speed of EW (a_2), m/s	347.4	347.5	348	348	357	415
Temp behind CW (T_2), K	300	300	300.2	302	318	427
Transient time, Δt_{UC}, mSec	0.288	0.288	0.287	0.285	0.264	0.175
Transient time, Δt_{OC}, mSec	0.288	0.288	0.287	0.287	0.280	0.241
p_2/p_1, CW	1.00002	1.0002	1.002	1.02	1.2	3
p_4/p_2, EW	1.00002	1.0002	1.002	1.02	1.25	3.8

It can be seen that below the fluctuating pressure of 160 dB (in RED in Table 2) in the domain of Classic Acoustics, CW strength and velocity are almost the same as EW, the IFF velocity ΔU is very low (< 0.52 m/s) and is linearly proportional to the strength of CW and EW. On the other hand, above 160 dB in the domain of gas pulsations, CW strength and velocity are very different from EW (CW is always faster than EW), and the IFF velocity ΔU is very high (302 m/s at 200 dB), much higher than the average flow velocity of screw compressors. This embodies the non-linearity or nature of the gas pulsations that the transient effect is dominant and demands an aerodynamic approach, such as shock tube theory or characteristics method [2,5,6] rather than traditional Thermodynamic approach.

The transient times in Table 2 also show that the time it takes to wave compress and induce IFF is much shorter than the time it takes to rotor compress (also called internal compression) the gas. In other words, transient events are driven by CW/EW waves as dynamic forces while internal traditional compression cycle is driven by rotors which are typically much slower. For example, for a screw running at high RPM of 15000 with a 4x6 lobe combination, the pocket passing time is 1 mSec (=1/(15000x4/60)), > 0.175 mSec shown in Table 2.

3. Dynamic interactions of discharge pulsation – a quasi 1-d model

From the above analysis, we obtain the needed knowledge of gas pulsations at its generation source (= location of discharge valve opening): its physical nature and the magnitude, the location and moment of generation, and the velocity it propagates both upstream and downstream. Now we are going to use this information to determine or derive how it will interact with compressor cavity and downstream components such as piping system, silencers as shown in Figure 7 in order to produce the shape of the waveforms and magnitude of pressure and flow pulsations. The results are then compared with measurement data by various researchers as shown in Figures 2a and 2b.

3.1 The making of the discharge gas pulsation in a free downstream system – model Y

The simplest case of the discharge system for the pulsation making process is modelled as the following as illustrated in Figure 9a for UC mode ($\Delta P_{UC} = \Delta p_{41}$): the process starts with the sudden valve opening at point A as the compressor cavity with low gas pressure p_1 is opened to the downstream piping with high gas pressure p_4. There isn't any obstruction downstream. The cavity displacement (modelled by a piston moving always to the right shown as yellow in Figure 9a) velocity U_0 and volume are assumed to be much smaller than the speed of sound of the gas at discharge ($U_0 \ll a$) and the downstream volume ($Lc \ll Ld$). A pressure tab is installed at point B downstream to sense the local pressure and flow variations within one period (= t_{cycle}, from the moment of one valve opening to the next opening).

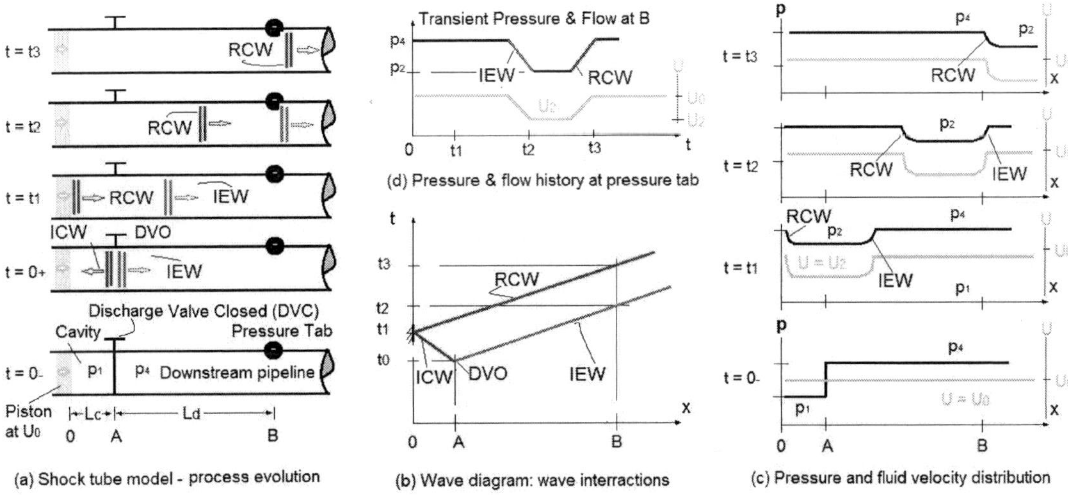

Figure 9. The making of the discharge gas pulsation for Model Y - UC mode.

According to the Transient Pulsation Generation Rule II, the moment of the discharge valve opening (DVO) triggers the generation of gas pulsations in the form of CW-IFF-EW at the point A as shown in Figures 9a-9b at $t=0^+$. A fan of the incident CW (ICW) would sweep into the cavity at the speed of wave W_{ICW} to compress the cavity gas from pressure p_1 to p_2 and then it is reflected at $t=t_1$ upon reaching the end wall (where $U=U_0$, displacement velocity), producing a reflective CW (RCW) that further compress the gas from pressure p_2 to p_4. While the generated fan of the incident EW (IEW) sweeps downstream at the speed of wave W_{IEW} causing a backflow into the cavity at velocity ΔU ($=U_0-U_2$) and reducing the downstream pressure to p_2. Both downstream going IEW and RCW are then sensed by the pressure tab at point B at the moments of $t=t_2$ and $t=t_3$ respectively, as shown in Figures 9a-9b-9c during t= 0 to t_3. The RCW would follow IEW travelling downstream, increasing the downstream pressure back to p_4 and stalling the flow to velocity $U=U_0$ as sensed by the pressure tab at point B at. For the purpose of analyzing and comparing with experimental data, the pressure and flow velocity history at point B is also expressed in time domain, as illustrated in Figure 9d, where a single pressure pulse produced by the combined actions of the IEW and RCW waves induces a single flow pulse in sync.

The time for fast moving waves to travel from 0→1→2→3 in model-Y can be calculated by the following formula:

$$t_1 = Lc / W_{IEW} \tag{8}$$

$$t_2 = Ld / W_{ICW} \tag{9}$$

$$t_3 = t_1 + (Lc+Ld) / W_{RCW} \tag{10}$$

The same discharge system (Model Y) can be also modeled for an OC mode ($\Delta P_{OC} = \Delta p_{41}$), as illustrated in Figure 10a where the gas pressure in cavity p_4 is higher than the downstream piping pressure p_1. Using the same Transient Pulsation Generation Rule II, a fan of the incident EW (IEW) would sweep into the cavity at the speed of wave W_{IEW} to expand the cavity gas from pressure p_4 to p_2 and then it is reflected at $t=t_1$ upon reaching the end wall (where $U=U_0$), producing a reflective EW

(REW) that further expand the gas from pressure p_2 to p_1. While the generated fan of the incident CW (ICW) sweeps downstream at the speed of wave W_{ICW} causing a forward flow out of the cavity at velocity ΔU ($=U_2+U_0$) and increasing the downstream pressure to p_2. Both downstream going ICW and REW are sensed by the pressure tab at point B at the moments of $t=t_2$ and $t=t_3$ respectively, as shown in Figures 10a-10b-10c during t= 0 to t_3. The REW follows ICW traveling downstream, decreasing the downstream pressure back to p_1 and stopping the flow to velocity $U=U_0$ as sensed by the pressure tab at point B. The pressure and flow velocity history at point B is also expressed in time domain, as illustrated in Figure 10d, where a single pressure pulse produced by the combined actions of the ICW and REW waves induces a single flow pulse in sync.

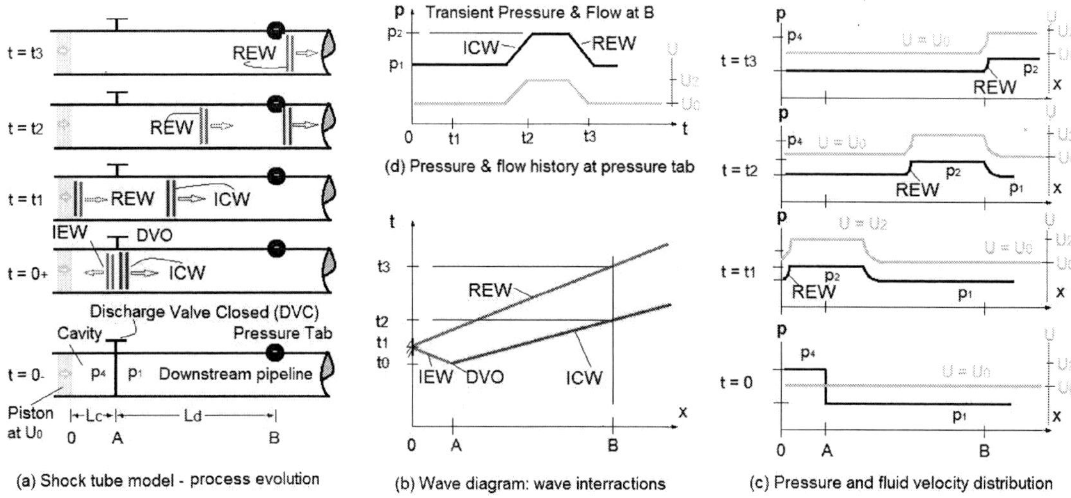

Figure 10. The making of the discharge gas pulsation for Model Y - OC mode.

3.2 The making of the discharge gas pulsation in an obstructed downstream – model Z

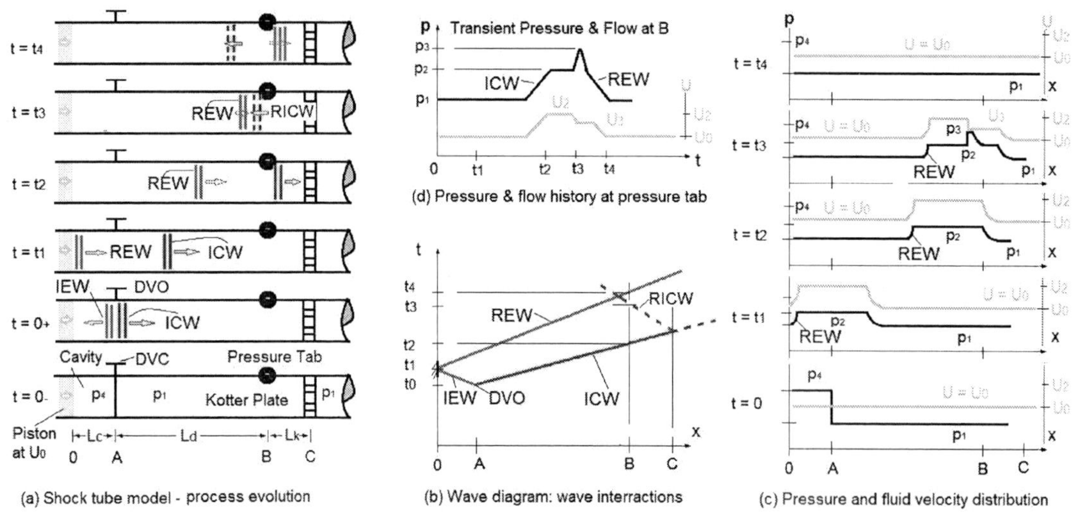

Figure 11. The making of the discharge gas pulsation for Model Z - OC mode.

The Model-Y assumption that there isn't any obstruction downstream of the compressor discharge isn't always true. Figure 7 shows an example of the reality with a downstream Kotter plate followed by a silencer. Kotter plate is basically a perforated plate that is often used to dampen the gas pulsations by blocking (or reflecting) a portion of the waves going downstream and transforming low frequency flow pulse into multiple high speed jets with a wide range of higher frequency noises. The selection from a wide range of the open-to-close area ratio determines the degree of the wave reflection and jet velocity. The silencer behind is often a reactive or combo type that can be regarded as a second stage dampening equipped with perforated tubes, reflective chambers and absorptive materials tackling both low and high frequency noises at the same time. For simplicity, only a Kotter plate (K-plate) is incorporated into model-Z, as shown in Figure 11a for the OC mode while all other assumptions for model-Y remain the same.

For Model-Z, the gas pulsation interaction with the compressor cavity and downstream K-plate has the same process during t=0-t_2 when the ICW passes through the point B and before hitting the K-plate as shown in Figures 11a-11b-11c. Because of the presence of K-plate, portion (say 50%) of the ICW is reflected at t=t_3 upon reaching the plate. Due to the obstruction of the K-plate, the induced flow is slowed down (say by half) at K-plate, producing a weakened reflective ICW (RICW) represented by the dashed wave fronts in Figure 11a that further compress the gas from pressure p_2 to p_3 and slow down the induced flow behind the wave front before colliding with the rightward going REW. The wave collision (refer to [12] for details) would strengthen the rightward moving REW (due to slowed flow by RICW) while weaken the leftward going RICW, as shown in Figures 11a-11b-11c during t= t_3-t_4. The resulting pressure and flow velocity history sensed by the pressure tab at point B can be illustrated in time domain in Figure 11d.

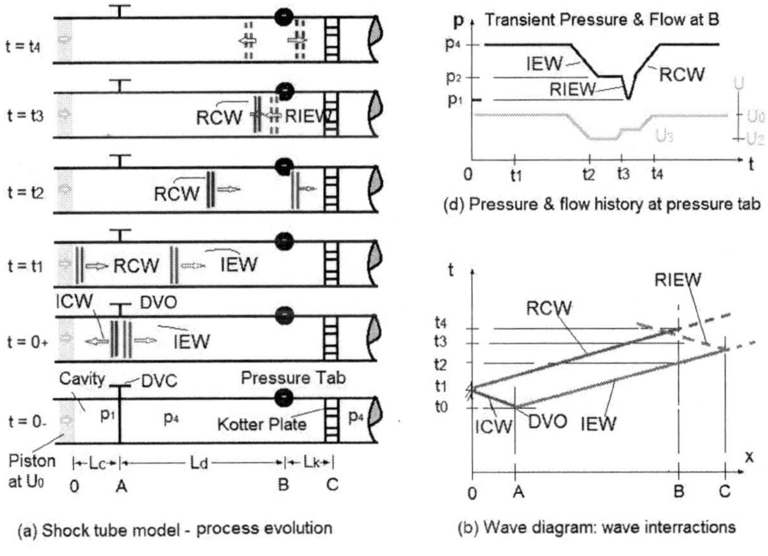

Figure 12. The making of the discharge gas pulsation for Model Z - UC mode.

The same discharge system for Model Z can be also modelled for the UC mode as illustrated in Figure 12a where the gas pressure in cavity p_1 is lower than the downstream piping pressure p_4. Using the same analysis as for Model Y, we could derive similar wave diagrams as shown in Figure 12b and the pressure and flow velocity history sensed by the pressure tab at point B as illustrated in time domain in Figure 12d.

3.3 Features of the dynamic process of gas pulsation making

Though Model-Y is a very simple case, it demonstrates some of the fundamental features of a dynamic gas pulsation making process for a typical screw discharge system. First of all, the source of ICW and

IEW is simply the re-distribution of the pre-opening UC pressure difference Δp_{41} that is being suddenly released at the discharge opening at point A and split into a pair of moving forces as IEW and ICW that are recombined (IEW and RCW) and teamed together to move and stop the flow when they sweep through downstream, resulting in a pulse flow (IFF). Secondly, the synchronization of the pressure and flow pulses in Figure 9d illustrates the dynamics of the transient production of the gas pulsation with the wave fronts IEW and RCW acting as the motive forces driving and stalling the instantly induced flow IFF between two wave fronts when they sweep through in the form of IEW-IFF-RCW. Thirdly, the repeat of the dynamic cycle shown in Figure 9d will produce a train of the pressure and flow pulses that ride on a main flow moving downstream at average velocity of U_0 (rotor displacement velocity) with the ICW and IEW pair shuttling through the main flow at the speed of waves to maintain the pulsed flow energized. This is in direct contrast to a steady state pipe flow where a static pressure difference stationary to the wall (that is, the driving force is stationary) is driving a continuous flow downstream. The last but not the least insight is that the transient gas pulsation event after initiation would be very difficult to control because there are three fast moving targets (CW, IFF, EW) bouncing around in different directions at high velocities that have to be dealt with.

3.4 Magnitude and waveform of the gas pulsation in comparison with previous investigators

From the above analysis and the Transient Pulsation Generation Rule III, it is shown that the magnitude of gas pulsations at the source is directly proportional to the pressure offset between the compressor cavity and outlet as expressed by the UC or OC or mathematically related to Δp_{41} by Equations (1) to (3). While the measured gas pulsations downstream of the discharge are related to its source value, they are affected by a few other design parameters nevertheless. In other words, the split of Δp_{41} into three different components at the source, CW-IFF-EW, so that each will interact with downstream system or each other by the ways of reflecting, inducing, colliding or merging, resulting in different types of gas pulsations in magnitude and waveform. For example, depending on whether the operating discharge conditions is a UC or OC, the split of the same magnitude Δp_{41} into CW and EW could have different induced flow directions as shown by the difference in Figures 9 and 10. Or the degree of the downstream obstruction determines whether the original CW and EW will be passed downstream or partially reflected to further strengthen the pulse amplitude, as demonstrated by the different magnitude between Figures 10 and 11 for the same OC mode. Another important parameter (not discussed in this paper) is the area expansion ratio from cavity to outlet that will weaken the strength of gas pulsations from the source to downstream pipeline. This is the reason the above models are called Quasi-1D as it can incorporate cross area changes when waves or flow travelling downstream. For a typical industrial screw with a "discharge port to outlet pipe" area expansion ratio around 3-5 times, plus the split of Δp_{41} into CW and EW, the downstream pulsation level is reduced 6-10 times from its generation source, and is roughly in the range of 10-15% of the source strength defined by Δp_{41}.

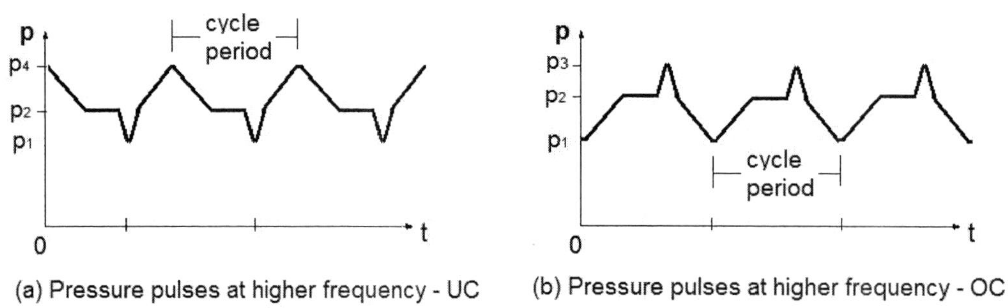

Figure 13. The pulses of the discharge gas pulsation for Model Z at higher frequency.

The geometrical parameters, such as locations of the pressure tab and K-plate, will affect the shape of the pulses while the speed of the compressor decides the duration of the cycle period or pulse frequency in relation to the pulse shape. For example, when compressor speed is increased so that more pulses are produced per period of time, the waveforms shown in Figures 11d and 12d for Model-Z will become pulses shown in Figures 13a and 13b for UC and OC modes respectively. It is interesting to compare the derived waveforms shown in Figures 13a and 13b with the measured waveforms in Figures 2a and 2b. The predicted waveform is basically a saw-tooth shape, same as those from both Fujiware [7] and Mujic [8]. However, we know now from the above analysis that the rising side of the pressure pulse corresponds to a CW action with a steeper slope (faster velocity) while the falling side of the pressure pulse corresponds to an EW action with a gentle slope when these waves sweep by the fluid at the pressure tab location at the speed of waves. There are also pressure ripples on different sides of the pulse from both the prediction and Mujic results, which can be seen as the secondary interactions such as reflection, colliding/merging from various boundaries inside the discharge system under UC or OC mode. However, the 1D assumption neglects a lot of influences from the reality and can be improved by the more precise CFD modelling in the future.

4. Conclusions and recommendations for future work

The use of sections to divide the text of the paper is optional and left as a decision for the author. Where the author wishes to divide the paper into sections the formatting shown in table 2 should be used.

4.1 Conclusions

The understanding of the physics and mechanism for the generation of gas pulsations and subsequent interactions with compressor system is fundamental and will help to provide valuable insight to developers of more precise CFD calculations and a comprehensive algorithm to predict gas pulsation and excited vibration and noise at the higher multiples of the PPF. Instead of using traditional thermodynamic approaches, this paper applies the transient aerodynamics from the shock tube theory to the transient discharge phase of the screw compressors to derive various scenarios of gas pulsations and compare them with the results of other researchers. It is concluded:

1. The most dominant source of gas pulsation at the screw discharge is directly caused by either an over compression (OC) or a under compression (UC) suddenly opening to compressor outlet, resulting in a combination of large amplitude compression waves (CW) and expansion waves (EW) accompanied with an induced fluid flow (IFF) in an inseparable formation of CW-IFF-EW.
2. The making of the discharge gas pulsation are demonstrated from the moment that the split of Δp_{41} into three different components at the source, CW-IFF-EW, through its formation process in which each component will interact with downstream system or each other by ways of reflecting, inducing, colliding or merging, resulting in different types of gas pulsations as observed downstream. These dynamically formed pressure and flow pulsations are comparable with measurement data by previous researchers with good agreement regarding the shape of the waveforms and order of magnitude.
3. The pressure and flow pulses are always in sync which illustrates the dynamics (Newton's Law of Motion) of the transient production of the gas pulsation with the wave fronts EW and CW acting as the motive forces driving and stalling the instantly induced flow IFF between two wave fronts when they sweep through in the form of EW-IFF-CW.
4. The methodology of the transient analysis with the shock tube analogy provides a pedagogic tool in understanding the physics of the transient phase of gas pulsation generation and subsequent formation downstream. As such it can provide valuable insight to developers of more precise CFD calculations and a comprehensive algorithm of gas pulsation predictions in the future.

4.2 Future work

The most important takeaway of this paper is the realization that the transient gas pulsation event after initiation would be very difficult to control because there are three fast moving targets (CW, IFF, EW) bouncing around in different directions and at high velocities within the system. Based on this insight, a new pulsation control method called Shunt Pulsation Trap or SPT [13,14] is proposed to target and take advantages of the fast moving waves in a parallel configuration that is advantageous to the traditional serial dampening schemes as analyzed in [15]. It is hoped that the principle of the SPT control strategy can be applied to the screw compressor design so that future generations of screw compressors can be simpler in structure, smaller in size and smoother in running than those of today.

5. References

[1] Smith, D., Pulsation, Vibration and Noise Issues with Wet and Dry Screw Compressors, *Proceedings of The 40th Turbomachinery Symposium*, 2011, p. 1709-202.

[2] Huang, P., Gas Pulsations: A Shock Tube Mechanism. *The 2012 International Compressor Engineering Conference at Perdue,* 2012

[3] Rotary Type PD Compressors for General Refinery Services, *API Standard 619 (5th Edition)* – 2010

[4] Linkamp, A.; Brümmer, A.: Calculation of discharge pressure pulsations of a screw compressor using the one-dimensional method of characteristics. *8th International Conference on Compressors and their Systems,* London 2013

[5] Linkamp, A.; Brümmer, A.: Experimental and theoretical investigation of the excitation of discharge pressure pulsations by a screw compressor with variable inner volume ratio. In: VDI (Hg.): *9th International Conference on Screw Machines. Dortmund. VDI, S.* 107–123, 2014

[6] Hauser, J.; Beinert, M.: CFD analysis of pressure pulsation in screw compressors – Combine theory with practice. *8th International Conference on Compressors and their Systems,* 2013

[7] Fujiwara, A.; Sakurai, N.: Experimental analysis of screw compressor noise and vibration. *The 1986 International Compressor Engineering Conference at Purdue,* 1986

[8] Mujic, E.; Kovacevic, A.; Stosic, N.; Smith, I. K.: Noise generation and suppression in twin-screw compressors. *Proceedings of the Institution of Mechanical Engineers, Part E: Journal of Process Mechanical Engineering 225 (2), S.* 127–148. DOI: 10.1177/1464419311403875, 2011

[9] Huang, P.; Yonkers, S.; Hokey, D.: Screw pulsation generation and control: a shock tube mechanism. *8th International Conference on Compressors and their Systems,* London 2013

[10] Huang, P.: Under Compression: An Isochoric or Adiabatic Process? *The 2012 International Compressor Engineering Conference at Purdue,* 2012

[11] WiSTL Shock Tube Calculator: http://silver.neep.wisc.edu/~shock/tools/gdcalc.html

[12] Rudinger, G., 1969, Non-steady Duct Flow: Wave Diagram Analysis, *Dover,* New York

[13] Huang, P., Yonkers, S., Hokey, D., Gas Pulsation Control Using a Shunt Pulsation Trap. *The 2014 International Compressor Engineering Conference at Purdue,* 2014

[14] Huang, P., Yonkers, S., Screw Compressor with a Shunt Pulsation Trap, *US Pat# 9,151,292,* 2015

[15] Huang, P., Groll, E., Yonkers, S., Parallel vs. Serial Noise Suppression Method for Positive Displacement (PD) Compressors & Internal Combustion Engines (ICE). *The 2018 International Compressor Engineering Conference at Purdue,* 2018

Acknowledgments

The author would like to thank Professor Brümmer for his introduction and invitation to the ICSM 2018 conference.

IOP Conf. Series: Materials Science and Engineering **425** (2018) 012022 doi:10.1088/1757-899X/425/1/012022

Notation

a	speed of sound
CW	compression wave, pulsation component
DVC	discharge valve closed
DVO	discharge valve open
EW	expansion wave, pulsation component
IFF	induced fluid flow, pulsation component
IEW	incident EW
ICW	incident CW
L	length
OC	over compression
ρ	gas density
p	gas pulsation pressure
Δp_{41}	gas pulsation source or measure of OC/UC
P	absolute gas pressure
PD	positive displacement
Q	volume flow rate
REW	reflected EW
RCW	reflected CW
SPT	shunt pulsation trap
t	time
UC	under compression
U	fluid flow velocity
ΔU	IFF velocity
V	PD compressor volume
W	wave velocity

Subscripts

1	initial low pressure in shock tube
2, 3	post opening pressures in shock tube
4	initial high pressure in shock tube
A, B…	locations
Inlet	compressor inlet
Outlet	compressor outlet

Numerical study on screw machines with large helix angles

Yang Lu, Ahmed Kovacevic and Matthew Read

City, University of London, Centre for Compressor Technology, London, UK

E-mail: yang.lu.4@city.ac.uk

Abstract. Modelling and performance calculation of screw machines with large helix angles such as a single and multiphase screw pumps by use of Computational Fluid Dynamics is challenging. The numerical procedures explained in literature are based on the 3D numerical meshes generated by series of 2D transverse cross sections which allows mesh to either follow the helix or be perpendicular to the rotor axis. This allows generating a conformal mesh. However, if the rotor helix angle is large, the cell skewness becomes prohibitively large which introduces errors in numerical simulation. The paper firstly attempts to generalize the generation method of rotor profiles with emphasis on producing a normal rack and rotors profiles. Then it introduces the method which uses series of 2D numerical meshes in the planes normal to each of the rotors and rack in order to decompose the working domain in two sub-domains. By this means it is possible to achieve 3D numerical mesh with extended capability of mesh refinement in clearances and alignment of the mesh to the main and leakage flows. However, special treatment is required to provide conformal interface between two moving meshes and with the casing. It is expected that it will greatly benefit accuracy and ease of performance calculation using a number of CFD solvers. In addition it is expected that it will allow generation of various different screw machine configurations like single screw machines or machines with conical rotors.

1. Introduction

Single and twin screw machines are positive displacement rotary machines comprising a meshing pair of rotors contained in a casing, which together form a working chamber whose volume depends on the angle of rotation. The performance and efficiency of screw machines are mostly related to the rotor profiles and machining precision. Generally, there are three different methods to generate rotor profiles namely, generation from a given curve of a main rotor, generation from a meshing line and generation from a rack on a transverse or on a normal plane. The envelope method is the basis for the design of rotor profiles of screw rotors. The envelope gearing method of the screw rotors is similar to the method used for helical gears. Litvin[1] described the process of coordinate transformation in matrix representation and implemented it in the generation of gear and rack cutter. Stosic[2] used the envelope gearing method to derive a general meshing condition for crossed helical gears which is then used to generate the profile of a hobbing or cutting tool. Stosic and Hanjalic[3]presented a rack based procedure to generate various lobe combination rotors. Refer to the rack-generated profile method of Stosic, Yu-ren[4] explicitly redefined the rack which including two normal-equidistant trochoids using at least one control parameter to each compound curve, which contributes to the flexible and adjustable rack. Then a method using a rack defined in the normal plane to design the twin screw rotor profiles was proposed by him[5]. The advantage of this method is that the normal circular pitch is same for main and gate rotor as the helix angle varies.

In order to enable performance calculation by CFD, the working domain between the casing and the rotors has to be mapped with a numerical grid. A 3D flow domain between the casing and two rotors is discretized by a number of 2D cross sections. In each of these cross sections, the flow domain is divided in two subdomains for easier grid generation. Practically, three methods can be used to decompose a 2D cross section of the flow domain, namely, differential division line, transverse rack and a normal rack. Kovacevic[6] was the first to use the analytical rack in the transverse cross section for decomposition of the 2D cross section of the flow domain as shown in Figure 1. Recently, as shown by Sham[7, 8], algebraic grid generation is used for a variable profile rotors with parallel axis using transverse rack. In this procedure, the split line is the rack segment stretching between the bottom and top cusp point. In his case, each 2D cross section has a different rotor profile while the position of transverse planes can also vary depending on the lead of the rotors. The two O blocks together form a composite grid. Voorde et al [9]constructed a block structured mesh using similar decomposition but has applied gradient lines and the potential lines in the transverse planes to obtain numerical solution. This often leads to a large gradient of volumes around the CUSP.

In the case of machines with non-parallel axes such as single screw compressors or twin screw compressors with large helix angle, cells formed from the consecutive 2D transverse planes may become too skewed for conservative solution. The large helix angle can cause a twist of the mesh too large for reliable grid generation. Despite a significant increase in the number of papers published recently in the area of grid generation for screw machines, very few journal papers have been published on grid generation of screw machines with large helix angles. In this paper, the proposed normal rack grid generation method will be proposed to decompose the fluid domain using normal planes. Grid Generation in this 2D normal cross section is expected to be much easier and would be directly applicable for screw machines with large helix angles. This paper proposes use of a normal rack in order to separates flow domains of the male and female rotors.

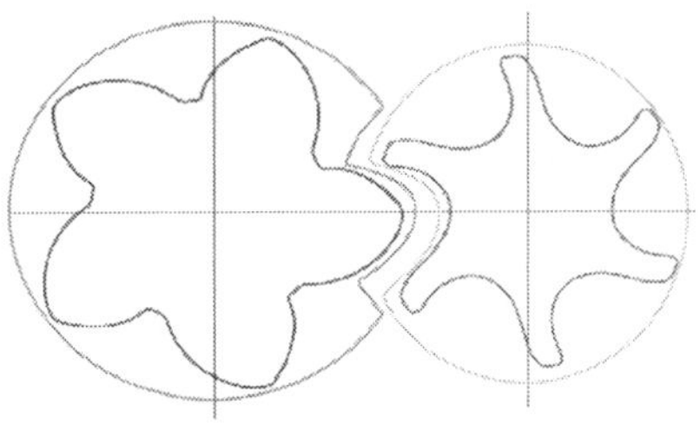

Reproduced from Kovacevic 1998

Figure 1. Transverse rack as a splitting curve.

2. Rotor generation

The performance of a screw compressor mainly depends on the design of rotor profiles and machining precision for their manufacturing. Nowadays, screw rotors are manufactured mainly by milling, grinding and hobbing tools for the rough and finish machining. Form tools for rotor manufacture can be regarded as crossed helical gears on non-parallel and non-intersecting axes system with a uniform lead such as in the case of form hobbing, or with no lead as in formed milling or grinding. The rotor profiles of twin screw compressors are usually with parallel and non-intersecting axes. Generally, there are three different mathematical methods for rotor profiles generation, namely, generation from a given main or gate rotor profile curve, generation from a meshing line and generation from a specifying rack on the transverse or normal plane. The envelop method is the basics for generation of conjugate rotor profiles.

IOP Conf. Series: Materials Science and Engineering **425** (2018) 012023 doi:10.1088/1757-899X/425/1/012023

It states that two rotors transforming in prescribed motion can generate or envelope each other under necessary conditions.

2.1. Applied coordinate systems

To start the procedure of rotor profiling, the coordinate system must be defined firstly. Single screw rotors and their cutting tools are normally defined in a non-parallel axes coordinate system while the twin screw rotors have parallel axes.

2.1.1. Non-parallel axes coordinate system

The coordinate system of non-parallel and nonintersecting axes is given in Figure 2. The shortest distance between the two axes is C and the angle between them is Σ. Consider coordinate system $S_1(X_1, Y_1, Z_1)$ and $S_2(X_2, Y_2, Z_2)$ that are rigidly connected to main rotor and gate rotor respectively and coordinate system $S_1(X_1, Y_1, Z_1)$ is the global coordinate system. And the coordinate system $R(X_{01}, Y_{01})$ rotates around Z_1 with main rotor and the coordinate system $R(X_{02}, Y_{02})$ rotates around Z_2 with gate rotor, which are local coordinate system. This global and local coordinate systems can be used to generate the single screw rotors and form cutter profiles from the rotor profiles.

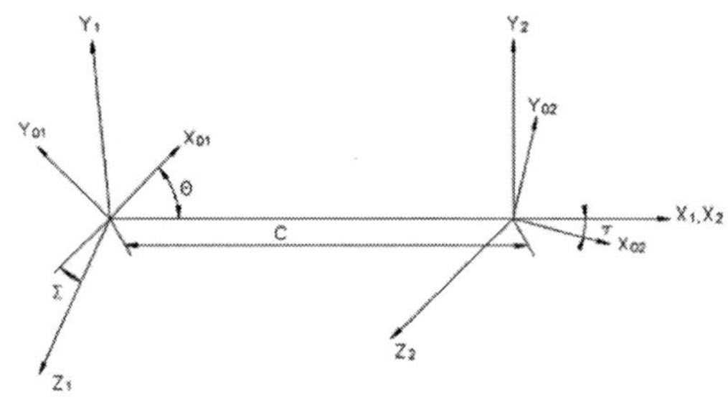

Figure 2. Coordinate systems of non-parallel and non-intersecting axes.

2.1.2. Parallel axis coordinate system.

When the angle between two axes is $\Sigma = 0$, the parallel and intersecting axes coordinate system can be obtained and the profiles of main and gate twin screw compressor rotors can be designed and generated in this coordinate system as shown in Figure 3. r_{1w} and r_{2w} are the radii of the pitch circle of main rotor profile and gate profile respectively. So the axes distance is $C = r_{1w} + r_{2w}$.

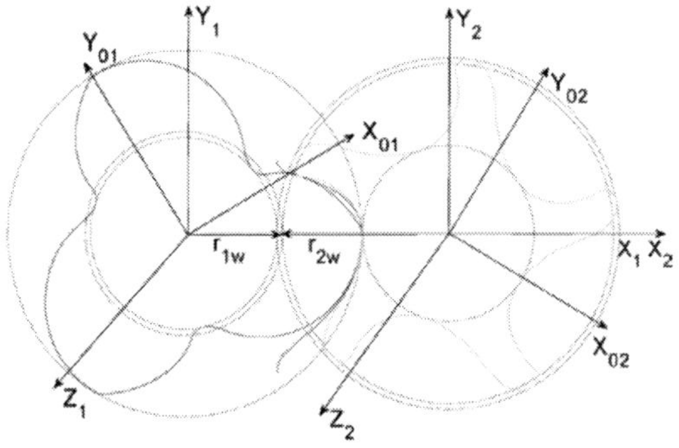

Figure 3. Coordinate systems of parallel and intersecting axes.

2.2. *Coordinate transformation*

With the coordinate system, the matrix of coordinate transformation provide the mathematical methods for generating profiles of the rack from the rotor profiles and conversely, generating the rotor profiles from the rack. With the coordinate systems showed in Figure 2, the coordinate transformation in transition from the main rotor profile coordinate S_1 to the gate rotor profile coordinate S_2 is based on the matrix equation:

$$\mathbf{r}_2 = M_{21}\mathbf{r}_1 = M_{2p}M_{pf}M_{f1}\mathbf{r}_1 \tag{1}$$

$$\mathbf{r}_1 = \begin{bmatrix} x_1 \\ y_1 \\ z_1 \\ 1 \end{bmatrix}, \mathbf{r}_2 = \begin{bmatrix} x_2 \\ y_2 \\ z_2 \\ 1 \end{bmatrix}, \mathbf{M}_{pf} = \begin{bmatrix} 1 & 0 & 0 & -C \\ 0 & \cos(\Sigma) & -\sin(\Sigma) & 0 \\ 0 & \sin(\Sigma) & \cos(\Sigma) & 0 \\ 0 & 0 & 0 & 1 \end{bmatrix}$$

$$\mathbf{M}_{2p} = \begin{bmatrix} \cos(\tau) & -\sin(\tau) & 0 & 0 \\ \sin(\tau) & \cos(\tau) & 0 & 0 \\ 0 & 0 & 1 & 0 \\ 0 & 0 & 0 & 1 \end{bmatrix}, \mathbf{M}_{f1} = \begin{bmatrix} \cos(\theta) & -\sin(\theta) & 0 & 0 \\ \sin(\theta) & \cos(\theta) & 0 & 0 \\ 0 & 0 & 1 & 0 \\ 0 & 0 & 0 & 1 \end{bmatrix}$$

Where $r_1(t,\theta)$ and $r_2(t,\theta,\tau)$ are the generating and generated surfaces respectively and t is a profile parameter while θ and τ are motion parameters of main rotor profiles and gate rotor profile respectively. M_{2p} and M_{f1} are rotational matrices corresponding to rotation with the Z_1 and Z_2 axes respectively and M_{pf} is a translational matrix corresponding to the transformation from coordinate system $S_1(X_1,Y_1,Z_1)$ to $S_2(X_2,Y_2,Z_2)$. Here these matrices are used to generate the rack from the given rotor profiles.

3. Rack generation

The rack line is the curve representing a rotor with infinite radius. Namely, 3D rotors and rack surface do not change when the cutting plane changes. Only if they are cut with the particular plane, they form either transverse or normal 2D projections of rotors and rack. The rack generation procedure for screw rotor profiles was firstly proposed by Rinder[10]. He used the rack as a generating curve and both the main and the gate rotor profiles can be generated from the same rack in the same coordinate system. The rack line can mesh with the main rotor profile and gate rotor profile simultaneously. In addition to that convenience, a rack generation procedure offers the possibility to design various lobe combinations rotors using the same rack. As shown in the Figure 4, the main rotor profile and gate rotor profiles are both generated from the rack between the main and gate rotor profiles and the main rotor with three lobes shown in the left side is able to mesh with the gate rotors with five lobes or seven lobes shown in the right side. The red line between two rotor profiles is the rack line. The rack can be defined in the transverse plane system which is called transverse rack and in the normal plane system which is called normal rack. Yu-ren[5] proposed normal-rack generation method (NRGM) to design all the tangent continuous profile segments explicitly in the normal section to generate the conjugated rotors which can be utilised for various lobe combinations, centre distances even the different helix angles of the main and gate rotors.

IOP Conf. Series: Materials Science and Engineering **425** (2018) 012023 doi:10.1088/1757-899X/425/1/012023

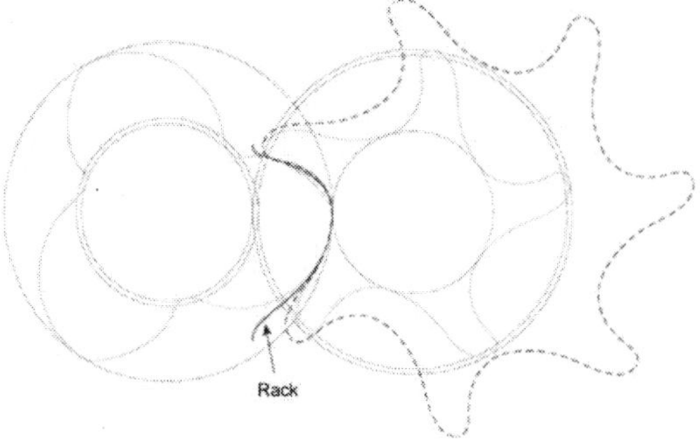

Figure 4 Rotor profiles of 3/5 and 3/7 with the same rack profile.

3.1. The transverse plane and the normal plane

The transverse plane is perpendicular to the rotor shaft. Usually the rotor profiles are generated in the transverse plane. The normal plane is perpendicular to the vector of the helix in the pitch circle. As showed in Figure 5, the transverse plane is indicated by the blue line and the normal plane is indicated by the red line. The angle between the transverse plane and normal plane is the helix angle β. For grid generation, if the transverse plane is used for the grid generation, then the transverse plane is unique for both rotors and rack. However, if the normal plane is used for grid generation, in that case each rotor will have a series of its own normal planes and the rack will have its own normal plane.

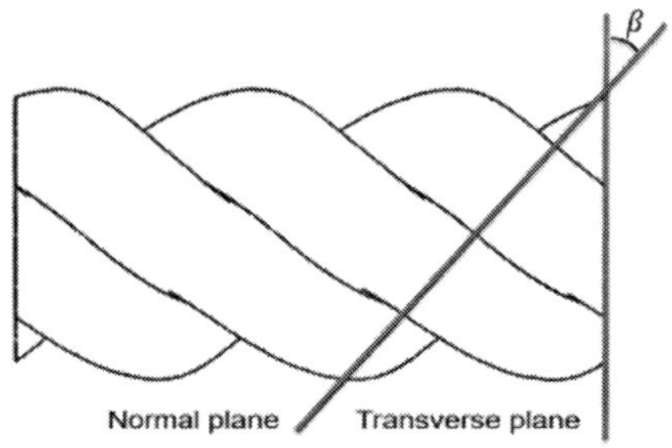

Figure 5. Transverse plane and normal plane.

3.2. Transverse rack

The transverse rack can be generated from the main rotor profile. The main rotor profile coordinate in system $R(X_{01}, Y_{01})$ is $r_1\ (x_{01}, y_{01})$ and the gate rotor profile coordinate in system $R(X_{02}, Y_{02})$ is $r_2\ (x_{02}, y_{02})$. For the parallel axes, since the angle between two axes is $\Sigma = 0$, gear ratio $i = \theta/\tau$ and $k = 1 + 1/i$. If the generated rotor profile is the main rotor, the meshing gate profile equations in the transverse plane can be obtained according to the Equation 1 as:

$$\begin{bmatrix} x_{02} \\ y_{02} \\ 0 \\ 1 \end{bmatrix} = \begin{bmatrix} \cos(\tau) & -\sin(\tau) & 0 & 0 \\ \sin(\tau) & \cos(\tau) & 0 & 0 \\ 0 & 0 & 1 & 0 \\ 0 & 0 & 0 & 1 \end{bmatrix} \begin{bmatrix} 1 & 0 & 0 & -C \\ 0 & 1 & 0 & 0 \\ 0 & 0 & 1 & 0 \\ 0 & 0 & 0 & 1 \end{bmatrix} \begin{bmatrix} \cos(\theta) & -\sin(\theta) & 0 & 0 \\ \sin(\theta) & \cos(\theta) & 0 & 0 \\ 0 & 0 & 1 & 0 \\ 0 & 0 & 0 & 1 \end{bmatrix} \begin{bmatrix} x_{01} \\ y_{01} \\ 0 \\ 1 \end{bmatrix}$$

Equation 2 can be obtained by simplification of the above equation as:

$$\begin{cases} x_{02} = x_{01}\cos(k\theta) + y_{01}\sin(k\theta) - C\cos(\frac{\theta}{i}) \\ y_{02} = -x_{01}\sin(k\theta) + y_{01}\cos(k\theta) + C\cos(\frac{\theta}{i}) \end{cases} \tag{2}$$

Rack coordinates can be calculated from Equation 2 if the gear ratio i tends to infinity:

$$\begin{cases} x_{0r} = x_{01}\cos(\theta) + y_{01}\sin(\theta) - r_{1w}\theta \\ y_{0r} = -x_{01}\sin(\theta) + y_{01}\cos(k\theta) \end{cases} \tag{3}$$

The rack can be regarded as the rotor with infinite radius which means the gear ration i tends to infinity. The main rotor profile and gate rotor profile can be generated from the rack profile $r_{0r}(x_{0r}, y_{0r})$ respectively as:

$$\begin{cases} x_{01} = x_{0r}\cos(\theta) - (y_{0r} - r_{1w}\theta)\sin(\theta) \\ y_{01} = x_{0r}\sin(\theta) + (y_{0r} - r_{1w}\theta)\cos(\theta) \end{cases} \tag{4}$$

$$\begin{cases} x_{02} = -x_{0r}\cos(\theta) - (y_{0r} - r_{2w}\theta)\sin(\theta) \\ y_{02} = x_{0r}\sin(\theta) + (y_{0r} - r_{2w}\theta)\cos(\theta) \end{cases} \tag{5}$$

By solving the meshing equation and the rack profile equations, the transverse rack profile can be obtained. The rack surface between two rotors can be obtained by projecting the transverse rack to the rack normal plane as shown in Figure 6. The rack surface always lies between the main rotor and gate rotor.

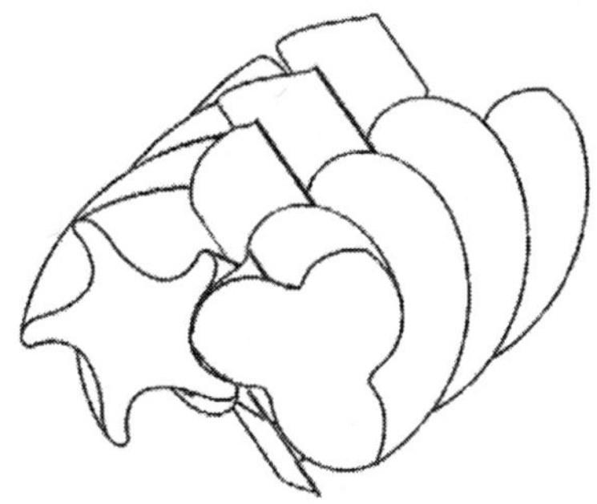

Figure 6. Rack surface.

3.3. Normal rack

The normal projection of the rack surface called the normal rack $r_{nr}(\theta, \beta)$ can be generated analytically from projecting the rack surface onto the normal plane. The normal rack profile equations are as follows:

$$\begin{cases} x_{nr} = x_{0r}\cos(\beta) \\ y_{nr} = y_{0r} \\ z_{nr} = x_{0r}\sin(\beta) \end{cases} \tag{6}$$

IOP Conf. Series: Materials Science and Engineering **425** (2018) 012023 doi:10.1088/1757-899X/425/1/012023

Where r_{nr} is the normal rack profile and β is the helix angle at the pitch circle. The normal rack for the main rotor and gate rotor in the normal plane is as shown separately in Figure 7.

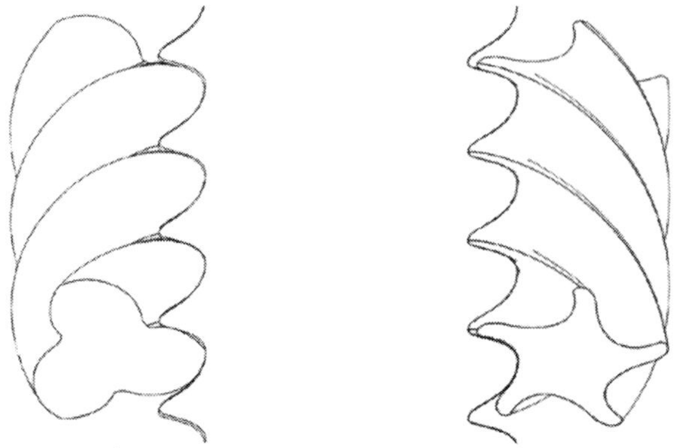

(a) (b)
Figure 7. Normal rack of (a) main rotor and (b) gate rotor.

4. Domain decomposition

The grid generation process of screw machine is performed in the Cartesian coordinate system introduced at the beginning of the paper and starts from the process of replacing the working chamber with a set of grid points. The decomposition of the flow region between the casing and two rotors consists of dividing the region into a series of 2D cross sections. In practise three methods can be utilised to decompose a cross section of the working domain, namely, transverse differential division curve, transverse rack and the normal rack. So far the grids of the fluid domain are generated in transverse planes. A different method using the normal rack to split the fluid domain in the normal cross section is proposed in this paper. It is more convenient if the main rotor and gate fluid domains are cutting by their normal planes respectively, so the grids can be generated in the normal planes. This process will make the grid lines between the consecutive planes nearly normal to the planes which will completely render or at least reduce the skewness of the grids. The 2D cross sections are then combined together to construct the full 3D fluid domain representing the main and gate fluid domains.

4.1. The fluid domain

The fluid domain of a screw rotor is helical type volume generated by the simultaneous revolving of the interlobe space around the rotor axis and translation along the axis. And the interlobe space can be separated to two domains according to the normal rack splitting surface. And then the rack line appears in the normal cross section. The outer boundary is the composition of the normal rack and the casing circle and the inner boundary is the rotor profile. To visualize the changing of rotor domain, two sub-domains are established as shown in Figure 8. The sub-domain around one lobe of the main rotor denoted by the green colour and the sub-domain in one interlobe of the gate rotor denoted by the red colour are meshing with each other and the unique rack surface is between the two rotors.

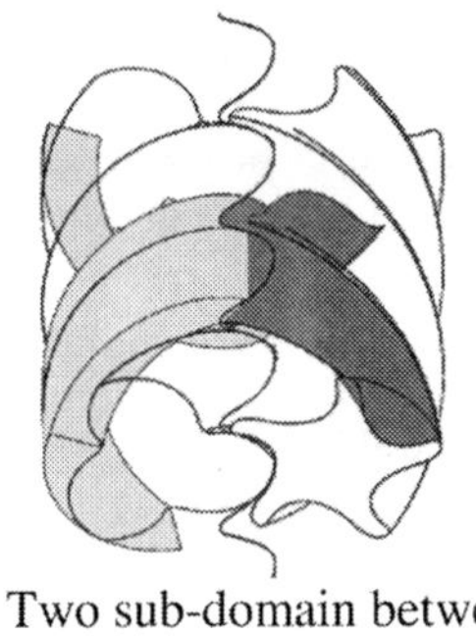

Figure 8. Two sub-domain between the rotors.

4.2. Decomposition method

The fluid domain between the rotors and casing is represented by a number of cross sections using the normal planes of the main rotor and gate rotor respectively. Then the normal rack line and the casing will form the outer boundary of the fluid domain while the rotor profile acts as the inner boundary. Figure 9 shows the sub-domain of the gate rotor cut by 20 consecutive normal planes to show how is the fluid domain changing. The orientation of the profile is aligned with the orientation of normal planes. The fist cross section is cut out at the end of the domain. Following that, the six normal cross sections are identical because the gate and the main sub-domains are apart. Once the two sub-domains interface, the normal rack replaces the outer circle as the outer boundary gradually cutting in and then out after which the domain returns to the shape formed by the outer circle and the rotor. If all of these 2D normal planes are meshed with a numerical grid, then a three dimensional fluid domain can be obtained by combining meshes in two dimensional cross sections. There are many advantages of the domain decomposition in normal planes. Firstly, the meshes in the clearance between rotors and casing can be refined by adjusting the density of nodes in required sections and the alignment of the mesh to the main and leakage flows can be achieved. Secondly, orthogonality of the mesh can be easily obtained. All of this advantages can contribute to a robust and efficient simulation of twin screw machines using a number of CFD solvers.

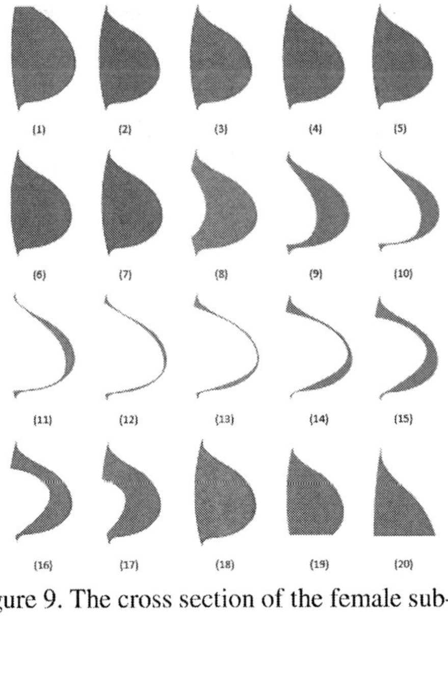

Figure 9. The cross section of the female sub-domain.

5. Conclusions

In this paper, the analytical racks of screw rotor in transverse plane and normal plane are firstly introduced. Then it introduces a method which uses normal rack to decompose the fluid domain between the twin screw rotors and the casing into two sub-domains in which numerical meshes can be generated in the planes normal to each of the rotors. By this means it is possible to achieve 3D numerical mesh with extended capability of mesh refinement in clearance and alignment of the mesh to the main and leakage flows. This method can also be used in the single screw and non-parallel axis compressors like internal conical rotary screw compressor. Application of the method will be demonstrated in the case of screw machines with large helix angles as used in single and multiphase screw pumps from the simulated and experimental results.

Acknowledgments
This research was supported by China Scholarship Council and the Centre for Compressor Technology at City, University of London.

References

[1] Litvin F L and Fuentes A 2004 *Gear Geometry and Applied Theory* (New York: Cambridge university press)
[2] Stosic N 1998 On gearing of helical screw compressor rotors *Proc. Inst. Mech. Eng.* 212 p587
[3] Stosic N and Hanjalic K 1996 General Method for Screw Compressor Profile Generation *Int. Compressor Engineering Conf. at Purdue* (West Lafayette, Indiana, USA) pp 157-162
[4] Wu Y and Fong Z 2009 Optimization design of an explicitly defined rack for the generation of rotors for twin-screw compressors *Mechanism and Machine Theory* **44** pp 66-82
[5] Wu Y and Fong Z 2008 Rotor Profile Design for the Twin-Screw Compressor Based on the Normal-Rack Generation Method *Journal of Mechanical Design* **130**
[6] Kovačević A 2002 Three-Dimensional Numerical Analysis for Flow Prediction in Positive Displacement Screw Machines Degree of Dorctor of Philosophy In: *School of Engineering*: (City University London)
[7] Rane S and Kovacevic A 2017 Algebraic generation of single domain computational grid for twin screw machines. Part I. Implementation *Advances in Engineering Software* **107** pp 38-50
[8] Rane S and Kovačević A 2017 Application of numerical grid generation for improved CFD analysis of multiphase screw machines *IOP Conf. Series: Materials Science and Engineering* **232**
[9] Voode J V, Vierendeels J and Dick E 2004 A grid generator for flow calculationsin rotary volumetric compressors. In: *European Congress on Computational Methods in Applied Sciences and Engineering* (Finland)
[10] Rinder L 1987 Screw rotor profile and method for generating. (United States)

Numerical and Experimental Analysis of Transient Flow in Roots Blower

S H Sun[1,2]*, A Kovacevic[2*], C Bruecker[3], A Leto[2,4], G Singh[2] and M Ghavami[2]

[1]Xi'an University of Technology, No.5, South Jinhua Road, 710048, CHINA.
[2]Centre for Compressor Technology, City University London, EC1V0HB, UK.
[3]Sir Richard Oliver Chair in Aeronautical Engineering, City University London, EC1V0HB, UK.
[4]Faculty of Mechanical Engineering, Džemal Bijedić University of Mostar, University Campus, 88104, Bosnia & Herzegovina.
* Corresponding Author.
Email: sun.shuaihui@city.ac.uk, a.kovacevic@city.ac.uk.

Abstract. The performance of rotary positive displacement machines highly depends on the operational clearances. It is widely believed that computational fluid dynamics (CFD) can help understanding internal leakage flows. Developments of grid generating tools for analysis of leakage flows by CFD in rotary positive displacement machines have not yet been fully validated. Roots blower is a good representative of positive displacement machines and as such is convenient for optical access in order to analyse internal flows. The experimental investigation of flow in optical roots blower by phase-locked PIV (Particle Image Velocimetry) performed in the Centre for Compressor Technology at City, University of London provided the velocity field suitable for validation of the simulation model. This paper shows the results of the three-dimensional CFD transient simulation model of a Roots blower with the dynamic numerical grids generated by SCORG and flow solution solved in ANSYS CFX flow solver to obtain internal flow patterns. The velocity fields obtained by simulation agree qualitatively with the experimental results and show the correct main flow features in the working chamber. There are some differences in the velocity magnitude and vortex distribution. The flow field in roots blower is highly turbulent and three-dimensional. The axial clearances should be included, and the axial grids should be refined in the simulation method. The paper outlines some directions for future simulation and experimental work. The work described in this paper is a part of the large project set to evaluate characteristics of the internal flow in rotary positive displacement machines and to characterize leakage flows.

1. Introduction

Rotary positive displacement machines are widely used in many industrial fields. Depending on the application they may contain one or more rotating elements and a stator. Typical representatives of a single rotor machine are progressive cavity pumps and single screw compressors. Twin rotor machines are more common. These can be designed either with straight lobes as in roots blowers and gear pumps, or with helical lobes used in screw compressors, expanders and pumps. Screw machines can handle single phase fluids in the form of a gas, vapour or liquid or multi-phase fluids mixed from any combination of single phase fluids and solids and may operate above or under atmospheric pressures. Liquid and multiphase pumps are often configured with multiple rotors. In all these machines, gaps between rotating and stationary parts have to be maintained in order to allow a safe and reliable operation but are desired to be minimal in order to reduce leakage flows, which play critical role in the performance. The challenge is to maintain the size of the gaps due to deformations of the machine elements which could be caused by thermal of physical loads.

Many researchers have studied leakage flows through clearance gaps in rotary positive displacement machines both experimentally and numerically. Numerical methods are mostly based either on chamber modelling [1], or computational fluid dynamics (CFD) model [2, 3]. In chamber models, it is usually assumed that the momentum change in the main domain is negligible due to the internal energy being dominant while the velocity of the leaking fluid is obtained based on the assumption of the isentropic flow through the nozzle. A CFD model allows more accurate calculation of velocities both in the main domain and in the leakage paths by numerically solving governing conservation equations such as mass, energy and momentum. This is of course subject to availability of an accurate numerical mesh which can capture both, the main flow domain and clearances. The latest developments in grid generation for screw machines described in detail in Rane et al. [4, 5] have led to the mesh which can be used in all flow calculations and for most rotary positive displacement machines. This grid generation methods allows use of any commercially available CFD solvers. The size of the mesh, the speed of its generation as well as the speed of calculation by commercial solvers is suitable for industrial application. However, it is yet not fully validated if it sufficiently accurately captures flow in clearances.

Numerical procedures for calculation of performance using either chamber models or 3D CFD are usually validated by measurements of the integral parameters such as the total mass flow rate and power as shown in recent studies by Kovacevic and Rane [6]. These indicate that the clearance flow is mostly well captured. However, unless the local velocities are measured, the leakage models cannot be fully validated. In addition, even the velocity distribution in the main flow of a rotary positive displacement machine has not been studied in detail experimentally. Therefore, for the full validation of numerical calculations it is required to obtain accurate measurements of the flow field both in the main working domain and in the clearance gaps of a rotary positive displacement machine.

Attempts were made in the past to evaluate local flows in screw compressors using optical methods. Firstly, Sachs [7] studied the gas flows in the tip gap of a static flat rotor screw compressor experimentally. The Toppler schlieren method was applied to obtain the swirls image in the leakage gap. The influence of pressure ratio, gap shape, gap height and moving boundary on the schlieren image were investigated. And the mass flow rate and velocity near the gap were compared. The boundary layer separation, swirls and Mach wave were also shown. The LDV method was recommended to obtain quantitative results in the gap without interference with the leakage flow. The Laser Doppler Velocimetry (LDV) measurements in the working chamber close to the discharge and in the discharge chamber of an oil free compressor were reported by Gueratto *et al.* [8] while the measurement of the flow in the suction chamber of an oil free compressor with water injection were reported by Kovacevic *et al.* [9]. However, neither of these managed to give quantitative velocity values within the working domain of the compressor and in the clearances. Screw machines with helical rotors are difficult to modify in order to provide optical access for the assessment of flow. The alternative is a roots blower which has straight lobes and similar characteristics of the flow as screw machines.

Roots blowers usually have two straight rotors with an involute profile and two or more lobes in each. The rotors rotate in the opposite directions within the casing and form a working chamber between the rotors and casing. Roots blowers do not have internal compression as the volume of the chamber remains constant while rotating. The increase of the pressure is external to the rotors due to the backflow from the high-pressure side. Recently, Liu *et al.* [10, 11] and Sun *et al.* [12, 13] established the CFD simulation model of the roots blower and validated it by measurements of mass flow rate and pressure distribution. In addition, the leakage flow in a two-lobe roots blower was predicted using a stationary mesh in CFD and the results were compared with the experimental leakage mass flow rate [14]. The development of experimental techniques to determine velocities using optical methods have been reported recently by Sun *et al.* [15].

In this paper, the three-dimensional CFD transient simulation model of a Roots blower was established using the dynamic grids prepared by SCORG and was solved in ANSYS CFX flow solver to obtain the flow patterns. The experimental investigation of flow in optical roots blower by phase-locked PIV (Particle Image Velocimetry) performed in the Centre for Compressor Technology at City, University of London [15] was used to validate the simulation model. The simulation and experimental velocity fields are analysed and compared. The work described in this paper is a part of the large project set to evaluate characteristics of the internal flow in rotary positive displacement machines and to characterize leakage flows with the objective to lead to further improvements in 3D CFD analysis of leakage flows in rotary positive displacement machines and ultimately lead to the improvement in the performance of rotary positive displacement machines.

2. Simulation model

2.1. Physical Model

This paper studies a roots blower with two-lobed rotors mounted on parallel shafts that rotate in opposite directions to transfer air as working fluid. The physical model is shown in Figure 1. The suction and discharge pipes connect to the inlet and outlet domains respectively. The air passes into the roots blower through the inlet and fills into the suction chamber formed by the casing and rotor. The female rotor rotates clockwise, and the reference crank angle when the female rotor is vertical is defined as 0°. Three types of gaps are recognised in the roots blower, namely the tip gap, the interlobe gap and the axial gap between the side of the lobes and the casing. The main dimensions are shown in Table 1.

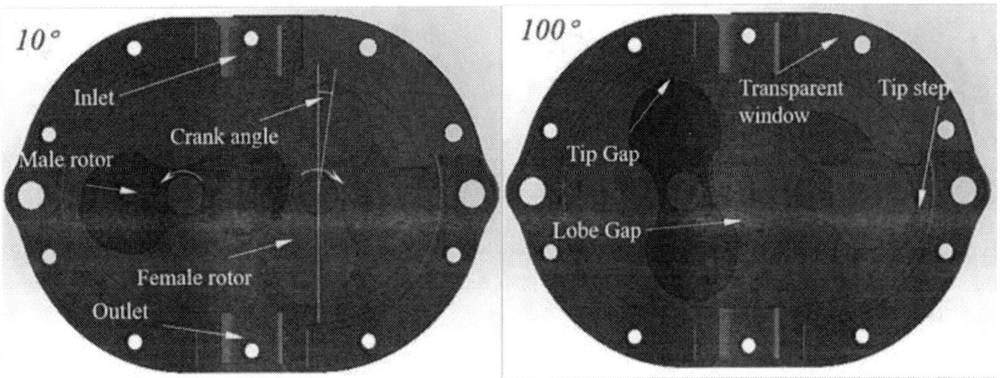

Figure 1. Physical model of roots blower at crank angle 10 ° and 100 °.

Table 1. The main parameters of the roots blower.

Items	Specification	Items	Specification
Diameter of the rotor/mm	101.3	Tip gap/mm	0.1
Axis distance/mm	63.12	Interlobe gap/mm	0.17
Rotor length/mm	50.5	Axial gap/mm	0.15
Displacement volume/ l/rev	0.4618	Width of tip step/mm	6.4

2.2. Grid generation

Stationary fluid domains of the inlet and outlet chambers can be extracted from a CAD model of the roots blower. For the transient simulation, rotor domains move and deform with the rotation of rotors. Figure 2 shows the fluid domain and the generated grids of rotors and ports. The moving rotor grids (left in Figure 2) were generated by in-house grid generation software SCORG. The rotor to casing non-conformal mesh was generated using algebraic transfinite interpolation and numerical smoothing. This preserves the shape of the tip and side steps on the lobes (see Figure 1). To check the grid independence, four levels of rotor grids were generated. Different grid levels used for calculations are shown in Table 2. The angular divisions for one chamber are 180, which means that there are 180 angular positions for single interlobe rotation. For the two-lobe Roots blower, the gird files were generated with the angle interval of 1°. And the time step was calculated out using the following equation.

$$\Delta t = \frac{60}{anglular\ divisions * Z * n} \quad (1)$$

where Z is the number of lobes, and n is the rotation speed (rpm). Inversely, the angular divisions can be calculated out when the simulation time step is given. As shown in the right of Figure 2, the stationary grids consist of inlet grids, outlet girds and axial-gap grids which have six layers of grids in each axial gap. The lengths of suction pipe and discharge pipe are three and nine times of the pipe diameter. Hexahedral grids with 889405 nodes for the stationary grids were generated in ANSYS-Mesh and used in all cases. The rotor grids and the stationary grids are combined and connected with interfaces.

Table 2. Grid levels and number divisions.

Level	Circumference	Radial	Angular	Axial	Number of grid nodes in Rotors	Y^+ at the centre line of tip step
Grid-1	200	14	180	30	37200	1.8
Grid-2	302	22	180	30	861304	1.3
Grid-3	400	30	180	35	1785600	1.1
Grid-4	480	36	180	38	2772560	1.0

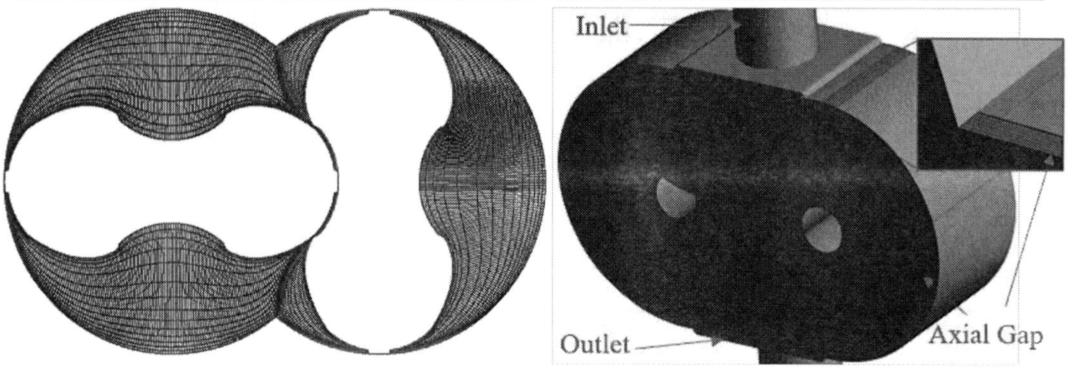

Figure 2. The rotor grids and the whole grids.

2.3. Simulation model boundary condition setting

The Ansys -CFX solver was used for the calculation of the roots blower. The working fluid was air as ideal gas. The high-resolution scheme was used for the advection term while the second order backward Euler scheme was used for the transient term. The SST (Shear Stress Turbulence) turbulence model was adopted in the calculation. The Y^+ at the centre line of the lobe tip step was checked and was lower than 2 in all cases, as shown in Table 2. The time step was 3.592E-4 as determined by equation (1) for the rotor speed of 464rpm. The number of iterations was 10 for every time step. The static pressure was used respectively for the inlet and outlet opening boundaries. The pressure at the inlet and outlet was given as measured ones during the experiment.

3. Experimental setup

The optical roots blower was mounted in the test rig which allows measurements of pressures, mass flow rates, power etc. at variable operating conditions. The rig also allows use of PIV laser technique for acquiring velocities inside the Roots blower. The layout of the test rig is shown in Figure 3a. The equipment used for measurements was introduced in detail in the paper published earlier by the authors [15]. The test was performed at the operating condition shown in the Table 3. The operation speed and pressure ratio are far lower than the design parameters of the prototype (5275rpm, pressure ratio 2.6). The work presented in this paper is the first step of the research to understand the mechanisms of leakage flows. The optical window, shown in Figure 3b, is made of perspex which can sustain only limited temperatures which was achieved by low speeds and pressures. The test at the practical operating conditions will follow in future.

Table 3. Operating conditions for the roots blower tests.

Inlet pressure/bar	Inlet temperature/K	Outlet pressure/bar	Outlet temperature/K	Pressure ratio	Speed/ rpm	Mass flow rate/ kg/s
0.975	300.6	1.045	306.2	1.072	464	0.00103

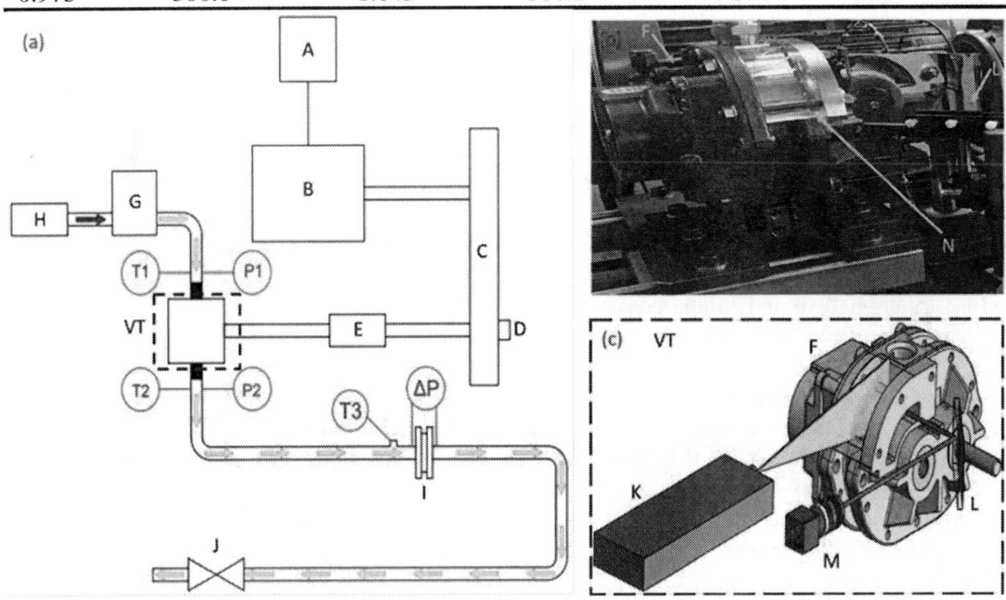

Figure 3. Roots blower test rig and diagram of visualizing test.

A - inverter, B – electromotor, C- pulleys, D – shaft encoder, E- torque meter, F – roots blower, G – smoke tank, H – smoke machine, I – orifice plate, J – valve, K – double pulse laser, L – surface mirror, M – double shutter camera, N – Position of Laser plane, T_1 – suction temperature transducer, P1 – suction pressure transducer, T2 – discharge temperature transducer, P2 – discharge pressure transducer, T3 – upstream temperature transducer, ΔP – differential pressure across an orifice plate

The optical window was designed to allow the optical access through two locations. One window is positioned radially while the other is on the side of the machine. The side window is used for collecting measurements through the reflection image in the mirror (L) in the Figure 3b. The Figure 3c is the 3D CAD model of standard PIV system for the test. The standard PIV system mainly consisted of double pulse laser (K, Dantec Dual power 200-15) and double shutter camera (M, Flow sensor E0) which has a full resolution of 2048*2048 pixels. The smoke particles made of liquid containing the glycol by a smoke machine (H) are used to seed the air. The smoke tank (G) was used to distribute the smoke homogeneously. The mixture of air and smoke then enters the roots blower and passes through the working chamber. A Nd:YAG laser with maximum 200mJ at the wave length of 532nm was used to illuminate the smoke particles. The laser was mounted on the laser arm and the beam was transformed to the laser sheet by using appropriate lenses at the end of the laser arm. The laser sheet plane (N) was 0.5cm inside the chamber from the window, as shown in Figure 3b. The test system uses a synchronizer to take the recording at a fixed crank angle, which is called a "phase lock". The shaft encoder shown in Figure 3a, was used to synchronise laser and camera to the required position of the lobe which was recorded by a Timer box. Once the lobe reached the required crank angle, the first laser pulse was emitted to light up the flow field. Then the camera takes the first picture. After a short interval of time of 2-50us which was dependent on the maximum velocity in the flow field, the second laser pulses, and the camera takes the second picture. This pair of pictures can be processed to calculate velocity of particles. Because the rotor has two symmetrical lobes, one completed circle including suction and discharge process takes up 180 °. The full cycle was divided into six steps with intervals of 30 °. The crank angles of shooting position are 10, 40, 70, 100, 130 and 160° respectively. Figure 1 shows the rotor positions at the crank angle of 10 and 100 °. The PIV results were processed using Dantec Dynamic studio by the adaptive PIV approach.

4. Results and discussion

4.1 Grid independence

Four cases with different grid density have been solved using Ansys CFX solver at the operation condition in Table 3. The cycle average mass balance between the rotor suction and discharge was achieved within 3% in all cases. Figure 4 shows the variation of the mass flowrate of the roots blower at various rotor grid refinements. The mass flow rate difference between the third level of the refinement and the fourth one is 4.0 %, which is far lower than the difference between the first level and second level. Therefore, the third level grids are regards as the enough fine grids. Because the interface between the two rotors in rotor to casing grids will influence the mass flow rate of the rotary machine [6], the casing to rotor grids which have the same divisions and nodes number with those of rotor to casing grids of Grid-3 were generated to be used in the simulation model whose results are selected for further analysis and compared with the experimental results.

4.2 Pressure variation and Mass flow rate

Figure 5 shows the static relative pressure variation in the working chamber formed by the female rotor and the casing. The reference pressure to compute this variation is 1bar. It is suction process when the crank angle moves from 0 to 180 °, and the discharge process corresponds to the 180 to 360 °. The pressure begins to increase at 170 ° and reach to maximum value at 200 °. It is mainly because of the tip step on the lobe which makes the suction chamber to close earlier and opens later, and finally extends the compression process and reduces the leakage flow during the process. There are two similar periodic pressure fluctuations during this process, and the second one begins at 270°. During every pressure fluctuation process, the pressure fluctuates sharply at the beginning and damps at the end, which has the inverse relation with the outlet mass flow rate shown in Figure 6. When the mass flow rate increases, the pressure decreases. If the crank angle reaches 270 °, the working chamber formed by the male rotor and the casing connects with the discharge chamber, so the outlet mass flow rate falls as shown in Figure 6, and the pressure in the female chamber increases, which leads to the second periodic pressure fluctuation.

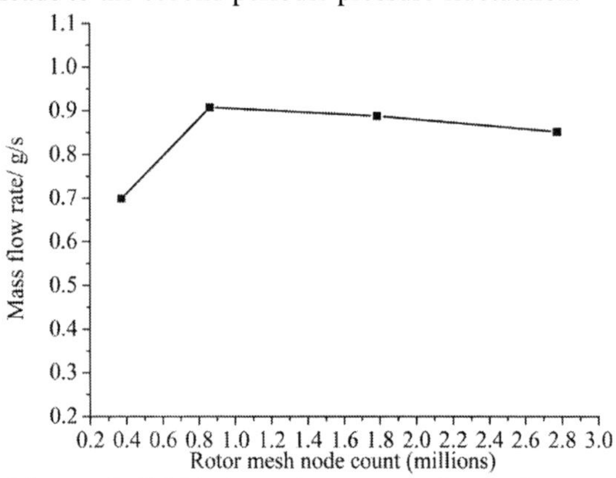

Figure 4. Grid independence analysis of a Roots blower rotor computational domain.

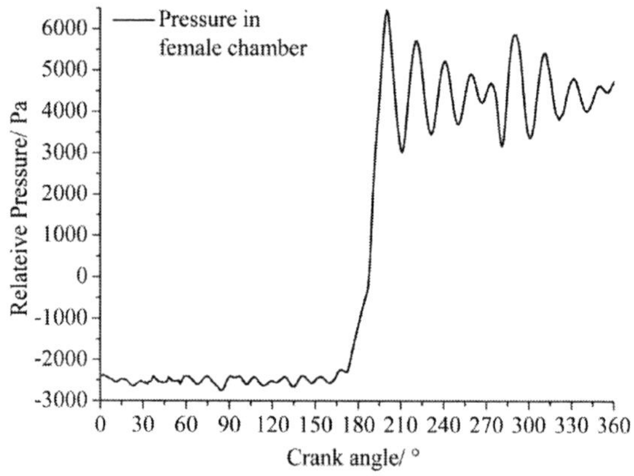

Figure 5. Pressure variation in working chamber.

The mass flow rate through the inlet and outlet surface were calculated and shown in Figure 6. The pink dash dot line represents the zero-mass flow rate line, and the lines below this line indicate that the fluid flows back. The blue dot line represents the mean value of the suction flow. The mean value of the inlet mass flow rate is 0.982g/s, which is 4.6% lower than the corresponding experimental value in Table 3. The reason for the deviation is caused by the measure error of orifice plate when it works under low pressure difference. In Figure 6, the total cycle has four repeated suction and discharge process. Because this roots blower has two same rotors which have two symmetric lobes respectively, the suction or discharge chamber formed by male rotor is symmetrical with that formed by the female rotor and has a period of 180 °. Hence, four periodic suction and discharge process appear, which also results in the two cycles of discharge pressure as shown in Figure 5. When the female lobe is at its vertical position, the crank angle is 0 °, as shown in Figure 1. Both inlet and outlet mass flow rate fluctuate with the crank angle. The outlet mass flow rate has higher pulsations than the corresponding inlet value. The reverse flow often happens at the outlet surface if the mass flow rate is negative. The maximum backflow rate of 5.9 g/s occurs at 15° and195° when the closed working chamber formed with female rotor connects with the discharge chamber. At that moment, the high pressure working fluid flows back into the working chamber, and the pressure in the working chamber increases quickly to the discharge pressure.

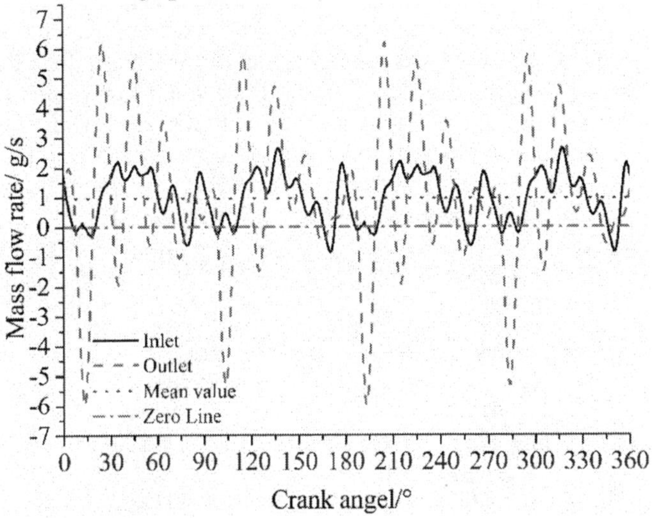

Figure 6. Inlet and outlet mass flow rate.

4.3 Comparison between the experimental and simulation results

Figure 7 shows the original figures from the standard PIV test at the shooting position of 10 ° and 100 °. The tracked particles can be distinguished easily in both pictures. At the crank angle of 10 °, the windows show the front of the lobe which has just closed the suction. At the crank angle of 100°, the window shows the area behind the lobe connected to the suction chamber. Two shadows caused by the bolts installed in the window (marked as 1 and 2) are visible in the frames, one at the top left of the window and another on the right side of the window which reduce the visible flow field.

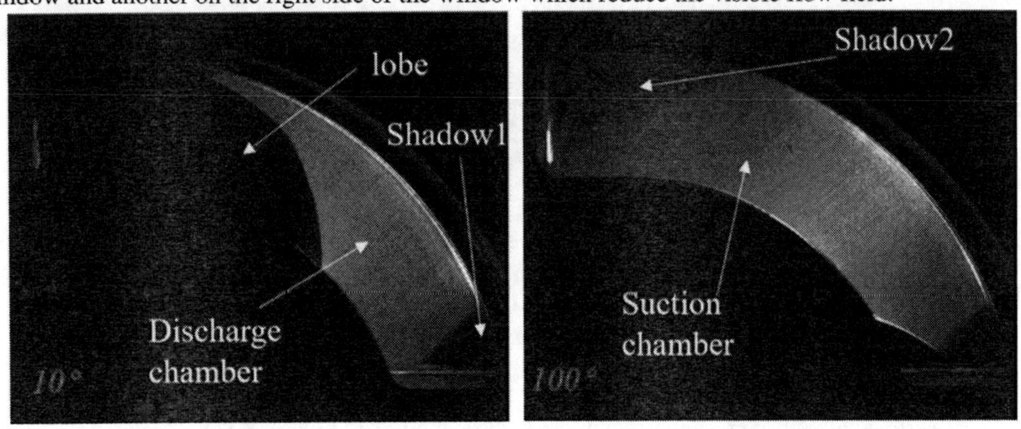

Figure 7. Original pictures of tracer particles at crank angle positions 10° and 100°.

The simulation and experimental results are compared in Figure 8. The velocity fields at three positions are displayed and compared. The colour bars of experimental and CFD results show the velocity magnitude in the plane of the laser sheet. The experimental results include the transient velocity flow field and average values obtained by averaging the transient velocity vectors from all 120 measured cycles. The simulation results are obtained from the results of next rotation cycle after the simulation was converged. The velocity vectors were calculated in the laser sheet plane. The black frame superimposed on the CFD results represents the border of visible area of the PIV pictures, so that the flow field in the frame can be compared with the measured velocity field. Comparing the simulation velocity field with the averaged PIV results, the fluid mainly flows downward at 10°; at 100°, the fluid in the two images flows from the bottom to the top and then is divided into two regions of flow movement. At 130°, most of fluid in the black frame moves towards the casing side. Hence, the simulation velocity fields have the similar flow features with the averaged PIV results. However, there are still obvious difference between the simulation and experimental average results. Firstly, the velocity magnitude of the simulation is larger than the experimental values. Also, the simulation velocity flow fields are more turbulent than the averaged PIV values which do not have the flow pattern in the same area A and C of the corresponding simulation images. In the simulation results of 100 °, the back flow in area B is caused by the tip leakage flow whose velocity decreases along the casing wall. The back flow and the main flow mixes and forms a low-velocity stripe, as shown in area A. The low-velocity area C results from the same reason. The experimental results do not show the same flow pattern, which implies that the simulation tip leakage flow may be overestimated.

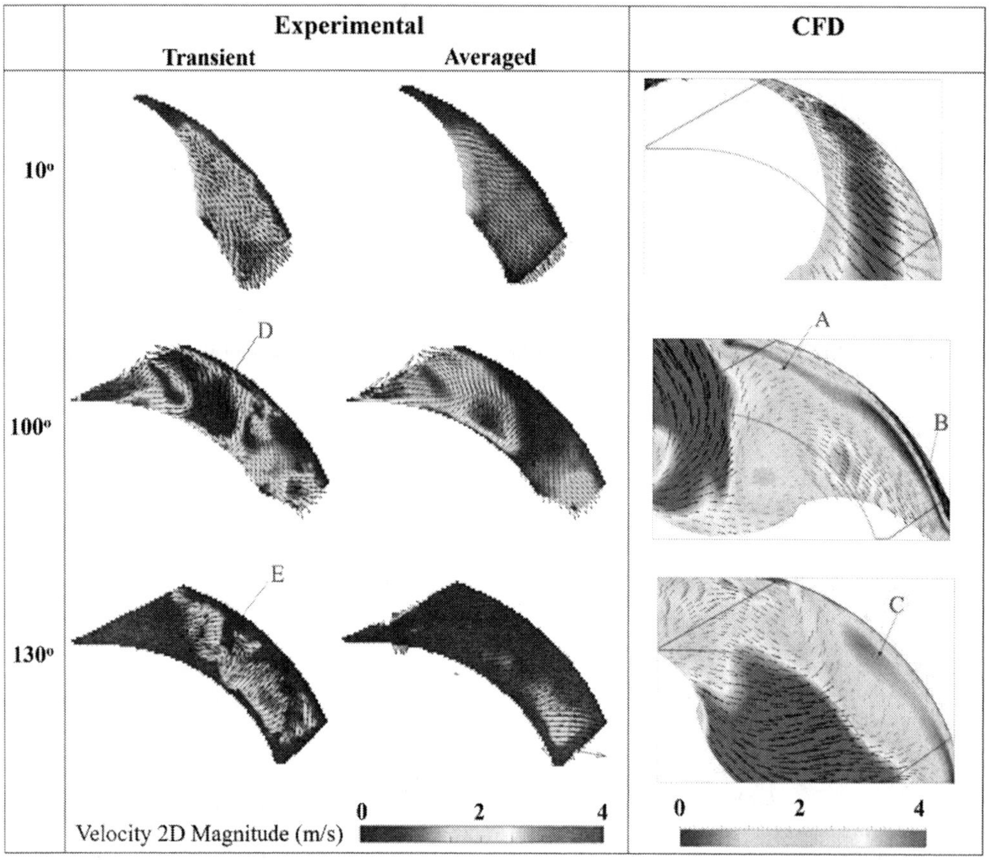

Figure 8. Comparison between the PIV results and CFD results at various crank angle positions.

The experimental transient velocity fields are also shown in Figure 8. The maximum velocity magnitude of the transient flow is close to that of the simulation velocity. And the main flow patterns in transient flow field are more similar to simulation flow field. But the transient flow field has more turbulence characteristics such as the vortex in areas D and E. Hence, the simulation results show the correct main flow features in the working chamber, but they have differences in the velocity magnitude and vortex distribution.

Usually, it is difficult to compare the transient PIV results with the URANS (Unsteady Reynolds-averaged Navier–Stokes) simulation results because the simulation model calculates the time-average velocity and adopts the Reynolds stress transport equations to impose the influence of turbulence on the time-average velocity. Some researchers use the simulation results of LES (Large Eddy Simulation) to compare with the transient PIV results [16, 17]. There is an indication that the simulation results agreed well with simulation results.

Many researchers compared the RANS or URANS simulation results with the time-average PIV results. Mortensen *et al.* [18] compared the simulation results of a rotor stator mixer from k-e, SST k-w and RSM (Reynolds stress model) with the time-average PIV results and found that simulation results over-predicted the dissipation rate of the TKE (turbulent kinetic energy) and indicated that the realizable *k-e* model was better than the other two turbulence models. Zou *et al.* (2016) used the standard *k-e* model to simulate the two-phase flow in a pump mixer and compared the simulation results with the PIV results, the deviation velocity error changes from 16.5% to 33%. Ryan *et al.* [19] adopted the SST-SAS (Scale adaptive simulation) turbulence model in the simulation of a sonolator liquid whistle and compared the simulation results with PIV time-averaged results. The simulation velocity agreed well with the experimental one, but the turbulence parameters matched with the experimental one poorly. Kurec *et al.* [20] had done simulation with standard *k-w*, SST *k-w* and SST-SAS turbulence model in a pressure exchange passage and validated the simulated results with PIV results. The comparison showed that the SST-SAS model can provide best prediction of the three turbulence models, but the simulation results are more turbulent than the PIV time-average results. Both Ryan *et al.* [19] and Kurec *et al.* [20] hinted that the CFD results needed to be averaged with different cycles because the PIV time-average results almost removed most of the turbulence features of the transient flow. Therefore, for our simulation, the CFD time-average results of several cycles will be firstly done to check if the simulation results can be improved, and then the SST-SAS model will be used in our simulation. LES simulation will be the last one because it requires a large amount of computational resources.

4.4 Analysis of the main velocity field

The experimental results can only capture a part of the domain while the simulation model can provide the whole 3D flow fields which can help to investigate the physical nature of the internal flow in a Roots blower. Figure 9 shows the whole simulation flow fields at two different axial planes. The flow fields in the left column are on the laser plane as shown in Figure 3. The crank angles are 10 ° and 100 ° which are same with the first two angles in Figure 8, so the downstream and upstream of the partial velocity fields in Figure 8 can be obtained. The velocity field in the right column are on the central plane which locates at the centre between the two side walls. Figure 10 displays the velocity flow fields in the cross sections along the axial direction. The cross section (a), is located in the mid-point between the centres of two rotors when the crank angle is 10 °; and the cross-section (B) which is also marked in Figure 9 is located at the one third of radius of the female lobe when the crank angle is 100 °. In addition, the axial positions of the central plane and the laser plane are shown in both cross sections of Figure 10. From Figure 9, the velocity direction in area A and C on the laser plane are opposite to that on the central plane. The reason is that the velocity field is three dimensional.

In Figure 10, two vortices with opposite sense of rotation exist on both cross sections. The vortices lead to the downward flow at the centre and upward flow near the sidewall, so the fluid at area A and C in the laser plane flows upward. The formation of the two vortices shown in Figure 10 is due to the side leakage through the axial clearance where the leakage velocity is very high, such as the area J. In the Figure 10a, the vortex L is bigger than the vortex K. The asymmetrical vortex distribution is mainly caused by the interlobe leakage flow. Most of the interlobe leakage flows right and makes the vortex L bigger than vortex K. But the reason why the interlobe leakage flows right under the symmetrical axial boundary condition is not clear. The unsteady simulation may contribute to this phenomenon.

Hence, the velocity field in roots blower is highly turbulent three-dimensional flow. The axial leakage clearance should not be neglected, or else the velocity field on the laser plane in Figure 8 will certainly

have big deviations with the PIV test results. Furthermore, the axial grids should be refined to improve the accuracy of simulation results. and the laser plane can move along the axis and measure the flow field at several planes including the central plane to validate the three-dimensional flow field. The situation is the same at the area B in Figure 9 where the fluid flows downward, but the fluid in central plane at the same area flows toward to the lobe. Apparently, there is one source D which results in the upward flow which influence the velocity direction at the area B of central plane. It is also caused by the three-dimensional flow.

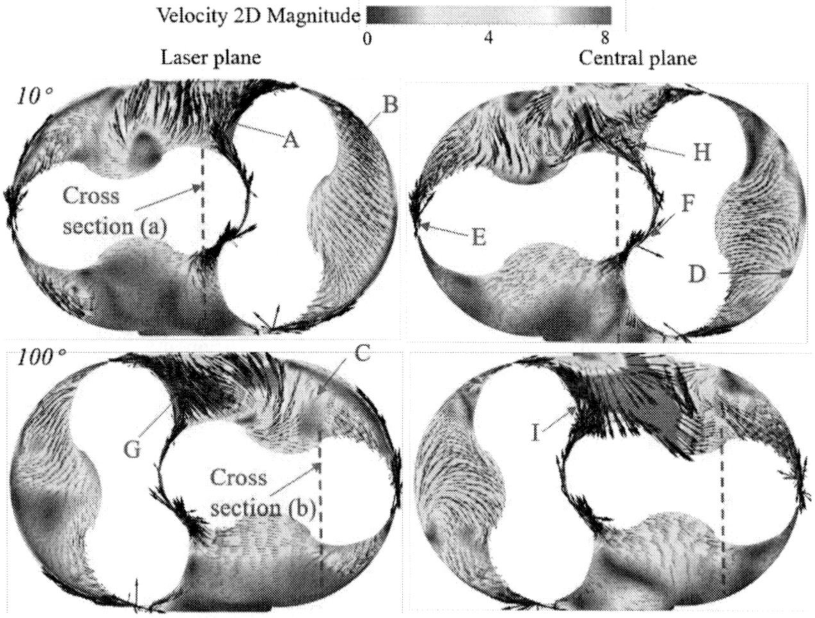

Figure 9. Simulation velocity fields at two different axial planes.

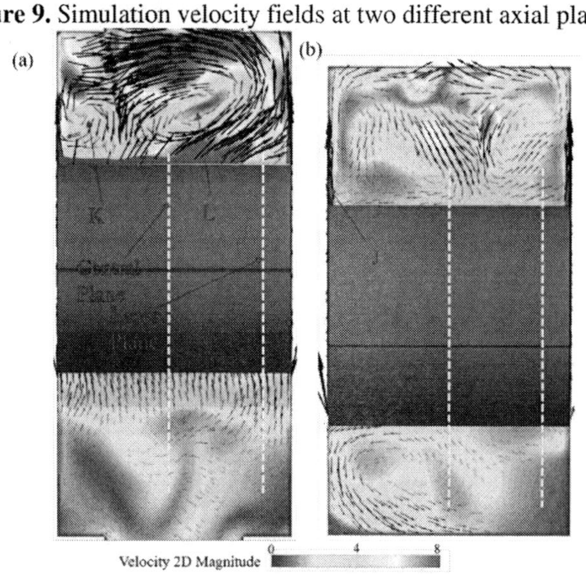

Figure 10. Simulation velocity field on the cross sections a and b as indicated in Figure 9.

The tip and interlobe leakages have different influence on the flow field as shown in Figure 9. The tip leakage has important influence on the downstream flow field of laser plane where it causes the low-velocity stripe while it has slight influence on the flow filed of central plane. It is because the main flow velocity on the central plane is bigger than that on the laser plane.

The maximum velocity in tip gap E is about 50m/s while the maximum velocity in the lobe gap F is 80m/s, which means the leakage flow rate through the lobe gap is far higher than that through the tip leakage. The interlobe leakage obviously influences the main flow pattern, as shown in Figure10a. In the laser plane, the interlobe leakage velocities have the same direction as the main flow, so it reinforces and accelerates the main flow such as the flow field G. However, in the central plane, the interlobe leakage flow velocities have the opposite direction with the main flow, so their influences are supressed. The leakage flow velocities reverses and the vortices H and I generate on the central plane.

5. Conclusions

The three-dimensional CFD transient simulation model of a Roots blower was established to obtain the internal flow field. The standard PIV test was conducted on the optical roots blower with transparent windows to validate the simulation model. The outlet mass flow has higher pulsations than the inlet mass flow. The maximum back flow happens when the working chamber connects to the outlet. The simulation velocity fields have similar features to the experimental results and show the correct main flow direction in the working chamber, but they show differences in the velocity magnitude and vortex distribution. Two asymmetric vortices exist on the cross section which leads the downward flow on the central plane and upward flow on the laser plane. The side leakage flow through the axial clearance has an important effect on the formation of the two vortices while the interlobe leakage makes the two vortexes asymmetrical. Therefore, the flow field in roots blower is highly turbulent and three dimensional. The axial clearance should not be neglected, and the axial grids should be refined, in order to ensure thc correctness of the simulation model. The interlobe leakage is larger than the tip leakage and affects the flow field significantly. It reinforces the main velocity field of the side flow (laser plane) and suppresses the flow in the central plane.

The future work will be conducted both on the improvement of the simulation model and the PIV test. The SST-SAS model will be used in simulation to improve the accuracy further. And the LES simulation will be considered at last. For the PIV test, the laser plane will be moved within the available domain to validate the prediction of the simulation model. And the microscope lens will be used to investigate the leakage flow. The operation speed and the pressure ratio will increase gradually, according to the quality of test results. The final objective of this research is to improve the 3D CFD analysis of leakage flows and ultimately lead to the improvement in the performance of rotary positive displacement machines.

References

[1] Stošić N, Smith I K and Kovačević A 2005 *Screw Compressors: Mathematical Modelling and Performance Calculation* (Berlin: Springer-Verlag)

[2] Kovacevic A 2005 Boundary adaptation in grid generation for CFD analysis of screw compressors *International Journal for Numerical Methods in Engineering* **64** 401-26

[3] Kovačević A, Stošić N and Smith I K 2007 *Screw compressors - Three-dimensional computational fluid dynamics and solid fluid interaction* (Berlin: Springer-Verlag)

[4] Rane S 2015 Grid Generation and CFD analysis of variable Geometry Screw Machines. (London: City, University of London)

[5] Rane S and Kovacevic A 2017 Algebraic generation of single domain computational grid for twin screw machines. Part I. Implementation *Advances in Engineering Software* **107** 38-50

[6] Kovacevic A and Rane S 2017 Algebraic generation of single domain computational grid for twin screw machines Part II – Validation *Advances in Engineering Software* **109** 31-43

[7] Sachs R 2002 Experimental investigation of Gas flows in screw machines In: *Faculty of Mechanical Engineering* (Dortmund: University of Dortmund)

[8] Guerrato D, Nouri J M, Stosic N, Constantine A and Smith I K 2007 Flow measurements in the discharge port of a screw compressor *Proceedings of the Institution of Mechanical Engineers, Part E: Journal of Process Mechanical Engineering* **222** 201-10

[9] Kovacevic A, Arjeneh M, Rane S, Stosic N and Gavaises M 2014 Flow Visualization at Suction of a Twin Screw Compressor. In: *International Screw Compressor Conference 2014,* (Dortmund: VDI Verlag Gmbh)

[10] Liu X M, Lu J, Gao R H and Xi G 2013 Numerical Investigation of the Aerodynamic Performance Affected by Spiral Inlet and Outlet in a Positive Displacement Blower *Chinese Journal of Mechanical Engineering* **26** 957-66

[11] Liu X M and Lu J 2014 Unsteady Flow Simulations in a Three-lobe Positive Displacement Blower *Chinese Journal of Mechanical Engineering* **27** 575-83

[12] Sun S K, Zhao B, Jia X H and Peng X Y 2017 Three-dimensional numerical simulation and experimental validation of flows in working chambers and inlet/outlet pockets of Roots pump *Vacuum* **137** 195-204

[13] Sun S K, Jia X H, Xing L F and Peng X Y 2018 Numerical study and experimental validation of a Roots blower with backflow design *Engineering Applications of Computational Fluid Mechanics* **12** 282-92

[14] Ashish M. Joshi, David I. Blekhman, James D. Felske, John A. Lordi and C.Mollendorf J 2006 Clearance Analysis and Leakage Flow CFD Model of a Two-LobeMulti-Recompression Heater *International Journal of Rotating Machinery* **2006** 1-10

[15] Sun S H, Kovacevic A, Bruecker C, Leto A, Ghavami G, Rane S and Singh G 2018 Experimental Investigation of the Transient Flow in Roots Blower. In: *24th International Compressor Engineering Conference,* (Chicago: Purdue university)

[16] Ji B, Luo X W, Arndt R E A, Peng X and Wu Y 2015 Large Eddy Simulation and theoretical investigations of the transient cavitating vortical flow structure around a NACA66 hydrofoil *International Journal of Multiphase Flow* **68** 121-34

[17] Liu M, Gao Z, Yu Y, Li Z, Han J, Cai Z and Huang X 2018 PIV experiment and large eddy simulation of turbulence characteristics in a confined impinging jet reactor *Chinese Journal of Chemical Engineering*

[18] Mortensen H H, Arlov D, Innings F and Håkansson A 2018 A validation of commonly used CFD methods applied to rotor stator mixers using PIV measurements of fluid velocity and turbulence *Chemical Engineering Science* **177** 340-53

[19] Ryan D J, Simmons M J H and Baker M R 2017 Determination of the flow field inside a Sonolator liquid whistle using PIV and CFD *Chemical Engineering Science* **163** 123-36

[20] Kurec K, Piechna J and Gumowski K 2017 Investigations on unsteady flow within a stationary passage of a pressure wave exchanger, by means of PIV measurements and CFD calculations *Appl. Therm. Eng.* **112** 610-20

Acknowledgement

The author would like to thank Howden Compressors for allowing authors to use and modify one of the Howden Roots Blowers in this experiment.

The authors gratefully acknowledge financial support from China Scholarship Council (CSC) to allow the researcher from China to perform this work at City, University of London.

Authors are thankful to Mr Robert Jaryczewski and Dr Ivan Zadrazil from Dantec for help in setting and demonstrating their equipment used for the part of this work.

IOP Conf. Series: Materials Science and Engineering **425** (2018) 012025 doi:10.1088/1757-899X/425/1/012025

Comparison of thermodynamic efficiency between constant, dual and multiple lead rotors for an industrial air screw compressor

Matthias Utri[1], Andreas Brümmer[1], Jan Hauser[2]

[1]Chair of Fluidics, TU Dortmund University
Leonhard-Euler-Str. 5, 44227 Dortmund, Germany
[2]Compression Technologies and Services, Ingersoll Rand
Steinbrinkstr. 1, 461485 Oberhausen, Germany

E-mail: matthias.utri@tu-dortmund.de, andreas.bruemmer@tu-dortmund.de, jhauser@irco.com

Abstract. The goal of the European Union funded project "MOTOR" (Multi-ObjecTive design Optimization of fluid eneRgy machines) is the optimisation of fluid energy machines by using isogeometric analysis. This paper deals with the optimisation of the geometry of screw compressors. Screw compressors are usually built using helical rotors with constant lead. However, variable rotor lead holds the promise of higher efficiency due to an optimised progression of chamber volume, clearance, and discharge areas. This is especially true for high pressure applications and for working fluids with low isentropic exponent. This paper describes the optimisation of a variable lead rotor pair for an industrial air screw compressor. The rotors are divided into a discrete number of segments of constant lead, named multiple lead. Using a high number of rotor segments allows a design that approaches a uniformly varying lead. Thermodynamic simulation of the machines is performed using multi-chamber simulation with the simulation tool "KaSim". The thermodynamic simulation is coupled with a Nelder Mead algorithm, which evaluates the efficiency of the machines and optimises the progression of rotor lead. The achievable efficiency of this optimised multiple lead machine is compared with constant and dual lead machine configurations. Results of this study show that the highest efficiency can be achieved by using dual lead rotors.

1. Introduction

The shape-optimisation of fluid energy machines, e.g. turbines and compressors, is a challenging task due to the complexities in geometry and numerical simulation involved. The industrial design process is usually performed using CAD systems, whereas the numerical simulation of fluids and solids is performed using CAE tools. The independent development of the respective research fields led to different data formats, requiring a conversion between them, which is always linked to an approximation and therefore to a loss in accuracy. Therefore, the vision of the Horizon 2020 project MOTOR is to reduce the approximation error resulting from the conversion steps and, at best, to eliminate it completely by linking all CAE tools involved in the design and development pipeline to a common and highly accurate master geometry model. To achieve this vision, MOTOR makes use of novel CAx technologies like isogeometric Analysis (IgA) [1], which tries to adopt the same mathematical approach (B-Splines or NURBS) to describe the geometry in the design and the simulation tools. This paper focusses on the optimisation process of the rotor wrap angle for an industrial air screw compressor. The optimisation approach is shown in figure 1. The rotor profile is described by a two-dimensional point cloud, which is used to generate a zero-dimensional chamber model. This chamber model is used to perform a thermodynamic simulation for user-specified boundary conditions. The results are input to an optimiser, which adjusts the geometric parameters in order to increase efficiency. The simulation and optimisation loop is repeated until the objective of maximum efficiency is realised, a converged case, or no such result is found within a specified number of iterations, a failed case. A meaningful chamber model simulation should provide reasonably precise calculation of mass flow exchanges in clearance and port areas. The determination of these flows is pictured on the right side of figure 1, where the geometry of the screw machine is used to perform more detailed flow simulation of the flow paths. The full IgA-based design and analysis of screw machines is not yet possible with the available CAE tools, but essential concepts from the IgA approach have been integrated into the workflow to realise the concept of a master geometry model from which both geometry parameterisations for IgA simulations and highly accurate computational meshes for classical CFD simulations can be easily generated. The construction of the master geometry model by the IgA approach is addressed in Hinz et al. [2] and the generation of computational meshes is presented by Möller et al. [3]. The utilisation of these meshes in CFD simulation is performed by Utri et al. and presented in [4]. Shamanskiy and Simeon present a coupling to the solid parts of the machine in order to estimate thermal expansion [5]. This paper focusses on the description of the left side of the presented optimisation approach, where the chamber model based optimisation of the screw machine geometry is performed.

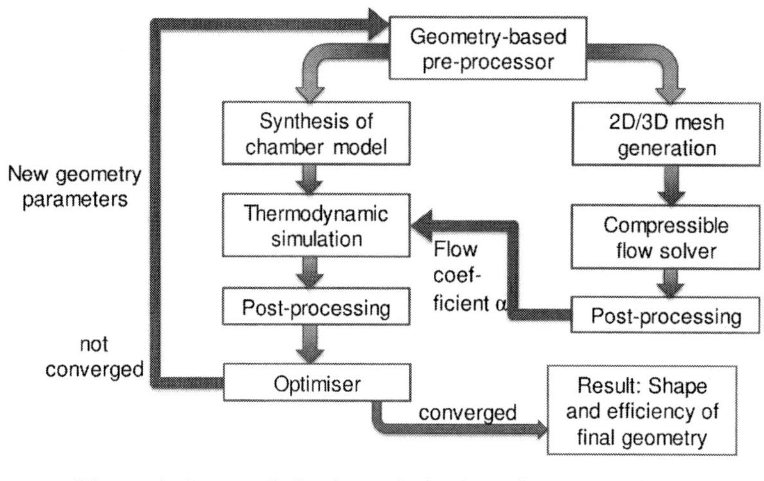

Figure 1. Approach for the optimisation of screw machines

Figure 2. Constant (left), dual (middle) and variable (right) lead

Industrial screw compressors are usually built with constant rotor lead (figure 2 left). However, recent experience reveals that use of a non-constant rotor lead has a positive impact on compressor efficiency. Probably the easiest way for a practical realisation of this new kind of machine is the combination of two different lead segments on one rotor (so called "dual-lead", figure 2 middle), a configuration investigated by Fost [6] and by the authors [7, 8]. Other contributions [9, 10] examine a continuous decrease in lead between the low and high pressure ends of the rotor (figure 2 right). Both approaches show promising results in terms of efficiency. In this paper, an optimisation of the rotor lead distribution is performed for an industrial air compressor in order to reveal which concept is most effective and to give practical advice for screw compressor design.

2. Concept for variable rotor lead

The lead of a screw machine rotor has great impact on important geometric properties (clearances and discharge port area) and should be matched to the intended compression task. High rotor lead (low wrap angle) usually implies a rapid compression, lower clearance areas and maximum possible chamber volume. However, it is associated with a small discharge port area. In contrast, low rotor lead (high wrap angle) leads to a longer duration of the working cycle, larger clearances and a beneficially large discharge port area. The wrap angle, φ, is defined as the angle of twist between front and rear ends of the rotor. The concept of variable rotor lead presented in this paper is to define a rotor using discrete segments with different leads in order to realise an appropriate balance of advantages of high and low lead configurations. An example of a multiple lead rotor is shown in figure 3. This rotor consists of three lead segments of equal length. The overall wrap angle is defined as the sum of the wrap angles of the separate rotor segments; the wrap angle associated with a rotor pair is usually that of the male rotor:

$$\varphi = \sum_{i=1}^{n} \varphi_i \tag{1}$$

The following investigations are limited to rotors of constant overall length with individual lead segments of equal length, as shown in figure 3. A multiple lead rotor with a large number of segments is almost equivalent to a rotor with a continuous change in rotor lead. The number of rotor segments and the overall wrap angle are varied in this study.

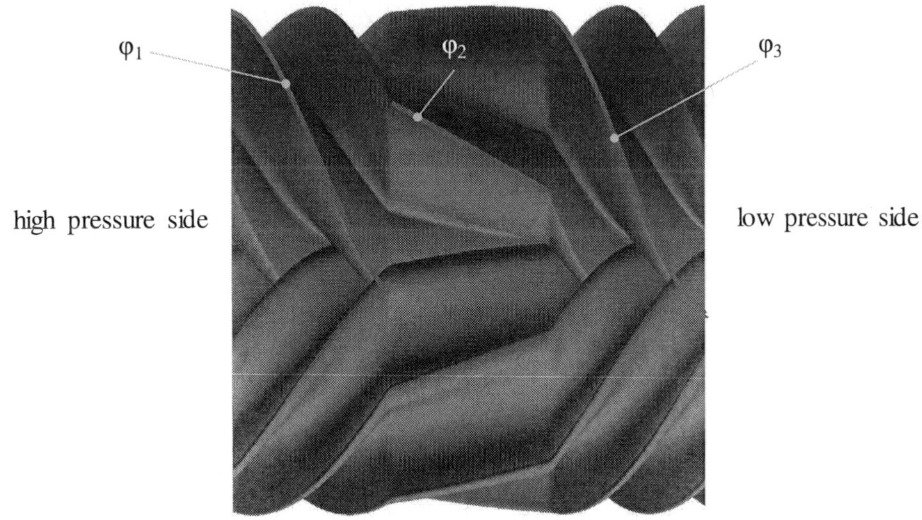

Figure 3: Screw machine with three lead segments (n = 3)

3. Chamber model simulation

Thermodynamic simulation is performed using the multi-chamber simulation tool "KaSim" [11]. The simulation requires a chamber model which contains all relevant geometric properties of the screw compressor under examination. This requires a detailed analysis of the geometry of the screw machine, this being performed in a series of two-dimensional cross sections which are then combined to describe the three-dimensional machine under investigation. It is the rotation angle-dependent progression of volumes and areas that are provided to the thermodynamic analysis.

Chamber model simulation is based on mass and energy conservation with the assumptions of an adiabatic process with a homogeneous distribution of pressure and temperature inside the working chambers. The simulation used here considers all working chambers simultaneously and has been used and validated for several use cases. The simulation process is divided into two consecutive steps: a change of chamber volume and an exchange of mass in the form of leakage and port flows. Mass flow is determined assuming a theoretical isentropic, frictionless flow $\dot{m}_{th}$. This flow is then adjusted using a flow coefficient, α, afterwards to approximate real flow conditions, equation (2):

$$\dot{m} = \alpha \cdot \dot{m}_{th} \tag{2}$$

The flow coefficient depends on the geometry of the flow path and on the thermodynamic states within the connected volumes. Determination of the flow coefficient for a certain geometry is discussed in more detail in [4].

The results of the chamber model simulation are, amongst others, delivered mass flow, $\dot{m}$, and indicated power, P_i. These can be used together with the isentropic power, P_s, which would be necessary for an isentropic compression, to evaluate the indicated isentropic efficiency of the compressor:

$$\eta_{is} = \frac{P_s}{P_i} = \frac{\dot{m} \cdot \Delta h_s}{P_i} \tag{3}$$

This efficiency is used by the optimiser to adjust the wrap angle parameters of all segments.

4. Optimisation principle

The geometry of the screw machine is optimised for application specific, user specified boundary conditions. In addition to the wrap angle parameters, the compressor's internal volume ratio, defined in equation (4), is used as a factor in the process.

$$v_i = \frac{V_{max}}{V_{compr,end}} \tag{4}$$

The internal volume ratio affects the duration of the compression and discharge phases as well as the size of the discharge port and needs to be selected based on screw machine geometry and application conditions. For a given set of rotor parameters, the optimisation of v_i is an easy search for a global maximum in efficiency. An example of this is shown in [8]. In contrast, optimisation of wrap angle is less trivial. The free wrap angle parameters of all segments lead to thousands of possible rotor geometries. An optimisation loop is used in order to reduce the necessary number of simulations to find the best rotor geometry. The optimisation algorithm used is the downhill simplex algorithm [12], also called the Nelder Mead algorithm. This gradient free method can optimise problems with multiple input parameters.

5. Industrial air compressor configuration and boundary conditions

The industrial compressor under investigation is a single-stage wet-running air compressor in the medium power range. Oil is injected into the compressor and serves to seal clearances. The compressor under examination can operate at a maximum outlet pressure of $16 \cdot 10^5$ Pa and runs at a circumferential speed of up to about 50 m/s. The boundary conditions, which remain constant for all simulations, are shown in table 1. Since the machine is wet-running, the heat of compression is partly absorbed by the injected fluid. Instead of an additional simulation of this fluid, the isentropic exponent of the homogeneous representation of the actual fluid conditions used in the simulation is adjusted so that the outlet temperature matches the temperature measured in experiments. Atmospheric conditions prevail at the inlet. The operating parameters selected for the study are a compression ratio of 10:1 with the compressor operating at its maximal circumferential speed. The rotors have an L/D-ratio of 1.6 and use a profile with four male and six female rotor lobes.

Table 1. Constant boundary conditions

Parameter [dimension]	Value
Fluid parameters:	
- Specific gas constant [J/kg/K]	287
- Isentropic exponent [-]	1.1
- Compressor inlet temperature [K]	293
- Compressor inlet pressure [Pa]	10^5
- Compressor outlet pressure [Pa]	10^6
Machine parameters:	
- Male rotor diameter [mm]	150
- Rotor length [mm]	238
Operating parameters:	
- Circumferential speed [m/s]	50

Table 2 shows parameters varied in this study. The number of lead segments is varied between 1 (which is equivalent to constant rotor lead) and 15. The wrap angle values of each segment are determined by the optimiser. The overall wrapping of the rotors is also varied and restricts the lead distribution in accordance with equation (1). The optimisation process is performed for different clearance heights and flow coefficients, which affect mass flow through clearances and ports. The flow coefficients are assumed to be the same for all clearances and port areas and do not vary with geometry or operating conditions. Two different clearance heights are investigated; these are applied to the inter-lobe, housing and front clearance. The blowhole area is determined based on the profile geometry and rotor lead.

Table 2. Varied parameters

Parameter [dimension]	Minimum / maximum value
Number of segments n [-]	1 / 15
Overall wrap angle φ [°]	200 / 350 (variable and dual lead) 200 / 500 (constant lead)
Clearance height h [mm]	0.025 / 0.075
Flow coefficient α [-]	0.5 / 0.7

6. Optimisation results

6.1. Optimisation of multiple lead - variation of the number of segments

Optimisation is performed by varying the number of segments used to define the multiple lead rotor set in order to show the global optimal geometry. Dual lead rotors are treated in a later section and are not comparable to multiple lead rotors with two segments since the segment length is different. Figure 4 shows optimisation results where indicated isentropic efficiency is maximised with the parameters being overall wrap angles and number of segments. This study is further defined by use of the smallest values of clearance heights and flow coefficient. Figure 5 shows the related rotor geometries for an exemplary wrap angle of φ = 275°; figure 6 shows the corresponding progressions of volume and optimised discharge area. The related indicator diagrams are shown in figure 7.

For a single rotor segment (which represents constant rotor lead), wrap angle has a large impact on the efficiency of the compressor. In general, outlet area is small for low wrap angles, leading to significant discharge throttling. For greater wrap angles, discharge throttling is reduced; the increased clearance areas do not affect efficiency significantly since clearance height and flow coefficient are set to small values in all simulations. A significant increase in efficiency is possible by splitting the rotor into two equal length segments (n = 2). The corresponding optimised rotor geometry, shown in figure 5, provides 80 percent of the overall wrap angle on the high pressure side. As shown in figure 6, the maximum chamber volume is independent of the wrap angle progression; rather, it is only influenced by the overall wrap angle [8]. When comparing constant lead with two lead segments, the two segments configuration leads to a strong impact on volume curve and discharge area. The reduction in chamber volume during compression is much greater for the optimised variable lead screw machine due to the low wrap angle of the low pressure side segment. As a result, the chamber volume when discharge begins is reached at earlier rotation angles, leading to an increased discharge area and longer discharge duration, even though the value of the optimised internal volume ratio is increased compared to constant lead. Together with the flat progression of the volume curve during discharge, this leads to a lower discharge volume flow rate and reduces the risk of throttling. This relation can be explained by referring to the indicator diagrams, shown in figure 7, which are used to calculate the indicated compressor power. The diagram shows that the highest pressure inside the screw machine is found with constant rotor lead due to discharge throttling. The optimised internal volume ratio of this machine opens the working chamber to the high pressure port long before reaching the high pressure conditions. The resulting under-compression is less detrimental in terms of efficiency than the discharge throttling which would occur for a higher internal volume ratio.

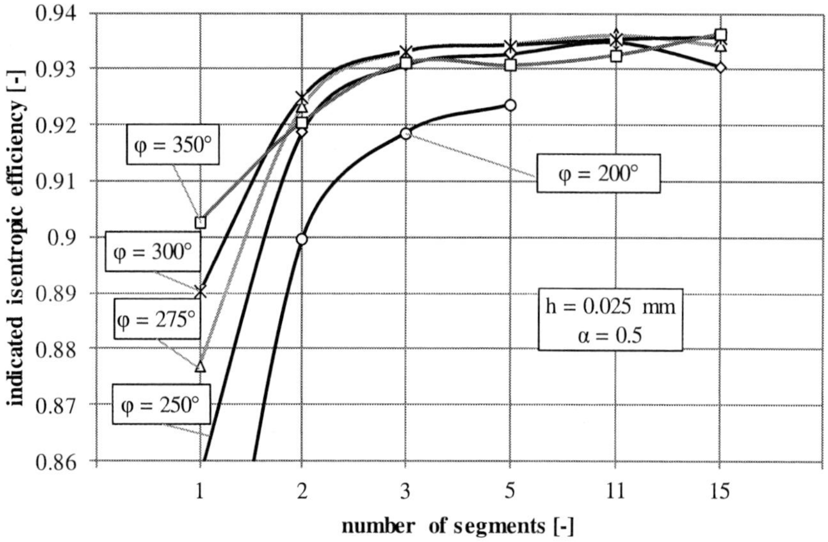

Figure 4: Indicated isentropic efficiency of the optimised screw compressors for a variation of segment numbers and overall wrap angle

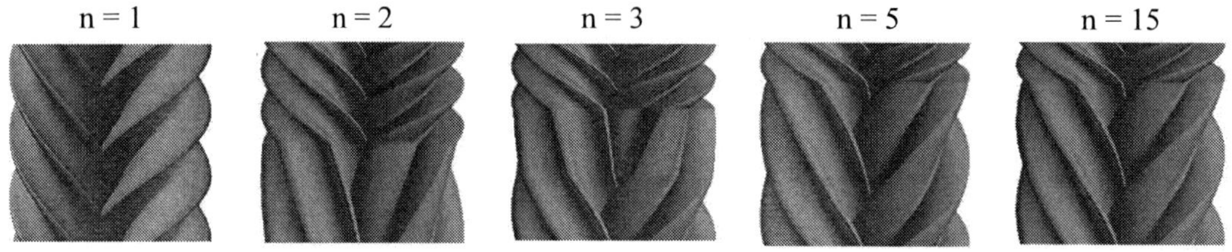

Figure 5: Optimised screw machines with $\varphi = 275°$ for different segment numbers (top: high pressure side, bottom: low pressure side)

For three rotor segments efficiency almost reached its maximum value. For $\varphi = 275°$ the rotor consists of a segment with high wrap angle on the high pressure side, a medium wrap angle on the low pressure side and a low wrap angle in between. In comparison with the rotor consisting of two segments, the medium wrap angle on the low pressure side leads to a decrease in throttling during suction. The low wrap angle of the middle segment maintains a fast compression and the high wrap angle on the high

IOP Conf. Series: Materials Science and Engineering **425** (2018) 012025 doi:10.1088/1757-899X/425/1/012025

pressure side leads to a gentle volume curve progression, so that the internal volume ratio can be increased without causing throttling during discharge.

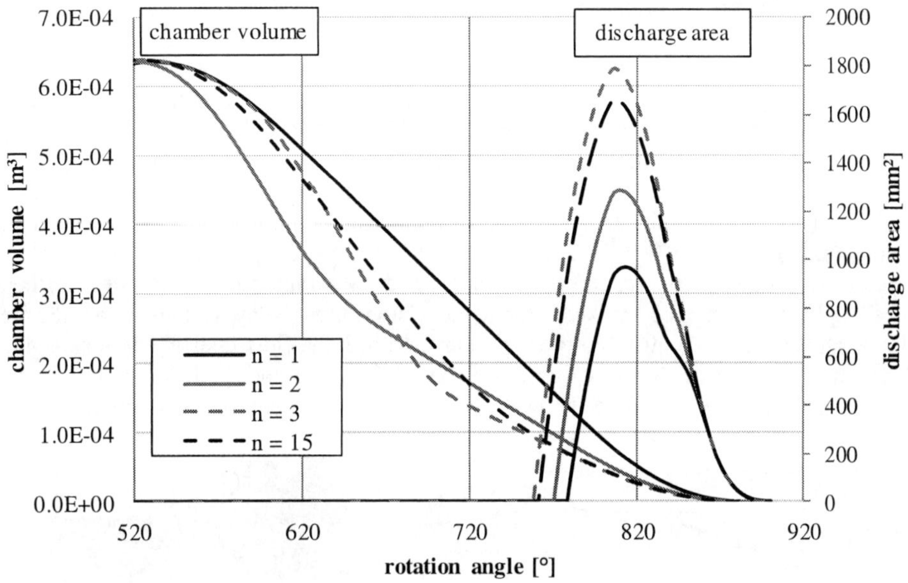

Figure 6: Chamber volume and discharge area for a fixed overall wrap angle of φ = 275° for the optimised screw machines for different numbers of segment

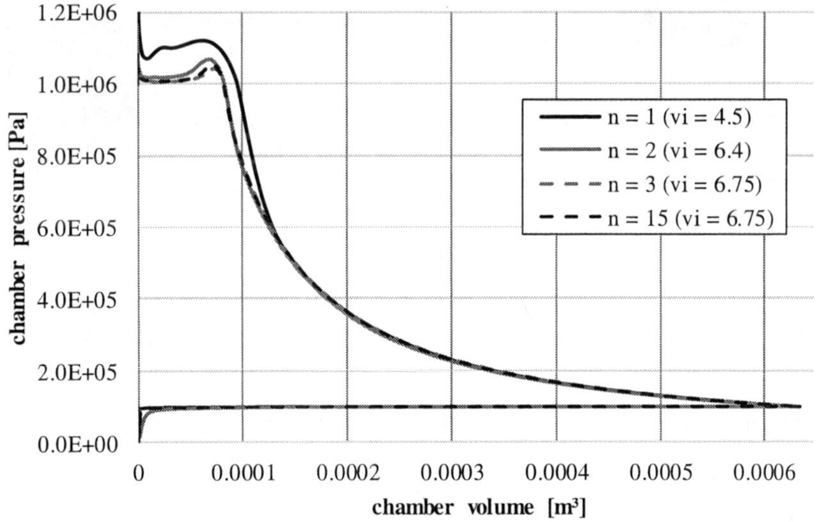

Figure 7: Indicator diagrams for the optimised screw machines (including optimised v_i) for different numbers of segments for a fixed overall wrap angle of φ = 275°

A further increase of the number of segments (n = 5 and n = 15) leads to almost the same results and reveals a characteristic shape of the optimised rotors. Although a continuous change in rotor lead (as shown in figure 2) from high to low pressure side would be possible as a result, the optimised rotors consist mainly of two wrap angle parts, a high wrapping on the high and a low wrapping on the low pressure side. The results emphasise that it is sufficient in terms of efficiency to build a rotor consisting of two rotor parts with different lengths. The shape represents the "dual lead" rotors, which the authors previously examined for different fields of application [7, 8].

6.2. Optimisation of dual lead - variation of flow coefficient and clearance height

An examination of multiple lead in the last section revealed that properly selected dual lead rotors can provide optimum performance. Since two rotor parts with different lengths are enough to optimise efficiency, it is not necessary to optimise the screw machine with completely variable lead; that reduces the number of influencing parameters and simplifies the optimisation process. With respect to this finding, further results for non-constant lead are shown for dual lead setups only (two parts on one rotor with different lengths).

The optimal wrap angle and the resulting efficiency of screw compressors are strongly influenced by the clearance heights which can be realised. Practical decisions must be made considering the clearance heights which can actually be realised in practise. Furthermore, the flow coefficient is constant in this study, so that the influence of the flow coefficient should be examined. Clearance height and flow coefficient are therefore varied in order to show the possible enhancement for other boundary conditions.

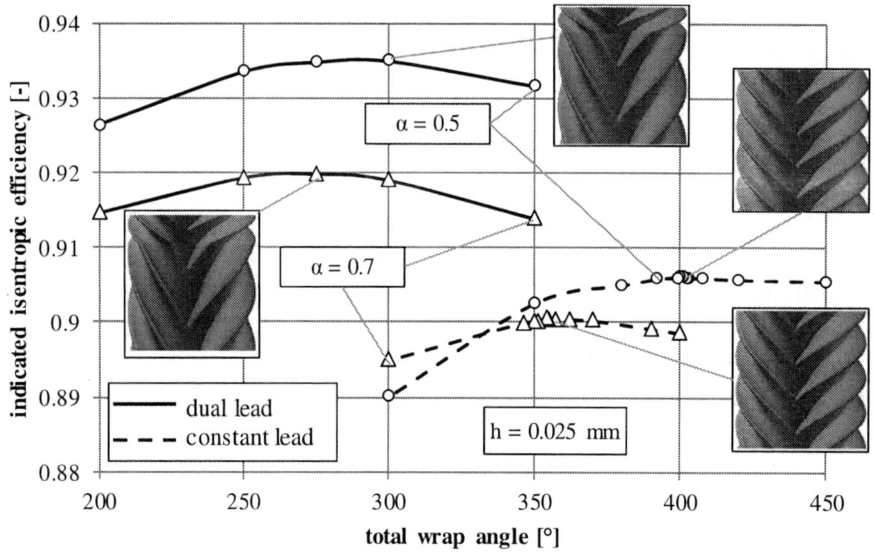

Figure 8: Indicated isentropic efficiency of the optimised screw compressor as a function of rotor wrap angle for two flow coefficients for constant and dual lead (clearance height h = 0.025 mm, v_i optimised for each point)

Figure 8 shows results for different flow coefficients with fixed small clearance heights. With higher flow coefficients port and clearance mass flows are increased. Due to higher port mass flow, the compressor is less at risk of discharge throttling, for which reason the optimal wrap angle decreases from 400 to 350 degrees for constant lead and from 300 to 275 degrees for dual lead. Due to higher clearance mass flow, maximal achievable efficiency is generally lower. By using the dual lead setup, discharge throttling can be reduced for lower overall wrap angles, for which reason the dual lead setup increases efficiency by three percentage points for the lower flow coefficient and by two percentage points for the higher flow coefficient. In addition to the higher efficiency, the dual lead setup increases the mass flow rate due to the lower overall wrap angle; this relation is explained in more detail in [8].

Figure 9 shows corresponding results for a higher clearance height, which again decreases overall efficiency. In comparison to the results from figure 8, the change in rotor lead of the dual lead rotors is less intense. The enhancement in efficiency which can be achieved by using dual lead rotors is reduced and, as explained before, the optimal wrap angle values are reduced. The diagram emphasises that using dual lead rotors is especially beneficial where low clearance heights can be achieved e.g. in the manufacturing process or by using fluids with favourable properties, since the enhancement is reduced for higher flow coefficients.

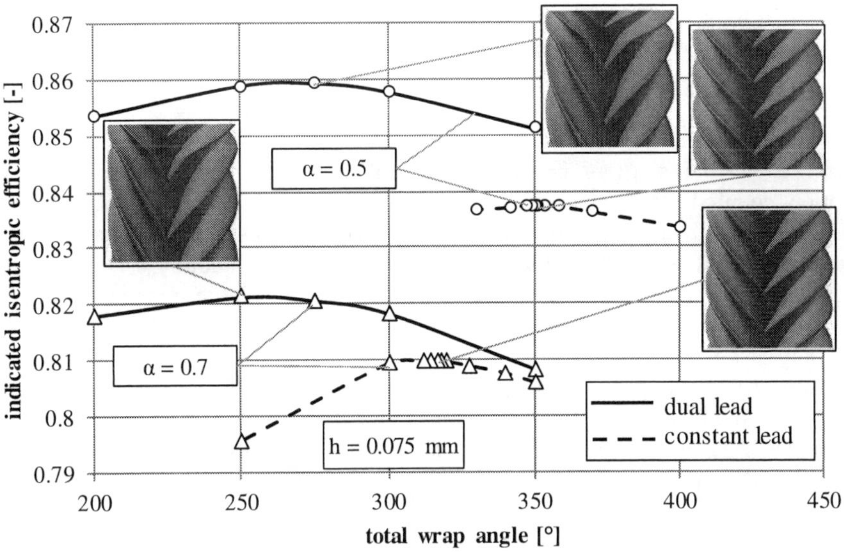

Figure 9: Indicated isentropic efficiency of the optimised screw compressor as a function of rotor wrap angle for two flow coefficients for constant and dual lead (clearance height h = 0.075 mm, v_i optimised for each point)

7. Manufacturing of prototype and outlook

The enhancement in efficiency suggested in this paper will be validated by experiment. Therefore, we have taken into account the optimisation results concerning wrap angles and lengths of the rotor parts in design of a dual lead prototype. We plan to build and test an industrial air compressor with dual lead rotors, where the casing remains unchanged. Manufacturing dual lead rotors is a challenging task since they cannot be manufactured by grinding in one piece. Instead, the two lead parts need to be

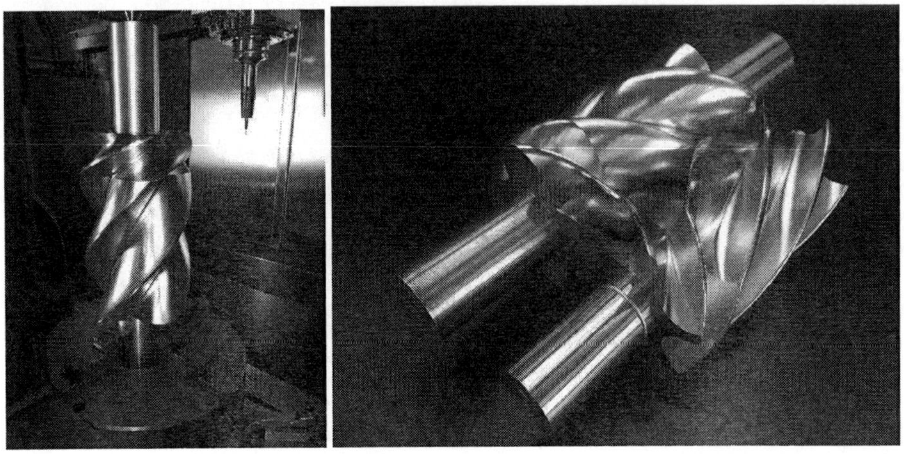

Figure 10: Dual lead male rotor before finishing with a two millimeter milling cutter on a five-axis milling machine (left) and finished dual lead rotor pair prototype made from aluminium (right)

manufactured separately and joined together afterwards. This is realised by grinding one part on the shaft and the other part on a ring. Another possibility to manufacture dual lead rotors is by milling on a five-axis milling machine, figure 10 left. We have used this approach in the workshops of TU Dortmund University to produce dual lead rotor prototypes. Manufacturing was first performed on aluminium rotors in order to investigate the possible dimensional accuracy. The prototype rotor pair (figure 10 right) shows promising results for the manufacturing of steel rotors, a step which is currently under way. The steel rotors will be investigated in a test rig at the facilities of GHH Rand in the near future and compared to the simulation results presented in this paper.

8. Conclusion

The present study examines the efficiency improvement that can be achieved in an industrial wet-running air compressor by using non-constant rotor lead. The optimisation is performed by chamber model simulation, coupled with a Nelder Mead optimisation algorithm. To realise variable rotor lead, the rotors are sectioned into discrete rotor segments. Each segment has a constant lead, so that an almost continuous progression of wrap angle distribution is possible by simulating a large number of rotor segments (multiple lead). The results reveal that dual lead rotors represent the optimised shape of variable lead. These rotors consist of two parts with different rotor lead and length. The increase in efficiency in comparison to constant rotor lead is up to three percentage points for the compressor under examination. However, this improvement needs to be opposed with the higher manufacturing costs of the dual lead rotors. The improvement depends to a great extent on the clearance heights which can be realised in manufacturing. A first dual lead rotor pair prototype made from aluminium has been realised. A steel rotor pair is currently manufactured and will be investigated in experiments in the near future.

References

[1] Cottrell J A, Hughes T J R and Bazilevs Y 2009 *Isogeometric Analysis* John Wiley & Sons, Ltd
[2] Hinz J P, Möller M and Vuik C 2018 Spline-Based Parameterization Techniques for Twin-Screw Machine Geometries *Int. Conf. on Screw Machines 2018*
[3] Möller M and Hinz J P 2018 Isogeometric analysis framework for the numerical simulation of rotary screw machines. I. General concept and early applications *Int. Conf. on Screw Machines 2018*
[4] Utri M, Höckenkamp S and Brümmer A 2018 Fluid flow through housing clearances of dry running screw machines using dimensionless numbers *Int. Conf. on Screw Machines 2018*
[5] Shamanskiy A and Simeon B 2018 Isogeometric simulation of thermal expansion for twin screw compressors *Int. Conf. on Screw Machines 2018*
[6] Fost C 2003 *Ein Beitrag zur Verbesserung der Kammerfüllung von Schraubenmotoren* Dissertation University of Dortmund
[7] Utri M and Brümmer A 2016 Improvement of the Efficiency of Twin-Screw Refrigeration Compressors by means of Dual Lead Rotors *Int. Compressor Engineering Conf. at Purdue 2016* paper 1428
[8] Utri M and Brümmer A 2018 *Journal of process mechanical engineering* Opportunity for Variable Rotor Lead in Dry Running Screw Compressors depending on Dimensionless Numbers (accepted paper to be published)
[9] Rane S, Kovacevic A, Stosic N and Kethidi M 2014 Deforming grid generation and CFD analysis of variable geometry screw compressors *Computers & Fluids* **99** pp 124-141
[10] Huang, C-Y 2015 *New Twin Screw Compressor Design by Deviation Function Method* Dissertation University of California
[11] Kauder K, Janicki M, Rohe A, Kliem B and Temming J 2002 Thermodynamic Simulation of Rotary Displacement Machines *VDI-Berichte 1715: Schraubenmaschinen 2002* pp 1-16
[12] Nelder J A and Mead R 1965 A simplex method for function minimization *The Computer Journal* 7 issue 4 pp 308-313

Nomenclature

h	Clearance height	(m)
Δh	Specific enthalpy difference	(J/kg)
$\dot{m}$	Mass flow	(kg/s)
n	Number of segments	(-)
P	Power	(W)
V	Volume	(m^3)
v_i	Internal volume ratio	(-)
α	Flow coefficient	(-)
η	Efficiency	(-)
φ	Overall wrap angle	(°)
φ_i	Wrap angle of segment i	(°)

Subscript

compr,end	End of compression
i	Indicated
max	Maximal
s	Isentropic
th	Theoretical

Acknowledgement

The work leading to these results has received funding from the European Community's Horizon 2020 Programme (2014-2020) under grant agreement no 678727. The opinions expressed in the document are of the authors only and no way reflect the European Commission's opinions. The European Union is not liable for any use that may be made of the information.

IOP Conf. Series: Materials Science and Engineering **425** (2018) 012026 doi:10.1088/1757-899X/425/1/012026

Analytical and numerical prediction of the flow and performance in a claw vacuum pump

J Willie

Chief Engineer, Innovation and Technology, Industrials Group, Gardner Denver Schopfheim GmbH, Johann-Sutter-Straße 6+8, 79650 Schopfheim, Germany

E-mail: James.Willie@gardnerdenver.com

Abstract. This paper presents the use of a 0-D thermodynamic chamber model and 3D Computational Fluid Dynamics (CFD) to predict the performance of a claw vacuum pump. The 0-D model can be used for pump sizing during the initial design or concept phase. Parameters such as the compression ratio (π), the built-in volume ratio (θ), the volumetric efficiency (η_V), the isentropic efficiency (η_{is}), the volume compression ratio (v_i) and the intermediate pressures (p_z, p_i) are essential to the performance of such machines. To determine these parameters, the volume-curve of the pump is determined as a function of the rotational angle. Using this model, it is possible to enhance the pump performance. For example, by decreasing the carry-over volume and by limiting or preventing over and under compression which can lead to an increase in the power consumption. This paper presents the results of the 0-D thermodynamic chamber model followed by the results of the 3D CFD simulations and validation with measurement data. After the benchmarking, CFD was used to improve the aerodynamic design of the pump discharge tube which led to a decrease in the power consumption of the pump compared to the baseline case. The diffuser angle was decrease to limit the boundary layer size and flow separation and recirculation at the discharge. When left unchecked, it can lead to large pressure drop and to an increase in the power consumption. Using CFD, other phenomenon such as such as pressure jump, which is commonplace in such machines, can be prevented. Otherwise, it can lead to shock wave generation and to noise in the pump.

1. Introduction

Modeling and simulation are playing an increasing role in the development process at Gardner Denver (GD). This reduces development time by reducing expensive tests in the laboratory. The availability of computing power at a relatively lower cost also makes modeling larger cases using finer meshes possible in industry. This paper presents one such example of how a 0-D chamber model and Computational Fluid Dynamics (CFD) can be used to help in the development and optimization process of vacuum pumps. Specifically, a claw vacuum pump is presented in this work. Based on the current knowledge of the author very little is available in the literature about these types of pumps, although they are used extensively in the vacuum industry. They are commonplace in the vacuum industry but the ultimate vacuum they can reach is limited by the temperature of the gas and the thermal loads on the rotors and bearings and sealing. The flow inside these pumps are influenced by many factors, for example, the opening and closing of the discharge contour, rotors profile and form, profile and housing gaps, the carried-over volume or dead volume, over-and under compression, the sealing, and so on. The tiny gaps inside these machines means that rotor to rotor contact or rotor to housing contact is possible.

IOP Conf. Series: Materials Science and Engineering **425** (2018) 012026 doi:10.1088/1757-899X/425/1/012026

This may damage the rotors or increase the pump power consumption. In the beginning of the design process the pump needs to be sized and to do this the 0-D model presented can be used. After the concepts development, 3D CFD is used to model the flow inside the pump including 3D effects like turbulence, etc. Flow optimization to reduce the pressure drop and increase the pumping speed or reduce the pump power can be performed using CFD. In this work, for example, engineering optimization is performed by changing the discharge tube geometry with the aim of lowering the pump power. The final goal of this work, which is not reported here, is to do conjugate heat transfer (CHT) analysis in order to determine the thermal loads on the rotors. The determined thermal and pressure loads can be exported to a structural code like Ansys Mechanical [1] to do Fluid-Structure Interaction (FSI) simulation to determine the deformation of the rotors and hence the hot gaps inside the pump.

Some applications of claw vacuum pumps include the following: Chemical industry, Environmental engineering (Aeration, drying, dust extraction, etc.), Industrial applications, Medical industry, Packaging industry, Pneumatic conveying, Post-press applications, Printing press, Wood working industry, and so on.

In the following, the 0-D thermodynamic chamber model is presented. This is followed by a description of the CFD set up used. After this, the measurement setup is presented followed by the results and discussions. Finally, the conclusions and future work are presented.

2. Thermodynamic chamber model

The schematics of a claw pump and the chamber volumes used in the 0-D thermodynamic chamber models are shown in Figure 1. The claw pump consists of two rotors in the form of a claw that rotate in opposite direction with one rotor (referred to as the male) bigger than the smaller rotor referred to as the female rotor. Both rotors are driven at the same speed with the female rotor transporting a larger volume of the working fluid, usually air. The torque generated and hence the power it consumes is therefore larger than that generated by the male rotor.

The working principle is shown in six steps in Figure 1(b) and described as follows [2]. In steps 1 and 2, air is taken into the pump. In step 3, the two closed volumes are transported by the rotors in an isochoric process that is assumed to be isobaric in the 0-D model. This is, however, not the case in reality because some compression takes place during the transport phase as shown in Figure 2. This diagram shows the p-V plot obtained by monitoring the pressure trace inside the pump at a normalized suction pressure of 0.36 in the CFD simulation compared to that obtained using the 0-D model. The power consumption is equal to the area bounded by the respective curve and the p-axis. In the chamber model the isochoric transport phase is assumed to be isobaric and therefore corresponds to only one point, namely, the point of maximum volume in the diagram. In the CFD simulations a rise in pressure occurs during the transport phase because of the gap flow and the heating inside the machine due to internal compression. The isothermal assumption used in the 0-D model is therefore only for simplicity. The CFD results show that we have over-compression at this suction pressure, even though it corresponds to our design point. The actual internal compression starts in step 4 and the carried-over volume is transported to the suction side. In step 5, the outlet begins to open, indicating that the internal compression is completed. Finally, in step 6, the volume is released at atmospheric pressure. A silencer is usually connected at the discharge to limit the pulsation caused by the lobe passing frequencies of the pump and the flow induced noise inside the pump at higher frequencies.

Based on the description above the equations of the 0-D thermodynamic model can be derived as presented below [2]. In the presented model equations the gap flows are neglected as well as heat transfer. The intermediate pressure p_z is given as a function of the chamber volumes and the suction pressure (p_S) as shown in equation 1:

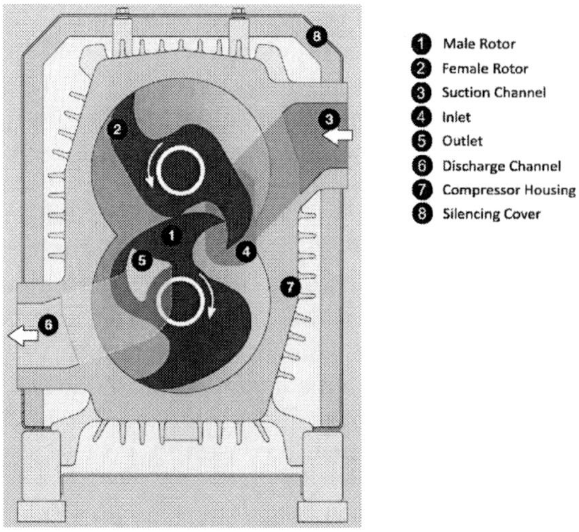

(a) Schematic of a claw pump

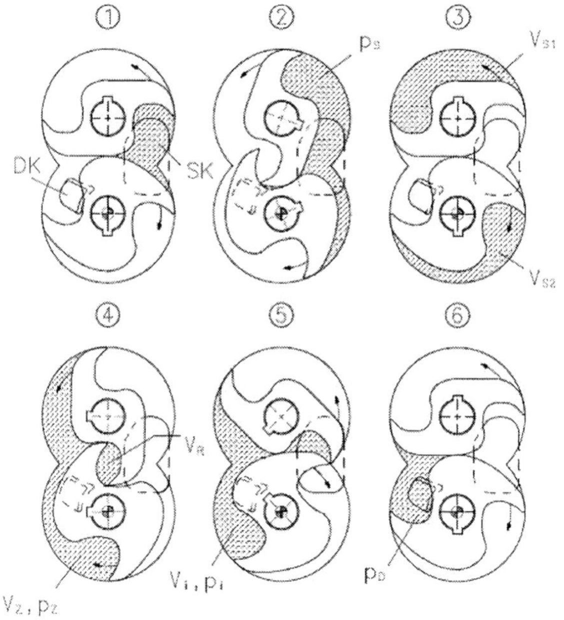

(b) Schematic of the chamber volumes used in model equations

Figure 1. Schematic of the claw vacuum pump and the chamber volumes used in the thermodynamic 0-D equations

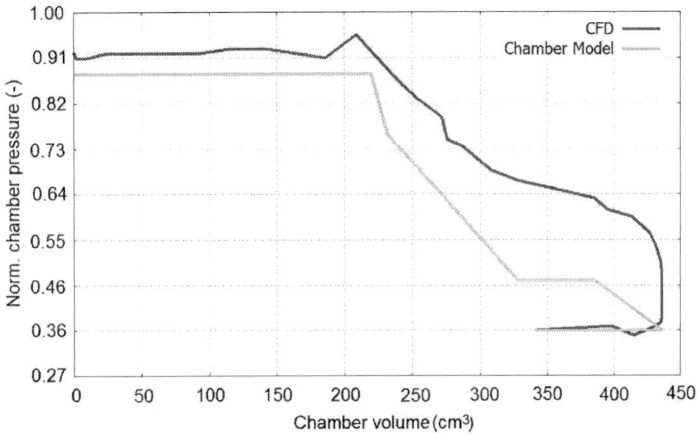

Figure 2. p-V Diagram of a typical claw pump operating at a normalized suction pressure of 0.36

$$p_z = p_S \left(\frac{V_{S1} + V_{S2}}{V_Z + V_R} \right) \tag{1}$$

The volume of the carried-over mass, V_{RS}, can be written as:

$$V_{RS} = V_R \left(\frac{p_z}{p_S} \right)^{\frac{1}{\kappa}}, \tag{2}$$

where V_R is the carried-over volume. Using this volume flow, the theoretical or ideal pumping speed of the pump can be expressed as:

$$\dot{V}_{th} = z(V_{S1} + V_{S2} - V_{RS})N, \tag{3}$$

where z is the number of rotor teeth and N is the rotational speed in Hz or 1/s. The maximum internal compression pressure just before the opening of the outlet is:

$$p_i = p_z \left(\frac{V_z}{V_i} \right)^{\kappa} = p_S \left(\frac{(V_{S1} + V_{S2})V_Z}{(V_Z + V_R)V_i} \right)^{\kappa} \tag{4}$$

The internal compression π is defined as:

$$\pi = \frac{p_i}{p_S} \tag{5}$$

Similarly, the internal volume ratio θ is given by:

$$\theta = \pi^{1/\kappa} = \left(\frac{(V_{S1} + V_{S2})V_Z}{(V_Z + V_R)V_i} \right) \tag{6}$$

The power taken by the pump is defined by the integral $\oint V dp$. Using the relationships above the power needed for isentropic compression (with no under or over-compression) and also not accounting for the partial compression of the carried-over volume is given by:

$$P_{is} = \frac{\kappa}{\kappa - 1} p_S \dot{V}_{th} \left[\left(\frac{P_D}{p_S} \right)^{\frac{(\kappa-1)}{\kappa}} - 1 \right] \tag{7}$$

If the pump is run at an off-design point (with under or over-compression), the theoretical isentropic power consumption becomes:

$$P^*_{is} = \frac{\kappa}{\kappa - 1} p_S \dot{V}_{th} \left(\left(\frac{P_i}{p_S} \right)^{\frac{(\kappa-1)}{\kappa}} - 1 \right) + (p_D - p_i)\dot{V}_i \tag{8}$$

Alternatively, equation 8 can be written as:

$$P^*_{is} = p_S \dot{V}_{th} \left[\frac{\kappa}{\kappa - 1} \left(\frac{1}{\kappa} \theta^{\kappa-1} - 1 \right) + \frac{p_D}{p_S} \frac{1}{\theta} \right] \tag{9}$$

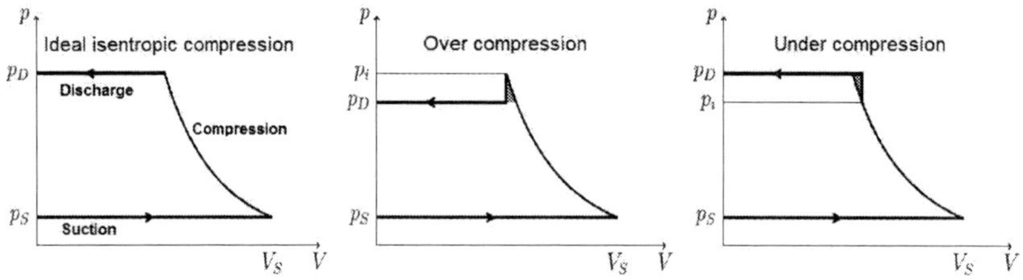

Figure 3. Schematic showing under and over-compression [3]

The power consumption calculated with equations 7 to 9 can be shown in the p-V diagram in Figure 3. Generally, three cases can be distinguished. The ideal case shown on the left where the pressure inside the pump housing at the discharge p_i is equal to the discharge pressure p_D when the discharge opens and the compression is isentropic. In the middle plot and just before the discharge opens the chamber pressure is greater than the discharge pressure and when the discharge opens the chamber pressure drops suddenly from p_i to p_D. The shaded area corresponds to the extra work that must be done because of the over-compression. In the sketch on the right the chamber pressure p_i is less than the discharge pressure before the discharge port is opened and it is referred to as under-compression. When the discharge opens a rapid flow is generated from the discharge tube to the chamber until we have an equalization of pressure. The extra work done as a result of the under-compression is shaded in the p-V diagram.

One possible deduction from equation 9 can be done by looking at the condition $\frac{\partial P^*_{is}}{\partial p_S} = \dot{V}_{th} \left[\frac{\kappa}{\kappa-1} \left(\frac{1}{\kappa} \theta^{\kappa-1} - 1 \right) \right] = 0$. This can be interpreted as the condition that must be satisfied in order for the power consumption of the pump to remain constant at all suction pressures.

Using the models above, performance parameters like the volumetric efficiency (η_V) and the isentropic efficiency (η_{is}) can be computed if the shaft power, $\mathrm{P_{shaft}} = 2\pi \mathrm{TN}$, where T is shaft torque, is known. The ideal volumetric efficiency (from the 0-D model) is given by:

$$\eta_{Vi} = \frac{\dot{V}_{th}}{\dot{V}_S}, \tag{10}$$

where $\dot{V}_S$ is the ideal volume flow rate given by:

$$\dot{V}_S = z(V_{S1} + V_{S2})N \tag{11}$$

Analogously, the real volumetric efficiency (η_{Vr}) is computed using the measured volume flow rate and given by:

$$\eta_{Vr} = \frac{\dot{V}_{measured}}{\dot{V}_S}, \tag{12}$$

The ideal volumetric flow rate does not depend on the suction pressure and therefore is constant where as the real volumetric flow rate depends on the suction pressures and decreases as the gap flow increases at lower suction pressures. For the volumetric efficiency computed from CFD, the numerator in equation 10 or 12 is replaced by the volume flow rate computed from CFD, which is obtained by dividing the average mass flow rate at the inlet or discharge by the average density.

The isentropic efficiency is computed using:

$$\eta_{is} = \frac{P_{is}}{P_{shaft}}, \tag{13}$$

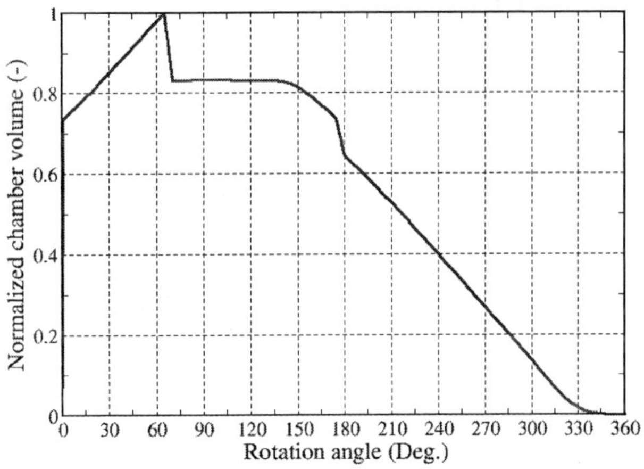

Figure 4. : Plot of the chamber volume of a claw pump

In case there is under or over compression, which is usually the case, the isentropic power P_{is}^* in equations 8 or 9 is used. In the initial or concept phase of the design, the shaft power may not be known. In this case CFD can be used to determine the pump power, which is less than the shaft power due to the extra power needed to run the gears and bearings and fan, in case there is a fan connection for cooling. Estimating the fan power and the losses due to the bearings and gears, it will be possible to predict the shaft power and hence the isentropic efficiency.

2.1. Volume curve generation

To be able to generate the chamber model presented in the previous section the volume curve as a function of the rotational angle is needed. One way of doing this is to use a CAD tool, in this case, Inventor Professional. The volume curve is generated for the claw pump from 0° to 360° using an angular increment of 5° as shown in Figure 4. With this, it is possible to determine the volumes at the various stages of the pump shown in Figure 1(b) and hence the model parameters presented in section 2. The volume curve begins at suction followed by transport, followed by the compression and the discharge. In one cycle, the process of the carried-over volume occur twice as expected.

3. CFD case setup

The simulated geometry of the claw pump is generated using Inventor Professional. By using this CAD tool and ANSYS Design Modeler [1] the fluid volume was extracted from the CAD model and exported as an stl file to CONVERGE Studio [7]. As a pre-processor, CONVERGE Studio was used. It is important to note that the cartesian mesh generation in this tool is automated. Therefore, only the boundaries and the numerical parameters have to be specified. The simulations of the unsteady, turbulent flow were carried out using the Reynolds Average Navier Stokes (RANS) and the Pressure Implicit with Splitting Operators (PISO) solver implemented in the CONVERGE solver. As an equation of state the ideal gas law was chosen because it depicts the behavior of air very well for moderate pressures. The mesh size was influenced by the base grid, adaptive mesh refining based on temperature and velocity gradients and fixed embedding for the geometry were used. For the base grid, 8 mm was chosen. The simulations are done in parallel using MPI and the CONVERGE solver using a Linux Solver with 8 cores and 256 GB RAM. For turbulence modeling, the $k-\varepsilon$ model is used with the law of the wall using the standard wall function [6]. Post-processing is done using EnSight.

Table 1. Flow regimes

Continuous flow	0	<	Kn	<	0.01
Sliding flow	0.01	<	Kn	<	0.1
Knudsen flow	0.1	<	Kn	<	10
Free molecular flow	10	<	Kn		

In doing the case setup it is important to consider the type of flow and this is described by its Knudsen number Kn that is defined by $Kn = \frac{\lambda}{l_{char}}$, which is the ratio of the mean free path length λ to the characteristic length l_{char} [5]. The different type of flows are shown in Table 1 [4]. In a sliding flow the collisions between gas particles still occur often but the no-slip condition of continuum mechanics is no longer valid. Inside a Knudsen (transitional) flow collisions between two molecules are about as often as collisions between molecules and walls. The Navier-Stokes equation can only be applied to the continuous and sliding flow [5]. It is important to point out that the solver took longer at lower suction pressures due to the larger velocity gradients in the flow at these pressures.

For the cases reported here the solid rotors and pump housing are not included in the simulation. The gap modelled is therefore the cold gap. To model the hot gap, sealing or gap models are used in CONVERGE Studio to decrease the axial gaps on the drive and the non-drive sides and the radial gaps between the pump housing and the rotors and that between the rotors. To account for the thermal expansion of the rotors and their effect on the gaps inside the machine a combination of Conjugate Heat Transfer (CHT) and Fluid-Structure-Interaction (FSI) simulations are needed.

Figure 5 shows a sketch of the pump setup used in the CFD and in the measurement. It shows the suction side denoted by using the subscript S and the discharge side shown by using the subscript D. The suction valve and the silencer are not included in the CFD simulations.

3.1. Measurement setup

For measurements, the setup shown in Figure 5 is used. It shows the pump being measured and temperature and pressure sensors are used to measure the suction pressure and temperature and the discharge pressure and temperature. The shaft torque is also measured using a torque meter. The volume flow rate or pumping speed is also measured for various suction pressures upto 75 mbar abs. The point of ultimate vacuum (almost zero mass flow rate) was not reached because of the high thermal loads on the rotors and sealing that can result in rotor-to-housing or rotor-to-rotor contact. The measurement was carried out at a speed of 60 Hz or 3600 rpm, but it is possible to run the machine at 50 Hz or 3000 rpm or at other speeds using a variable frequency drive (VFD).

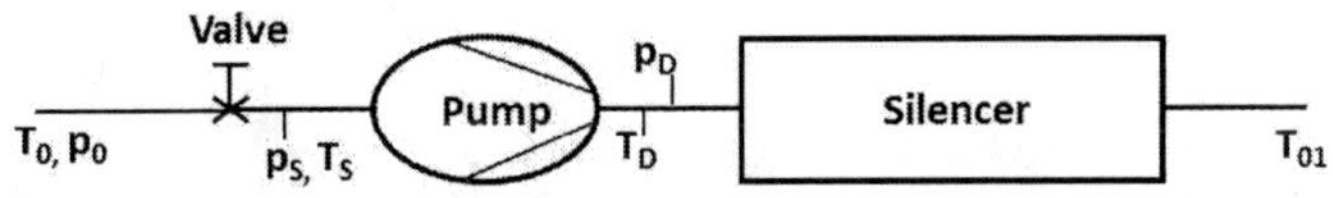

Figure 5. : Schematic of the setup used in the CFD simulations and measurements

4. Results and discussions

To be able to use CFD and the 0-D thermodynamic model in the design process benchmarking is required. To do this, a baseline case is selected and simulations are run for this case and validated with measurement data. Sealing, which helps with predicting the hot gaps inside the machine are used in CONVERGE Studio to ensure that the right losses are being predicted by the solver. When the machine is designed, the gaps included are the cold gaps but in reality the machine heats up when it runs and the thermal expansion of the rotors and the housing will change the radial and axial gaps inside the machine.

In the following, the baseline case results and validation with measurement data are presented. Grid dependency studies are performed to ensure that the results obtained are physical. This is followed by results to predict the entire performance map of the baseline machine simulated. In each case, validation is done with measurement data. Comparison to the 0-D model, which is the ideal case, is also done. Results for the pumping speed, pump power and the volumetric efficiency and isentropic efficiency are presented.

Table 2. Comparison of measurement and simulated data.

Parameter	CFD	Measurement	% error
Norm. pump power (-)	0.91	1.0	9
Norm. mass flow rate(-)	1.0	0.91	10
Norm. female rotor avg. torque (-)	1.0	-	-
Norm. male rotor avg. torque (Nm)	0.56	-	-

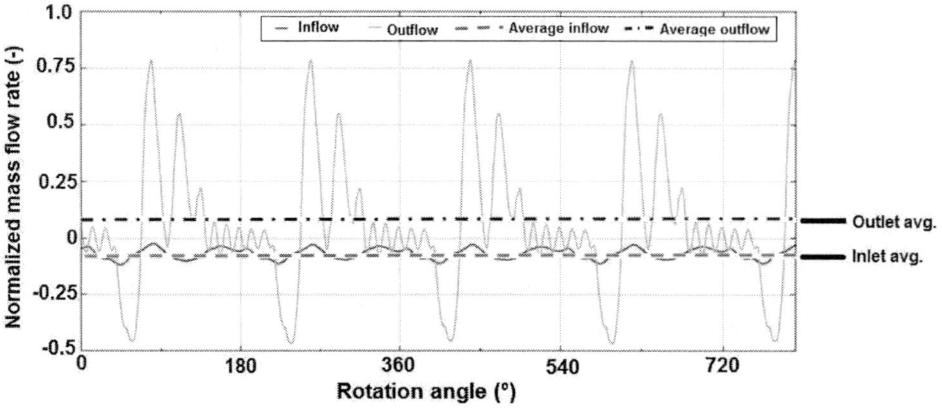

(a) Mas flow rates at inlet and discharge

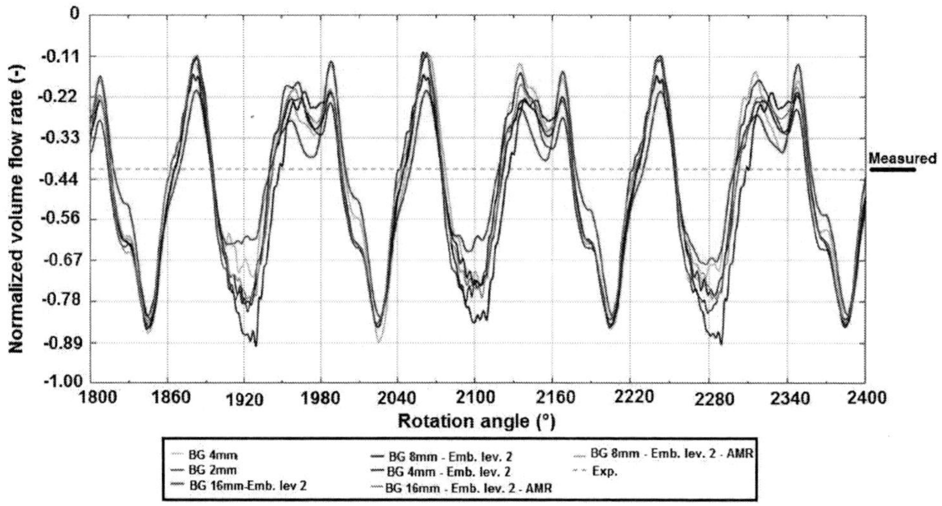

(b) Volumes flow rates at inlet

Figure 6. Plots of the mass flow rate and the volume flow rates at the inlet and discharge computed from CFD

4.1. Grid dependency studies

For grid independent studies, an operating point with a suction pressure of 150 mbar abs. is selected. The mass flow rates and hence the volume flow rates and the male and female rotors power are monitored and compared to measurement data as shown in Figure 6 and Table 2. In Figure 6, the normalization is done using the maximum mass flow rate and the maximum volume flow rate plotted from CFD. In Table 2, the pump power is normalized using the measured pump power. For the mass flow rate in Table 2, the average mass flow is determined from CFD and this value is used in the normalization. The torques are normalized using the average female rotor torque. Even though we are operating in vacuum, the principle of mass conservation at the boundaries of our pump applies. The mass entering must equal the mass leaving the pump. Due to the small gaps inside such machines and the high velocity gradients, especially at low suction pressures, the flow is simulated as compressible. The base grid and the level of embedding and AMR (Adaptive Mesh Refinement), based on the temperature and the velocity gradients were

IOP Conf. Series: Materials Science and Engineering **425** (2018) 012026 doi:10.1088/1757-899X/425/1/012026

changed until a mesh independent solution was obtained [7]. The goal was to obtain a mesh independent solution while at the same time limiting the computational cost. Because the solver is semi-implicit, it is possible to change the Courant-Friedrichs-Lewy (CFL) number depending on the gradients in the flow. The mesh parameters are specified and the meshing itself is done automatically, which allows for more flexibility. For boundary conditions, pressure inlet and pressure outlet are used. The mass flow is computed by the solver as a result. Constant wall temperatures are specified and heat transfer within the fluid and between the fluid and the wall is considered but the solid rotors and pump housing are not included in the simulation.

In Table 2, it is important to note that the female rotor torque is greater than the male torque as expected because it transport a larger volume of fluid than the male. The difference between the computed pump power and the measured is due to the fact in the measurement the torque is that of the input shaft which runs the gears and overcomes other losses in the bearings, sealing, etc. which are not accounted for in the CFD simulations.

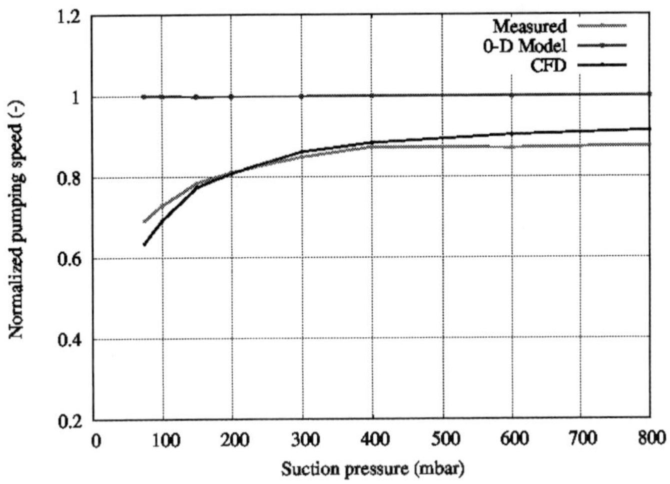

Figure 7. : Measured and predicted claw pump normalized pumping speed verses suction pressure

4.2. Performance prediction and optimization

Performance prediction is crucial in the design of vacuum pumps. The right motor and pumping speed must be predicted as well as critical parameters like bearings temperature, air discharge temperature, cooling needed, etc., before the prototype is developed and tested. In the following, the performance predicted for a typical claw pump is presented. Using the 0-D model presented in section 2 it is possible to do such prediction. CFD can also be used with a greater degree of accuracy because it includes the losses due to gaps inside the machine as well as the 3D and dynamic effects due to the rotation of the rotors and the interaction between the various chambers. In both cases presented, validation is done with measurement data. In Figure 7, the normalized pumping speed, which is normalized using the maximum computed pumping speed from the 0-D model is presented. The measured curve is closer to the predicted CFD curve than to the 0-D model since the latter does not account for the gap flow or losses inside the machine. The CFD prediction is better in terms of the trend at higher suction pressure than at lower suction pressure because of the high gradients associated with the flow at lower suction pressure.

The plot of the pump power normalized using the maximum shaft power measured is shown in Figure 8. The CFD model does not account for the power consumed by the gears and bearings and other losses accruing in the sealing but the trend is similar between the CFD and measurements results, unlike the pumping speed. If the power versus suction pressure is plotted for the 0-D model, it would show a linear increase from high suction pressure to low suction pressure since the over-compression at the high suction pressure that leads to high power consumption is not accounted for in the 0-D model.

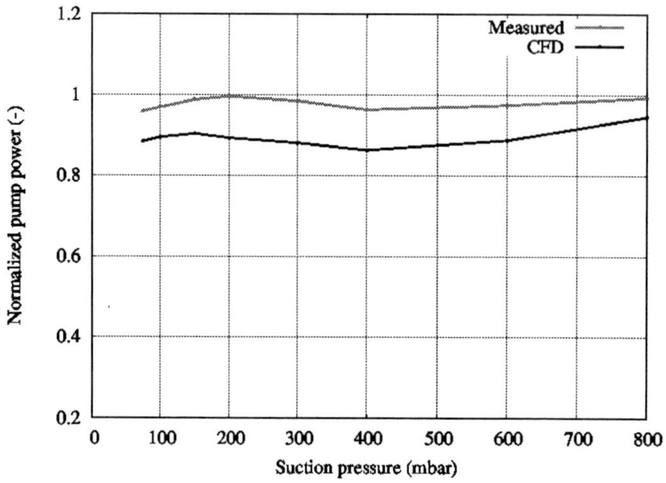

Figure 8. : Performance curve of the power of a claw pump: Measured versus predicted

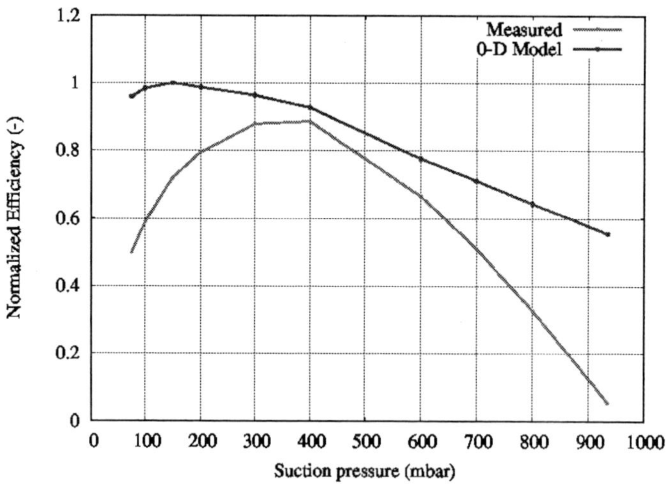

Figure 9. : Performance curve of the isentropic efficiency of a claw pump: Measured versus predicted

In Figure 9, the plot of the pump isentropic efficiency is shown. It shows similar trends: lower on the high pressure side due to over compression and it decreases on the low pressure side due to under-compression and increase in leakages due to high velocity gradients in the flow. It is maximum at the design point where there is no under or over compression. For the measured isentropic efficiency equation 13 is used. The efficiency of the 0-D model is computed by using P^*_{is} in equation 13 instead of P_{is} and P_{shaft} is obtained by using the measured torque and the rotational speed.

One of the goals of this work was to carry out an engineering optimization of the pump investigated. As an example, the discharge tube is optimized using 3D CAD in combination with 3D CFD. The pump power relative to the baseline (benchmark) case is computed and presented in Figure 10. In this case, the optimization parameter that was selected for minimization was the pump power since pressure drop will not be appropriate because of the pressure inlet and outlet boundary conditions used. Clearly, design 3b is the best design since it offers the highest power saving achievable of the cases considered. The worst case was design 3 since it has the highest power consumption relative to the baseline case.

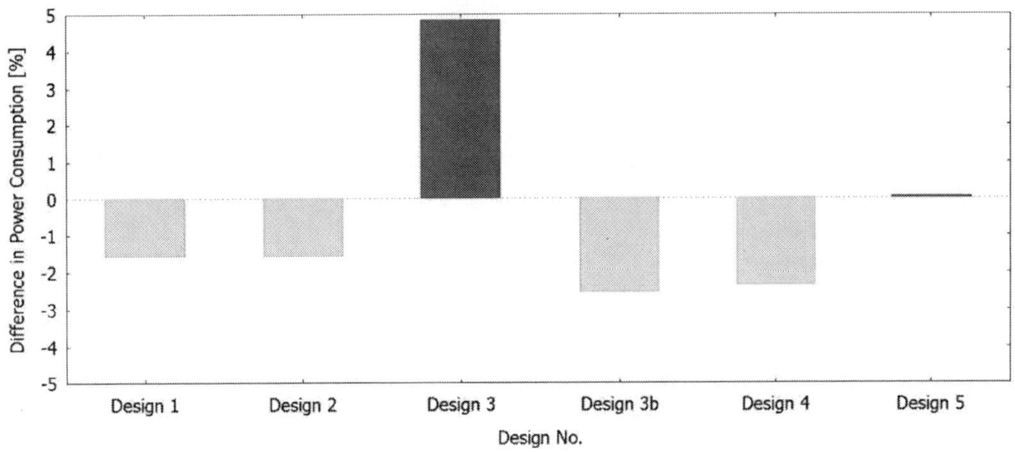

Figure 10. : Optimization of the discharge tube of claw pump

5. Conclusions

This paper presents the successful simulation of the flow inside a claw vacuum pump using 3D CFD. Simple 0-D thermodynamic model as well as 3D CFD results and validation with measurement data are presented. Using the 0-D thermodynamic model the compression ratio (π), the built-in volume ratio (θ), the volume compression ratio (v_i) and the intermediate pressures (p_z, p_i) as well as the ideal volume flow rate were determined during the concept phase. 3D CFD was used to determine more accurately the performance parameters like the pumping speed and the pump power. An existing claw pump with measurement data is used as a benchmark for validating the CFD process after which engineering optimization of the pump is done. The discharge tube is optimized with the goal of reducing the pump power as presented. The following are planned for future work:

- Using CFD with heat transfer to be able to predict the temperature of the rotors and housing of the pump. The heat flux and pressure distribution on the rotors can be used for computing the deflection or deformation of the rotors. This information is useful in sizing the gaps inside the pump. With this it will be possible to determine the hot gaps inside the pump.

- Flow optimization to reduce leakages and extend the vacuum range of the pump
- Extension of the chamber model to include flow losses or leakages and heat transfer

Nomenclature

θ	Internal volume ratio	[-]
κ	Isentropic exponent	[-]
π	Internal compression ratio	[-]
ρ	Density	[kg/m^3]
$\dot{m}$	Mass flow rate	[kg/s]
T	Torque	[Nm]
N	Rotational speed	[1/s or Hz]
p	Pressure	[Pa]
p_D	Discharge pressure	[Pa]
p_i	Pressure before opening of the outlet	[Pa]
p_S	Suction pressure	[Pa]
p_z	Pressure when the carried-over volume is separated	[Pa]
P_{is}	Isentropic or ideal power	[W]
P_{is}^*	Power with under or over-compression	[W]
V	Volume	[m^3]
V_i	Volume before opening the outlet	[m^3]
V_R	Carried-over volume	[m^3]
V_{RS}	Expanded carried-over volume	[m^3]
V_{S1}	Volume of chamber 1	[m^3]
V_{S2}	Volume of chamber 2	[m^3]
$\dot{V}$	Pumping speed or volume flow rate	[m^3/s]
V_{th}	Theoretical volume flow rate or pumping speed	[m^3/s]
V_S	Ideal volume flow rate or pumping speed	[m^3/s]
η_V	Volumetric efficiency	[%]
η_{is}	Isentropic efficiency	[%]
z	Number of teeth	[-]
λ	Free mean path length	[m]
l_{char}	Characteristic length	[m]

Acknowledgment
The authors would like to thank Dr. David Rowinski of Convergent Science for his invaluable support. His valuable contribution is acknowledged.

References
[1] ANSYS 2018 *DesignModeler User Guide*, Release 18.0 (ANSYS, Inc.)
[2] Faragallah W H 2008 *Vakuum Pumpen (Gastransfervakuumpumpen)*, (Sulzbach: Verlag und Bildarchiv w.H. Faragallah)
[3] Fister W 2013 *Fluidenergiemaschinen: Physikalische Voraussetzungen, Kenngrößen, Elementartstufen der Strömungs-und Verdrängermaschinen*, (Berlin: Springer-Verlag, Band 1)
[4] Frohn A 1979 *Einführung in die kinetische Gastheorie: Studienbuch für Studierende der Ingenieurwissenschaften, der Physik und der physikalischen Chemie*, (Akademische Verlagsgesellschaft)
[5] Müller R 2013 *Spaltströmung mit Wärmeübertragung in Vakuumpumpen*, PhD thesis. Universität Kaiserslautern
[6] Pope S B 2000 *Turbulence Flows*, (Cambridge University Press, ed 1)
[7] Richards J K, Senecal P K and Pomraning E 2017 *Converge 2.4 Manual*, Madison, WI (Convergent Science, Inc.)

IOP Conf. Series: Materials Science and Engineering **425** (2018) 012027 doi:10.1088/1757-899X/425/1/012027

Limits of one dimensional modeling of rarefied Couette Poiseuille clearance flow in vacuum pumps

Christopher Huck, Andreas Brümmer

Chair of Fluidics, TU Dortmund University, Leonhard-Euler Str. 5, 44227 Dortmund Germany

E-mail: `christopher.huck@tu-dortmund.de`, `andreas.bruemmer@tu-dortmund.de`

Abstract. Clearance flows are the main loss mechanism in dry running positive displacement vacuum pumps. In order to calculate the operation of those pumps a detailed knowledge of the clearance mass flow rates is crucial. The dimensions of such pumps and the large pressure range of the operating points require a wide range of gas rarefaction to be taken into account. The clearance flow can be described by a combined Couette Poiseuille flow due to the pressure gradient between two chambers and the rotation of the rotary pistons. A clearance of variable cross section is investigated and in order to determine the mass flow rate, the DSMC (direct simulation Monte Carlo) method is used. Results in the slip and transition flow regime are compared to experimental findings and a good agreement can be found. In addition, a one dimensional theory is presented based on a linear superposition of Poiseuille and Couette flow rates. The aim of the present work is to determine the limits of the one dimensional theory concerning the geometry of such clearances. For this, the boundary conditions circumferential speed, inlet and outlet pressure are varied.

1. Introduction

Dry running vacuum pumps have become an important part of vacuum systems due to their oil-free working principle. The need for a clean vacuum, for instance in the semiconductor industry, has pushed the development of dry running positive displacement pumps, such as screw and roots pumps (see Ref. [1]). Due to contactless operation of the pistons, the sealing between the working chambers is realized by small clearances. Regarding Figure 1 a) the effective mass flow rate of a screw machine can be calculated by the delivered mass flow rate $\dot{m}_d$ of a working chamber driven by the rotation of the pistons and reduced by the clearance mass flow rate $\dot{m}_{cl}$. Clearances typically appear between both, rotary pistons and the housing in such machines. The geometry of the surrounding boundary varies in dependence of the shape of the rotors and the housing. Calculating the thermodynamical operation of such pumps, the knowledge of these clearance mass flow rates is essential, which represent the main loss mechanism.

In the current work a clearance geometry formed by a plane contour on one side and a contour with a radius R on the other side shall be regarded, which is shown in Figure 1 b). The clearance height is defined by h_{min}. Due to the rotation of the rotors a moving boundary is present. This is considered by a circumferential speed $U = \omega R$ for the contour of radius R. The clearance has a finite width w, which is constant in the flow direction x. The thermodynamic state of the connected working chambers is defined by the static pressure and static temperature at the inlet (p_{in}, T_{in}) and outlet (p_{out}, T_{out}).

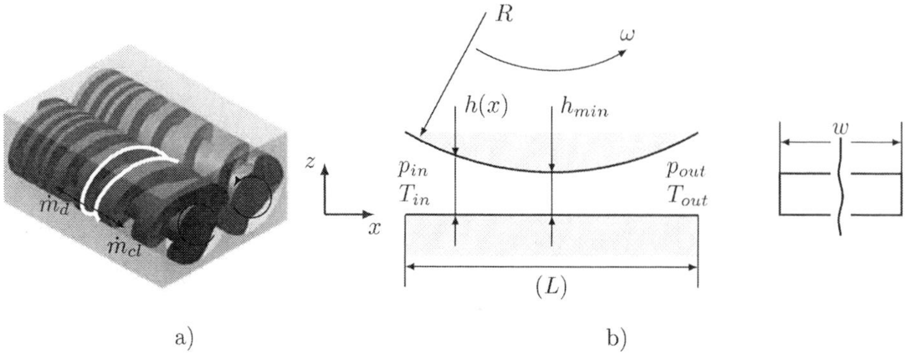

Figure 1. Screw vacuum pump and investigated clearance geometry.

Screw vacuum pumps are deployed as fore vacuum pumps and the suction pressure ranges from atmospheric state down to only a few Pascal, meaning that the pressure ranges in more than five decades. In addition, the circumferential speed of the tip of the rotary piston can reach up to 80 m/s. In this context two basic flow problems can be identified in order to predict the mass flow rate in such clearances. The Couette flow, which is driven by the movement of the rotary piston and the Poiseuille flow, driven by the pressure gradient between two working chambers. In addition, the prediction of the mass flow rate is complicated by the fact, that rarefaction effects of the clearance flow have to be taken into account, due to the large pressure range. The degree of rarefaction is expressed by the Knudsen number, which can be described by the ratio of the mean free path of the molecules and a characteristic length (the minimal clearance height h_{min}). Typically all flow types appear in these machines, ranging from the continuum flow to the free molecular flow.

The rarefied gas flow in channels, tubes and clearances has been studied extensively in the field of microelectromechanical systems during the past decades. These systems often operate at atmospheric flow conditions with very small length scales (see Ref. [2]). Thus, the range of Knudsen numbers which has to be dealt with is comparable to clearance flows in vacuum pumps. Regarding the literature various approaches modeling rarefied flows are present. While in the continuum flow regime normally the Navier-Stokes equations are used applying a no-slip boundary condition on the walls, this approach fails at higher Knudsen numbers. Therefore in the so called slip flow regime a slip boundary condition at the walls can be introduced to improve the results of the Navier-Stokes equations (Ref. [3]). The application of a first order slip model fails to predict the Knudsen minimum, so that a second order model has to be introduced (see Ref. [2]). If the Navier-Stokes equation fail at moderately high Knudsen numbers, one can calculate the rarefied flow on the basis of the Boltzmann equation, but the solution is complicated because of the collision term (see Refs. [4] and [5]). A widely used method for the solution of the collision term is the Bhatnagar-Gross-Krook model (BGK-model) described in Ref. [6]. The idea of the model is that due to the collision term the distribution function tends to equilibrium state. Often this is combined with a linearization of the Boltzmann equation. Cercignani uses this method to determine the volume flow rate between two parallel plates and compares it with experimental data (see Ref. [7]). The method is suitable to predict the Knudsen minimum (Ref. [8]). Beside the BGK-model different models for the collision term exist, like the S-model (Ref. [9]) in order to represent the correct Prandtl number. Sharipov uses this model to calculate a Poiseuille and thermal creep flow through a long tube (Ref. [10]) and parallel plates (Ref. [11]). A method for solving the Boltzmann equation statistically is provided by the direct simulation

Monte Carlo (DSMC) method (see Ref. [4]), which considers a discrete set of particles, each owning a position and a velocity. Key feature of the method is the decoupling of molecular movement and intermolecular interactions. The method has been used widely in the context of microdevices, see Refs. [12], [13], [14], [15] and [16].

As described earlier, two basic flow problems have to be handled for calculating the mass flow rate in clearances of vacuum pumps, the Couette flow and the Poiseuille flow. For both flow problems various investigations can be found in the literature and a summary is located in Ref. [17]. The one dimensional approach to calculate a combined Couette Poiseuille flow is described in Ref. [17]. The aim of the current work is to determine the limits of applicability of the one dimensional approach due to assumptions of this flow model. In a first step results of the DSMC method are compared to those obtained by experiment in order to verify the DSMC method. Following this, the radius R is varied for the presented clearance for different Knudsen numbers and boundary velocities using the one dimensional approach. Results are then compared to simulations using the DSMC method in order to evaluate the range of applicability of the one dimensional approach.

2. Theoretical approaches

In order to achieve a universal declaration of applicability, it is useful to describe the flow by dimensionless numbers. Assuming a flow of constant temperature $T(x) = T_{in} = T_{out}$ five dimensionless numbers are needed for the investigated geometry, naming the Knudsen number Kn, the pressure ratio Π, the dimensionless circumferential speed U/c_m, the ratio of the radius to the minimal clearance height R/h_{min} and the normalized mass flow rate C_0.

The Knudsen number using the model of hard sphere molecules is defined as

$$Kn = \frac{\mu \cdot c_m \cdot \sqrt{\pi}}{2 \cdot h_{min} \cdot p_{in}}, \tag{1}$$

μ is the viscosity and

$$c_m = \sqrt{\frac{2 \cdot k \cdot T_{in}}{m}} \tag{2}$$

is the most probable molecular speed with the molecular mass m and the Boltzmann constant k.

$$\Pi = \frac{p_{in}}{p_{out}} \tag{3}$$

is the pressure ratio. The results are given in the dimensionless mass flow rate

$$C_0 = \frac{\dot{m}}{\dot{m}_{mol}} = \frac{\sqrt{\pi} \cdot c_m}{p_{in} \cdot w \cdot h_{min}} \cdot \dot{m} \tag{4}$$

with $\dot{m}_{mol}$ being the mass flow rate through an orifice with equal cross sectional area at molecular flow into a perfect vacuum.

2.1. One dimensional approach

The one dimensional approach is described in Ref [17] and a brief summary shall be given in the current work. The aim is to calculate the mass flow rate $\dot{m}$ of a clearance for given boundary velocity U, inlet pressure p_{in} and outlet pressure p_{out}. The temperature of the gas and the boundaries is assumed to be equal $T_{in} = T_{out} = T$.

According to Refs. [18] and [19] the clearance is assumed to be long (i.e. $max(h) \ll L$) so that the local dimensionless pressure gradient

$$\xi_P = \frac{h}{p} \cdot \frac{\mathrm{d}p}{\mathrm{d}x}, \quad |\xi_P| \ll 1 \tag{5}$$

is considered to be small in any cross section. Following Refs. [18] and [20] a superposition of the Couette and Poiseuille flow is assumed leading to the mass flow rate in any cross section

$$\dot{m} = \frac{h \cdot w \cdot p}{c_m} \cdot \left(2 \cdot G_C \cdot \frac{U_x}{c_m} - G_P \cdot \xi_P \right) , \tag{6}$$

where $U_x = U_x(x)$ is the local wall velocity of the moving boundary in flow direction. G_C is the reduced flow rate of the plane Couette flow, which can be calculated to the constant value $G_C = 1/2$ due to the symmetry of the Couette flow for a clearance of infinite width w (see Ref. [20]). The reduced flow rate G_P of the Poiseuille flow depends on the local Knudsen number and the local aspect ratio h/w. The determination of G_P is described in Ref. [17] and values are extracted from Ref. [21], [22], [23] and [24]. Eq. 6 then can be solved using a finite difference scheme, also described in Ref. [17].

2.2. DSMC method

The DSMC method provides a statistical tool for determining the solution of the Boltzmann equation. The method, developed by Bird [4], is a particle-based simulation tool, according to the kinetic theory of gases. The key feature of the DSMC method is, that one particle represents a set of molecules and that the movement of particles and intermolecular collisions are decoupled. Figure 2 shows the complete algorithm sequence, as it is implemented in the open source software package OpenFOAM.

The algorithm starts with a grid generation and initialization of particles according to a given uniform number density of the flow field. A velocity and inner energy is assigned to every particle according to a given uniform temperature. The main loop starts with the calculation of the movement of every particle. According to a given time step new positions are calculated and potential collisions with boundaries are handled according to the underlying model. Particles which leave the simulation domain via the inlet or outlet have to be removed. To be able to simulate an inflow and outflow at a given pressure and temperature, new particles have to be inserted at the inlet and outlet. For this, a new boundary condition according Ref. [4] was implemented in the OpenFoam software package. Next, the particles are indexed to every cell, as is necessary to calculate the intermolecular collisions for every cell separately. The collision algorithm itself is based on a non-deterministic approach, using the probability of collisions. The NTC-method (no time counter method), developed by Bird [25], allows the calculation of the correct number of collisions. For every time step the macroscopic flow fields are calculated and are then time-averaged in order to minimize the statistical noise. This procedure is repeated N_{av} times until the end of the averaging interval is reached. All averaged fields are written and reset and the next averaging interval begins. The procedure is repeated until a flow of steady state is reached, regarding the mass flow at the inlet and outlet. In order to reduce the statistical noise, a subsequent longer averaging run is necessary.

Within this study, the flow of air in the clearance geometry presented in Figure 1 is investigated. In all cases the length L is chosen such, that the pressure gradient at the inlet and outlet of the clearance is small. The gas model is composed as a mixture of nitrogen and oxygen with the appropriate fraction of number density. The used values for the relaxation collision number, the mass of molecule, the variable hard sphere (VHS) diameter, the number of inner degrees of freedom and the viscosity index for each molecule type is given in Table 1. For the intermolecular collisions the Larsen-Borgnakke-VHS model is used, described in Ref. [4]. Wall collisions are handled according a diffuse reflection of the molecule and in case of a moving boundary, the wall velocity is superposed.

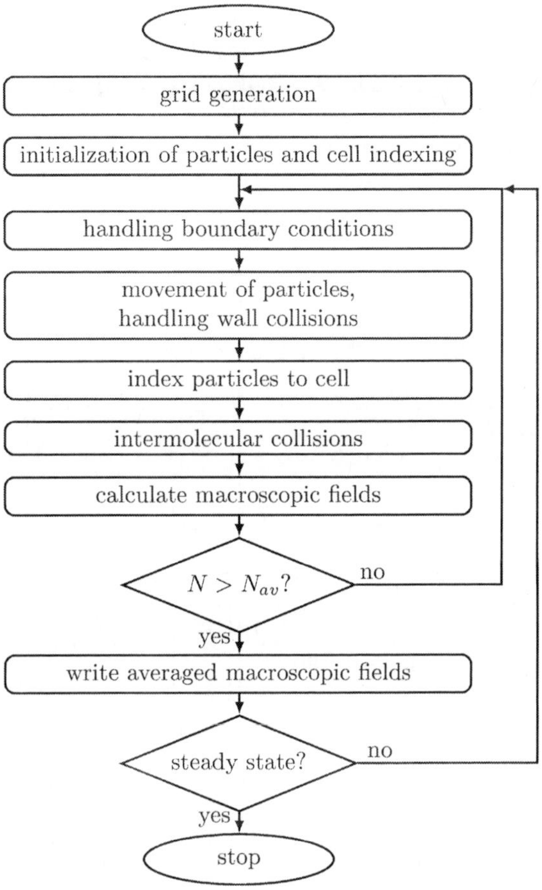

Figure 2. DSMC-Algorithm

Table 1. Model of gas (according Ref. [4]) and boundary conditions used in the DSMC method.

inflow/ outflow boundary conditions	specifying: p, T	
wall interaction model	diffuse (Maxwell)	
Binary collision model	Larsen-Borgnakke-VHS	
relaxation collision number	5	
gas	N_2	O_2
fraction number density	77.7 %	22.3 %
mass of molecule	$46.5 \cdot 10^{-27}\ kg$	$53.12 \cdot 10^{-27}\ kg$
VHS diameter at 273 K	$4.17 \cdot 10^{-10}\ m$	$4.07 \cdot 10^{-10}\ m$
number of inner degrees of freedom	2	2
viscosity index	0.74	0.77

3. Comparing theoretical approaches to Experimental results

In order to validate the simulations of the DSMC method a comparison to experimental results is carried out. For this, an experimental setup is used with a clearance geometry according to Figure 1, where the upper contour is realized by a rotatable shaft. It is possible to adjust the inlet and outlet pressure p_{in} and p_{out} and the circumferential speed U. A complete description of the setup and the method of measurement can be found in Ref. [17]. Figure 3 shows experimental results and simulations of the DSMC method for the dimensionless mass flow rate as a function of the Knudsen number. The dimensionless circumferential speed U/c_m is varied for constant pressure ratio $\Pi = p_{in}/p_{out} = 8$ and geometry $R/h_{min} = 250$. Measurements are conducted in the transition, slip and continuum flow regimes. Simulations are carried out in the transition and slip flow regimes. The smallest Knudsen number calculated by the DSMC method is $Kn = 0.044$ because for smaller values the computational time gets very large.

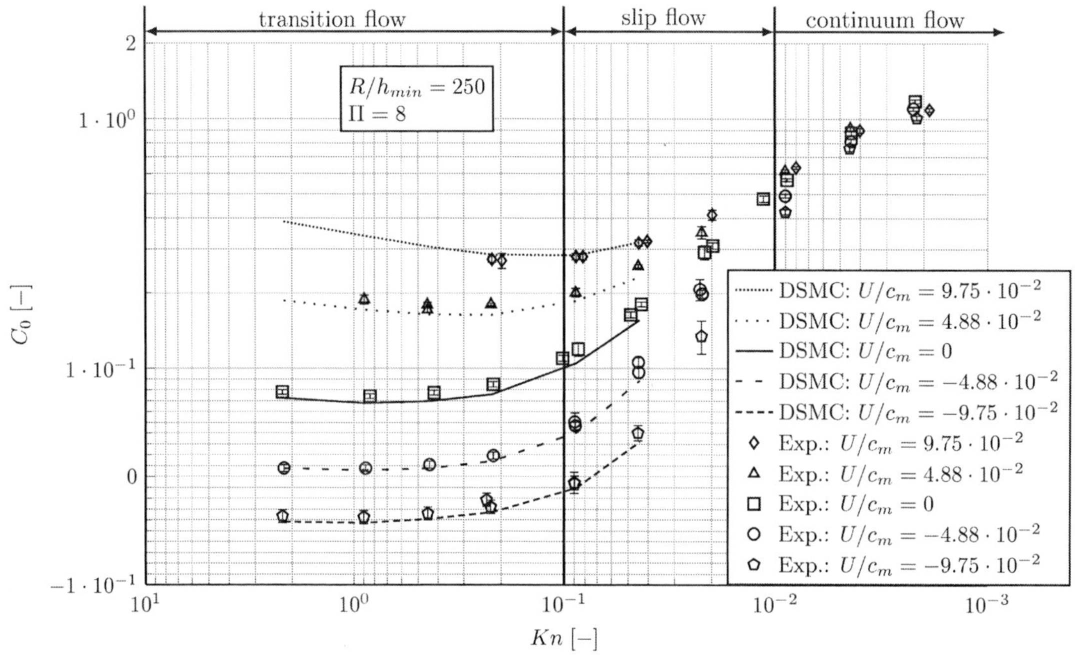

Figure 3. Dimensionless mass flow rate as a function of the Knudsen number. A variation of the dimensionless circumferential speed U/c_m is shown for simulations using the DSMC method and results are compared to experimental findings from Ref. [17].

Regarding the experimental results for a stationary boundary $U/c_m = 0$ a characteristic dimensionless mass flow rate as a function of the Knudsen number can be observed. For small Knudsen numbers $0.004 < Kn < 0.05$ in the continuum and slip flow regime a linear decrease of the dimensionless mass flow rate with increasing Knudsen number can be observed. This is due to a higher influence of the friction in the flow, which is laminar in this regime. For very small Knudsen numbers $Kn < 0.004$ the dimensionless mass flow rate slightly tends to a constant value due to a choked flow in the clearance. For greater Knudsen numbers $Kn > 0.05$ a slip of the gas flow at the wall occurs, so that the effect of increasing friction is reduced. In the transition flow regime rarefaction effects get significant, and the Knudsen minimum can be found for a Knudsen number of $Kn \approx 0.8$. For greater Knudsen numbers the normalized mass flow rate increases again, until it reaches the mass flow rate for a molecular mass flow rate, which is independent of the Knudsen number. Regarding a positive dimensionless circumferential speed $U/c_m = 4.88 \cdot 10^{-2}$ and $U/c_m = 9.75 \cdot 10^{-2}$ the dimensionless mass flow rate is shifted towards higher values due to the Couette flow, which operates in the same direction as the Poiseuille flow. The highest impact can be observed in the transition flow regime and gets very small in the continuum flow regime. Here the flow is highly influenced by the Poiseuille flow. A negative circumferential speed $U/c_m = -4.88 \cdot 10^{-2}$ and $U/c_m = -9.75 \cdot 10^{-2}$ shifts the normalized mass flow rate towards lower values. Again, the highest impact can be found in the transition flow regime. For a Knudsen number smaller than $Kn < 0.08$ and a negative circumferential speed of $U/c_m = -9.75 \cdot 10^{-2}$ a negative dimensionless mass flow rate can be observed.

Comparing the simulation results of the DSMC method with the experimental results a very good agreement can be found. The DSMC method is able to depict the normalized mass flow rate for the shown Knudsen numbers and circumferential speeds. Regarding a negative circumferential speed, the simulation using the DSMC method is within the confidence interval of the experimental results, except for a Knudsen number $Kn = 0.044$. Here, the relative error is within 20%. Regarding the results without circumferential speed and a positive circumferential speed $U/c_m = 4.88 \cdot 10^{-2}$, the normalized mass flow rate obtained by the DSMC method is smaller than the experimental results for all calculated Knudsen numbers. Here, the relative error is within 13%. For a positive circumferential speed $U/c_m = 9.75 \cdot 10^{-2}$ the results of the DSMC method are within the confidence interval of the experimental results, again.

4. Variation of geometry

As described in section 1, the aim of the current work is to evaluate the applicability of the one dimensional approach regarding different geometries. Therefore, the geometrical parameter R/h_{min} is varied and the calculated dimensionless mass flow rates of the one dimensional approach are compared to those, obtained with the DSMC method. Further on, the relative deviation is shown, which is defined as

$$f = \frac{C_{0,one\ dim.} - C_{0,DSMC}}{C_{0,DSMC}} \cdot 100\% \tag{7}$$

Figure 4 shows the dimensionless mass flow rate as a function of R/h_{min} for three different Knudsen numbers, which are located in the transition and slip flow regimes. No circumferential speed $U/c_m = 0$ is present in this case and the pressure ratio is $\Pi = 10$. Reducing the geometrical parameter R/h_{min} a rise of the normalized mass flow rate occurs for all three Knudsen numbers. A smaller radius of the clearance results in a smaller effective clearance length in the flow direction and yields less frictional losses in the flow. This is depicted by the DSMC method and the one dimensional approach as well. It has to be stated, that the normalized mass flow rate using the one dimensional approach offers higher values compared to the DSMC method in all cases and diverges for smaller values of R/h_{min}. Regarding a Knudsen number $Kn = 4.4$ the relative deviation is $f = 7.5\,\%$ for $R/h_{min} = 250$ and rises to $f = 28\,\%$ for $R/h_{min} = 16$. A similar process can be observed for smaller Knudsen numbers, starting with a relative deviation $f = 3\,\%$ for $R/h_{min} = 250$ rising to $f = 10\,\%$ for $R/h_{min} = 16$ for a Knudsen number $Kn = 0.44$ and $f = 3.3\,\%$ for $R/h_{min} = 250$ rising to $f = 18.5\,\%$ for a Knudsen number $Kn = 0.044$. It can be stated in general, that the deviation of both methods is small in the region of the Knudsen minimum and rises for higher and smaller Knudsen numbers. The diverging results of both methods for smaller values R/h_{min} can be explained by an increasingly influence of a two dimensional flow. With a smaller radius R, the z-component of the flow has a higher impact, which can not be depicted by the one dimensional approach, but the DSMC method can. It can be stated, that a great increase in the deviation f can be observed for $R/h_{min} < 100$ for all Knudsen numbers and no circumferential speed.

Figure 5 shows the dimensionless mass flow rate as a function of R/h_{min} for three different Knudsen numbers, a positive circumferential speed $U/c_m = 9.75 \cdot 10^{-2}$ and a pressure ratio $\Pi = 10$. Due to the positive Couette flow, a higher normalized mass flow rate can be observed compared to the results without circumferential speed for all Knudsen numbers. Again, the deviation f rises for smaller values of the geometrical parameter R/h_{min}, due to a higher

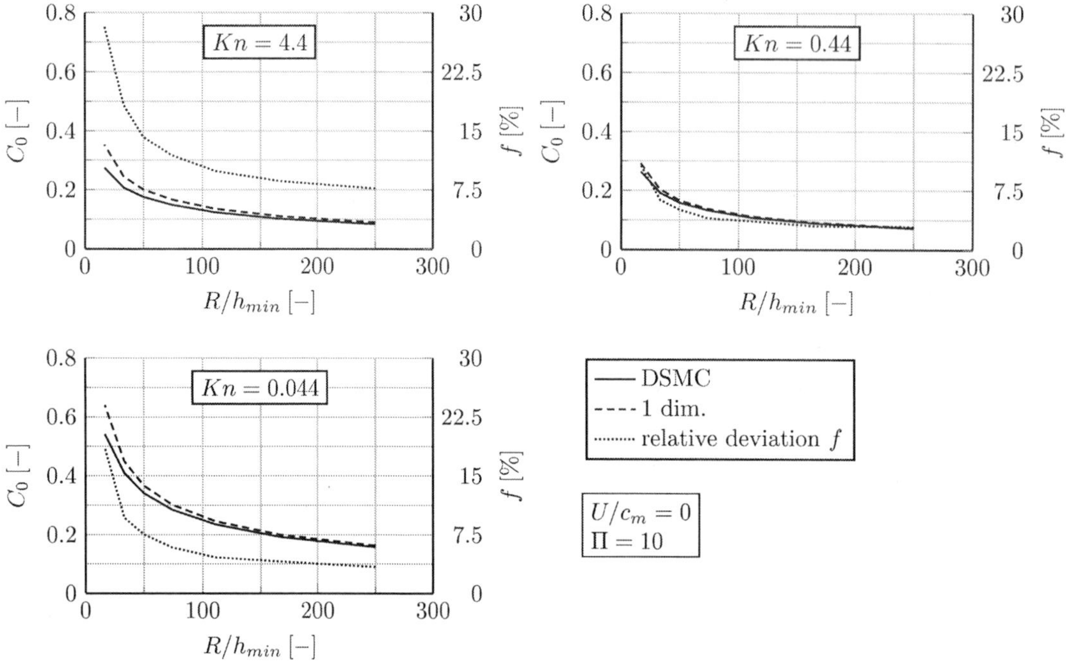

Figure 4. Dimensionless mass flow rate as a function of the ratio R/h_{min} for different Knudsen numbers and a boundary velocity $U/c_m = 0$ and a pressure ratio $\Pi = 10$. Results of one dimensional theory are compared with the DSMC method and relative deviation f is given.

influence of the z-component of the flow. Regarding a Knudsen number $Kn = 4.4$, a reduction of the geometrical parameter R/h_{min} only has a small impact on the normalized mass flow rate. Results of the DSMC method indicate a constant normalized mass flow rate so that a high influence of the Couette flow can be stated for this Knudsen number. A slight rise in the normalized mass flow rate can be observed for smaller values of R/h_{min} using the one dimensional approach. For smaller Knudsen numbers $Kn = 0.44$ and $Kn = 0.044$ a rise in the normalized mass flow rate can be observed for smaller values of R/h_{min}. For $R/h_{min} > 100$ the deviation f is smaller than 4% for all Knudsen numbers.

Figure 6 shows the dimensionless mass flow rate as a function of R/h_{min} for three different Knudsen numbers, a negative circumferential speed $U/c_m = -9.75 \cdot 10^{-2}$ and a pressure ratio $\Pi = 10$. Due to a negative Couette flow, lower normalized mass flow rates can be observed compared to the results without circumferential speed. For smaller values of the geometrical parameter R/h_{min} an increase of the normalized mass flow rate can be observed for all Knudsen numbers. Again, a smaller radius R decreases the effective length of the clearance and thus the frictional losses of the Poiseuille flow. For a Knudsen number $Kn = 0.044$ an increase of the deviation f of both methods for smaller values of the geometrical parameter R/h_{min} can be observed again. In contrast to this, for higher Knudsen numbers $Kn = 0.44$ and $Kn = 4.4$ the deviation rises for smaller values R/h_{min}. Because the absolute values in the normalized mass flow rates tends to zero the deviation f is large. Here, the Poiseuille and the negative Couette flow nearly compensate. But the absolute deviation is small again for $R/h_{min} > 100$.

IOP Conf. Series: Materials Science and Engineering **425** (2018) 012027 doi:10.1088/1757-899X/425/1/012027

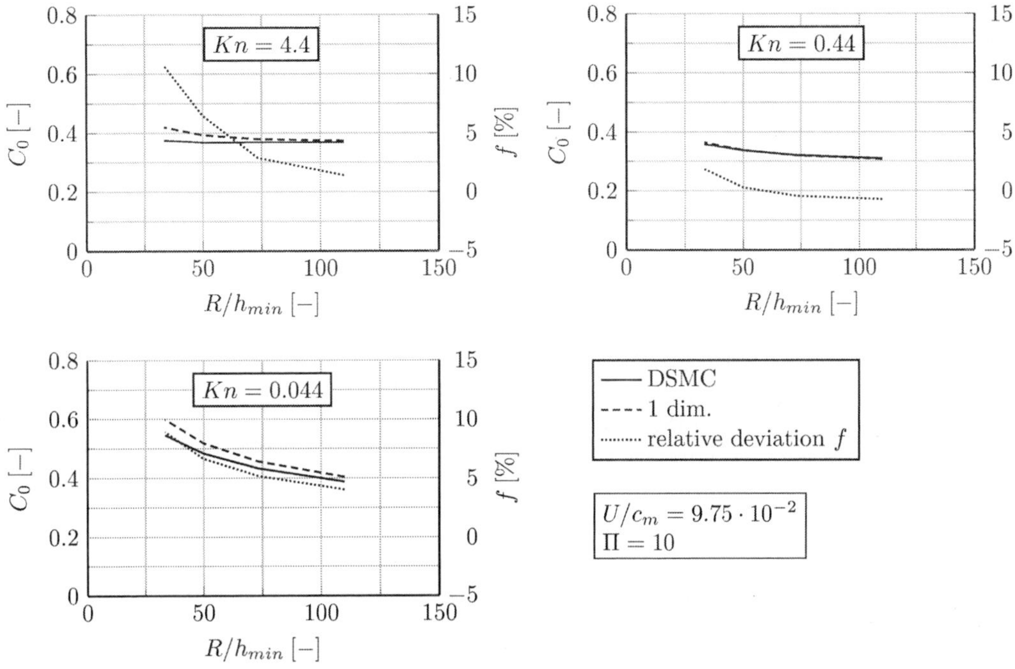

Figure 5. Dimensionless mass flow rate as a function of the ratio R/h_{min} for different Knudsen numbers and a boundary velocity $U/c_m = 9.75 \cdot 10^{-2}$ and a pressure ratio $\Pi = 10$. Results of one dimensional theory are compared with the DSMC method and relative deviation f is given.

5. Conclusion

In the present paper, the combined Couette Poiseuille flow of a rarefied gas is investigated using a one dimensional approach and the DSMC method. A comparison of the normalized mass flow rate obtained by the DSMC method and experimental results show, that the DSMC method is able to depict the combined Couette Poiseuille flow. A variation of the geometrical parameter R/h_{min}, the circumferential speed U/c_m and the Knudsen number is conducted. In summary the one dimensional approach seems to provide reasonable results for geometrical parameters $R/h_{min} > 100$.

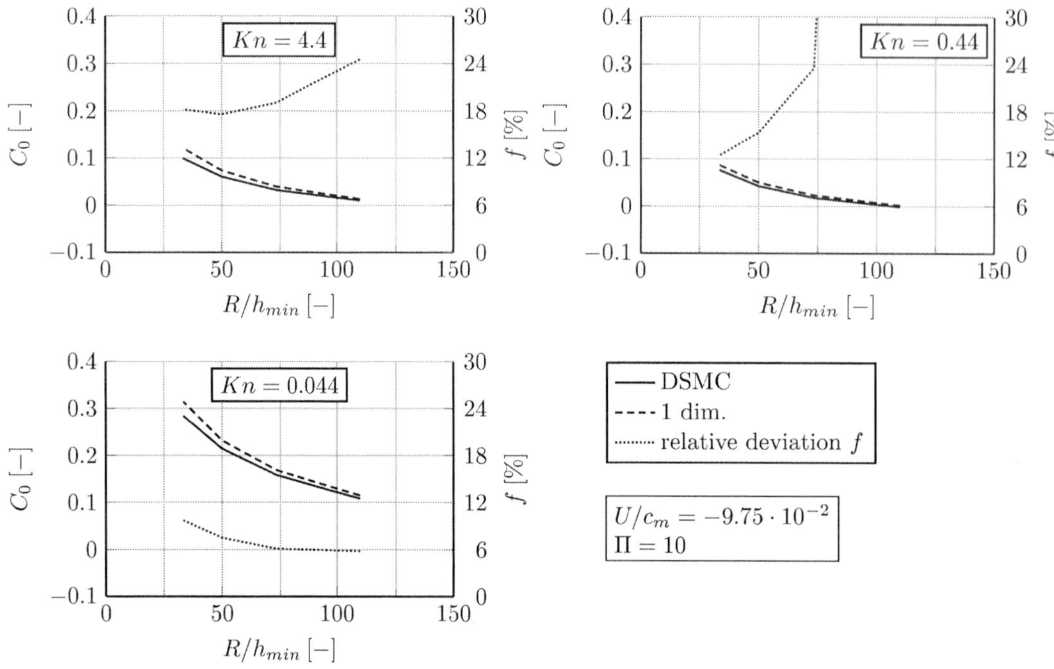

Figure 6. Dimensionless mass flow rate as a function of the ratio R/h_{min} for different Knudsen numbers and a boundary velocity $U/c_m = -9.75 \cdot 10^{-2}$ and a pressure ratio $\Pi = 10$. Results of one dimensional theory are compared with the DSMC method and relative deviation f is given.

Acknowledgments

Supported by the Federal Ministry for Economic and Energy (BMWi) on the basis of a decision by the German Bundestag, AiF, IGF No.: 18140 N/1.

Gefördert durch:

Bundesministerium
für Wirtschaft
und Energie

aufgrund eines Beschlusses
des Deutschen Bundestages

References

[1] Jousten K 2010 *Wutz - Handbuch der Vakuumtechnik* vol 10 (Wiesbaden: Vieweg+Teubner)
[2] Karniadakis G, Beskok A and Aluru N 2005 *Microflows and Nanoflows - Fundamentals and Simulation* (New York: Springer Science+Business Media, Inc.)
[3] Arkilic E B, Schmidt M A and Breuer K S 1997 *J. Microelectromech. Syst.* **6** 167
[4] Bird G A 1994 *Molecular Gas Dynamics and the Direct Simulation of Gas Flows* (Oxford: Clarendon Press)
[5] Shen C 2005 *Rarefied Gas Dynamics - Fundamentals, Simulations and Micro Flows* (Berlin, Heidelberg: Springer Science+Business Media, Inc.)
[6] Bhatnagar P L, Gross E P and Krook M 1954 *Phys. Rev.* **94** 511
[7] Cercignani C and Daneri A 1963 *J. Appl. Phys.* **34** 3509
[8] Loyalka S K and Hamoodi S A 1990 *Phys. Fluids A* **2** 2061
[9] Shakov E M 1968 *Fluid Dyn.* **3** 142
[10] Sharipov F 1996 *J. Vac. Sci. Technol., A* **14** 2627
[11] Sharipov F 2002 *Eur. J. Mech. B. Fluids* **21** 113
[12] Piekos E S and Breuer K S 1995 *30th AIAA Thermophysics Conference*
[13] Wu J and Tseng K 2001 *Comput. Fluids* **30** 711
[14] Wang M and Li Z 2004 *Int. J. Heat Fluid Flow* **25** 975
[15] Chong X 2006 *Phys. Fluids* **18** 127104–1
[16] White C 2013 *Benchmarking, Development and Applications of an Open Source DSMC Solver* Phd thesis University of Strathclyde
[17] Huck C, Pleskun H and Brümmer A 2018 *J. Vac. Sci. Technol., A* **36** 031606 1–10
[18] Sharipov F, Fahrenbach P and Zipp A 2005 *J. Vac. Sci. Technol., A* **23** 1331
[19] Graur I and Ho M T 2014 *Vacuum* **101** 328
[20] Fukui S and Kaneko R 1988 *J. Tribol.* **110** 253
[21] Titarev V A and Shakhov E M 2010 *Comput. Math. Math. Phys.* **50** 1221
[22] Zahid W A, Yin Y and Zhu K Q 2007 *Microfluid. Nanofluid.* **3** 55
[23] Sharipov F 1999 *J. Vac. Sci. Technol., A* **17** 3062
[24] Alexander F, Garcia A and Alder B 1994 *Phys. Fluids* **6** 3854
[25] Bird G A 1989 *Rarefied gas dynamics: Theoretical and computational techniques; International Symposium, 16th* 211–226

IOP Conf. Series: Materials Science and Engineering **425** (2018) 012028 doi:10.1088/1757-899X/425/1/012028

Optimisation of screw spindle vacuum pumps with variable rotor pitch regarding load-lock operation

T. Jünemann[1] and A. Brümmer[2]

Chair of Fluidics, Faculty of Mechanical Engineering, TU Dortmund University, Germany

E-mail: [1]timo.juenemann@tu-dortmund.de, [2]andreas.bruemmer@tu-dortmund.de

Abstract. Screw spindle vacuum pumps are characterised by high suction speed and the ability to achieve high pressure ratios in a dry running process. If the dimensions of the pump are comparable to the volume of the system's receiver, evacuation only takes a few seconds. For this reason, a typical application for screw spindle vacuum pumps is the evacuation of a load-lock chamber in a predefined time during a clocked production process. In order to reduce power consumption, screw spindle vacuum pumps with an internal change of volume are used almost exclusively. These pumps are commonly with variable pitch rotors. In this article, a new approach for optimisation of variable pitch rotors in consideration of the requirements of the load-lock process is presented. A dimensionless energetic efficiency is defined based on an idealised load-lock process, this parameter is maximised by the optimisation algorithm. The rotor pitch progression, an important factor in pump performance, is approximated by segments of constant pitch at the suction and discharge sides with a linear decrease in between. The overall wrap angle, the minimum rotor pitch, the crown circle diameter, the root circle diameter, and the circumferential speed are constant for the optimisation, during which the rotor length and load-lock pressure are varied. Optimisations are carried out for double-lobed cycloid profiles. The optimisations result in selection of internal volume ratios which are significantly below a theoretical calculated value for the idealised lock-lock process. This is explained by the internal clearances and details of compression process. The studies reveal a positive influence of the rotor length on the load-lock efficiency, this is because the proper rotor length leads to an uniform compression. It is also shown that the wrap angle of the segment on the suction side should be kept low.

1. Introduction

Evacuation of a load-lock chamber during a clocked production process is a common application for vacuum pumps. In such a process a given volume is evacuated until the target pressure in the receiver is reached. The operating point of the vacuum pump is changing continuously from atmospheric pressure to the target pressure. Therefore, the whole pressure range needs to be taken into account in the design process. In recent years dry-running screw spindle vacuum pumps have replaced multi-stage-roots pumps and oil-lubricated rotary vane pumps because they are tolerant to dirt and small amounts of liquid [1]. In order to reduce compression power, screw spindle vacuum pumps with an internal change of volume are used almost exclusively. In this context, an ideal volume ratio can be identified for every operating point. The ideal volume ratio depends on the pressure ratio, internal leakage and heat exchange within the positive displacement machine. The most common solution is an change of volume during the transport phase, which can be achieved in various ways. This article deals with the solution based on a variable rotor pitch, an option now available due to advances in manufacturing [2]. In [3] variable rotor pitch and other options affecting internal compression are discussed. It is shown that a continuous, uniform progression of chamber volume is advantageous since this leads to an uniform heat distribution and a low maximum component temperature. Although there are disadvantages regarding to the volumetric efficiency and the ultimate pressure. Pfaller [4] presented an evolutionary multi-criteria optimisation approach to optimise variable pitch rotors for mass flow and power consumption , when only a single operating point is taken into account. The influence of manufacturing methods on the mismatch of the profile shape of vacuum pumps with variable rotor pitch and therefore on the volumetric efficiency of these pumps is discussed in [5]. Utri [6] investigated the potential of a variable pitch for dry running screw compressors and identified an opportunity to increase the efficiency for fluids with low isentropic exponents and in applications with high pressure ratios. In [7] Utri uses a Nelder-Mead-simplex optimisation to identify the best rotor pitch and internal volume ratio for dual pitch rotors. The aim of work the current work is to present a suitable approach to optimise the rotor pitch of a screw spindle vacuum pump for a load-lock process.

2. Working principle and characteristics of screw spindle vacuum pumps

Screw spindle vacuum pumps are two shaft rotary positive displacement machines with a large wrap angle, which transport the trapped gas axially from the inlet to the outlet by a rotation of the rotors (Figure 1). A large wrap angle is necessary to reduce the internal clearance back flow and therefore to achieve the high pressure ratio, common for such vacuum pumps [2].

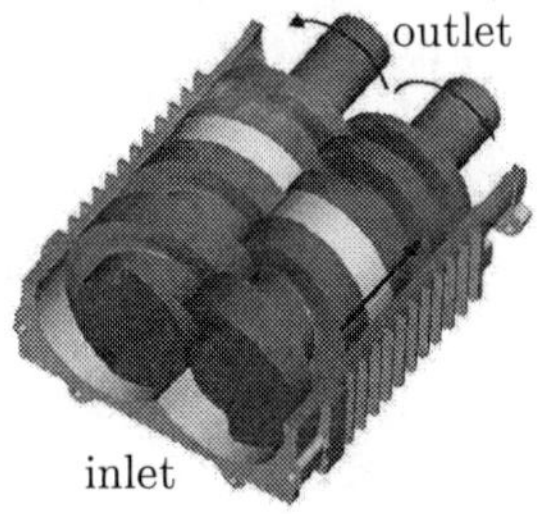

Figure 1. Operating principle of a screw vacuum pump [1]

Screw spindle vacuum pumps are commonly used as forepumps and are therefore compressing against atmospheric pressure ($p_{out} = p_{at}$). The inlet pressure ranges from $10^{-3}\,hPa$ to $10^{3}\,hPa$ for the most of the time. The suction speed S is the pump's delivery volume flow rate, referred to the thermodynamic state at the inlet. Another important characteristic is the inner power consumption P_i, which is strongly influenced by the volume ratio v_i, defined as the ratio of the volume of the first closed working chamber at the inlet to the volume of the last closed working chamber at the outlet.

$$v_i = \frac{V_{in}}{V_{out}} \qquad (1)$$

The theoretical process of a single operating point is defined by assuming no internal leakages and the following change of state:

(i) isobaric suction at inlet pressure
(ii) isothermal compression
(iii) isochoric adjustment to the outlet pressure

IOP Conf. Series: Materials Science and Engineering **425** (2018) 012028 doi:10.1088/1757-899X/425/1/012028

(iv) isobaric discharge at outlet pressure

The density work is defined as the ratio of inner power consumption P_i and the suction speed S:

$$w_i = \frac{P_i}{S} \tag{2}$$

It is calculated for the theoretical process by

$$w_i^{th} = p_{in} \cdot \left(\frac{\Pi}{v_i} + (\ln(v_i) - 1) \right) \tag{3}$$

which is minimised by choosing the volume ratio equal to the pressure ratio ($v_i = \Pi$), which corresponds to an adjusted operating point. The minimal value is given by:

$$w_{i,ideal}^{th} = p_{in} \cdot \ln(\Pi) \tag{4}$$

A theoretical work similar to Eq. 3 is presented by Dreifert [1], who assumes an isentropic internal compression.

In order to evaluate the efficiency of an operating point of a screw spindle vacuum pump, two dimensionless quantities are commonly used. The volumetric efficiency λ_v, given by the ratio of the suction speed S to the suction speed S_{th} of a theoretical ideal machine with no leakage:

$$\lambda_v = \frac{S}{S_{th}} \tag{5}$$

The isothermal efficiency η_T is defined by the ideal theoretical density work (Eq. 4) divided by the real density work w_i^{real} at the operating point:

$$\eta_T = \frac{w_{i,ideal}^{th}}{w_i^{real}} \tag{6}$$

3. Load-lock operation and quantities

In a load-lock process, like it is shown in Fig. 2, a receiver of volume V is evacuated from atmospheric pressure p_{at} to a desired load-lock pressure p_{LL}. The time required - the load-lock time t_{LL} - is normally required to be short. The inner power consumption P_i of the vacuum pump depends on the current inlet pressure and defines the work $W_{i,LL}$ which is needed for the evacuation. An efficient load-lock process is characterised by a low inner work $W_{i,LL}$. This process parameters are evaluated as follows where the temperature inside the receiver is assumed to be constant $T = 273\,K$.

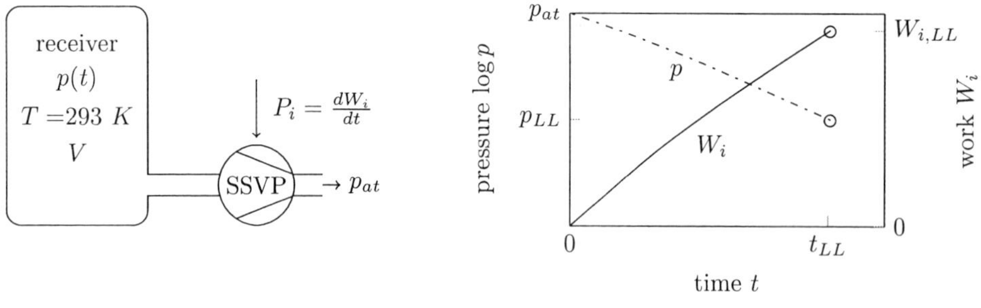

Figure 2. Sketch of a load-lock process

The load-lock time is calculated for a given suction speed characteristic $S(p)$ as shown below, assuming an ideal gas:

$$t_{LL} = \int_0^{t_{LL}} dt = -V \int_{p_{at}}^{p_{LL}} \frac{1}{S(p) \cdot p} dp \tag{7}$$

Using this equation the load-lock work $W_{i,LL}$ can be calculated using the known characteristics of inner power consumption $P_i(p)$ and suction speed $S(p)$:

$$W_{i,LL} = \int_0^{t_{LL}} P_i(t)dt = -V \int_{p_{at}}^{p_{LL}} \frac{P_i(p)}{S(p) \cdot p} dp = -V \int_{p_{at}}^{p_{LL}} \frac{w_i(p)}{p} dp \tag{8}$$

The density load-lock work is defined as the ratio of the load-lock work $W_{i,LL}$ to the volume V of the receiver:

$$w_{i,LL} = \frac{W_{i,LL}}{V}. \tag{9}$$

If the theoretical process (Eq. 3) is assumed at every operating point, Eq. 8 can be rearranged to

$$w_{i,LL}^{th} = \frac{W_{i,LL}^{th}}{V} = p_{at} \left(\frac{\ln(\Pi_{LL})}{v_i} + (ln(v_i) - 1) \left(1 - \Pi_{LL}^{-1}\right) \right) \tag{10}$$

which can be minimised with the following volume ratio

$$v_{i,LL,ideal}^{th} = \ln(\Pi_{LL}) \cdot \left(1 - \Pi_{LL}^{-1}\right)^{-1}, \tag{11}$$

where Π_{LL} is the ratio of the atmospheric pressure p_{at} to the load-lock target pressure p_{LL}. The minimised density load-lock work $w_{i,LL,ideal}^{th}$ for a theoretical load-lock process is given by the combination of Eq. 11 and Eq. 10.

$$w_{i,LL,ideal}^{th} = p_{at} \cdot \left(1 - \Pi_{LL}^{-1}\right) \cdot \ln \left(\frac{\ln(\Pi_{LL})}{1 - \Pi_{LL}^{-1}} \right) \tag{12}$$

Results from application of equation 10 are shown for a range of load-lock pressures in Figure 3 where the minima are marked. Starting at an volume ratio of $v_i = 1$, the density load-lock work at first strongly decreases to a minimum with increasing volume ratio, then more gradually increases. Especially in the case of low load-lock pressures, the volume ratio can be chosen less than the optimal parameter without having a great impact on the load-lock work. A reduced volume ratio reduces the maximum power consumption in case of over-compression and improves the volumetric efficiency and ultimate pressure of the pump.

The theoretical optimal evacuation work for a given load-lock pressure is known, and this can be used to define a dimensionless load-lock efficiency $\eta_{T,LL}$. This efficiency is defined as the ratio of the ideal work to the real density inner load-lock work. In the following, this quantity is optimised.

$$\eta_{T,LL} = \frac{w_{i,LL,ideal}^{th}}{w_{i,LL}^{real}} \tag{13}$$

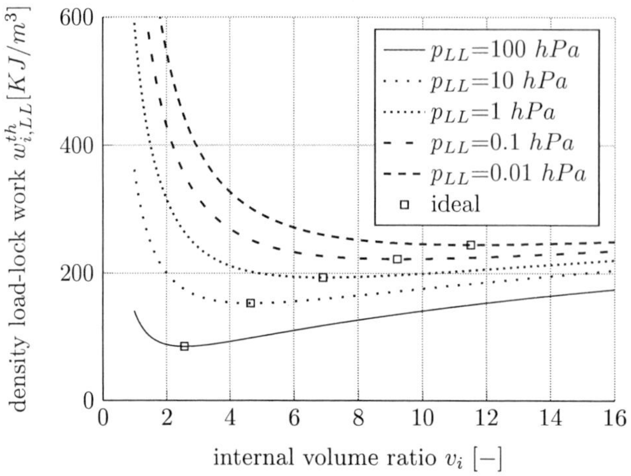

Figure 3. Density load-lock work $w^{th}_{i,LL}$ for theoretical load-lock processes as function of the internal volume ratio v_i

4. Geometric characteristics of cycloid profile rotors

For the zero-dimensional simulation of rotary displacement pumps, it is necessary to define the working chamber and the adjacent clearances. Nadler [8] describes an approach based on the analysis of different transverse planes of a screw machine. The analysis requires the clearance points and chamber areas, which need to be defined numerically by searching the points of minimal distance between the rotor flanks for various rotor orientations. The results are combined with the rotor pitch to create a phase-depending progression of chamber volume and clearance dimensions.

The front section of a typical pump rotor pair is shown in Figure 4. The cycloid geometry used on these rotors is well known in conjunction with tooth flanks of transmissions [9]. The segments of a cycloid flank are the curves traced by a point on the rim of a roll circle rolls on the inside (hypocycloid) and the outside (epicycloid) of a pitch circle. In the case of rotary positive displacement pumps, the tooth flank needs to be modified at the root to ensure a closed interlobe line. Therefore, an envelope of the tooth tip is created [10]. Since the interlobe line of cycloid profiles is well known from transmission technology [9], the analysis of the front section can be simplified using the analytical solution. For every front section position up to three interlobe clearance (ILC) points can be identified. The radial point (1) is the contact between the crown and the root circle. The flank-to-tip point (2) occurs when the tooth tip is contacting the envelope of the opposite rotor. The flank-to-flank point (3) results from the contact of the profile flanks. The combination of all clearance points for several front section positions results in the complete interlobe line, which is highlighted in bold in Figure 4. The housing clearance (4) is defined in the middle of the tooth tip.

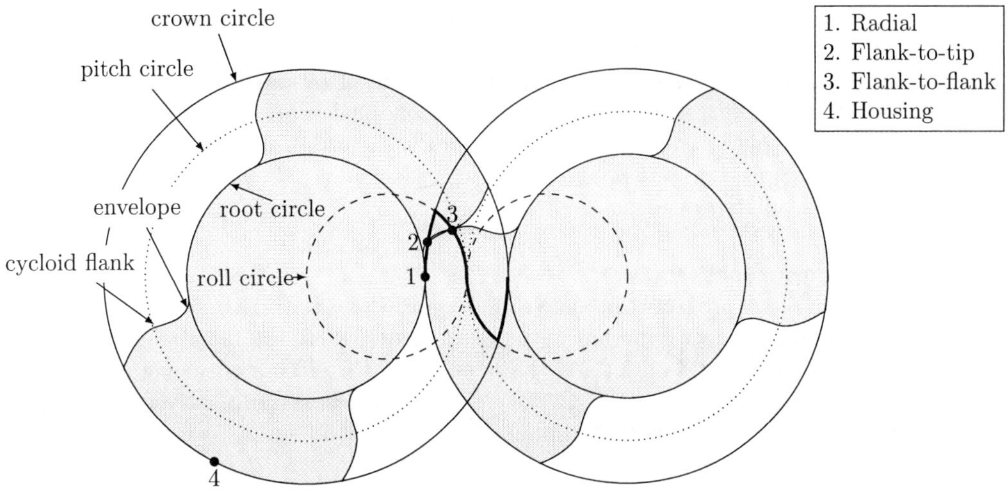

Figure 4. Clearance points in the front section of a screw spindle vacuum pump

5. Simulation approach

In consideration of the optimisation task a computationally efficient thermodynamic model is needed. The approach used here is taken from Rohe [11], who approximates the spindle pump as series of pumping stages (Figure 5). Rohe used this model to investigate the effects of internal leakages, external leakages and internal volume ratio on the suction speed characteristic of a screw spindle vacuum pump [11]. Later, Pfaller used the same approach to optimise the rotor pitch of screw spindle vacuum pumps for a single operating point [4]. In this model, every pumping stage i transports the trapped mass m_i of the chamber i to the next chamber $i+1$, while the internal leakages are lead to back flow $\dot{m}_{cl}$ from higher pressure chambers to lower pressure ones. Based on the conservation of mass and energy and the ideal gas law, the distribution of pressure p_i and the mass flow rate $\dot{m}_d$ can be calculated. The volumes V_i of the working chambers are determined for a constant rotor position (see [4] and [11]).

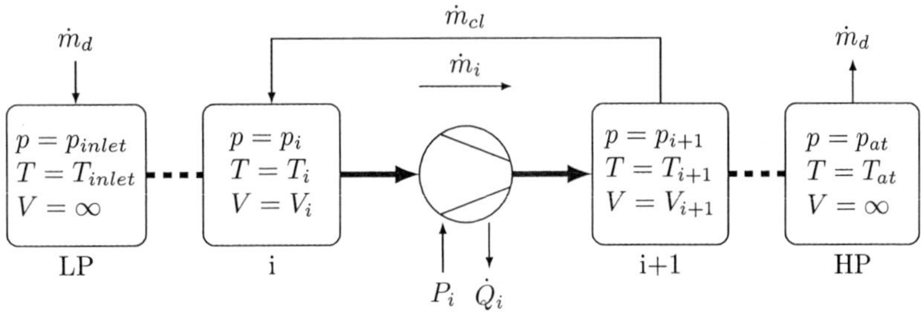

Figure 5. Sketch of the thermodynamic model (according to [11] and [4])

In order to calculate the internal leakages of the vacuum pump for the full range of pressure and referred to effects of gas rarefaction, the results of the experimental investigations of Wenderott [12] are used. Additionally, the influence of moving boundaries is taken into account by using the theoretical investigations of Stratmann [13]. Regarding the results of [11] and [13], the heat transfer between the transported gas and the pump components is modelled as complete heat exchange. For this, an isentropic compression is calculated followed by a cooling of the gas to the inlet temperature T_{Inlet} in every pumping stage. The inner power consumption of the screw spindle vacuum pump is therefore equal to the sum of all partial heat flows $\dot{Q}_i$ of every pumping stage [4]. In order to solve the integral for load-lock work (eq. 8) the characteristics of mass flow and inner power consumption are calculated for several discrete inlet pressures. In the context of the optimisation one pressure point per decade is calculated. The values between these points are interpolated linearly.

6. Optimisation approach

A generalised rotor pitch progression, shown in Figure 6, is used in the optimisation process. The rotor pitch s is plotted as a function of the wrap angle, where the rotor is divided into three sections, inlet, compression and outlet. At the inlet, the rotor pitch s_{in} is constant for the wrap angle φ_{in}. In the next section ($\varphi_{compress}$), the rotor pitch decreases linearly from the inlet s_{in} to the outlet parameter s_{out} and finally remains constant in the outlet section φ_{out}.

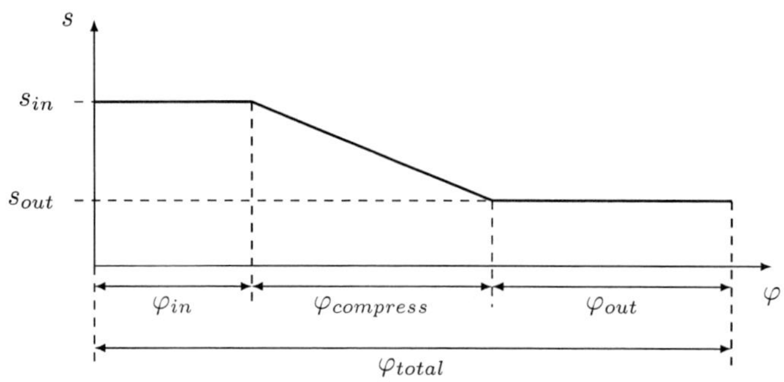

Figure 6. Simplified rotor pitch progression for the optimisation task

The rotor pitch progression needs to be optimised for the best dimensionless evacuation work (Eq. 13) while satisfying four constraints. rhe first constraint is that the specified total rotor wrap angle needs to be realised:

$$\varphi_{in} + \varphi_{compress} + \varphi_{out} = \varphi_{total} \tag{14}$$

The second constraint is the rotor length, which is equal to the integral of the rotor pitch progression:

$$\int_0^{\varphi_{total}} s(\varphi)d\varphi = L \tag{15}$$

The third constraint takes possibilities of manufacturing into account by requiring a minimum gap $h_{chamber,min}$ between the rotor flanks of one rotor. This is ensured by defining a minimum rotor pitch:

$$s(\varphi) \geq s_{min} \tag{16}$$

The last constraint defines a minimum wrap angle for the inlet and outlet side which should be greater than the wrap angle of load change. In the case of double-lobed rotors, this is given by:

$$\varphi_{in}, \varphi_{out} \geq 450° \tag{17}$$

As a result of the last constraint the internal volume ratio can be calculated as the ratio of the inlet and the outlet pitch:

$$v_i = \frac{s_{in}}{s_{out}} \tag{18}$$

IOP Conf. Series: Materials Science and Engineering **425** (2018) 012028 doi:10.1088/1757-899X/425/1/012028

In conclusion, the optimisation problem can be characterised as single objective and constrained. A local minimum of an unconstrained problem with one objective can be determined by several algorithms. In this case the down-hill simplex algorithm [14] is used. This approach was used in the context of rotor pitch optimisation in [7]. A solution to handle the bound constraints is given in [15] and [16]. In order to handle the non-linear constraints an augmented lagrangian algorithm is used, as described in [17] and [18]. At least one global minima should be found, this can be achieved using the monotonic basin hopping algorithm [19]. All algorithms are provided by the pagmo [20] and NLopt [21] optimisation libraries. In the context of the optimisation some parameters are not varied. These are listed in Table 1. The geometry of the front section is kept constant. Additionally the ratios of crown circle diameter to clearance heights are assumed to be constant. For all rotor pitch progressions a constant total wrap is chosen. All simulations are done with a constant circumferential speed and with dry air. The optimisations are done for several ratios of rotor length and crown circle diameter and for several load-lock pressures (Table 2).

Table 1. Constant optimisation boundary conditions

name	symbol	value
Gas		dry Air
Circumferential speed	u	60 m/s
Wrap angle	φ_{total}	2160°
Crown circle diameter	D_{CC}	150 mm
Number of lobes	z	2
Root circle ratio	D_{CC}/D_{RC}	1.7
Ratio of pitch and roll circle	D_{PC}/D_{Roll}	2
Housing clearance height ratio	D_{CC}/h_{HC}	600
Flank-to-flank clearance height ratio	$D_{CC}/h_{ILC,FP}$	700
Flank-to-tip clearance height ratio	$D_{CC}/h_{ILC,FT}$	700
Radial clearance height ratio	$D_{CC}/h_{ILC,R}$	800
Min. gap between rotor flanks	$D_{CC}/h_{chamber,min}$	17.0

Table 2. Varied optimisation boundary conditions

name	symbol	value
Dimensionless rotor lengths	L/D_{CC}	$2 \ldots 3.5$
Load-lock pressure	p_{LL}	$10^{-3} \ldots 10^{2}$ hPa

7. Results

Figure 7 shows the optimised internal volume ratio for different load-lock pressures p_{LL} and various dimensionless rotor lengths L/D_{CC}. Results are compared with the solution for an ideal process given by Eq. 11. It can be seen that the optimised volume ratios for the real process are significant below the parameters for the ideal process. Additionally, it can be noticed that the optimum volume ratios are increasing with decreasing load-lock pressure until they reach a maximum value. As discussed on the example of the ideal process (Figure 3) the internal volume ratio can be reduced relative to the theoretical best point without a considerable impact on the efficiency. However, an increasing volume ratio increases internal leakage and the ultimate pressure, which has been explained in detail by [11]. So the ideal volume ratio is always a compromise between volumetric efficiency and an ideal compression. Even if higher volume ratios could be realised for larger dimensionless rotor lengths, this turns out to be disadvantageous. For an increasing dimensionless rotor length the volume ratio converges to a maximum which is still clearly below the theoretical ideal value. Additionally, it should be noticed that the curve for $L/D_{CC} = 2$ is limited by its maximal possible volume ratio $v_i = 3$. With higher volume ratios, the desired load-lock pressure can not be achieved, thus the volume ratio decreases below the maximum. Since machines with larger rotor lengths have better abilities for sealing, these have higher volume ratios for low load-lock pressures. A load-lock pressure of $p_{LL} = 10^{-3}$ could not be achieved for $L/D_{CC} = 2$.

The optimised isothermal load-lock efficiency is shown in Figure 8. For all dimensionless rotor lengths the optimised efficiency is decreasing with a decreasing load-lock pressure. This is explained by the higher pressure ratios inside the pump and the resulting internal leakages. The isothermal load-lock efficiency is increasing with larger dimensionless rotor lengths. This is explained by the uniform compression of larger rotors which is discussed below.

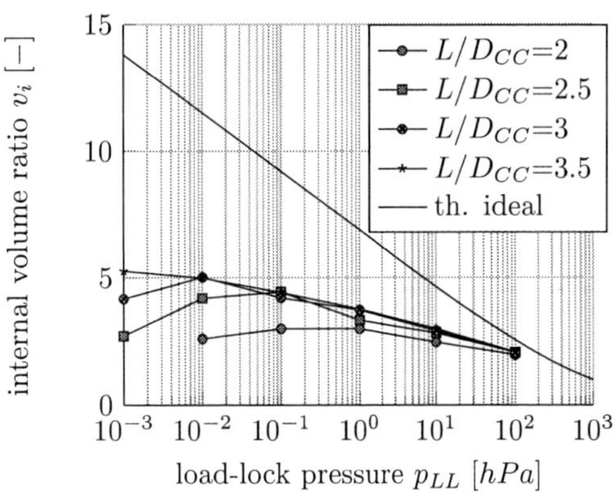

Figure 7. Optimised internal volume ratio as function of the load-lock pressure for various dimensionless rotor lengths L/D_{CC}

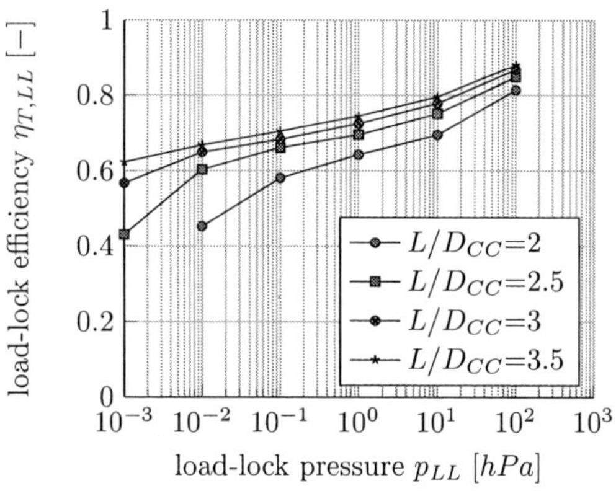

Figure 8. Optimised isothermal load-lock efficiency as function of the load-lock pressure for various dimensionless rotor lengths L/D_{CC}

The positive effect of a larger dimensionless rotor length is explained considering the characteristics of isothermal (Eq. 6) and volumetric (Eq. 5) efficiency in Figure 9 and 10, which is for several L/D_{CC}-ratios optimised for a load-lock pressure of $p_{LL} = 1\,hPa$. The characteristics of volumetric efficiency have a progression typical for screw spindle vacuum pumps with inner volume ratio. In the range of $1\,hPa$ to $20\,hPa$ the maxima occur. For higher inlet pressures the characteristic is decreasing. For lower inlet pressures the volumetric efficiency is decreasing to zero delivery. If the characteristics for different dimensionless rotor lengths are compared, an advantage of higher rotor length can be observed again. A change of the dimensionless rotor lengths by 0.5 effects the ultimate pressure by about half a decade. Additionally a high L/D_{CC}-ratio is advantageous for high inlet pressures $> 10^1\,hPa$. The biggest improvement occurs by an increase of the L/D_{CC}-ratio from 2 to 2.5.

For each dimensionless rotor length L/D_{CC} the best point of isothermal efficiency (Fig. 10) occurs at an inlet pressure of about $3 \cdot 10^2\,hPa$. Because of the effect of over-compression, the efficiency decreases rapidly to zero for the highest inlet pressures. Starting from the best point, the efficiency is decreasing with decreasing inlet pressure and is almost zero at $10^{-1}\,hPa$. An increase of the dimensionless rotor length L/D_{CC} leads to an improvement of isothermal efficiency for almost all operating points. Only in the range of over-compression is a lower volume ratio advantageous, making use of short rotor lengths beneficial.

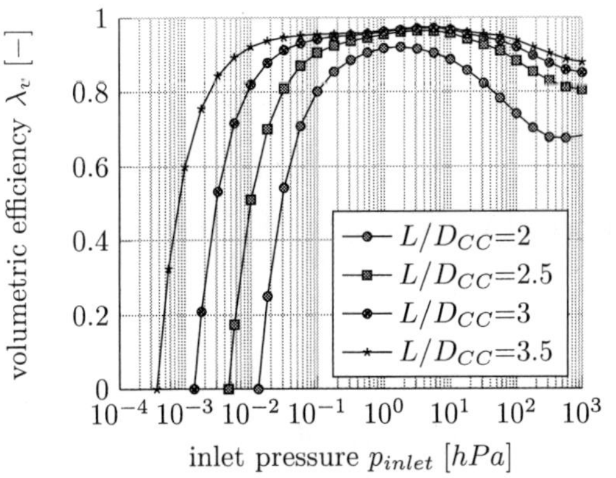

Figure 9. Characteristics of volumetric efficiency λ_L for various dimensionless rotor lengths L/D_{CC} (optimised for p_{LL}=1 hPa)

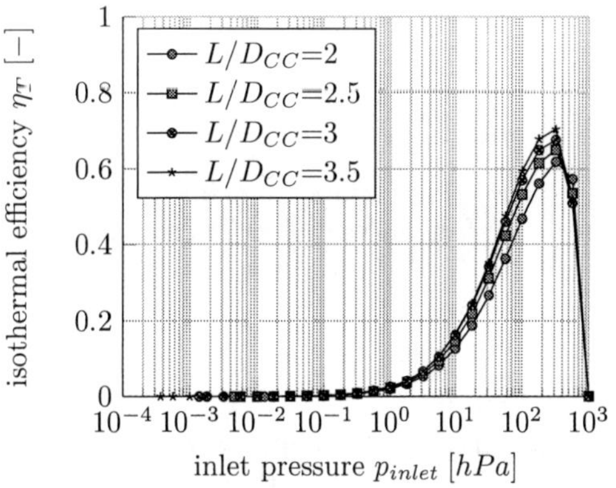

Figure 10. Characteristics of isothermal efficiency for various dimensionless rotor lengths L/D_{CC} (optimised for p_{LL}=1 hPa)

Finally the optimised rotor pitch progression is considered. Some example progressions for several dimensionless rotor lengths are shown in Figure 11 where all progressions are optimised for $p_{LL} = 1\,hPa$. All progressions start at high rotor pitch and decrease to a lower outlet pitch, which means an internal volume ratio. Notable is the wrap angle φ_{in} of the inlet segment which is equal to the minimum constrained angle $\varphi_{in,min} = 450°$ for all dimensionless rotor lengths. The wrap angle of the compression $\varphi_{compress}$ increases with the dimensionless rotor length. In the case of the lowest dimensionless rotor length $L/D = 2$ the upper limit of the volume ratio is reached and therefore no free rotor length is used for the compression phase. Based on the rotor pitch progressions the differences in the volumetric efficiency characteristics (Fig. 9) can be explained. A continuous change of the chamber volume leads to a low pressure gradient at the suction side which results in low leakage mass flow rates during the charging, leading therefore to a good volumetric efficiency. Additionally, a continuous compression more closely follows an isothermal path, this benefits the isothermal efficiency of longer rotors (Fig. 10). An isochoric transport phase, which means a high value of φ_{in}, does not occur since an isochoric pressure increase would increase the enclosed area of the pv-diagram, even though a high value of φ_{in} would improve the volumetric efficiency.

Figure 12 shows the influence of the load-lock pressure on the optimised rotor pitch progression for the dimensionless rotor length $L/D_{CC} = 3.5$. Again, it can be noticed that the wrap angle of the inlet segment is equal to the minimum allowed value $\varphi_{in,min} = 450°$ for all dimensionless rotor lengths. Comparable with the results for the variation of the rotor length, a large wrap angle for compression $\varphi_{compress}$ is preferred. For example, in the case of $p_{LL} = 10^2\,hPa$ the optimisation results in the maximum possible wrap angle $\varphi_{compress,max} = 1260°$.

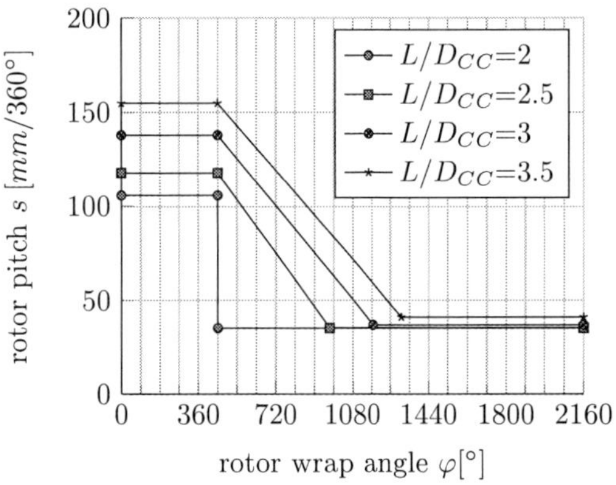

Figure 11. Influence of the dimensionless rotor length L/D_{CC} on the rotor pitch progression ($p_{LL} = 1\,hPa$)

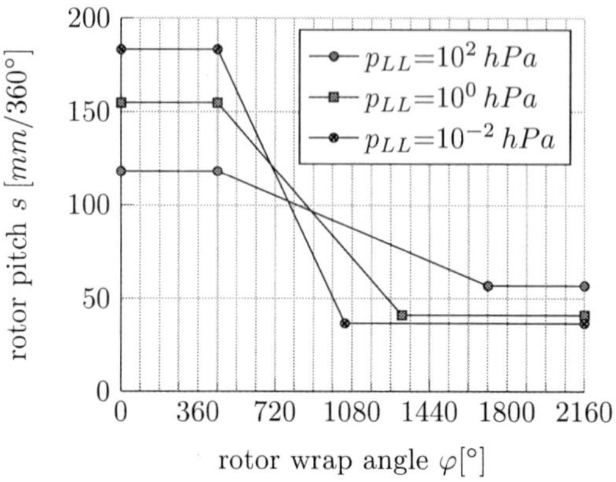

Figure 12. Influence of the load-lock pressure p_{LL} on the rotor pitch progression ($L/D_{CC} = 3.5$)

8. Conclusion

In the current article, a method to optimise the rotor pitch progression of screw spindle vacuum pumps for load-lock operations is presented. In this study an idealised load-lock process is used to define a dimensionless efficiencies. It is shown that the influence of the internal volume ratio on the load-lock efficiency decreases for high volume ratios. The optimisations of actual load-lock processes result in volume ratios below the theoretical values for the idealised process. This is explained by the negative influence of the volume ratio on the volumetric efficiency and the ultimate pressure. Additionally, a positive influence of the rotor length on the load-lock efficiency is identified. Based on the optimised rotor pitch progressions, it is shown that a uniform compression benefits the characteristics of isothermal and volumetric efficiencies over the entire range of suction pressure. In order to optimise the isothermal load-lock efficiency, the decrease of rotor pitch should start directly after the end of charging. All optimisations are performed for double-lobed rotors. Since single lobed rotors reveal a good potential for volumetric efficiency, these should be taken into account for future work.

References

[1] Dreifert T and Müller R 2014 *9th Int. Conf. Screw Mach.* (*VDI Berichte* vol 2228) pp 29–42
[2] Jousten K 2010 *Wutz - Handbuch der Vakuumtechnik* vol 10 (Vieweg+Teubner, Wiesbaden)
[3] Pfaller D, Stratmann D, Brümmer A and Kauder K 2007 *5th Int. Conf. Compr. Sys.* (iMechE) pp 214–254
[4] Pfaller D, Brümmer A and Kauder K 2010 *8th Int. Conf. Screw Mach.* (*VDI Berichte* vol 2101) pp 167–181
[5] Holmes C S, Rane S and Wang X 2017 Manufacturing method and performance assessment for variable lead vacuum rotors *IOP Conf. Series: Mater. Sci. Eng.* **232** 012022
[6] Utri M and Brümmer A 2016 *23th Int. Compr. Eng. Conf.* (Purdue) paper 1428
[7] Utri M and Brümmer A 2018 Opportunity for variable rotor lead in dry running screw compressors depending on dimensionless numbers *Proc. Inst. Mech. Eng.* E (accepted)
[8] Nadler K and Brümmer A 2010 *8th Int. Conf. Screw Mach.* (*VDI Berichte* vol 2101) pp 333–344
[9] Wittel H, Muhs D, Jannasch D and Voiek J 2015 *Roloff/Matek Maschinenelemente* (Wiesbaden: Springer Vieweg) chap 20, pp 726–727
[10] Huck C, Brümmer A and Sachs R 2014 *9th Int. Conf. Screw Mach.* (*VDI Berichte* vol 2228) pp 331–345
[11] Rohe A 2005 *Wärmehaushalt von Schraubenspindel-Vakuumpumpen* Ph.D. thesis Dortmund University
[12] Wenderott D and Kauder K 2002 *6th Int. Conf. Screw Mach.* (*VDI Berichte* vol 1715) pp 253–266
[13] Stratmann D 2010 *Kennfeldverhalten zweiwelliger Rotationsverdrängervakuumpumpen* Ph.D. thesis Dortmund University
[14] Nelder J A and Mead R 1965 A simplex method for function minimization *Comp. J.* **7** 308–313

[15] Box M J 1965 A new method of constrained optimization and a comparison with other methods *Comp. J.* **8**(1) 42–52

[16] Richardson J A and Kuester J L 1973 The complex method for constrained optimization *Commun. ACM* **16**(8) 487–489

[17] Conn A R, Gould N I M, and Toint P L 1991 A globally convergent augmented lagrangian algorithm for optimization with general constraints and simple bounds *SIAM J. Numer. Anal.* **28**(2) 545–572

[18] Birgin E G and Martnez J M 2008 Improving ultimate convergence of an augmented lagrangian method *Opt. Meth. Soft.* **23**(2) 177–195

[19] Wales D J and Doye J P K 1997 Global optimization by basin-hopping and the lowest energy structures of lennard-jones clusters containing up to 110 atoms *J. Phys. Chem.* A **101**(2) 5111–5116

[20] Biscani F, Izzo D and Märtens M 2018 Pagmo 2.7 URL `https://doi.org/10.5281/zenodo.1217831`

[21] Johnson S G 2018 The NLopt nonlinear-optimization package URL `http://ab-initio.mit.edu/nlopt`

IOP Conf. Series: Materials Science and Engineering **425** (2018) 012029 doi:10.1088/1757-899X/425/1/012029

Study on the performance prediction of dry twin screw vacuum pump

Jinliang Tuo[1], Bei Guo[1], Rui Wu[1], Yu Zhang[1], Xiaole Chen[1]
[1]Xi'an Jiaotong University, Xi'an, Shaanxi, China

E-mail： guobei@mail.xjtu.edu.cn

Abstract. Dry twin screw vacuum pump, which generates no oil and gas during operation, inherits the advantages of screw machinery, such as high reliability, stable medium conveying, small vibration, simple and compact structure, convenient operation, etc. which has been widely used in petrochemical and new or high-technology industries, like vacuum heat treatment, nuclear research, micro electro mechanical systems, nanotechnology, precision manufacturing and national defense technology. In this paper, the geometric feature of the twin screw vacuum pump such as contact line, meshing line and chamber volume between teeth is analysed, respectively. In order to study the micro working process within the dry twin screw vacuum pump, a test rig has been designed and built and the p-V indicator diagrams of the working process of dry twin screw vacuum pump have been recorded successfully under various working conditions. Based on the indicator diagrams, the working process within the dry twin screw vacuum pump is analysed in more detail, and especially, the effects of the suction pressure as well as rotational speed on the performance of dry twin screw vacuum pump, such as pumping speed and ultimate pressure, is thus investigated. The results presented in this paper are helpful to deepen understanding of the working process inside dry twin screw vacuum pump and offers a good reference for to improve design and optimize such machines.

1. Introduction

The dry twin screw vacuum pump, which generates no oil and gas during operation, is a novel positive-displacement device capable of ''pumping'' gas. Its core element consists of a pair of inter-meshing screw rotors rotating in opposite direction about parallel axes, and it removes gas from a sealed working chamber in order to leave behind a partial vacuum. The meshing of a pair of conjugated male and female rotors inside the housing of the dry twin screw vacuum pump allows suction, compression and discharge in the screw channel and the space inside the housing wall. Additionally, the dry twin screw vacuum pump inherits the advantages of screw machinery, such as high reliability, stable medium conveying, small vibration, simple and compact structure, convenient operation, etc. So, it is widely used in low and medium-low vacuum applications, such as semiconductor industry, metallurgical industry, nanomaterials industry, chemical industry, pharmaceutical industry and high-technology industry, et al.

In recent years, a lot of investigations have been done on the performance prediction of the dry twin screw vacuum pump under different operating conditions. Ohbayashia[1] established a method of the performance prediction and proposed a way to design the pump that satisfies specific requirements. The leaks flow through clearances between the screw rotor and housing, and clearances between two meshing rotors were estimated with the results based on the linearized BGK model. Pfaller[2] presented a method based on the evolutionary calculation algorithm for the energetic optimization of the dry twin screw vacuum pump, accordingly changing the pitch of a rotor at constant design volume. Feng[3] and Huang[4] analyzed the deformation of rotor and the performance of the dry twin screw vacuum pump. Compared with the experimental results, the calculated results obtained from the performance prediction method can agree well. Zhao[5] and Wang[6] derived a theoretical formula to calculate the leakage in the dry twin screw vacuum pump, but there was no experimental validation.

Although much research have been performed on the theoretical and experimental study of the dry twin screw vacuum pump, there is seldom analysis on the thermodynamic process of the dry twin screw vacuum pump based on the recorded p-V indicator diagram. In this paper, the geometric feature of the dry twin screw vacuum pump, such as contact line, meshing line and working volume between teeth is analyzed, respectively. In order to study the micro working process within the dry twin screw vacuum pump, a test rig has been designed and built and the p-V indicator diagrams of the working process of dry twin screw vacuum pump have been recorded successfully under various working conditions. Based on the indicator diagrams, the working process within the dry twin screw vacuum pump is analyzed in more detail, and especially, the effects of the suction pressure as well as rotational speed on the performance of the dry twin screw vacuum pump, such as pumping speed and ultimate pressure is thus investigated. The results presented in this paper are helpful to deepen understanding of the working process inside dry twin screw vacuum pump and offers a good reference for to improve design and optimize such machines.

2. Geometrical parameters

The dry twin screw vacuum pump is a type of fluid-extracting apparatus, which consists of two partly overlapping cylinders and two same-shaped meshing rotors. The two rotors rotate in opposite directions around their axes and are synchronized by a pair of gears and maintain only tiny clearances during operation. The rotor profile plays a significant role on the performances and applications of the dry twin screw vacuum pump since it permits the rotors to remain meshed with each other.

A novel rotor profile, which is completely conjugate, is proposed, as shown in Figure 1. Taking the male rotor as an example, it contains 5 curves: extended epicycloid *ab*, root circular arc *bc*, pseudo Archimedes *cd*, envelope curve of pseudo Archimedes *de* and head circular arc *ea*. The equations as well as design procedures of the rotor profile can be found in the paper by Yang[7] published in 2015.

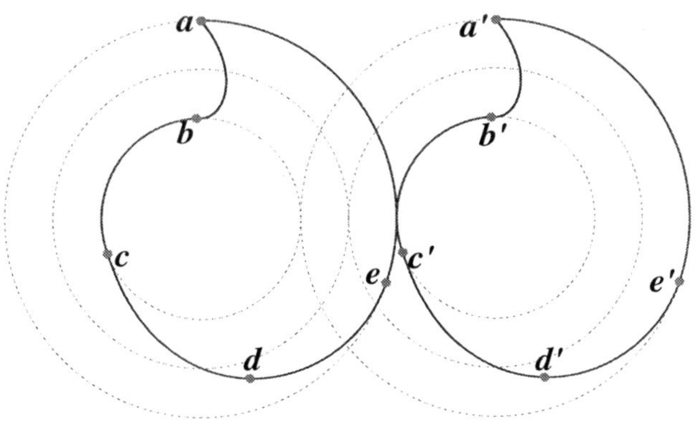

Figure 1. The profile of the rotors.

2.1. Rotor Parameters

The design parameters of the male and female rotors are shown in Table 1 commonly, because their parameters are completely the same. The rotor pitch is divided into two sections: one section of the pitch is 88mm, and the other section of the pitch is 60 mm.

Table 1. Geometry parameters of screw rotor

Item	Value
Outer radius/ *mm*	130
Root radius/ *mm*	66
Rotor length/ *mm*	340
First rotor pitch/*mm*	88
Second rotor pitch/ *mm*	60
First rotor length/ *mm*	200
Second rotor length/ *mm*	140

2.2. *Calculation of the sealing line*

The sealing line has a significant effect on the thermodynamic performance of the twin screw vacuum pump. The equation of the sealing line can be obtained by the simultaneous of meshing condition and the equation of the rotor profile. Figure 2 shows the sealing line of the rotor profile, where Z is the axial length of the rotor and Y is the upright length of the rotor. The section *LM* and *PQ* are generated by the meshing of point and the extended epicycloid; the section *MN* and *OP* are generated by the meshing of the head and root circular arc; the section *NO* is generated by the meshing of the pseudo Archimedes and the envelope curve of pseudo Archimedes.

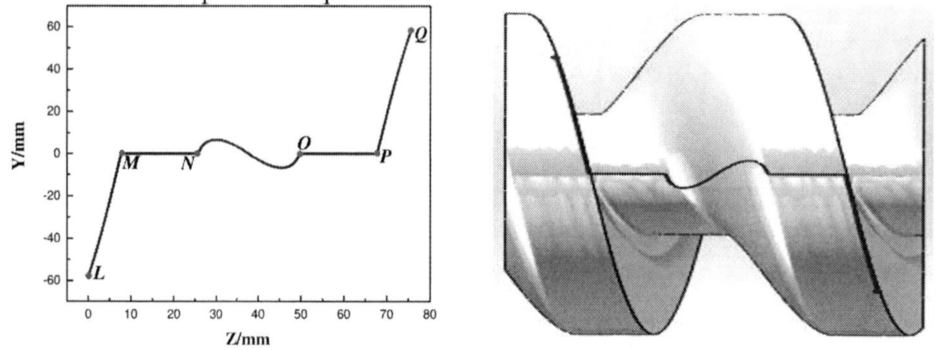

Figure 2. Sealing line.

2.3. *Calculation of chamber volume*

The operation of the twin screw vacuum pump is based on the change of the chamber volume enclosed by the male rotor, female rotor and the housing (working volume). The chamber volume of the twin screw vacuum pump can be expressed by:

$$A=\sum_{i=1}^{n}\frac{1}{2}\int_{t_i}^{t_{i+1}}\left[\dot{y}_i x_i-\dot{x}_i y_i\right]dt \tag{1}$$

Where A is the tooth area of the tooth, $x_i = x_i\left(t\right)$, $y_i = y_i\left(t\right)$, are the parametric equation of the i-th curve of the rotor profile, $\dot{x}_i = dx_i\left(t\right)/dt$ and $\dot{y}_i = dy_i\left(t\right)/dt$ are the derivative of the parameter, t_i and t_{i+1} are the range of parameter.

$$V_0=\frac{T}{2\pi}\int_0^{\theta}A\left(\theta\right)d\theta \tag{2}$$

IOP Conf. Series: Materials Science and Engineering **425** (2018) 012029 doi:10.1088/1757-899X/425/1/012029

Where V_0 is the chamber volume, T is the pitch of the rotor and θ is the rotation angle of the rotor.

The chamber volume of the dry twin screw vacuum pump versus the rotation angle of the male rotor is shown in Figure 3. Obviously, the maximum volume of the working chamber is 0.767 L. And the twin screw vacuum pump contains six processes during operation: suction, transportation, internal compression, transportation, compression and discharge.

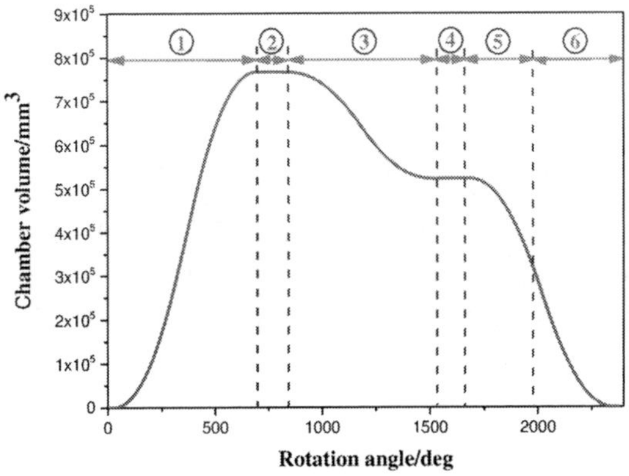

Figure 3. Chamber volume varies with the rotation angle. (① — suction, ② — transportation, ③ — internal compression, ④—transportation, ⑤—compression, ⑥—discharge)

3. Experimental setup

3.1. Test rig

A schematic diagram of the dry twin screw vacuum pump test rig is shown in Figure 4 and the photograph of which is shown in Figure 5. The dry twin screw vacuum pump prototype is driven by a variable-speed motor. A tacho-torquemeter with an uncertainty of ±0.1% reading is coupled between the motor and dry twin screw vacuum pump for measurements of the shaft speed and power. The variable rotation speed of the male rotor is obtained by a frequency-converter. The pressure in the test cover is measured by a PSG 500 type vacuum gauge made by Inficon. Its measurement range is from 0 Pa to 10^5 Pa and the accuracy is less than 2% of reading. The revolving type flow meter with an uncertainty of 0.5% is used to measure the vacuum pump flow rate. Data obtained from the measurement system mainly include pumping speed, ultimate pressure and pressures at various positions.

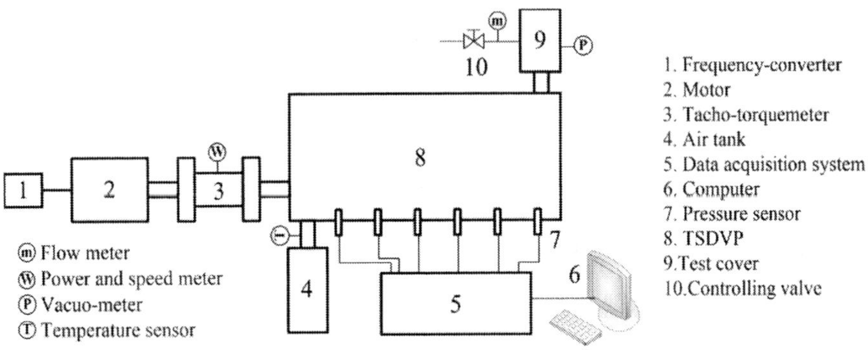

Figure 4. The schematic diagram of the test rig.

Figure 5. The photograph of the test rig.

3.2. Recording of the p–V indicator diagram

As we know, p–V indicator diagram is an important and efficient method to investigate the thermodynamic performance of displacement machines. In order to study the micro working process within the dry twin screw vacuum pump, a test rig has been designed and built. Firstly, the installment of the pressure sensors is discussed. Five GE pressure sensors and one Kulite pressure sensor are installed in consecutive positions in its housing to measure the p-V diagram of the studied dry twin screw vacuum pump. The model UNIK 5000 pressure sensors are made by GE Group and have an accuracy of less than ±0.04% of full scale and a response frequency of 3.5 kHz. The model XTL-140M pressure sensor is made by Kulite Group and has an accuracy of ±0.3% of reading and a response frequency of 10 kHz. Each pressure sensor has some overlap with the following pressure sensor in order to continuously monitor pressure changes in the compression process, as shown in Figure 6. However, since there is no position for installing the pressure sensor in the compressor discharge housing, the pressure in the late part of discharge process cannot be measured in this study. Therefore, for this part, the discharge pressure is considered approximately equal to the atmosphere pressure.

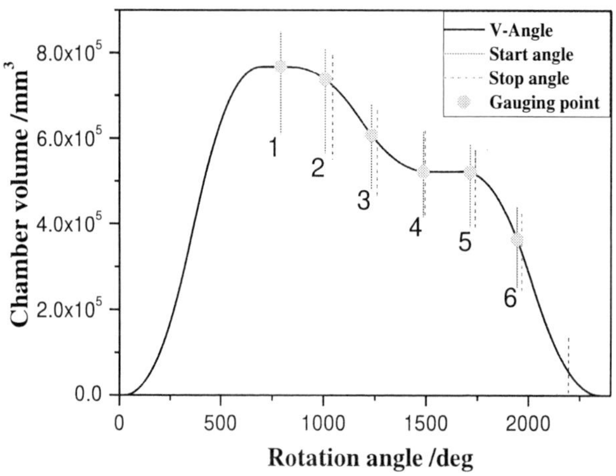

Figure 6. Location of sensors.

The installation of pressure sensors on the casing of dry twin screw vacuum pump is shown in Figure 7. The sealing ring can prevent the ambient gas from going into the working chamber. There should have some distance between the sensor film and the pressure-leading system to avoid the damage of sensor. The design of pressure-leading system can be referred to the following equations[8]:

$$\xi=2\sqrt{\frac{3}{\pi}\frac{\mu}{r^3}}\frac{\sqrt{lV}}{\rho c} \tag{3}$$

Where ξ is damp coefficient, μ is the dynamic viscosity, r is the radius of pressure-leading pipe, l is the length of leading pipe, V is the volume of the sensor pocket, ρ is the density, c is the speed of sound.

$$\omega_1=\sqrt{\frac{3\pi}{2}}\frac{rc}{\sqrt{lV}} \tag{4}$$

$$\frac{p_1}{p_2}=\frac{1}{\sqrt{\left[1-(\omega_2/\omega_1)^2\right]+4\xi^2(\omega_2/\omega_1)^2}} \tag{5}$$

$$\varphi=-\text{ac}\tan\left(\frac{2\xi(\omega_2/\omega_1)}{1-(\omega_2/\omega_1)^2}\right) \tag{6}$$

Where p_1 is the pressure of the sensor film, p_2 is the pressure within the working chamber, ω_1 is the angular frequency of the pressure signal of the sensor film, ω_2 is the angular frequency of the pressure signal within the working chamber.

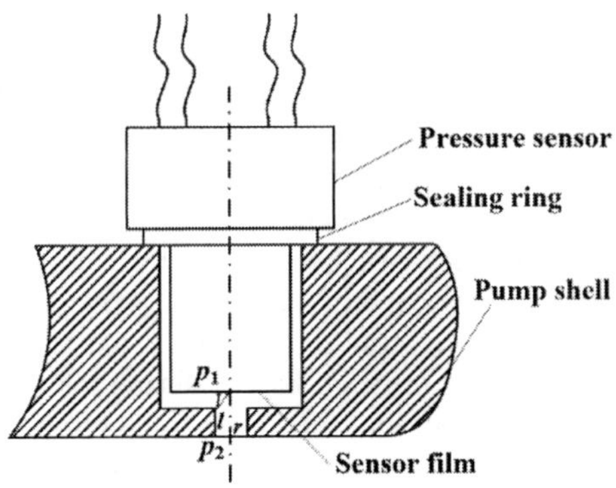

Figure 7. Installation of pressure sensors.

In general, with the increase of the angular frequency of the pressure signal within the working chamber ω_2 and decrease of damp coefficient ξ, the dynamic performance of the leading system increases and the distortion of the pressure signal decreases.

Secondly, in order to determine a phase position of the recorded pressure, a photoelectric sensor is used. When the dry twin screw vacuum pump working process begins, the photoelectric sensor will produce an impulse. Based on the pressure signal and the position signal, the pressure distribution versus the rotational angle of the rotor can be obtained. By transformation between the rotational angle of rotor and the volume of working chamber as explained by Xing[9], the diagram of pressure distribution versus chamber volume can be carried out. The pressure signal is collected and analyzed in the dynamic data acquisition system NI-PXIe1078. The model of data acquisition card is PXIe-6356. This card can provide 8-differential and 16-single ended measurement. The differential measurement is chosen in order to eliminate the impact of common-mode voltage on the signal. Because the signal to be measured is floating type, bias resistor should be applied to satisfy the bias current path requirement. In the experiment, one resistor connected between the negative input and the com terminal is required because the input signal of data acquisition card is DC-coupled and the impedance of pressure sensor is low. Moreover, the same isolated power source is used to supply the power of pressure sensors and the data acquisition card to improve accuracy of measurement. Figure 8 shows typical p-t measurements from the signal collecting and analyzing system when the compressor is operating at a suction pressure of 4kPa and rotation speed of 3000 r/min.

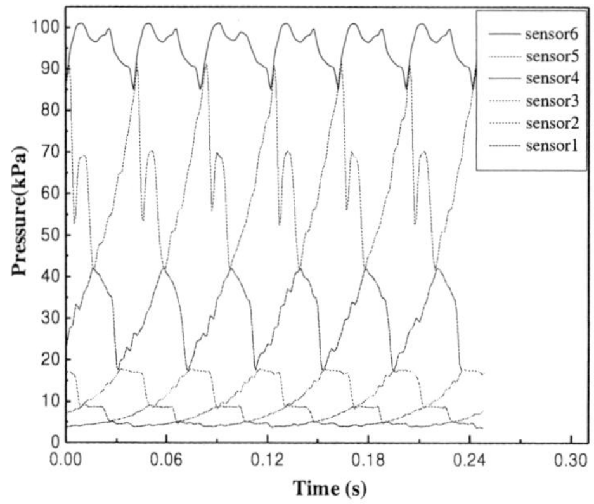

Figure 8. A typical p-t diagram obtained in the experiment.

4. Results and discussion

4.1. Effects on p–V indicator diagram

As we know, p–V indicator diagram is an important and efficient method to investigate the performance of displacement machines. The change of the gas pressure in the working chamber can be described in detail by the p–V indicator diagram. In addition, it can be used to estimate the indicated power consumption of the dry twin screw vacuum pump.

The p–V indicator diagrams of the dry twin screw vacuum pump with variable suction pressures are shown in Figure 9. Here, the rotation speed of the rotor is 3000 r/min and the suction pressures are 4kPa and 6kPa, respectively. It can be seen from Figure 9 that the dry twin screw vacuum pump with the suction pressure of 6kPa has a higher pressure almost during the whole working process than that of 4kPa. This can be explained by the fact that the larger the suction pressure is, the larger the back flow of gas is. It also can be seen that the pressures in the discharge process with variable suction pressures are almost equivalent.

It also can be seen that the area of the p–V diagram is different for different suction pressures. And the area for the case with 6kPa suction pressures is larger than that with 4kPa suction pressures. It is thought that the pressure within the working chamber becomes higher, causing the indicated power to rise up, as the suction pressure increases.

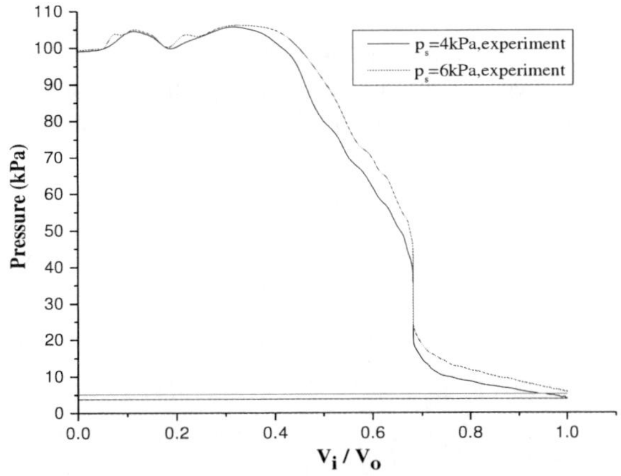

Figure 9. The diagrams of dry twin screw vacuum pump with variable suction pressures.

Figure 10 shows the p–V indicator diagrams of the dry twin screw vacuum pump at different rotation speeds of 3000r/min and 2700r/min, and a suction pressure of 4kPa. As seen from Figure 10, the pressure during the whole working process of rotation speed of 2700 r/min is higher than that of 3000 r/min. This is because of the fact that the heat generated during the whole working process of dry twin screw vacuum pump increases as the rotation speed increases and the temperature difference between the housing and rotors increases with the rotation speed, so the clearances between them become narrow, which results in smaller backflow occurs from the discharge chamber into working chamber and also lead to larger pressure increase. Moreover, the pressure level inside the working chamber under high speed is also higher than that under low speed. It's can be partly explained by the fact that a reduced fraction of a total slip passes through the internal clearance as rotational increases.

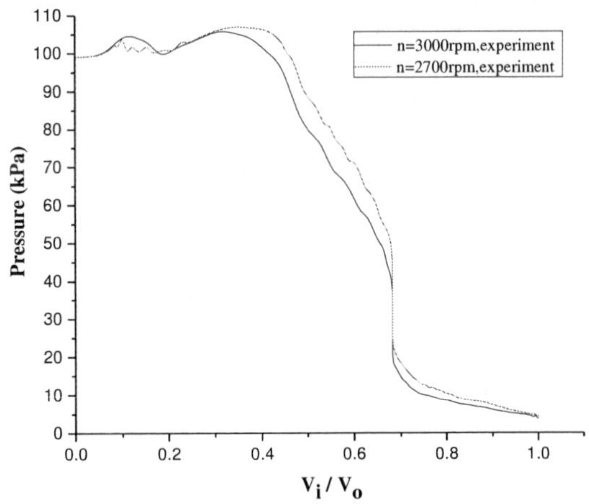

Figure 10. The diagrams of dry twin screw vacuum pump at variable rotation speeds.

4.2. Pumping speed

The pumping speed of the dry twin screw vacuum pump, which is based on the ISO 1607 standard[10], is measured by varying the frequency from 40 to 50 Hz with a 5 Hz interval. The changes of pumping speed with the rotation speed and suction pressure are shown in Figure 11. It can be seen that the pumping speed increases sharply when the suction pressure is less than about 200 Pa with the increase of the suction pressure and then decreases slowly when the suction pressure is in other ranges at the same rotation speed. Moreover, it can be seen that the pumping speed increases monotonously with the rises of the rotation speed at the same suction pressure. So, the increases of pumping speed with rotation speed can be explained by the increases of volumetric efficiency with rotation speed. Obviously, the suction pressure corresponding to the maximum value of the pumping speed is about 200 Pa.

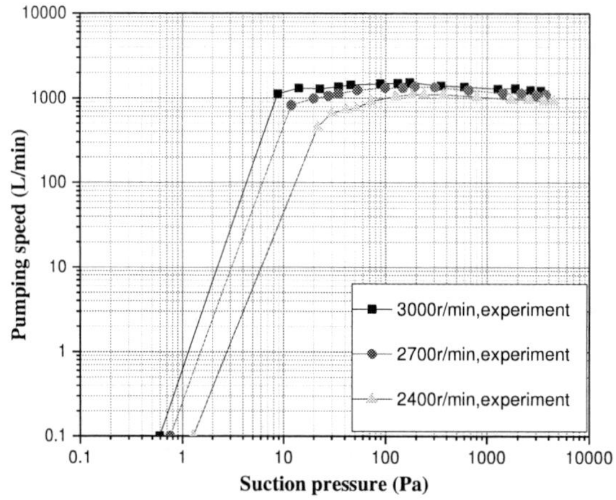

Figure 11. Pumping speed for different suction pressures.

4.3. Ultimate pressure

The experimental results of ultimate pressure for different rotation speeds are shown in Figure 12. It can be observed that the ultimate pressure goes down rapidly with the increment of speed in a certain range. As the speed is large enough, the changing rate of ultimate pressure with speed becomes is less and less. It can be explained by the two reasons: one is that the temperature of the rotors rises higher than that of the housing because the housing are cooled more easily than the rotors. The heat generated during gas compression phase increases as the rotation speed increases and the temperature difference between the housing and rotors increases with the rotation speed, and the clearances between them become narrow, which results in the back flow of gas lessen; two is that the external leakages go down with the increment of speed, resulting in an increase in the volumetric efficiency, so the ultimate pressure decrease. The minimum ultimate pressure of the dry twin screw vacuum pump is 0.7Pa.

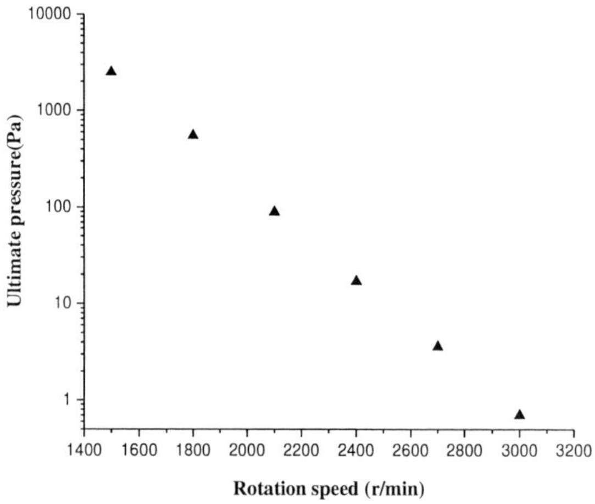

Figure 12. Ultimate pressure.

5. Conclusions

In order to study the micro working process within the dry twin screw vacuum pump, a test rig has been designed and built and the p-V indicator diagrams of the working process of dry twin screw vacuum pump under various working conditions have been recorded successfully. Based on the indicator diagrams, the working process within the dry twin screw vacuum pump is analyzed in more detail, and especially, the effects of the suction pressure as well as rotational speed on the performance of dry twin screw vacuum pump is thus investigated. Some main conclusions are drawn as detailed below:

(1) The suction pressure and rotational speed have a direct influence on the p–V indicator diagrams of the dry twin screw vacuum pump. The pressure curve during the whole working process has a higher level under the condition of higher rotational speed and higher inlet pressure.

(2) The pumping speed of the dry twin screw vacuum pump increases with the increase of the suction pressure at the same rotation speed. Also, it increases monotonously with the rise of the rotation speed at the same suction pressure.

(3) The ultimate pressure goes down rapidly with the increment of speed in a certain range. As the speed is large enough, the changing rate of ultimate pressure with speed becomes is less and less.

(4) The integrated experimental studies of the dry twin screw vacuum pump make it possible to create experimental data bank concerning pressure dependence in working chambers on rotors rotating angle. These results can be used for further development of the pump design and optimization with the help of mathematical model.

Acknowledgments

The authors would like to express appreciation to the Department of Compressor Engineering of the Xi'an Jiaotong University for supporting and encouraging our efforts.

References

[1] Ohbayashi T, Sawada T, Hamaguchi M, & Miyamura H 2001. Study on the performance prediction of screw vacuum pump. *Applied Surface Science*, **169**, pp 768–771.

[2] D. Pfaller, A. Brümmer, & K. Kauder 2011. Optimized rotor pitch distributions for screw spindle vacuum pumps. *Vacuum*, **85**, pp 1152-1155.

[3] Z H Fong, F C Huang 2006. Evaluating the interlobe clearance and determining the sizes and shapes of all the leakage paths for twin-screw vacuum pump. *Mechanical engineering science*, **220** Part C pp 499-506.

[4] FeiChuan H 2007. Study on the leakage evaluation of the non-conjugate rotor profiles for a twin screw vacuum pump. *National Chuang Cheng University*.

[5] Yu Z 2008. The structure and properties of the rotors of dry screw vacuum pump. *Northeastern University*.

[6] Qiong'e, W 2005. Study of design and performance of the rotor's curve of dry-running twin screw pump. *Zhejiang University*.

[7] Yang L, Bei G, Maofei G 2015. Study on design of rotor profile for the twin screw vacuum pump with single thread tooth. *IOP Conf. Ser.: Mater. Sci. Eng.* **90** 012007.

[8] Baolong W, Xianting L, Wenxing S, & Qisen Y 2007. Design of experimental bench and internal pressure measurement of scroll compressor with refrigerant injection. *International Journal of Refrigeration*, **30**, pp 179-186.

[9] Ziwen X 2000. Screw Compressors: Theory, Design and Application. *China Machine Press*, pp 71–78.

ISO 1607-1:1993, Standard Positive-Displacement Vacuum Pumps-Measurement of Performance Characteristics. Part 1: Measurement of Volume Rate of Flow (Pumping Speed). ISO 1607-2:1989 Standard Positive-Displacement Vacuum Pumps—Measurement of Performance Characteristics. Part 2: Measurement of Ultimate Pressure (International Standardization Organization).

IOP Conf. Series: Materials Science and Engineering **425** (2018) 012030 doi:10.1088/1757-899X/425/1/012030

Spline-Based Parameterization Techniques for Twin-Screw Machine Geometries

Jochen Hinz, Matthias Möller and Cornelis Vuik

Delft University of Technology, Faculty of Electrical Engineering, Mathematics and Computer Science, Delft Institute of Applied Mathematics, Van Mourik Broekmanweg 6, 2628 XE Delft, The Netherlands

E-mail: J.P.Hinz@tudelft.nl, M.Moller@tudelft.nl and C.Vuik@tudelft.nl

Abstract.
The fully automated generation of computational meshes for twin-screw machine geometries constitutes a mandatory aspect for the numerical simulation (and shape-optimization) of these devices but proves to be a challenging task in practice. Therefore, the successful generation of computational meshes requires sophisticated mathematical tools. Commercially available classical mesh generators can produce high quality meshes from no more than a description of the rotor contours. However, since we are particularly interested in numerical simulations using the principles of *Isogeometric Analysis*, a spline-based geometry description rather than a classical mesh is needed.

In this paper, we propose a practical approach for the automated generation of spline-based twin-screw machine geometry parameterizations in two dimensions. For this purpose, we adopt the principles of *Elliptic Grid Generation* and present a parameterization algorithm that is compatible with an automated simulation pipeline based on the principles of isogeometric analysis.

To demonstrate the proposed techniques, we apply them to an example geometry and present the resulting parameterizations. Finally, we give a qualitative explanation of how the discussed techniques can be utilized to generate geometry parameterizations in three dimensions and their applications to shape-optimization on a variable rotor-pitch.

1. Introduction

The generation of analysis-suitable meshes for twin-screw geometries constitutes the first step towards the numerical simulation and shape-optimization of twin-screw machines. However, the full automation of this process remains difficult, often contributing substantially to the total amount of labour hours and computational costs. On the one hand, this is caused by the inherent difficulty of generating a mesh from no more than a description of the boundary contours, and, on the other hand, further aggravated by the challenging characteristics of twin-screw geometries, such as tiny clearances and rapidly changing gap sizes (see figure 1). To the best of our knowledge, there are two commercially available classical structured mesh generators that only require a description of the geometry contours as input: *twin-mesh* [1] and *SCORG* [2], both being capable of exporting directly into *ANSYS CFX* format [3]. We are particularly interested in using Isogeometric Analysis (IgA) [4] techniques to perform shape-optimization on twin-screw machine geometries with variable rotor pitch. Even though the commercial mesh generators produce high-quality classical meshes, they cannot be used in an IgA-setting where a

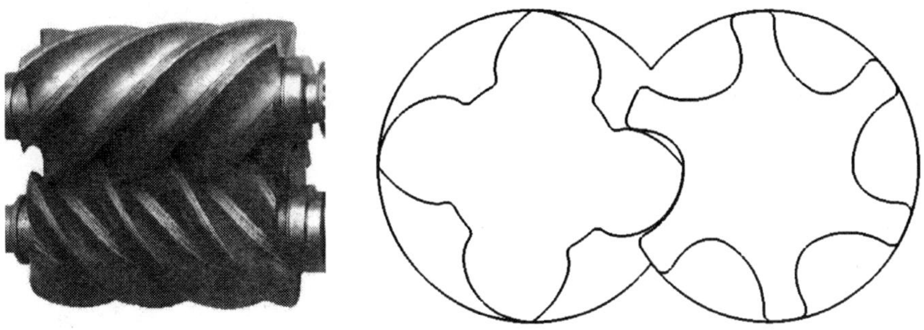

Figure 1. Rotary-screw compressor (Wikipedia, file: *Lysholm_screw_rotors.jpg*) (left) and a cross section showing the rotor profiles with casing (right).

spline-based parameterization of the target geometry is mandatory. This is the main motivation for the techniques presented in this paper.

The general idea of a spline-based parameterization is to build a continuous mapping operator $\mathbf{x} : \hat{\Omega} \to \Omega$ comprised of higher-order spline functions that maps the entire computational domain $\hat{\Omega}$ onto the target geometry Ω (or an approximation thereof). The spline-parameterization $\mathbf{x}$ is then directly used for an IgA-based simulation. Unlike in the classical case, application-specific features such as boundary layers are added after the geometry description has been completed. This is accomplished by performing knot-refinement on $\mathbf{x}$ (see section 2) and has a negligible impact on the overall computational costs.

A volumetric parameterization of the geometry is accomplished by generating a large number of planar parameterizations at various discrete rotational angles θ_i and stacking them in the z-direction. Hereby, it is desirable to achieve a degree of smoothness in the cell boundaries of the mesh corresponding to consecutive slices in order to achieve the same smoothness in the elements of the resulting volumetric parameterization. This translates to the requirement of smoothly varying control points of the mapping $\mathbf{x}$ as a function of the rotation angle θ. Therefore, it is furthermore desirable to, if possible, avoid topology changes in the planar slices, even though this may be a challenging task. Variable pitches are easily accomplished by a tighter or wider stacking of the discrete slices.

It should be mentioned that a spline-based description can be turned back into a classical mesh by performing a large number of function evaluations in $\mathbf{x}$ and connecting the resulting point cloud by linear edges. Hereby, the evaluation points determine the properties of the mesh and have to be chosen wisely. However, this topic will not be covered in this paper.

For the purpose of generating folding-free planar parameterizations using spline-functions, we adopt the principles of *Elliptic Grid Generation* (EGG) and present a numerical scheme that is suitable for an IgA-based computational approach. In section 2, we will briefly review the concept of (B-)spline functions while in section 3, we present the various possible topologies along with their advantages and disadvantages. In section 4, we discuss the principles of EGG and in sections 7 and 8, we will present the computational approach along with the parameterization strategy we employed for the results presented in section 9

2. B-Splines

B-splines are piecewise-polynomial functions that can be constructed so as to satisfy various continuity properties at the places where the polynomial segments connect. Their properties are determined by the entries of the so-called *knot vector*

$$\Xi = \{\xi_1, \xi_2, \ldots, \xi_{n+p+1}\}. \tag{1}$$

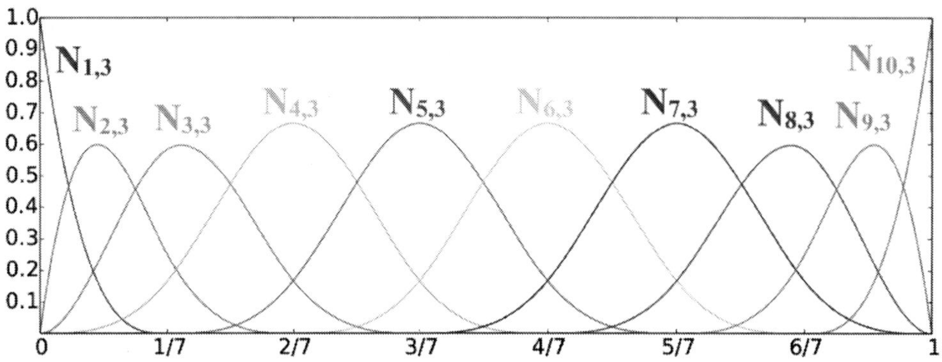

Figure 2. The univariate B-spline basis resulting from the knot-vector Ξ_3.

The knot vector is a monotone-increasing sequence of parametric values $\xi_i \subset [0, 1]$ that determine the boundaries of the segments on which the spline-basis is polynomial. Selecting some polynomial order p, the p-th order spline-functions $N_{i,p}$ are constructed recursively, utilizing the relation (with $\frac{0}{0} \equiv 0$)

$$N_{i,q}(\xi) = \frac{\xi - \xi_i}{\xi_{i+1} - \xi_i} N_{i,q-1}(\xi) + \frac{\xi_{i+q+1} - \xi}{\xi_{i+q+1} - \xi_{i+1}} N_{i+1,q-1}(\xi), \tag{2}$$

starting from

$$N_{i,0} = \begin{cases} 1 & \text{if } \xi_i \leq \xi \leq \xi_{i+1} \\ 0 & \text{otherwise} \end{cases}, \tag{3}$$

and iterating until $q = p$. The support of basis function $N_{i,p}$ is given by the interval $\mathcal{I}_{i,p} = [\xi_i, \xi_{i+p+1}]$ and the amount of continuous derivatives across knot ξ_j is given by $p - m_j$, where m_j is the multiplicity of ξ_j in $\mathcal{I}_{i,p}$. In practice, $\xi_1 = 0$ is repeated $p + 1$ times as well as ξ_{n+p+1} such that $\xi_1 = \ldots = \xi_{p+1} = 0$ and $\xi_{n+1} = \ldots = \xi_{n+p+1} = 1$. As a result, the resulting basis $\sigma = \{N_{1,p}, \ldots, N_{n,p}\}$ forms a non-negative partition of unity on the entire parametric domain $[0, 1]$, that is:

$$\sum_{i=1}^{n} N_{i,p}(\xi) = 1, \tag{4}$$

with

$$N_{i,p}(\xi) \geq 0, \tag{5}$$

for all spline functions $N_{i,p}$ [4]. Figure 2 shows the $p = 3$ B-spline basis resulting from the knot-vector

$$\Xi = \left\{0, 0, 0, 0, \tfrac{1}{7}, \tfrac{2}{7}, \tfrac{3}{7}, \tfrac{4}{7}, \tfrac{5}{7}, \tfrac{6}{7}, 1, 1, 1, 1\right\}. \tag{6}$$

The extension to bivariate spline bases is now straight-forward: given two univariate bases $\sigma_\Xi = \{N_1, \ldots, N_n\}$ and $\sigma_\mathcal{H} = \{M_1, \ldots M_m\}$, we build a bivariate basis $\Sigma = \{w_{i,j}\}_{(i,j) \in \{1,\ldots,n\} \times \{1,\ldots,m\}}$, by means of a tensor-product, where

$$w_{i,j}(\xi, \eta) = N_i(\xi) M_j(\eta). \tag{7}$$

The knots corresponding to the knot-vectors Ξ and $\mathcal{H}$ (without knot-repetitions) used to construct σ_Ξ and $\sigma_\mathcal{H}$, respectively, hereby become the boundaries of the polynomial segments, which can be regarded as the counterparts of classical elements.

We construct the mapping of a B-spline surface as follows:

$$\mathbf{x} = \sum_i \sum_j \mathbf{c}_{i,j} w_{i,j}, \tag{8}$$

where the $\mathbf{c}_{i,j} \in \mathbb{R}^2$ are referred to as the *control points*. We refer to the $\mathbf{c}_{i,j}$ with $i \in \{1, n\}$ or $j \in \{1, m\}$ as the *boundary control points* while the remaining $\mathbf{c}_{i,j}$ are called *inner control points*. As $\mathbf{x}$ is a linear combination of the $w_i \in \Sigma$, it will inherit the local continuity properties of the basis. This implies that many geometrical features can be better captured by a clever choice of the knot multiplicities in Ξ and $\mathcal{H}$.

An additional appealing feature of spline basis functions is the possibility of knot refinement. Let

$$f = \sum_i a_i N_{i,p} \tag{9}$$

be a function from the linear span of the spline basis $\sigma = \{N_{1,p}, \dots, N_{n,p}\}$ with corresponding knot-vector Ξ. We can refine σ by adding additional knots to Ξ, resulting in the refined basis $\tilde{\sigma}$. It can be shown that $\operatorname{span} \sigma \subset \operatorname{span} \tilde{\sigma}$. For an algorithm to prolong the a_i to the refined basis, see [4].

In the following, we shall drop the tensor-index notation and introduce a global index with $(i, j) \to i + (j-1)m$. The mapping then simplifies to

$$\mathbf{x} = \sum_i \mathbf{c}_i w_i. \tag{10}$$

For many geometries a parameterization with only one mapping operator is not possible, which is why several mappings that jointly parameterize the geometry have to be employed. The individual geometry segments that result from each of the mappings are referred to as *patches*.

3. Choice of Topology

As discussed in section 1, we would like to parameterize the geometry from figure 1 (right) for all rotational angles θ. The discrete angles θ_i then either correspond to the screw-machine at time-instances t_i in the planar case or to some cross-section of the screw-machine in the z-direction. A volumetric parameterization can therefore be acquired by parameterizing the geometry for a large number of θ_i and interpolating the planar cross-sections in the z-direction.

Figure 3 shows the possible topologies that come to mind. Here, black lines indicate boundaries at which the grid is held fixed and red lines indicate boundaries that slide along the grid. The various patches are indicated in different colors. Since the target geometry is of genus two (it has two *'holes'*), at least two patches are needed.

Even though figure 3 (right) takes advantage of the symmetries of the geometry, topology changes are unavoidable. For instance, since the blue patch bounded by the region surrounding the CUSP-points and the two rotor lobes (henceforth referred to as the *separator*) is static while the others rotate, patches (such as the grey patch) will eventually disappear, making a computational simulation with IgA-techniques difficult.

In figure 3 (center), two O-type patches are employed which lead to a sliding interface (in the separator region). Sliding interfaces are generally difficult to handle since element conformity is difficult to achieve. In an IgA-setting, the CUSP-points themselves pose an additional problem:

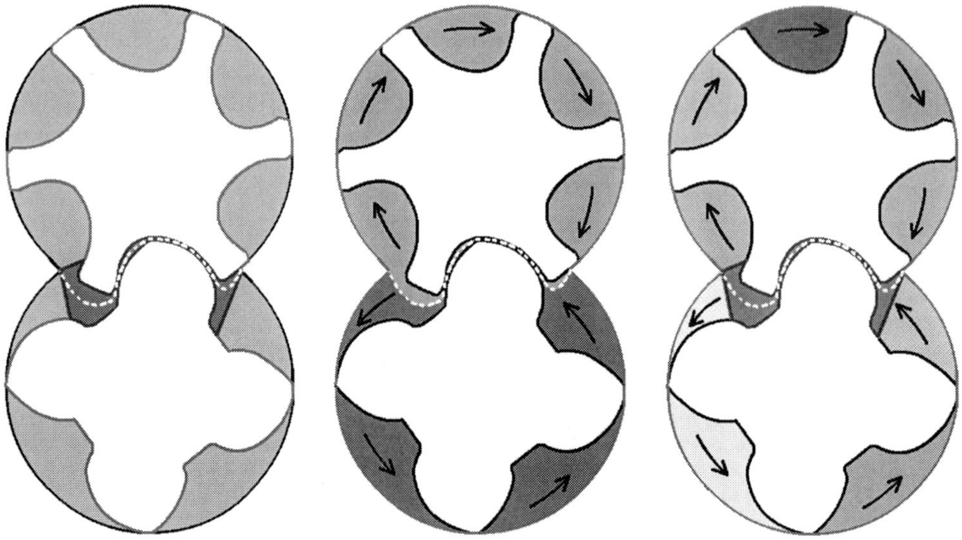

Figure 3. The various possible topologies for the parameterization of the fluid-part of the screw-machines.

as the patches slide along, the CUSP-point C^0-continuities can only be captured by adding repeated knots to the knot-vectors. If several cross sections are interpolated in the z-direction for a volumetric parameterization, they all have to be prolonged to the same knot-vector, soon leading to an infeasibly dense trivariate knot-vector.

Therefore, we are aiming for the topology of figure 3 (left) in which the patches are held fixed at the casings while the rotors slide along and the separator is parameterized with one static patch. Here, no topology changes are required and no sliding interface exists. In all cases, we need to generate the dotted white curve connecting the two CUSP-points, in order to parameterize the separator using two patches with mutual element conformity as well as conformity to the C-type patches.

4. Elliptic Grid Generation

Having discussed the aimed-for topology in section 3, we need a tool to generate analysis-suitable (i.e., bijective or folding-free) parameterizations at every discrete angle θ_i for geometries Ω from no more than a boundary description $\partial\Omega$.

As it is known to produce folding-free meshes in many applications, *Elliptic Grid Generation* (EGG) is among the most popular meshing approaches, especially in settings where computationally inexpensive algebraic methods such as [5] [6] and [7] fail due to the complexity of the target geometry. However, due to the higher chance of success, higher computational costs can be expected.

Assuming Ω is topologically equivalent to the unit quadrilateral $\hat{\Omega} = [0,1] \times [0,1] \subset \mathbb{R}^2$, a mathematical operator $\mathbf{x} : \hat{\Omega} \to \Omega$ that maps $\partial\hat{\Omega}$ onto $\partial\Omega$, and is furthermore folding-free exists and can be constructed. For this purpose, EGG imposes the Laplace-equation on the components of the inverse mapping $\mathbf{x}^{-1}$. Assuming the free topological variables are given by the tuple (ξ, η), the equation takes the form:

$$\begin{cases} \Delta\xi(x,y) &= 0 \\ \Delta\eta(x,y) &= 0 \end{cases} \quad \text{s.t. } \mathbf{x}^{-1}|_{\partial\Omega} = \partial\hat{\Omega}. \tag{11}$$

This system of equations is scaled in order to yield an equation for $\mathbf{x}$ that is suitable for a computational approach. The resulting equations read [8]:

$$\begin{cases} \mathcal{L}(x,y,x) &= 0 \\ \mathcal{L}(x,y,y) &= 0 \end{cases} \quad \text{s.t. } \mathbf{x}|_{\partial\hat{\Omega}} = \partial\Omega, \tag{12}$$

where

$$\mathcal{L}(x,y,f) = g_{22}f_{\xi\xi} - 2g_{12}f_{\xi\eta} + g_{11}f_{\eta\eta}, \tag{13}$$

with $g_{11}(x,y) = \|\mathbf{x}_\xi\|^2$, $g_{12}(x,y) = \mathbf{x}_\xi \cdot \mathbf{x}_\eta$ and $g_{22}(x,y) = \|\mathbf{x}_\eta\|^2$. Since the target space of $\mathbf{x}^{-1}$ is always convex, the bijectivity of the exact solution of (12) follows from the maximum principle [9]. As a result, we may conclude that any sufficiently accurate approximation $\mathbf{x}_h$ of $\mathbf{x}$ will also be bijective. This property distinguishes EGG from most other meshing techniques, which may tend to produce folded meshes, due to them being less grounded in mathematical theory.

4.1. Discretization

Traditionally, (12) is approximately solved using a finite-difference approach [8]. However, since this only yields a finite collection of grid points that can serve as the vertices for a classical mesh, we have to look for different options. We discretize the equations with FEM-techniques. First, we introduce the operator

$$\tilde{\mathcal{L}}(x,y,f) = \frac{\mathcal{L}(x,y,f)}{g_{11} + g_{22}}. \tag{14}$$

As a next step, we select a $p \geq 2$ bivariate B-spline basis $\Sigma = \{w_1, \ldots, w_n\}$ with global C^1-continuity (i.e., the $w_i \in \Sigma$ possess at least one continuous derivative in $\hat{\Omega}$). By Σ_0, we denote the subset of Σ consisting of inner basis functions, that is, the collection of $w_i \in \Sigma$ that vanish on $\partial\hat{\Omega}$ (corresponding to the inner control points from section 2). We discretize (12) as follows:

$$\forall w_i \in \Sigma_0 : \begin{cases} \int_{\hat{\Omega}} w_i \tilde{\mathcal{L}}(x_h, y_h, x_h) \mathrm{d}\boldsymbol{\xi} = 0 \\ \int_{\hat{\Omega}} w_i \tilde{\mathcal{L}}(x_h, y_h, y_h) \mathrm{d}\boldsymbol{\xi} = 0 \end{cases}, \quad \text{s.t. } \mathbf{x}_h|_{\partial\hat{\Omega}} \simeq \partial\Omega. \tag{15}$$

Denoting by $\mathcal{I}$ and $\mathcal{I}_0$ the index-sets of Σ and Σ_0, respectively, $\mathbf{x}_h$ will be of the form

$$\mathbf{x}_h = \sum_{i \in \mathcal{I}_0} \mathbf{c}_i w_i + \sum_{i \in \mathcal{I} \setminus \mathcal{I}_0} \mathbf{d}_i w_i, \tag{16}$$

where the $\mathbf{c}_i$ denote the inner control points and the $\mathbf{d}_i$ the boundary control points, which follow from a regression of the input point cloud and serve as to ensure that the mapping operator satisfies $\mathbf{x}_h|_{\partial\hat{\Omega}} \simeq \partial\Omega$ (see section 5). As the $\mathbf{d}_i$ are known, the objective is to find the $\mathbf{c}_i$ such that $\mathbf{x}_h$ satisfies (15). Equation (15) leads to a nonlinear root finding problem of the form

$$\mathbf{F}(\mathbf{c}) = \mathbf{0}, \tag{17}$$

where the vector $\mathbf{c}$ contains the unknown inner control points $\mathbf{c}_i$.

The scaling introduced in (14) has the advantage of allowing for a more scaleable convergence criterion $\|\mathbf{F}(\mathbf{c})\| \leq \epsilon$, that is, the value of ϵ that corresponds to a converged solution is less sensitive to the characteristic length-scale of the geometry (and can therefore be taken approximately equal in all cases). The choice of Σ and the optimal selection of the inner control points $\mathbf{d}_i$ shall be the topic of section 5, while the computational approach for solving (17) will be the topic of section 7.

IOP Conf. Series: Materials Science and Engineering **425** (2018) 012030 doi:10.1088/1757-899X/425/1/012030

5. Contour Approximation and Choice of Basis

Given four input point clouds corresponding to each boundary of Ω: $P_\alpha = \{\mathbf{p}_\alpha^i\}_{i=1}^{I_\alpha}$ with $\alpha \in \{s, e, n, w\}$, the selection of a suitable spline basis Σ and the corresponding boundary control points $\mathbf{d}_i$ (see section 2) constitutes a preliminary step before (17) is tackled computationally. For this purpose we select a coarse initial basis $\Sigma_\square$ (resulting from two coarse univariate knot-vectors) and four sets of monotone increasing parametric values $\{\xi_\alpha^i\}_{i=1}^{I_\alpha}$, each starting on $\xi_\alpha^1 = 0$ and ending on $\xi_\alpha^{I_\alpha} = 1$. Let $\mathbf{m}_s(s) = (s, 0)$, $\mathbf{m}_e(s) = (1, s)$, $\mathbf{m}_n(s) = (s, 1)$ and $\mathbf{m}_w(s) = (0, s)$. The objective is to select the $\mathbf{d}_i$ such that $\mathbf{x}_h(\mathbf{m}_\alpha(\xi_\alpha^i)) \simeq \mathbf{p}_\alpha^i$. To this purpose a least-squares regression is carried out by minimizing the functional

$$R(\partial\Omega, \mathbf{d}) = \frac{1}{2} \sum_{\alpha \in \{s,e,n,w\}} \sum_{i=1}^{I_\alpha} \left\| \mathbf{x}_h(\mathbf{m}_\alpha(\xi_\alpha^i)) - \mathbf{p}_\alpha^i \right\|^2 \tag{18}$$

over the vector of boundary control points $\mathbf{d} = (\ldots, \mathbf{d}_j, \ldots)^T$. Hereby, it is advisable to constrain the corner control points to the corners of the input control points in order to avoid mismatches. In practice, (18) can suffer from instabilities. This is usually a result of the local amount of DOFs exceeding the local amount of points. We add a small least-distance penalty term to (18) in order to improve the stability:

$$\tilde{R}(\partial\Omega, \mathbf{d}) = \frac{1}{2} \sum_{\alpha \in \{s,e,n,w\}} \left(\sum_{i=1}^{I_\alpha} \left\| \mathbf{x}_h(\mathbf{m}_\alpha(\xi_\alpha^i)) - \mathbf{p}_\alpha^i \right\|^2 + \lambda \int_{\gamma_\alpha} \left\| \frac{\partial \mathbf{x}_h}{\partial \boldsymbol{\xi}} \cdot \hat{\mathbf{t}} \right\|^2 \mathrm{d}s \right), \tag{19}$$

where $\hat{\mathbf{t}}$ is the unit tangent vector and γ_α the subset of $\partial\hat{\Omega}$ that corresponds to side α. Here $\lambda > 0$ is a small penalty factor whose value should be chosen small enough not to noticeably alter the outcome of (19) while avoiding instabilities.

After (19) has been minimized, the mismatch

$$r_{\alpha,i} = \left\| \mathbf{x}_h(\mathbf{m}_\alpha(\xi_\alpha^i)) - \mathbf{p}_\alpha^i \right\| \tag{20}$$

serves as a local refinement criterion. Whenever $r_{\alpha,i}$ exceeds the approximation tolerance $\mu > 0$, we place a knot in the center of the element corresponding to the knot-vector of side α that contains ξ_α^i in order to locally increase the resolution of the basis.

Above steps are repeated until the convergence criterion is reached. Upon completion, we are in the possession of the coarse- and fine-grid bases $\Sigma_\square$ and Σ with corresponding vector of boundary control points $\mathbf{d}_\square$ and $\mathbf{d}$. The fine-grid basis then constitutes the coarsest possible basis to properly resolve the boundary condition

$$\mathbf{x}|_{\partial\hat{\Omega}} = \partial\Omega. \tag{21}$$

The tuple $(\Sigma_\square, \mathbf{d}_\square)$ serves as a means to build initial guesses for the computational approach employed to solve (17) (see section 7).

It should be noted that the choice of Σ is purely based on its capabilities to properly approximate $\partial\Omega$. However, this does not mean that it yields sufficient resolution to appropriately approximate the solution of (12) at every point in $\hat{\Omega}$. Heuristically, folding due to insufficient accuracy is uncommon. Should it nevertheless happen, we refine Σ by adding knots to the knot-vectors and prolong $\mathbf{d}$ to the refined basis Σ_r. Equation (17) is solved with Σ_r to yield a better approximation at internal points of $\hat{\Omega}$. This step can be repeated until the approximation is sufficiently accurate and bijectivity is achieved.

IOP Conf. Series: Materials Science and Engineering **425** (2018) 012030 doi:10.1088/1757-899X/425/1/012030

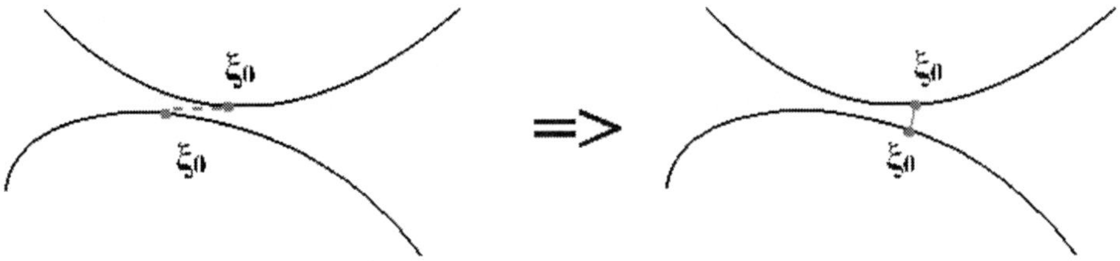

Figure 4. Skewed isolonies due to poorly parameterized boundary contours and improved parametric properties resulting from reparameterization. Near the tiny gaps, it is of major importance that the same parametric value is assumed on either side.

6. Choice of Parametric Values

The parametric values $\{\xi_\alpha^i\}_{i=1}^{I_\alpha}$ from section 5 have a profound influence on the parametric properties of the resulting parameterization and have to be chosen wisely. By default, the input point clouds are chord-length parameterized. Defining l_i recursively by

$$l_{i+1} = l_i + \|\mathbf{p}_{i+1} - \mathbf{p}_i\|, \tag{22}$$

starting with $l_1 = 0$ and ending on l_{I_α}, a chord-length parameterization corresponds to taking

$$\xi_\alpha^i = \frac{l_i}{l_{I_\alpha}}. \tag{23}$$

Using (23) ensures that the parametric velocity at the boundaries is (approximately) constant and therefore constitutes the default choice. In the presence of extreme aspect ratios (tiny gaps), however, we have found (23) to lead to dissatisfactory results. To ensure the quality of the resulting parameterization, we need to ensure that the (approximately) same parametric value is assumed on either side of the gaps (see figure 4).

For this purpose, we employ the matching algorithm proposed in [10] to two opposite point clouds P_w, P_e (or P_n, P_s) in order to match pairs of points that are too close. Let $\mathcal{I}_w = \{2, \ldots, I_w - 1\}$ and $\mathcal{I}_e = \{2, \ldots, I_e - 1\}$, upon completion of the matching, we are in the possession of a finite set of matched tuples

$$\mathcal{I}^m = \{(i, j) \mid \mathbf{p}_i \in P_w \text{ and } \mathbf{p}_j \in P_e \text{ have been matched}\}. \tag{24}$$

The tuples $(i, j) \in \mathcal{I}^m$, $i \in \mathcal{I}_w$, $j \in \mathcal{I}_e$ can be utilized to build a reparameterization function that improves the parametric properties of the mapping. By $\{\hat{\xi}_\alpha^i\}_{i=1}^{I_\alpha}$, we denote the default chord-length parameterization resulting from (23). We assign the parametric values $\{\hat{\xi}_w^i\}_{i=1}^{I_w}$ to the points $\mathbf{p}_i \in P_w$. Furthermore, we set $\xi_e^j = \hat{\xi}_w^i$ whenever $(i, j) \in \mathcal{I}^m$. The question remains what parametric value to impose on the unmatched points $\mathbf{p}_j \in P_e$. Given the set $\mathcal{I}^k = \{j \mid \xi_e^j \text{ is known}\}$ (note that $\xi_e^1 = 0$ and $\xi_e^{I_e} = 1$), we carry out a monotone cubic spline-interpolation [11] of the values $\{\hat{\xi}_e^i\}_i$, $i \in \mathcal{I}^k$ versus the known values $\{\xi_e^i\}_i$, $i \in \mathcal{I}^k$, to yield the monotone reparameterization function $\xi_e'(\xi)$. The ξ_e^i then follow from evaluating ξ_e' in the $\hat{\xi}_e^i$, that is

$$\xi_e^i = \xi_e'\left(\hat{\xi}_e^i\right). \tag{25}$$

In practice, reparameterization is a crucial ingredient for the successful parameterization of the target geometry and should therefore always be employed along with the numerical approach that is the topic of section 7.

IOP Conf. Series: Materials Science and Engineering **425** (2018) 012030 doi:10.1088/1757-899X/425/1/012030

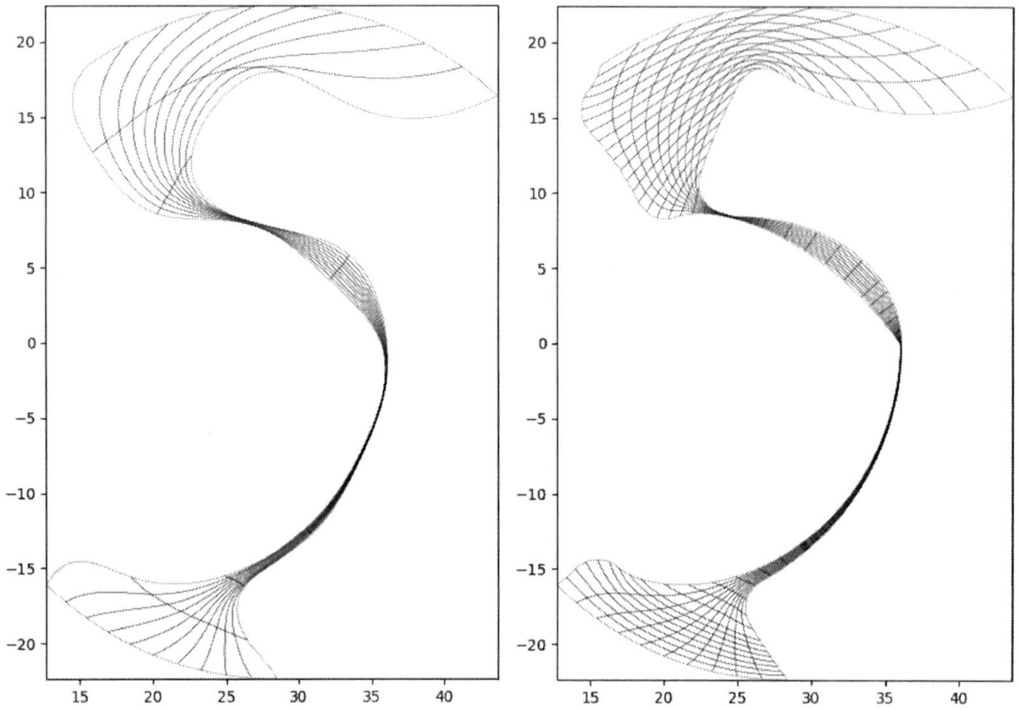

(a) Coarse-grid solution

(b) Fine-grid solution

Figure 5. Coarse- (a) and fine-grid solution (b) of a challenging input geometry.

7. Computational Approach

After completion of the regression from section 5, possibly in conjunction with reparameterization (see section 6), we are in the position to tackle the root-finding problem from equation (17). The point cloud regression yields the tuples $(\Sigma_\Box, \mathbf{d}_\Box)$ and $(\Sigma, \mathbf{d})$ of coarse- and fine-grid bases with corresponding boundary control points. We tackle (17) with a truncated Newton-approach. We first solve the coarse-grid problem $\mathbf{F}_\Box(\mathbf{c}_\Box) = \mathbf{0}$ using transfinite interpolation [5] to generate an initial guess $\mathbf{c}_\Box^0$. Both the coarse- and the fine-grid problem are of the form $\mathbf{G}(\mathbf{c}) = \mathbf{0}$, where $\mathbf{G}$ is nonlinear in $\mathbf{c}$. Given some initial guess $\mathbf{c}^0$, the new iterate is computed using the following recursive relation:

$$\left.\frac{\partial \mathbf{G}}{\partial \mathbf{c}}\right|_{\mathbf{c}=\mathbf{c}^i} \Delta\mathbf{c} = -\mathbf{G}(\mathbf{c}^i), \tag{26}$$

$$\mathbf{c}^{i+1} = \mathbf{c}^i + \delta\Delta\mathbf{c}. \tag{27}$$

Here $0 < \delta \leq 1$ is a truncation parameter whose optimal value is estimated from the current and updated tangents and residuals. Upon solution of the coarse-grid problem $\mathbf{F}_\Box(\mathbf{c}_\Box) = \mathbf{0}$, the coarse-grid solution $\mathbf{c}_\Box$ is prolonged to the fine-grid basis Σ and serves as an initial guess for the fine-grid problem $\mathbf{F}(\mathbf{c}) = \mathbf{0}$.

In practice, the coarse-grid problem typically converges after $4-6$ iterations, while the fine-grid problem requires an additional $2-3$ iterations. Thanks to the relatively small number of DOFs in $\Sigma_\Box$, the impact of the large number of required coarse-grid iterations is manageable.

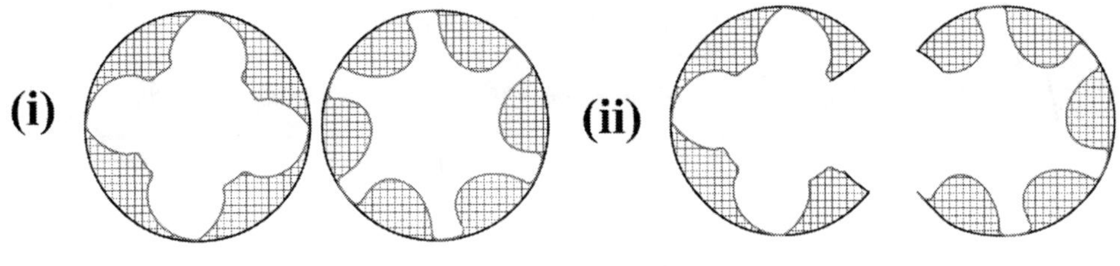

Figure 6. Steps (i) and (ii) of the parameterization strategy.

Since the required number of fine-grid iterations is approximately halved, the expected speed-up is $\sim$ 50%. Furthermore, the robustness of the approach is greatly improved: thanks to the high quality of the initial guess, failure of convergence is extremely uncommon. As an example, figures 7 (a) and (b) show the mapping corresponding to the coarse- and fine-grid solution of a challenging geometry. Here, we performed reparameterization (see section 6) on the western and eastern boundaries, keeping the western boundary chord-length parameterized while letting the eastern boundary float. Convergence on the fine grid is reached within 3 nonlinear iterations.

8. Application to Twin-Screw Machine Geometries

The computational approach discussed in section 7 constitutes the basic ingredient for the parameterization approach that will be the topic of this section. When no good initial guess is available, we will employ the hierarchical approach from section 7. Else, we will use the initial guess to solve the fine-grid problem right away. Selecting K uniformly-spaced discrete angles θ_k from the interval $[0, \pi/2]$, the objective is to compute a planar paramterization for every θ_k. The approach consists of the following steps:

(i) Generate separate O-type parameterization for the male and female rotors with casing for every θ_k.

(ii) Cut the O-parameterizations at the CUSP-points in order to produce two C-parameterizations for every discrete angle.

(iii) Combine the cuts with the male and female rotor parts to form a contour description of the separator.

(iv) Compute single-patch parameterizations for the separator at every discrete angle.

(v) Use the single-patch parameterized separators to generate curves, connecting the two CUSP-points and splitting the separator in half.

(vi) Generate parameterizations on the left and on the right of the splitting curves using the same knot vector as for the C-grids to acquire a conforming parameterization.

The key-steps are depicted in figures 6 and 7.

To generate the O-parameterizations from (i), we first generate exact (chord-length parameterized) cubic spline-fits through the input point clouds of the rotors and casings, using one of the FITPACK [12] routines. We evaluate both spline fits in N uniformly-spaced points over the parametric interval $[0, 1]$ and utilize the resulting point clouds to build male and female reparameterization functions η'_m and η'_f, respectively (see section 6). Hereby the casings are held chord-length parameterized (i.e., their reparameterization functions are simply given by the identity function) while we let the rotor parameterizations float.

At every discrete angle θ_k, we act with the canonical rotation matrix on the control points of the fitted spline-curves and reparameterize such that the CUSP-points always coincide with the same parametric values η_1 and η_2. Since the casings are chord-length parameterized, we

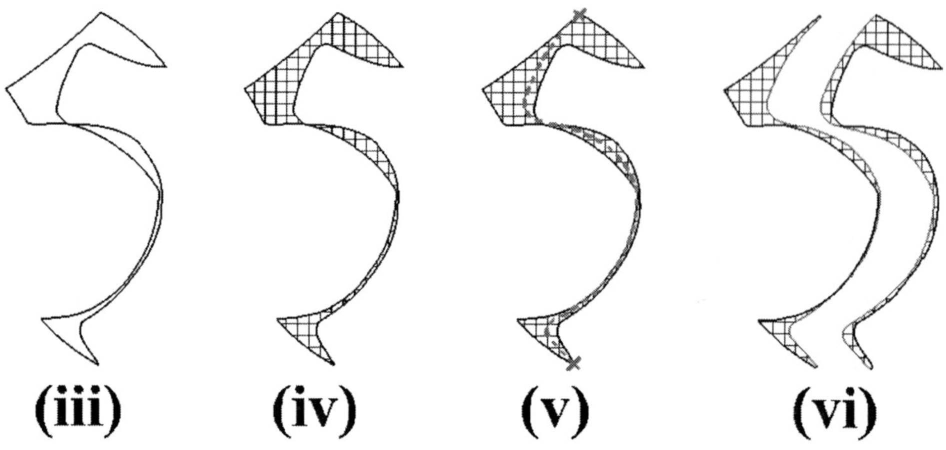

Figure 7. Steps (iii) to (vi) of the parameterization strategy.

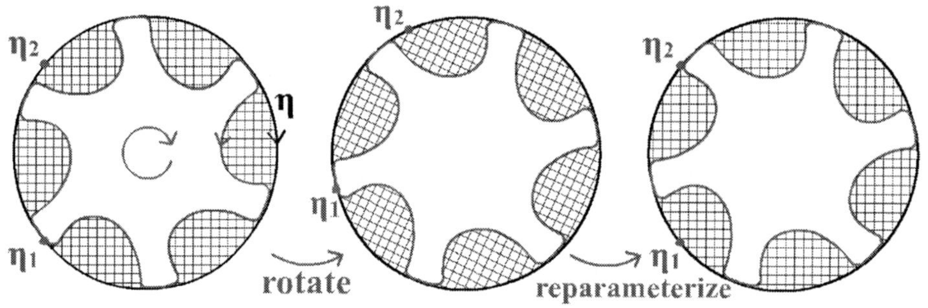

Figure 8. At every angle θ_k we act with the canonical rotation matrix on the control points of the spline-fit and reparameterize such that the CUSP-points always correspond to the same parametric values η_1 and η_2.

reparameterize simply by a shift of $-\theta_k/(2\pi)$ in the parametric domain. The rotor spline-fits have to be shifted by the same value in the reparameterized η'-domain. Therefore, we compute the shift in η by inverting $\eta'_{m,f}$ (see figure 8). As a next step, we utilize the shifted spline fits to generate a large number of uniformly-spaced points which serve as an input point cloud. We utilize the reparameterization functions $\eta'_{m,f}$, which we shift in a way similar to the spline fits, to assign parametric values to the rotor point clouds.

For the O-parameterizations, we utilize a $p \geq 2$-th order knot-vector which is periodic in the η-direction while disregarding the northern and southern boundaries. We compute parameterizations corresponding to θ_k, $k = 1, \ldots, 6$ utilizing the hierarchical approach from section 7. Upon completion, for each θ_{k+1}, we extrapolate the inner control points of $\theta_{k-6}, \ldots, \theta_k$ to θ_{k+1} and utilize them as an initial guess.
Upon completion of the K slices for each rotor, we add the $p+1$-times repeated knots η_1 and η_2, the η-values that correspond to the CUSP-points in $\hat{\Omega}$, to the knot vectors of the O-grids. This way, the two separate male and female rotor O-grids are each split into two parts: one C-grid and one grid which is discarded. By cutting in the domain, we avoid accidentally cutting the rotor lobes twice, as a result, the cuts are not (necessarily) straight.

Upon completion, our database is filled with left and right C-grid parameterizations for each

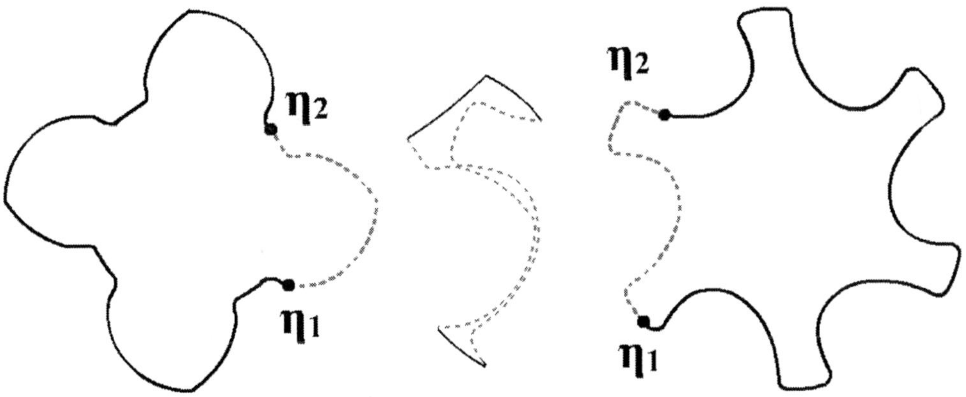

Figure 9. The exact rotor spline-fits are evaluated from η_1 to η_2 and the resulting point clouds are combined with the C-grid cuts

discrete angle θ_k. Having completed steps (i) and (ii), a description of the separator contours can be acquired by evaluating the exact rotor spline-fits from η_1 to η_2 (on both sides) and combining the resulting point clouds with the C-grid cuts (see figure 9).
The resulting geometries are parameterized with one patch. Instead of sequentially parameterizing all K slices, we start off by building parameterizations for each L-th slice (with $L \ll K$), keeping the western boundary chord length parameterized. Here, we employ the hierarchical approach from section 7. Let η'_k, denote the reparameterization function for the k-th slice. Upon completion of every L-th parameterization, we are in the possession of K/L single-patch parameterizations for the separator as well as a reparameterization function η'_k, $k = 1, L, 2L, \ldots$ for every L-th slice. We build reparameterization functions for the remaining slices by blending the available η'_k to achieve full smoothness of the parametetric properties in θ.

We use the K/L available single-patch parameterations and interpolate them in θ. The inner control points of the resulting interpolation function are extracted and serve as an initial guess for the remaining slices, in a way similar to the rotor O-grids.

Upon completion of all K-slices, we traverse $\hat{\Omega}$ from one CUSP-point to the other, leaving splitting-curves in our wake (see figure 10).
Their properties are fully determined by the path taken. To maximize the quality of the parameterizations in step (vi), we traverse $\hat{\Omega}$ such that the separator is split most-evenly on both sides of the small gaps. Upon completion, we are in the possession of a splitting-curve for all discrete angles θ_k. As a last step, we utilize the resulting database to parameterize the separator with two patches, one on each side of the splitting curve. Hereby, we employ the same computational approach as for the single-patch parameterized separator. To achieve boundary conformity between separator and C-grids, we employ the same knot-vector(s) in the ξ-direction.

9. Results

We have implemented the parameterization approach proposed in section 8 in the FEM python-library *Nutils* [13]. The geometry has been parameterized for $K = 200$ discrete angles over the interval $[0, \pi/2]$ (corresponding to a quarter rotation on the male rotor after which the initial position is again assumed). Figures 11 to 13 show the two rotor C-grids at $\theta = \theta_1$, $\theta = \theta_{75}$ and $\theta = \theta_{150}$.
We have used $L = 5$ to fill database with 40 single-patch parameterizations of the separator. The single-patch parameterized separator along with the computed splitting-curve is plotted for

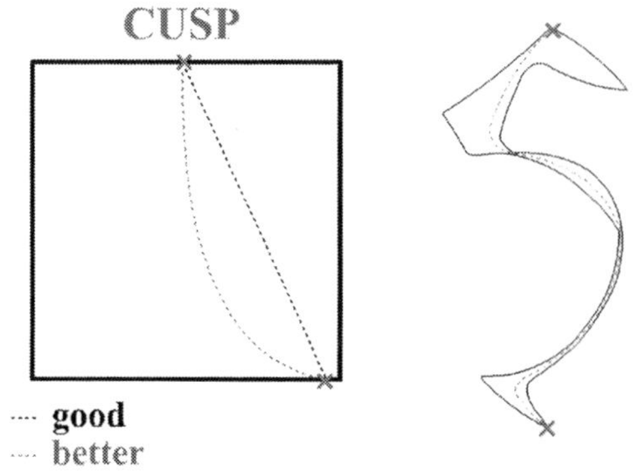

Figure 10. A splitting curve can be generated by traversing the computational domain from one CUSP-point to the other. The properties of this splitting curve is tuned by changing the path taken. We choose the path such that the separator is split most-evenly at the narrow gaps.

Figure 11. The two C-grids at $\theta = \theta_1$

$\theta = \theta_1$, $\theta = \theta_{75}$ and $\theta = \theta_{150}$ in figure 14. Here, we do not plot the isolines for improved visibility. Figure 15 shows the final geometry at $\theta = \theta_{100}$ and 17 figure (b) a zoom-in on the conforming separator showing the parametric properties by the splitting curve.
Finally, figure 16 shows the final geometry at $\theta = \theta_1$ and 17 (a) a zoom-in on the separator.

With $K = 200$, the computation of the rotor O-grids converges after 1 iteration as soon as enough slices for a 5-th order extrapolation are available. The 40 ($L = 5$) initial single-patch parameterized separators that we utilize to build an interpolation function typically converge within 3 iterations on the fine-grid using the hierarchical approach from section 7. Using the interpolation function to build an initial guess, for the remaining slices convergence is typically reached within 1 iteration with a maximum of 2. The same level of efficiency is achieved for the two-patch parameterized separator.

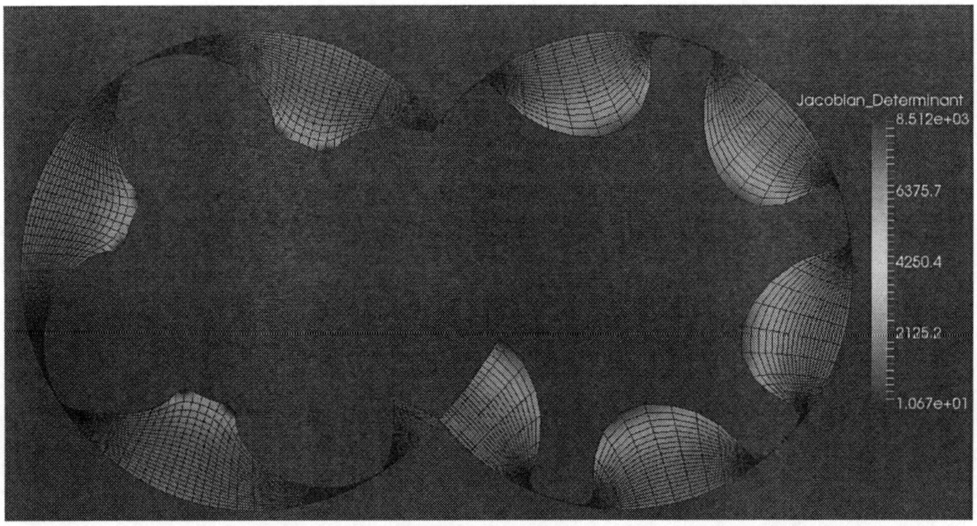

Figure 12. The two C-grids at $\theta = \theta_{75}$

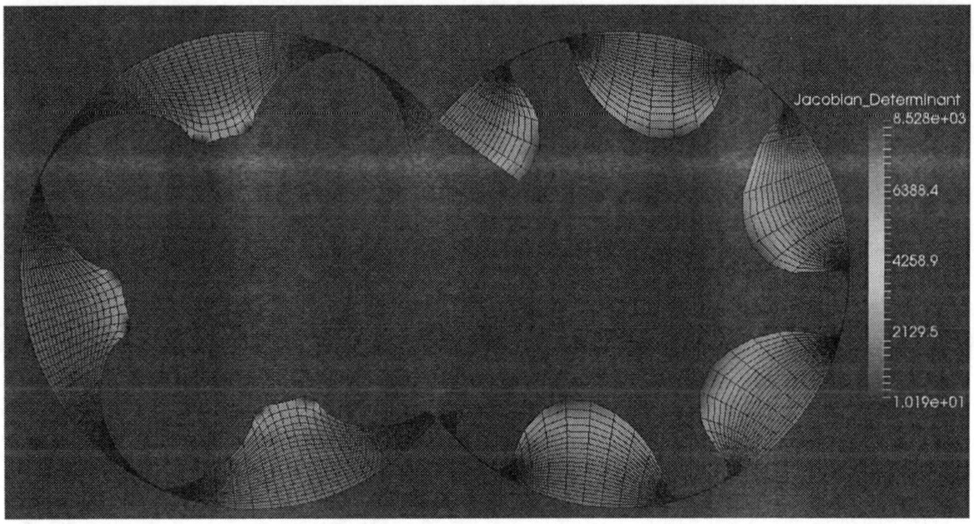

Figure 13. The two C-grids at $\theta = \theta_{150}$

10. Discussion

We have successfully implemented the approach proposed in section 8 utilizing the principles from sections 4 to 7. Here, inter- and extrapolation in the rotational angle θ greatly improves the efficiency since the number iterations is reduced to only 1 in almost all cases (against typically 3 without interpolation). The quality of the interpolation (and by that the expected number of required iterations) greatly depends on the value of K. Here, we used $K = 200$ which is a realistic number for an accurate flow-simulation. The reduction to only 1 iteration is remarkable since it implies that in this setting, EGG is nearly as efficient as algebraic parameterization techniques while yielding superior results.

The properties of the splitting curve, which are fully determined by the path taken in $\hat{\Omega}$, have a profound influence on the parametric properties of the two-patch parameterized separator as

IOP Conf. Series: Materials Science and Engineering **425** (2018) 012030 doi:10.1088/1757-899X/425/1/012030

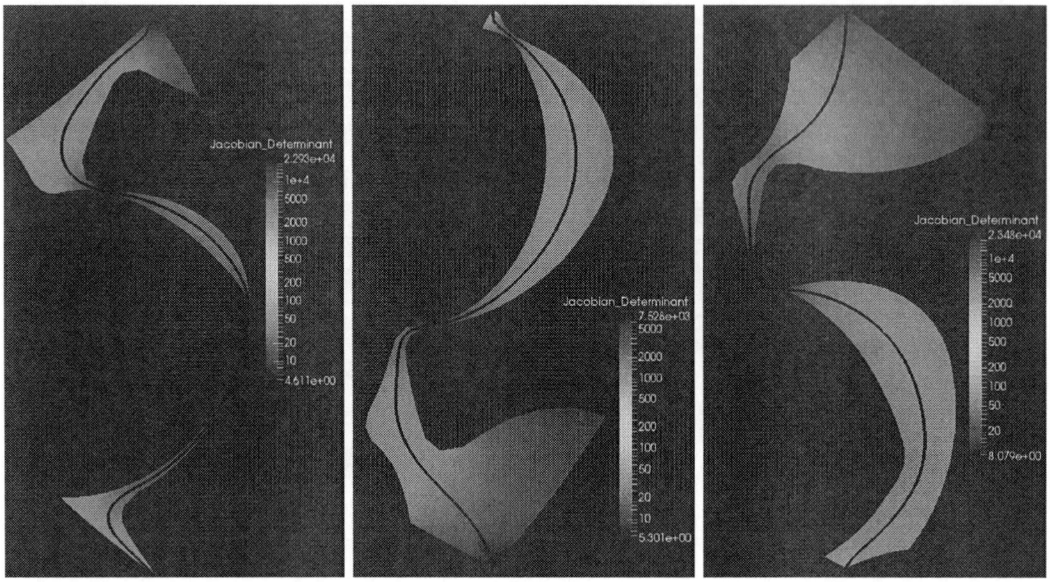

(a) $\theta = \theta_1$ (b) $\theta = \theta_{75}$ (c) $\theta = \theta_{150}$

Figure 14. The single-patch parameterized separator along with the generated splitting curve for various values of θ.

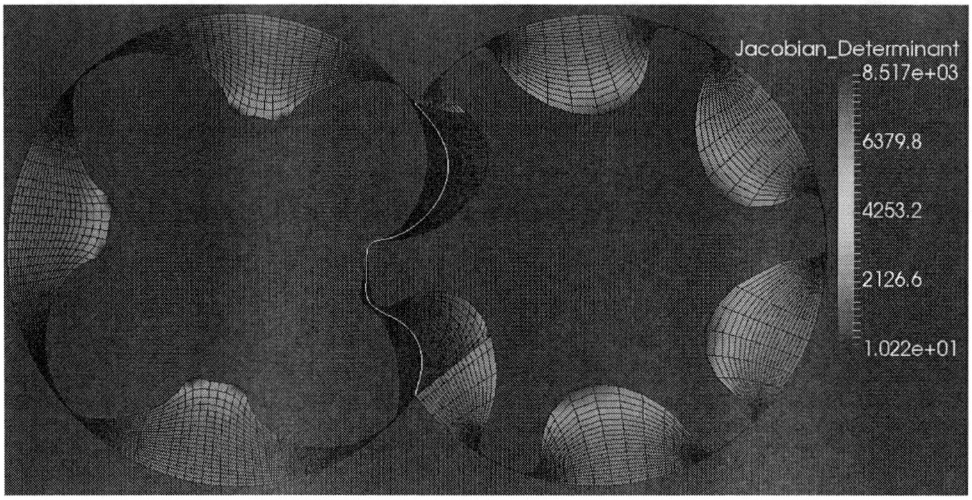

Figure 15. The two-patch parameterized separator along with the splitting curve at $\theta = \theta_{100}$

can be seen in figure 17. In (a), we clearly see the steep inter-element angles at the splitting curve interface. In (b) this is less pronounced. We conclude that an approximate halving of the separator by the splitting curve may not be the best quality criterion for all rotational angles θ and that further optimization is necessary. Hereby, it will be important to make a good trade-off between the steepness of the inter-elements angles at the splitting curve and C-grid interfaces which can, in turn, serve as a splitting curve selection criterion. A posteriori smoothing may be an option, too.

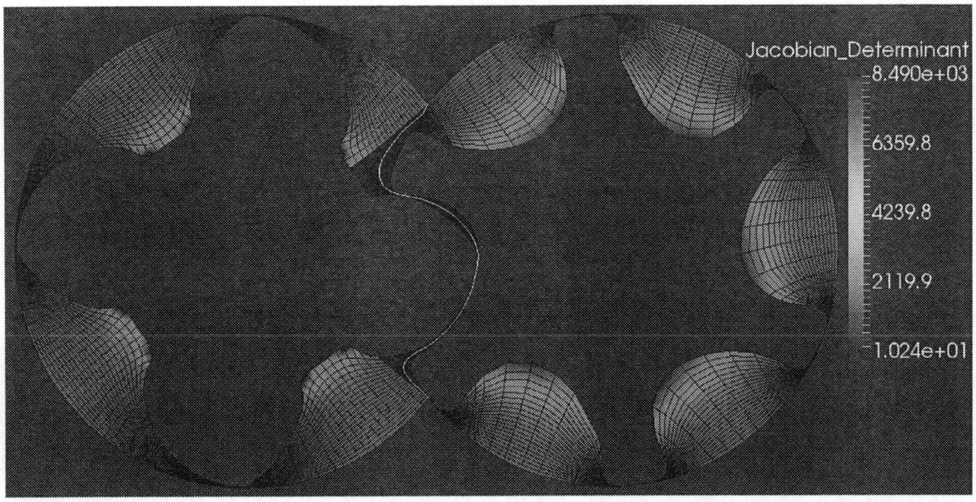

Figure 16. The two-patch parameterized separator along with the splitting curve at $\theta = \theta_1$

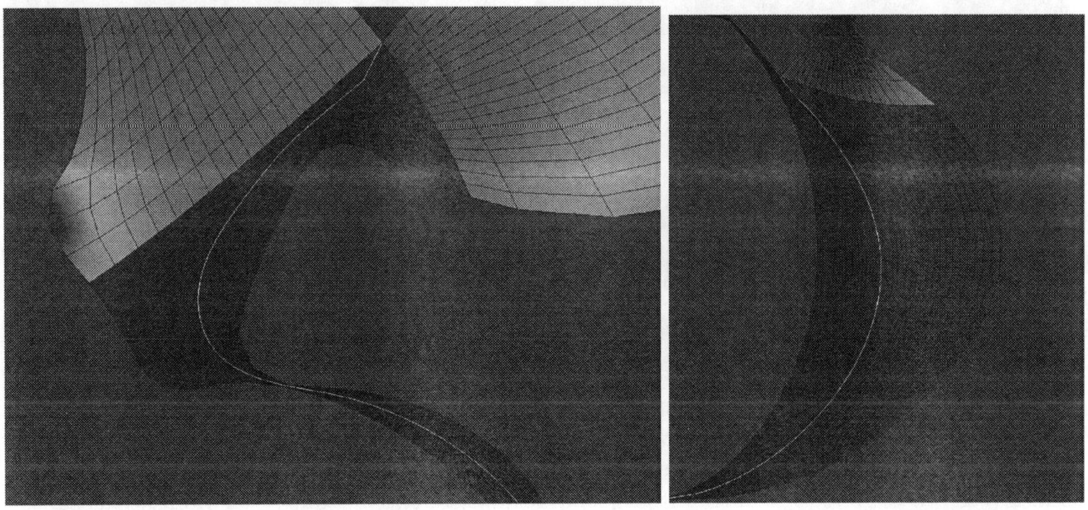

(a) $\theta = \theta_1$ (b) $\theta = \theta_{100}$

Figure 17. A zoom-in on the two-patch parameterized separator at $\theta = \theta_1$ (a) and $\theta = \theta_{100}$ (b).

10.1. Shape Optimization

As stated in section 1, we are particularly interested in performing shape optimization on a variable pitch-function. Given a particular rotor profile input, the objective is to minimize the objective function over the three-vector of shape parameters $\boldsymbol{\alpha}$ comprised of left- and right pitches $\dot{\theta}_{l,r}$ and the z-coordinate l_z at which the pitch changes. We base our approach on the observation that the planar slices at angle θ_k coincide with planar cross-sections of a volumetric parameterization in the z-direction.

The idea is to parameterize the geometry at a large number of discrete angles θ_k with angular increment $\Delta\theta$ and fill the database with a large number of planar slices $\mathbf{x}_h^{\theta}$. A parameterization

Figure 18. Part of a volumetric parameterization acquired by stacking a large number of planar slices.

for a particular configuration of $\boldsymbol{\alpha} = \left(\dot{\theta}_l, \dot{\theta}_r, l_z\right)^T$ is accomplished by a proper stacking of the slices in the z-direction. Hereby, a (locally) larger pitch will require a higher density of slices for an accurate description of the geometry, while a lower pitch allows for less slices. A database should therefore be generated with a slice-density that corresponds to the largest admissible pitch in the design space. Lower-pitched segments can be parameterized using a subset of the available slices $\mathbf{x}_h^\theta$. Since interpolation in the z-direction is a relatively cheap operation, a decent parameterization for a certain $\boldsymbol{\alpha}$ comes at a relatively low cost.

Figure 18 shows a segment of a volumetric geometry with constant pitch, generated by the stacking of a large number of planar slices in the z-direction.
Finally, figure 19 shows a segment of the separator with non-constant pitch along with a dotted red line to indicate the z-coordinate at which the pitch changes. This geometry has been constructed using the same planar slices as figure 18 but with a tighter stacking in the stronger-ptiched region.

11. Conclusion

In this article we presented a practical approach for the parameterization of twin-screw machine geometries with spline functions. For this, we adopted the principles of Elliptic Grid Generation and presented a computational approach that is compatible with the principles of Isogeometric Analysis. We presented automated boundary contour reparameterization techniques that further improve quality of the resulting parameterization.

We have successfully applied the approach to a twin-screw machine geometry. We have concluded that the parametric properties can be improved by optimizing the properties of the splitting curve that is a necessary ingredient for the parameterization of the separator. Finally, we have given a qualitative explanation of how the proposed techniques may be employed for a database-driven shape-optimization on a variable rotor pitch and presented an example of a

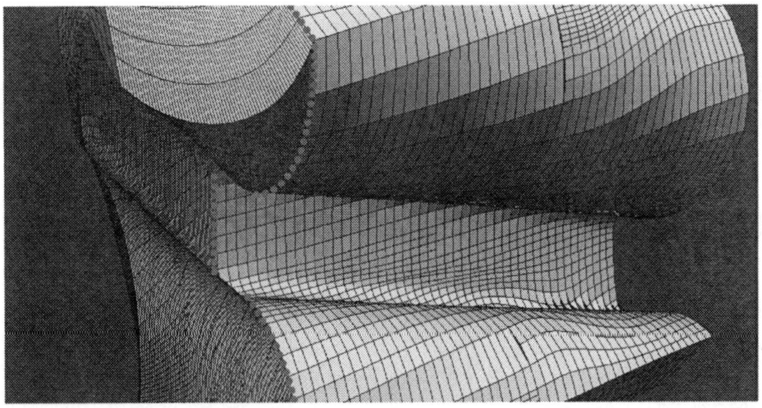

Figure 19. Part of a volumetric parameterization of the separator with non-constant pitch. The coordinate at which the pitch changes is indicated by the dotted red line.

volumetric parameterization resulting from the stacking of a large number of planar slices.

ACKNOWLEDGEMENTS

This project (MOTOR) has received funding from the European Unions Horizon 2020 research and innovation programme under grant agreement No 678727.

Bibliography

[1] Twin-Mesh Twin-mesh software `https://www.twinmesh.com/` [Online; accessed 2018-03-24]
[2] SCORG Screw compressor rotor grid generator `http://pdmanalysis.co.uk/scorg/` [Online; accessed 2018-03-24]
[3] ANSYS I Ansys cfx `https://www.ansys.com/products/fluids/ansys-cfx/` [Online; accessed 2018-03-24]
[4] Hughes T J, Cottrell J A and Bazilevs Y 2005 Isogeometric analysis: Cad, finite elements, nurbs, exact geometry and mesh refinement *Computer methods in applied mechanics and engineering* **194** 4135–4195
[5] Coons S A 1967 Surfaces for computer-aided design of space forms Tech. rep. DTIC Document
[6] Gordon W J and Hall C A 1973 Construction of curvilinear co-ordinate systems and applications to mesh generation *International Journal for Numerical Methods in Engineering* **7** 461–477
[7] Gordon W J and Thiel L C 1982 Transfinite mappings and their application to grid generation *Applied Mathematics and Computation* **10** 171–233
[8] Thompson J F, Soni B K and Weatherill N P 1998 *Handbook of grid generation* (CRC press)
[9] Castillo J E 1991 *Mathematical aspects of numerical grid generation* (SIAM)
[10] Hinz J, Möller M and Vuik C 2018 Elliptic grid generation techniques in the framework of isogeometric analysis applications *Computer Aided Geometric Design*
[11] Fritsch F N and Carlson R E 1980 Monotone piecewise cubic interpolation *SIAM Journal on Numerical Analysis* **17** 238–246
[12] Dierckx P 1995 *Curve and surface fitting with splines* (Oxford University Press)
[13] van Zwieten G, Verhoosel C, van Zwieten J, van Opstal T and Hoitinga W 2016 Nutils v2.0 URL `https://doi.org/10.5281/zenodo.822381`

IOP Conf. Series: Materials Science and Engineering **425** (2018) 012030 doi:10.1088/1757-899X/425/1/012030

List of Symbols

Symbol	Property
Ω	target geometry
$\partial\Omega$	boundary of the target geometry
$\hat{\Omega}$	computational domain (unit quadrilateral in $\mathbb{R}^n$)
$\mathbf{x}$	mapping operator from $\hat{\Omega}$ onto Ω
$\mathbf{x}_h$	numerical approximation of $\mathbf{x}$
Ξ	(univariate) knot-vector
σ	univariate spline basis
p	polynomial order
$N_{i,p}$	i-th spline function in σ with polynomial order p
Σ	bivariate spline basis
Σ_0	bivariate spline basis comprised of inner basis functions only
w_i	i-th spline function in Σ
θ	rotational angle
(ξ, η)	tuple of free topological variables
(ξ', η')	reparameterization function in ξ and η direction, respectively
$\mathbf{c}$	vector of inner control points
$\mathbf{d}$	vector of boundary control points
$\mathbf{R}(\partial\Omega, \mathbf{d})$	projection residual
K	number of planar slices
L	spacing between the slices used to build an interpolation function
$\boldsymbol{\alpha}$	vector of shape parameters

Subscript	Property
k	k-th index in the total number of planar slices K
(s, e, n, w)	south, east, north, west
(m, f)	male, female

IOP Conf. Series: Materials Science and Engineering **425** (2018) 012031 doi:10.1088/1757-899X/425/1/012031

Isogeometric simulation of thermal expansion for twin screw compressors

A Shamanskiy and B Simeon

Department of Mathematics, TU Kaiserslautern, Paul-Ehrlich-Str. 31, Kaiserslautern, Germany

E-mail: shamansk@mathematik.uni-kl.de, simeon@mathematik.uni-kl.de

Abstract. Isogeometric Analysis (IGA) is a recently introduced computational approach intended to breach the gap between the Finite Element Analysis and the Computer Aided Design worlds. In this work, we apply it to numerically simulate thermal expansion of oil free twin screw compressors in operation. High global smoothness of IGA leads to a more accurate representation of the compressor geometry. We utilize standard tri-variate B-splines to parametrize the rotors, while the casing is modeled exactly by using NURBS. We employ the Galerkin version of IGA to solve the thermal expansion problem in the stationary case. The results allow to estimate the contraction of the clearance space between the casing and the rotors. The implementation is based on the open source C++ library G+Smo.

1. Introduction

A twin screw compressor is a type of gas compressor with a great practical significance in various branches of engineering [1]. In this work, we focus on the oil free subclass of these machines. The absence of oil and its cooling effect leads to a much higher temperature rise than in oil injected compressors. The resulting thermal stress distorts the shape of the rotors, making it necessary to increase clearances in order to avoid contact between the rotors. By considering the effects of thermal expansion in numerical simulations [2], an estimate of the minimal safe distance can be acquired and the efficiency of compressors could be increased.

Here we propose an alternative approach to thermal expansion simulation with the use of Isogeometric Analysis (IGA), which is a modern computational method breaching the gap between the Finite Element Analysis and the Computer Aided Design worlds [3, 4]. NURBS-based IGA is particularly attractive for applications with rotor-type geometries and where spinning motion is involved, due to its ability to exactly represent circular arcs. Another key feature of IGA is its high global geometric smoothness which allows to achieve greater accuracy per degree of freedom when modeling intricate shapes. In this paper, we demonstrate one possible strategy of generating 3D IGA models of the rotors and the casing. We start from the point cloud representation of the rotors' profiles and approximate it with cubic B-spline curves. Using these approximations, the 3D tensor product B-spline parametrizations of the rotors are generated, while the casing model is built with NURBS. We then state a mathematical model of thermal expansion and use it to simulate thermal growth in the case of stationary rotors. The results are in a good agreement with the previous work, but much less computational resources are needed. We then use the results to estimate the contraction of the clearance space between the casing and the rotors.

We give now a short overview of different simulation methods and aspects for twin screw compressors. One of the most common approaches, 0D chamber-model simulations, is described in the work by Kauder [5]. For a systematic introduction to 3D CFD simulations we refer to Kovacevic et al. [6], while advanced CFD simulations including leakage flows are presented in [7, 8]. A novel efficient tool for fluid domain meshing is proposed by Hesse [9]. In a recent work by Utri [10] the optimization of screw machines with variable rotor pitch is addressed.

This paper is organized as follows. In Section 2, we describe the creation of geometric models for the rotors and the casing. Section 3 deals with the mathematical model of thermal expansion and demonstrates its use on a numerical example. In Section 4, we draw conclusions and outline current and future research directions.

2. Geometric Models

Both of the compressor rotors possess an intricate geometry (Fig. 1), a careful representation of which is the focal point of this article. An extremely small clearance height between the rotors and the surrounding casing requires a high degree of accuracy from the approach chosen to describe these objects. The conventional approach is the use of point clouds. To run numerical simulations, however, another representation is necessary. Within the framework of Isogeometric Analysis (IGA) the well-known tools from the world of Computer-Aided Design (CAD) - B-splines or NURBS - are utilized to model objects. However, unlike in CAD, not only the boundaries or surfaces of objects are modeled, but also their interior. The standard way to do this in IGA are tensor product B-splines.

Figure 1. Screw machine's rotors. Throghout this article they are refered to as the male (left, red) and the female (right, blue).

2.1. Tensor Product B-splines

To fix the notation, we shortly outline the construction of tensor product B-splines in the case of $d = 2$ dimensions. First we specify the polynomial degrees (p, q) and the horizontal and vertical knot vectors

$$\Xi = \{\xi_1 \leq \xi_2 \leq \ldots \leq \xi_{m+p+1}\}, \quad \Psi = \{\eta_1 \leq \eta_2 \leq \ldots \leq \eta_{n+q+1}\},$$

which contain non-decreasing parametric real values so that

$$0 \leq \mu(\Xi, \xi) \leq p + 1 \qquad \text{and} \quad 0 \leq \mu(\Psi, \eta) \leq q + 1$$

are the multiplicities of the parameter values in the knot vectors (the multiplicity $\mu(X, x)$ is zero if the given value x is not a knot in X). We write $M_{i,p}(\xi)$ and $N_{j,q}(\eta)$ for the univariate B-splines, which are computed by means of the recursion [11]

$$N_{j,0}(\eta) = \begin{cases} 1 & \text{for } \eta_j \leq \eta < \eta_{j+1}, \\ 0 & \text{otherwise}, \end{cases} \tag{1}$$

$$N_{j,q}(\eta) = \frac{\eta - \eta_j}{\eta_{j+q} - \eta_j} N_{j,q-1}(\eta) + \frac{\eta_{j+q+1} - \eta}{\eta_{j+q+1} - \eta_{j+1}} N_{j+1,q-1}(\eta) \tag{2}$$

for $j = 1, \ldots, n$. Fractions with zero denominators are considered zero. The same recursion applies to $M_{i,p}(\xi)$, replacing j, q in (1)–(2) by i, p for $i = 1, \ldots, m$.

A function $\mathbf{f}(\xi, \eta) : [\xi_{p+1}, \xi_{m+1}] \times [\eta_{q+1}, \eta_{n+1}] \to \mathbb{R}^2$ is called a bivariate tensor product B-spline function if it has the form

$$\mathbf{f}(\xi, \eta) = \sum_{i=1}^{m} \sum_{j=1}^{n} M_{i,p}(\xi) N_{j,q}(\eta) \mathbf{d}_{i,j} \tag{3}$$

with the de Boor control points $\mathbf{d}_{i,j} \in \mathbb{R}^2$, which form the control net associated to the parametric representation. In order to work with the unit square as parameter domain, combined with open knot vectors, we assume additionally $\xi_1 = \ldots = \xi_{p+1} = 0$ and $\xi_{m+1} = \ldots = \xi_{m+p+1} = 1$ at the beginning and at the end; analogously $\eta_1 = \ldots = \eta_{q+1} = 0$ and $\eta_{n+1} = \eta_{n+2} = \ldots = \eta_{n+q+1} = 1$.

In $d = 3$ dimensions, a trivariate B-spline tensor product function is generated analogously, which results in

$$\mathbf{f}(\xi, \eta, \psi) = \sum_{i=1}^{m} \sum_{j=1}^{n} \sum_{k=1}^{\ell} M_{i,p}(\xi) N_{j,q}(\eta) L_{k,r}(\psi) \mathbf{d}_{i,j,k} \tag{4}$$

with an additional set of univariate B-splines $L_{k,r}$ and control points $\mathbf{d}_{i,j,k} \in \mathbb{R}^3$.

2.2. Model Construction

The geometry was originally provided in the form of a point cloud where the male and the female rotors of SRM type are described with 2572 and 2292 points correspondingly, see Fig. 2. It was scaled to match the configuration from [2]. As a result, the axis distance is 80 mm, the clearance height is 44 μm between the casing and the rotors and approximately 100 μm between the rotors, the radius of the rotors is 50.97 mm. The difference in scales makes it challenging to come up with a B-spline approximation of rotors' profiles without a large number of control points. That is one of the reasons why point clouds, rather than CAD models, are usually used to describe these objects.

We consider a B-spline approximation accurate enough if it deviates from the point cloud for less than 10% of the clearance height. This accuracy can be achieved by using cubic splines with 73 control points for the male rotor and with 109 control points for the female rotors. Four sharp corners of the male rotor are modeled by increasing the knots multiplicity. For the technical details of B-spline curve approximation we refer to [11]. Two circular arcs forming the casing can be modeled exactly by NURBS. This ability to exactly represent circular shapes makes IGA particularly suitable for applications where spinning motion is involved.

Next we generate 2D parametrizations for the cross-sections of the rotors. We choose the Scaled Boundary approach [12] which is natural for rotary objects and allows to describe the interior of each rotor with a minimum number of patches, see Fig. 3. The outer layer is formed by "scaling" the boundary towards the circular patch in the center. This central patch is later used to construct a shaft. The parametrization is C^2-continuous inside the patches, but only C^0 between them. Both patches are obtained as a geometric mapping from a unit square. For

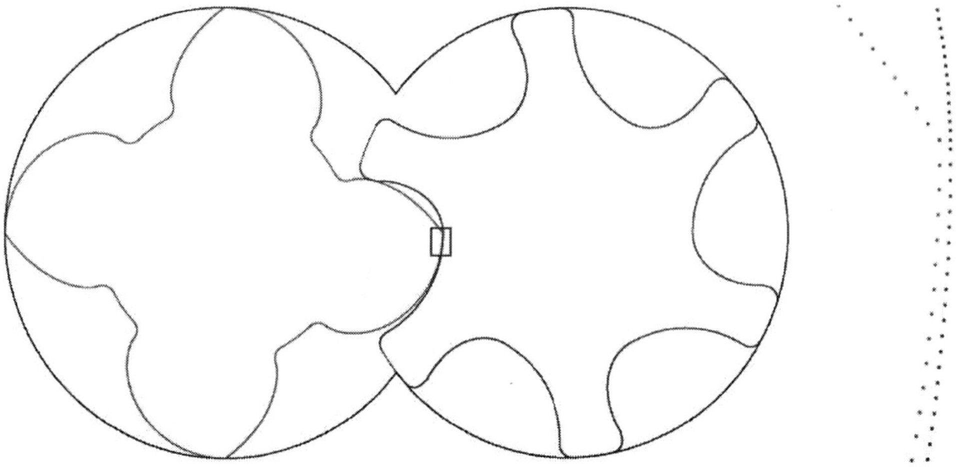

Figure 2. Point cloud describing the male and the female rotors as well as the surrounding casing. Magnified segment to the right clearly depicts the points and the clearance height.

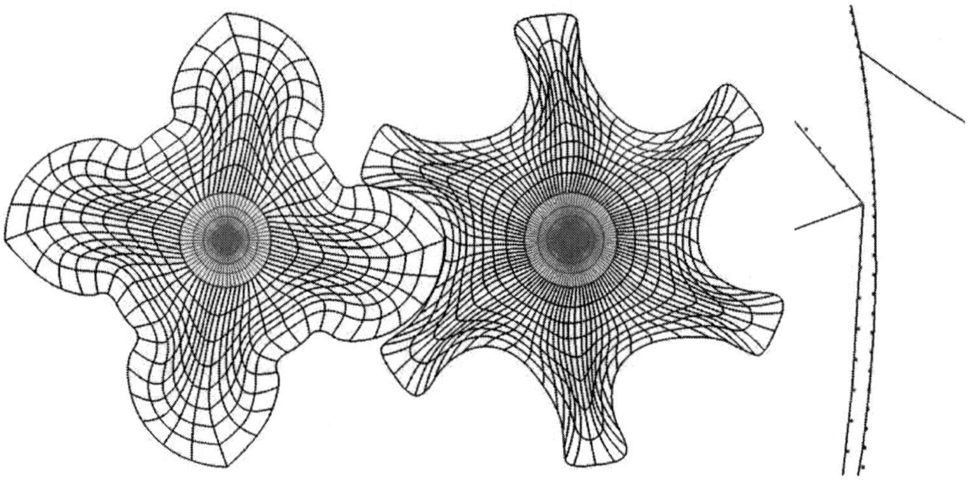

Figure 3. 2D scaled boundary parametrizations for the rotors. Original point cloud and its B-spline approximation (to the right, magnified).

the central patch this mapping is not bijective at the center. However, despite this singularity, this type of parametrizations is analysis-suitable.

Optionally, the planar parametrizations can be locally refined or coarsened using THB-splines [13]. By locally refining boundary layers of the rotors a finer resolution can be achieved there. This is especially attractive for the twin screw compressor example since the surfaces of the rotors require a finer resolution for analyzing and optimizing the mechanical properties. Figure 4 depicts a locally coarsened planar mesh. The B-spline approximation of the boundaries is maintained while less degrees of freedom are used in the interior of the rotors.

The final step is a 3D model. Each rotor is represented by a multi-patch geometry with 4 patches, see Fig. 5. The shaft has to be split into three parts so that the central part could have the matching boundary description with the inner surface of the outer layer. The control points for the outer layer are obtained by combining several layers of rotated control points from the

IOP Conf. Series: Materials Science and Engineering **425** (2018) 012031 doi:10.1088/1757-899X/425/1/012031

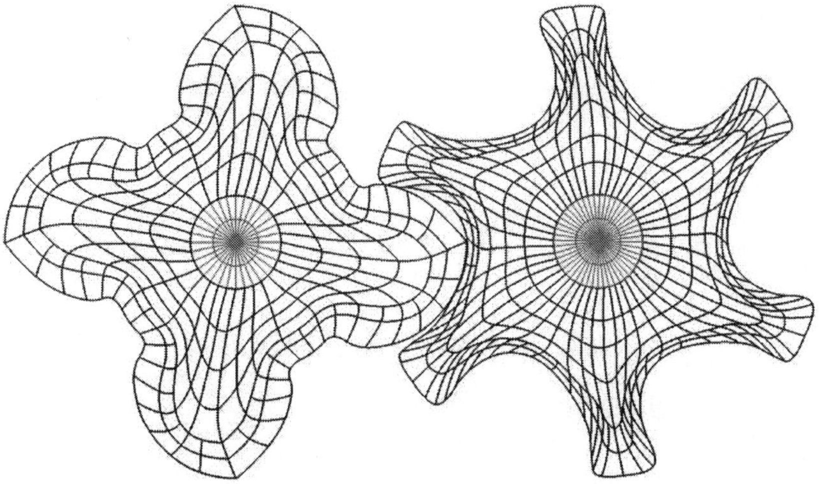

Figure 4. Local refinement with THB-splines allows to allocate degrees of freedom denser along the boundaries of the rotors than in their interior.

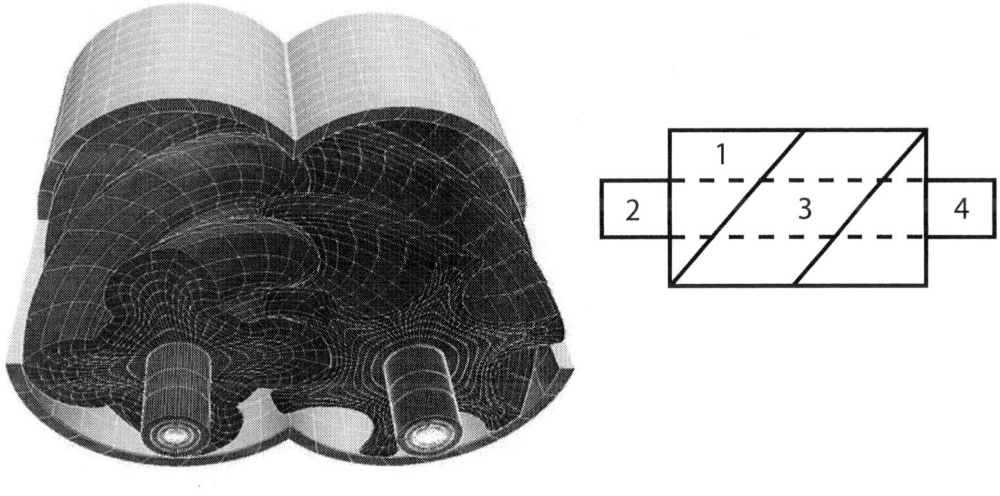

Figure 5. 3D models of the rotors and the casing. Multi-patch structure of the parametrizations: each rotor consists of 4 patches with matching boundary descriptions.

2D cross-section. 20 layers of control points are used for both rotors. Both rotors have a length of 168.3 mm, the male is modeled with a pitch of 300°, the female with -200°. The shaft patches 2 and 4 are 30 mm long and the casing is 6 mm thick.

3. Thermal Expansion

We consider here a very basic model of thermal expansion, which is nevertheless applicable for the case of screw compressors. It is based on the equations of linear elasticity, where the thermal growth is represented by an extra term. The use of a linear model is justified since only small deformations appear during the operation of the compressor, otherwise the machine would be blocked. However, these small deformations may still alter the shape of the clearing and change the efficiency of the compressor.

3.1. Mathematical Model

A body of isotropic material subject to a heat change from an initial temperature T_0 to a temperature T expands or contracts uniformly in all directions [14]. This can be modeled by the thermal strain

$$\varepsilon_T = \alpha(T - T_0)\mathbf{I}, \tag{5}$$

where α is the coefficient of thermal expansion of the material and $\mathbf{I}$ is the $d \times d$ identity matrix. The resulting deformation is described by the displacement field $\mathbf{u} \in \mathbb{R}^d$. To take into account the mechanical load as well we consider the linearized mechanical strain $\varepsilon(\mathbf{u}) := \nabla^{(s)}\mathbf{u}$, where $\nabla^{(s)}$ is the symmetric gradient. We define the total strain $\boldsymbol{\varepsilon}_S(\mathbf{u})$ as a superposition of both:

$$\boldsymbol{\varepsilon}_S(\mathbf{u}) = \boldsymbol{\varepsilon}(\mathbf{u}) + \boldsymbol{\varepsilon}_T = \frac{1}{2}(\nabla\mathbf{u} + \nabla\mathbf{u}^T) + \boldsymbol{\varepsilon}_T. \tag{6}$$

The total stress $\boldsymbol{\sigma}(\boldsymbol{\varepsilon}_S(\mathbf{u}))$ is related to the total strain $\boldsymbol{\varepsilon}_S(\mathbf{u})$ by Hooke's law with the Lamé parameters λ and μ:

$$\boldsymbol{\sigma}(\boldsymbol{\varepsilon}_S(\mathbf{u})) = \mathbf{C}\boldsymbol{\varepsilon}_S(\mathbf{u}) = 2\mu\boldsymbol{\varepsilon}_S(\mathbf{u}) + \lambda \mathrm{tr}(\boldsymbol{\varepsilon}_S(\mathbf{u}))\mathbf{I}. \tag{7}$$

With the total stress defined we can now write the system of linear elasticity equations for the domain $\Omega \subset \mathbb{R}^d$ as

$$-\operatorname{div}\boldsymbol{\sigma}(\boldsymbol{\varepsilon}_S(\mathbf{u})) = \mathbf{g} \text{ in } \Omega, \tag{8}$$

$$\mathbf{u} = \mathbf{0} \text{ on } \partial\Omega_D, \tag{9}$$

$$\boldsymbol{\sigma}(\boldsymbol{\varepsilon}_S(\mathbf{u})) \cdot \mathbf{n} = \mathbf{f} \text{ on } \partial\Omega_N. \tag{10}$$

Since both the divergence operator and Hooke's law are linear, we can move the thermal terms to the right-hand side of the equations (8) and (10). Keeping in mind a change of the sign for the outer normal, we rewrite the equations in the following form:

$$\begin{aligned} -\operatorname{div}\boldsymbol{\sigma}(\boldsymbol{\varepsilon}(\mathbf{u})) &= \mathbf{g} - \operatorname{div}\boldsymbol{\sigma}(\boldsymbol{\varepsilon}_T) \text{ in } \Omega, \\ \mathbf{u} &= \mathbf{0} \text{ on } \partial\Omega_D, \\ \boldsymbol{\sigma}(\boldsymbol{\varepsilon}(\mathbf{u})) \cdot \mathbf{n} &= \mathbf{f} + \boldsymbol{\sigma}(\boldsymbol{\varepsilon}_T) \cdot \mathbf{n} \text{ on } \partial\Omega_N. \end{aligned} \tag{11}$$

Now for any given temperature distribution T the displacement field $\mathbf{u}$ can be computed. The temperature itself is assumed to be given or can be acquired as a solution of Poisson's equation or the heat equation depending on whether the stationary or the transient problem is considered. We assume that the temperature change is small enough so that the material properties remain constant. Another assumption is that the domain deformation does not affect the temperature distribution inside. Then the problem is only one-way coupled with the temperature change driving the thermal expansion without feedback.

3.2. Galerkin IGA

Let $\mathcal{V} = \{\mathbf{f} \in H^1(\Omega)^d \mid \mathbf{f} = \mathbf{0} \text{ on } \partial\Omega_D\}$ be the trial solution and the weighting function spaces, which are assumed to be the same. Then the weak form of the system (11) is defined in the following way, which is equivalent to the Principle of Virtual Work:

$$\begin{aligned} &\textit{Find}\, \mathbf{u} \in \mathcal{V},\ \textit{such that} \\ &\int_\Omega \nabla^{(s)}\mathbf{u} : \mathbf{C}\,\nabla^{(s)}\mathbf{v}\,\mathrm{dx} = \int_\Omega (\mathbf{g} - \operatorname{div}\boldsymbol{\sigma}(\boldsymbol{\varepsilon}_T)) \cdot \mathbf{v}\,\mathrm{dx} \\ &\qquad + \int_{\partial\Omega_N} (\mathbf{f} + \boldsymbol{\sigma}(\boldsymbol{\varepsilon}_T) \cdot \mathbf{n}) \cdot \mathbf{v}\,\mathrm{ds}, \quad \textit{for}\, \forall\, \mathbf{v} \in \mathcal{V}. \end{aligned} \tag{12}$$

Here : denotes the Frobenius inner product of two matrices. For functions $\mathbf{u}, \mathbf{v} \in \mathcal{V}$, the bilinear form

$$a(\mathbf{u},\mathbf{v}) = \int_\Omega \nabla^{(s)}\mathbf{u} : \mathbf{C}\,\nabla^{(s)}\mathbf{v}\,\mathrm{d}\mathbf{x} \tag{13}$$

is well-defined, and even more, it is symmetric and coercive. Setting

$$l(\mathbf{v}) = \int_\Omega (\mathbf{g} - \operatorname{div}\boldsymbol{\sigma}(\boldsymbol{\varepsilon}_T)) \cdot \mathbf{v}\,\mathrm{d}\mathbf{x} + \int_{\partial\Omega_N} (\mathbf{f} + \boldsymbol{\sigma}(\boldsymbol{\varepsilon}_T)\cdot\mathbf{n}) \cdot \mathbf{v}\,\mathrm{d}\mathbf{s} \tag{14}$$

as linear form for the integration of the right hand side, the equation of the weak form (12) can be written as

$$a(\mathbf{u},\mathbf{v}) = l(\mathbf{v}). \tag{15}$$

The Galerkin projection of the weak form (12) in physical coordinates $\mathbf{x}$ replaces the infinite dimensional space $\mathcal{V}$ by a finite dimensional subspace $\mathcal{V}_h \subset \mathcal{V}$, with the subscript h indicating the relation to a spatial grid. Let $\boldsymbol{\phi}_1, \ldots, \boldsymbol{\phi}_K$ be a basis of $\mathcal{V}_h$, then the numerical approximation $\mathbf{u}_h$ is constructed as linear combination

$$\mathbf{u}_h = \sum_{i=1}^{K} q_i \boldsymbol{\phi}_i \tag{16}$$

with unknown real coefficients $q_i \in \mathbb{R}$.

Upon inserting $\mathbf{u}_h$ into the weak form (15) and testing with $\mathbf{v} = \boldsymbol{\phi}_j$ for $j = 1, \ldots, K$, one obtains the linear system

$$\mathbf{A}\mathbf{q} = \mathbf{r} \tag{17}$$

with $K \times K$ stiffness matrix $\mathbf{A} = (a(\boldsymbol{\phi}_i, \boldsymbol{\phi}_j))_{i,j=1,\ldots,K}$ and right hand side vector $\mathbf{r} = (l(\boldsymbol{\phi}_i))_{i=1,\ldots,K}$. Since the matrix $\mathbf{A}$ inherits the properties of the bilinear form a, it is straightforward to show that $\mathbf{A}$ is symmetric positive definite, and thus the numerical solution $\mathbf{q}$ or $\mathbf{u}_h$, respectively, is well-defined.

Let $\mathcal{S}$ denote the function space of tensor product B-splines defined in Section 2. In Galerkin-based IGA, we define $\boldsymbol{\phi}_i = \boldsymbol{\psi}_i \circ \mathbf{F}^{-1}$, $i = 1, \ldots, K$, as basis functions via the push forward operator with $\boldsymbol{\psi}_i \in \mathcal{S}^d : [0,1]^d \to \mathbb{R}^d$. The numerical approximation becomes, expressed in parametric coordinates, a vector-valued d-variate B-spline tensor product function (3).

3.3. Numerical example

We study now the magnitude of thermal expansion and its effect on the clearance height. To that end, it is enough to consider thermal growth of the rotors and the casing in a stationary case. We assume a known temperature distribution with the temperature growing linearly along the axial direction towards the high pressure section. In order to compare our results with [2], we use the similar temperature distribution from 70°C to 180°C. The temperature at the shafts' end is also set to 70°C. Both rotors are fixed at the low pressure end of the shaft, while the high pressure ends are allowed to move in the axial direction. In this setting both the deformation of the lateral surfaces and the axial elongation can be computed. The casing is fixed at the outer boundary and is allowed to expand inside. Its outer surface is set to 70°C while the same linear temperature distribution is assumed at the inner surface. For the material parameters we choose those of steel: the thermal expansion coefficient α is 12e-6 K^{-1}, Young's modulus is 209 GPa and Poisson's ratio is 0.3. The computational meshes consist of 14592 cubic elements for the male rotor, of 18944 cubic elements for the female rotor and, after one iteration of uniform refinement, of 5184 quadratic elements for the casing. On a single core the simulation took 190s including the solution of a linear system, where we use the BiCGStab solver with domain

IOP Conf. Series: Materials Science and Engineering **425** (2018) 012031 doi:10.1088/1757-899X/425/1/012031

decomposition preconditioning. The simulation code was implemented using G+Smo, a C++ library proving most of the geometric routines needed in IGA [15].

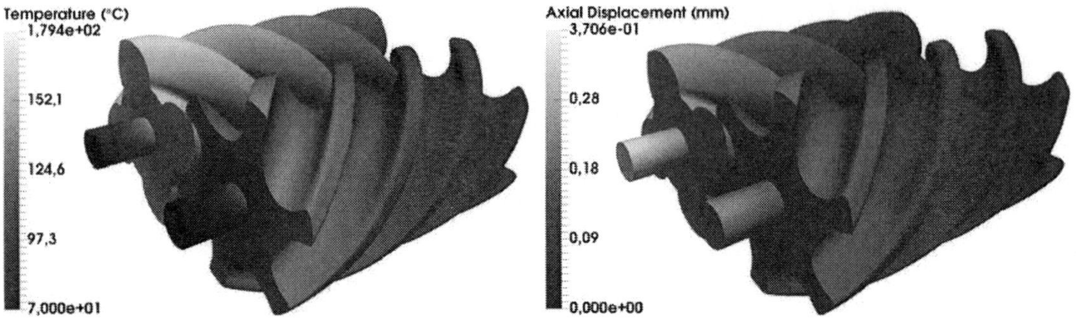

Figure 6. Temperature distribution for the compressor in operation (left). Axial elongation for the rotors (right).

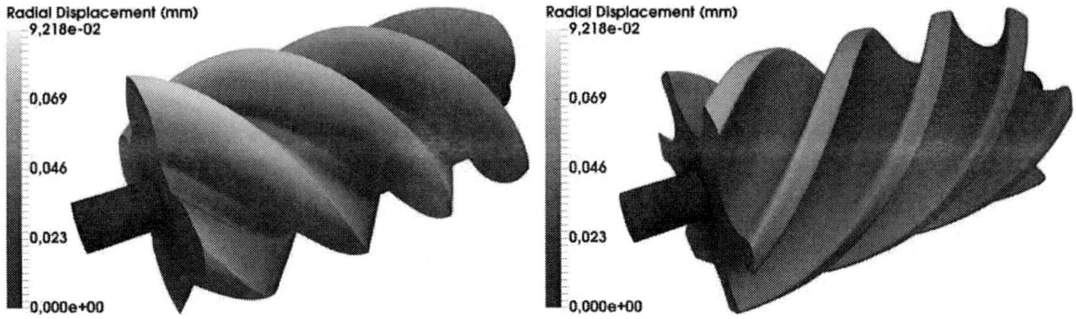

Figure 7. Radial expansion for the male rotor (left) and the female rotor (right).

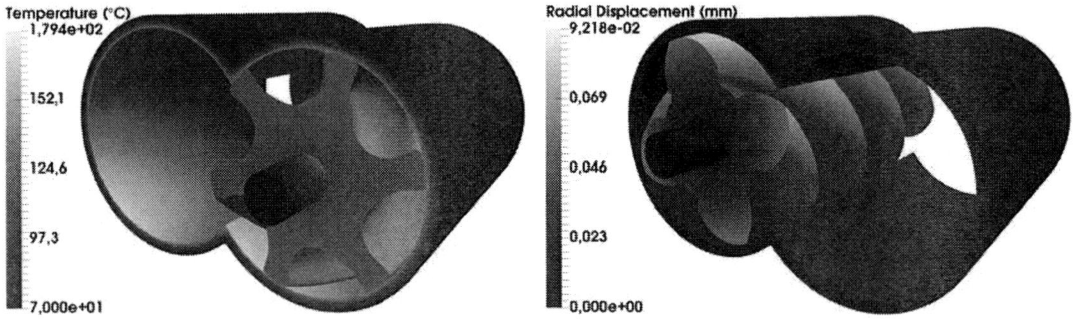

Figure 8. Simulation results with the casing. For the sake of visibility only one rotor is visualized. Temperature distribution (left) and radial expansion (right).

The results of the simulation are depicted in Fig. 6, 7 and 8. The simulation predicts an axial elongation of 317 μm for the male rotor, 370 μm including the shaft. For the female rotor a predicted axial elongation is 251 μm, 300 μm including the shaft. Predicted radial expansion is 92 μm and 79 μm for the male and the female correspondingly. The casing expands for 12 μm towards rotors. The results on radial expansion are in a very good agreement with [2], where approximately 300000 elements were used for each rotor. However, just like in the original paper they would mean a complete closing of the clearance between the casing and the rotors, which is 44 μm originally. Figure 9 shows a collection of magnified views on the clearance space in the 2D cross-section of the high pressure side of the rotors. Due to thermal expansion the clearance is completely closed between the casing and the rotors. Next to the sharp edge of the male the clearance space between the rotors is contracted by approximately 60%.

Figure 9. Clearance contraction. Axial expansion in 2D cross-section (left). Magnified views on the clearance space (right): between the male rotor and the casing (left); between the rotors (middle); between the female rotor and the casing (right).

4. Conclusions and future work

In this paper, we proposed to use Isogeometric Analysis to simulate thermal expansion of twin screw compressors. We demonstrated one possible approach to generating tensor-product B-spline parametrizations of the rotors and used them to solve a stationary thermal expansion problem. The results are consistent with the results from the previous works obtained with the help of the commercial software, but much less computational resources are necessary. However, further investigation is required to understand the excessive radial expansion which leads to a complete closing of the clearance space. In the future we plan to improve the model by using heat fluxes obtained from a coupled FSI simulation. This simulation will include the flow of the gas between the rotors and will take into account the deformation of the flow domain due to the thermal expansion. Furthermore, we plan to validate the simulation results on an experimental test rig with the help of our colleagues.

Acknowledgments

We are grateful to Andreas Brümmer and Matthias Utri for supplying us with the point cloud data of the rotors' geometry and for the fruitful discussions about the simulation setup. We are grateful to Matthias Möller and Jochen Hinz for valuable advice on the meshing strategy. This research is supported by the European Union under grant no. 678727 (Alexander Shamanskiy, Project MOTOR).

IOP Conf. Series: Materials Science and Engineering **425** (2018) 012031 doi:10.1088/1757-899X/425/1/012031

References

[1] Stosic N, Smith I and Kovacevic A 2005 *Screw compressors: mathematical modelling and performance calculation* (Springer Science & Business Media)
[2] Spille-Kohoff A, Hesse J and El Shorbagy A 2015 Cfd simulation of a screw compressor including leakage flows and rotor heating *IOP Conference Series: Mtrl. Sc. and Eng.* vol 90 (IOP Publishing) p 012009
[3] Cottrell J A, Hughes T J and Bazilevs Y 2009 *Isogeometric analysis: toward integration of CAD and FEA* (John Wiley & Sons)
[4] Hughes T J R, Cottrell J A and Bazilevs Y 2005 Isogeometric analysis: cad, finite elements, nurbs, exact geometry and mesh refinement *Comput. Meth. Appl. Mech. Engrg.* **194** 4135–4195
[5] Kauder K and Janicki M 2003 Adiabatic modelling and thermodynamic simulation of rotary displacement machines *IMECHE Conference Transactions* vol 4 (Professional Engineering Publishing; 1998) pp 511–520
[6] Kovacevic A, Stosic N and Smith I 2007 *Screw compressors: three dimensional computational fluid dynamics and solid fluid interaction* vol 46 (Springer Science & Business Media)
[7] Pascu M, Buckney D, Heiyanthuduwage M and Cook G 2014 Numerical investigation of the leakage flows in twin screw compressor rotors *Int. Compr. Eng. Conf.* (*Preprint* `http://docs.lib.purdue.edu/icec/2321`)
[8] Andres R, Hesse J, Babic H, Salecker U, Spille-Kohoff A, Nikolov A and Brümmer A 2016 Cfd simulation of a twin screw expander including leakage flows *Int. Compr. Eng. Conf.* (*Preprint* `http://docs.lib.purdue.edu/icec/2497`)
[9] Hesse J, Spille-Kohoff A, Hauser J and Schulze-Beckinghausen P 2014 Structured meshes and reliable cfd simulations: Twinmesh for positive displacement machines *International screw compressor conference, TU Dortmund*
[10] Utri M and Brümmer A 2017 Energy potential of dual lead rotors for twin screw compressors *IOP Conference Series: Materials Science and Engineering* vol 232 (IOP Publishing) p 012018
[11] Piegl L and Tiller W 1997 *The NURBS book* Monographs in visual communications (Springer)
[12] Arioli C, Shamanskiy A, Klinkel S and Simeon B 2017 Scaled boundary parametrizations in isogeometric analysis *arXiv preprint arXiv:1711.05760*
[13] Vuong A V, Giannelli C, Jüttler B and Simeon B 2011 A hierarchical approach to adaptive local refinement in isogeometric analysis *Computer Methods in Applied Mechanics and Engineering* **200** 3554–3567
[14] Gross D, Hauger W and Wriggers P 2011 *Technische Mechanik, Bd. 4: Hydromechanik, Elemente der Hheren Mechanik, Numerische Methoden* vol 4 (Springer)
[15] Jüttler B, Langer U, Mantzaflaris A, Moore S and Zulehner W 2014 Geometry + simulation modules: implementing isogeometric analysis *Proc. Appl. Math. Mech.* **14** 961–962 ISSN 1617-7061 special Issue: 85th Annual Meeting of the Int. Assoc. of Appl. Math. and Mech. (GAMM), Erlangen 2014

IOP Conf. Series: Materials Science and Engineering **425** (2018) 012032 doi:10.1088/1757-899X/425/1/012032

Isogeometric analysis framework for the numerical simulation of rotary screw machines. I. General concept and early applications

Matthias Möller and Jochen Hinz

Delft University of Technology, Faculty of Electrical Engineering, Mathematics and Computer Science, Department of Applied Mathematics, Van Mourik Broekmanweg 6, 2628XE Delft, The Netherlands

E-mail: {m.moller,j.p.hinz}@tudelft.nl

Abstract. This paper reports on the current status of an isogeometric modeling and analysis framework for rotary twin-screw machines that is being developed by an international consortium of academic partners within the EU-funded MOTOR project. The approach aims at combining accurate geometry modeling capabilities with modern high-performance computing techniques to enable efficient multi-physics simulations. The paper discusses the overall design philosophy of our isogeometric modeling and simulation pipeline and presents an isogeometric grid generator that makes it possible to integrate ANSYS CFX. Preliminary computational results for Sod's shock tube problem and an asymmetric SRM profile are shown.

1. Introduction

Rotary twin screw compressors are positive displacement (PD) machines that use the mechanical power of two counter-rotating rotors to suck in a fluid at the inlet, transport it in the chambers that are formed between the two rotors and reduce their volume in the course of the rotation, and finally release the compressed fluid with higher pressure at the exit. Likewise, screw expanders adopt the inverse working principle and recover mechanical power from the fluid that enters with high pressure at the suction side and is released with lower pressure at the outlet. Both PD machine types are nowadays widely used in industrial applications.

The accurate numerical analysis of screw machines and the simulation-based optimization of their design remains a challenging task. In the simplest case, high-speed turbulent compressible fluid flow through the chambers, heat exchange between the fluid and the rotors and the casing through the respective surfaces, and the resulting thermal expansion of the rotor profiles need to be taken into account, which leads to a mathematical model consisting of a coupled system of partial differential equations (PDEs) that is closed by one or more constitutive laws. The complexity of this multi-physics (MP) problem can be further enhanced by including additional physical phenomena like condensation and vaporization. The second major challenge stems from the complicated rotor geometries and the resulting fluid chambers that, by definition of the working principle of PD machines, vary in time, thereby giving rise to small gap sections.

Current best practice is to utilize special purpose mesh generation tools like TwinMesh^{TM} [1] and SCORG^{TM} [2] to pre-generate a sequence of computational meshes for the fluid and solid part of the domain at pre-selected time instances and apply commercial simulation packages like ANSYS CFX [3] and STAR-CCM+ [4] to perform multi-physics simulations using finite elements (FEM) and finite volumes (FVM) or a combination of both. This *mesh-based coupling approach* via the discrete unknowns of the numerical approximation is particularly attractive since it simplifies the combination of different simulation packages and makes the full potential of each of these tools immediately available for the numerical analysis of PD machines.

On the other hand, mesh-based coupling approaches have several conceptual limitations, which arise from the lack of a parametric master geometry model. This is not a particular problem of the numerical analysis of PD machines or of the particular mesh generation tools but it is inherent to the 'discreteness' of mesh-based coupling approaches.

Some of the major shortcomings that are frequently observed in practice are as follows:

1. Non-matching meshes at the MP domain interface make it difficult to exchange data between disciplines without introducing excessive interpolation errors. Moreover, the identification of 'neighboring' unknowns that communicate via the MP domain interface might become computationally expensive and calls for efficient search data structures.
2. Modeling of the MP domain interface as discrete surface mesh imposes an *a priori* maximum resolution on the geometry model and makes solution-adaptive mesh refinement beyond this level ineffective. Moreover, the lack of a parametric geometry model counteracts advanced computational approaches such as adjoint-based shape optimization or sensitivity analysis, which perform best if applied in the parameter space rather than on mesh points.
3. Combining the computational modeling of the geometry with the generation of high-quality computational meshes complicates this tasks and makes it particularly challenging to develop mathematical methods that can be proven to produce 'admissible' meshes.
4. Load-balancing strategies for distributed high-performance computing have no access to topology information of the geometry but need to schedule hundreds of thousands of individual cells to the different processors trying to minimize communication overhead.

In addition, TwinMesh^{TM} [1] and SCORG^{TM} [2] are particularly designed as pre-processings tools for ANSYS CFX^{TM} [3], which can only handle computational meshes consisting of straight-sided cells. Our interest is mainly in high-order methods (see Section 2.2 for the technical details of the adopted approach), which require computational meshes (actually, parametrizations) with curved element boundaries in order to represent complex geometries more accurately. High-order mesh generation is a topic of active research in itself and, to the best of our knowledge, neither TwinMesh^{TM} [1] and SCORG^{TM} [2] are capable of producing such meshes.

Geometry-aware parametrization-based multi-physics coupling. This work is the first in a series of publications that propose a novel approach for the numerical analysis of PD machines based on a parametric master geometry model with the ultimate aim to overcome the aforementioned shortcomings and to enable much more accurate predictions. Our approach builds on the concept of Isogeometric Analysis (IGA), which is a recent simulation technology that was introduced by Hughes et al. [5] in 2005 with the explicit vision to unify the worlds of Computer-Aided Design (CAD) and Computer-Aided Engineering (CAE). The central idea is to utilize the same mathematical language, namely, (non-uniform rational) B-Splines for describing the geometry in design and analysis and for seeking approximate solutions to the multi-physics problems.

The full realization of the envisaged isogeometric analysis framework is ongoing work that is carried out by an international consortium of academic partners and will be reported in subsequent publications. This paper describes the general concept including a short introduction to IGA and computational geometry modeling in Section 2. Section 3 discusses our recently developed isogeometric mesh generation tool, which makes it possible to couple any of the IGA applications that are provided by the open-source G+Smo (Geometry + Simulation Modules) library [6], which serves as common software base for all IGA-related developments, with ANSYS CFX [3]. This development is both meant as intermediate step to test the new approach with an established flow solver and to enable early adopters of this new technology to exploit the benefits of using a parametric master geometry model within an established simulation pipeline consisting of commercial solver packages. Preliminary computational results for fluid flow through an asymmetric SRM profile and a variation of Sod's shock tube benchmark problem [7] are reported in Section 4. Conclusions and an outlook on future research are given in Section 5.

2. Isogeometric Modeling and Analysis Pipeline

This section introduces the general concepts of our isogeometric modeling and simulation approach in an abstract mathematical formulation. Our aim is to highlight the conceptual differences of what we consider as *parametric master geometry model* to existing mesh generation techniques such as those adopted by TwinMesh^{TM} [1] and SCORG^{TM} [2], respectively.

2.1. Parametric curves

Mathematically speaking, the interfaces between MP domains are given by so-called *functional free-form surfaces*, which can be considered as two-dimensional manifolds that are exposed to the three-dimensional Cartesian space $\mathbb{R}^3$. These geometric objects are typically constructed by defining a parametric mapping $\mathbf{F}$ between the *parameter domain* $\hat{\Omega}$ and the *physical domain* Ω.

In order to illustrate the underlying principles of parametric curves and their algorithmic manipulation, consider the parametric curve depicted in Figure 1, which results from defining

$$\mathbf{F}(u) := \begin{pmatrix} \cos(2u)\cos(u) \\ \cos(2u)\sin(u) \end{pmatrix} \quad \text{where} \quad u \in \hat{\Omega} := [0,1]. \tag{1}$$

The so-defined mapping is bijective (one-to-one and onto), which means that the parameter value u^* of an arbitrary point (x^*, y^*) on the curve can be recovered from is inverse, that is

$$u^* = \mathbf{F}^{-1}(x^*, y^*). \tag{2}$$

The red circles in Figure 1 illustrate the location of the equidistantly distributed parameter values $u_i = i/9$, $i = 0, 1, \ldots, 9$ along the curve. Likewise, black dots stand for the non-equidistantly distributed parameter values $u'_i := \mathbf{G}(u_i)$ with bijective function $\mathbf{G} : [0,1] \to [0,1]$ defined as

$$u' = \mathbf{G}(u) := \sqrt{u} \quad \text{with} \quad u = \mathbf{G}^{-1}(u') = (u')^2. \tag{3}$$

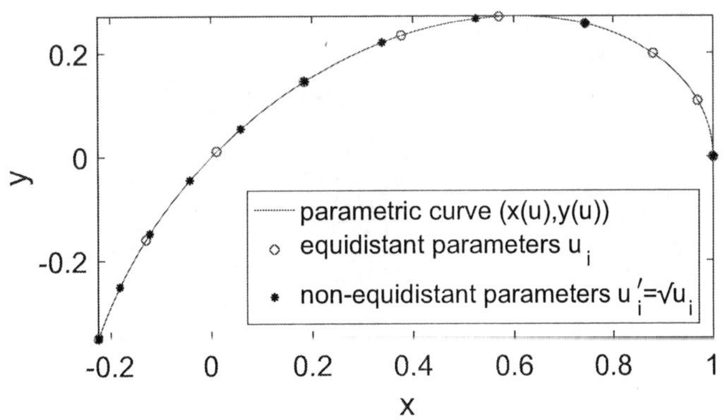

Figure 1. Illustration of a univariate parametric curves $\mathbf{F}(u) = (\cos(2u)\cos(u), \cos(2u)\sin(u))$ with equidistantly and non-equidistantly distributed parameter values $0 \leq u, u' \leq 1$.

We would like to emphasize that both $\mathbf{F}$ and $\mathbf{F} \circ \mathbf{G}$ represent the same *geometric* object but the so-called parametric speed at which the curve is traversed when varying u and u' between zero and one differs. Nonetheless, the bijectiveness of all involved mappings makes it possible to convert between both descriptions back and forth. This feature is particularly helpful when exchanging data between the solid and the fluid along a common interface curve.

For illustration purposes let (x^*, y^*) denote the point on the curve at which information should be exchanged. According to (2) and (3) both u^* and $(u^*)^2$ are mapped to the same physical location (x^*, y^*) under the mapping $\mathbf{F}$ and $\mathbf{F} \circ \mathbf{G}$, respectively.

2.2. Multi-patch parametrizations

Consider the benchmark configuration of a two-dimensional test rig depicted in Figure 2, which consists of an asymmetric SRM rotor profile (reddsih), a circular casing (greenish), and the fluid region (blueish) between the two solids. This benchmark has been studied experimentally by Sachs in [8] and numerically using ANSYS CFX^{TM} by Utri et al. in [9]. The (planar) physical domain that is depicted in Figure 2 (b) consists of 3×4 so-called *patches*, that is, geometric objects that can be mapped onto one of the unit squares defined in the parameter domain; cf. Figure 2 (a). The decomposition of the screw machine geometry into several patches is advisable in order to simplify the parametric treatment of the 'step' profile at the tip of of the rotor; cf. Figure 3. Here and below white numbers denote the patch number.

Mathematically speaking, there exist $k = 1, \ldots, 12$ distinct bijective mappings between the parameter domain $\hat{\Omega}_k = [0,1] \times [0,1]$ and the physical patch $\Omega_k \subset \mathbb{R}^2$ such that

$$\mathbf{F}_k(u_k, v_k) := \begin{pmatrix} x_k(u_k, v_k) \\ y_k(u_k, v_k) \end{pmatrix} \qquad \begin{matrix} (u_k, v_k) & \in \hat{\Omega}_k, \\ (x_k, y_k) & \in \Omega_k, \end{matrix} \tag{4}$$

and $(u_k, 0)$, $(u_k, 1)$, $0 \leq u_k \leq 1$ and $(0, v_k)$, $(1, v_k)$, $0 \leq v_k \leq 1$ are mapped onto the four boundary segments Γ_k^s, $s \in (S, N, W, E)$, which together form the boundary of the k-th patch.

Even though it is not mandatory for the method to work, it eases the implementation if the mappings of two adjacent patches, say, Ω_k and Ω_l that share the common boundary, say,

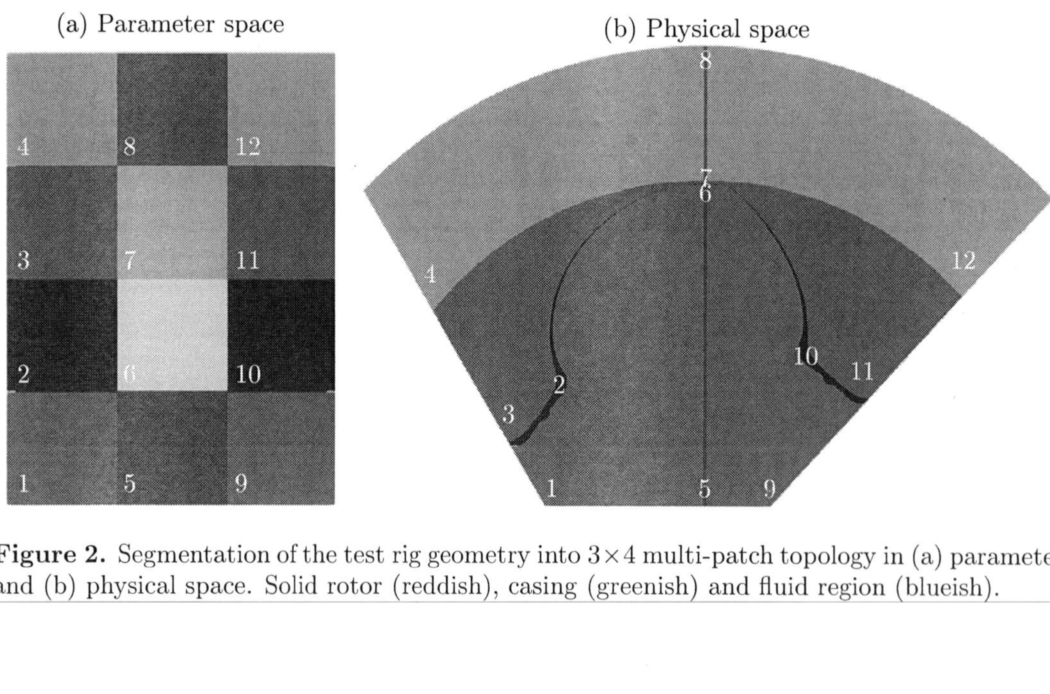

Figure 2. Segmentation of the test rig geometry into 3×4 multi-patch topology in (a) parameter and (b) physical space. Solid rotor (reddish), casing (greenish) and fluid region (blueish).

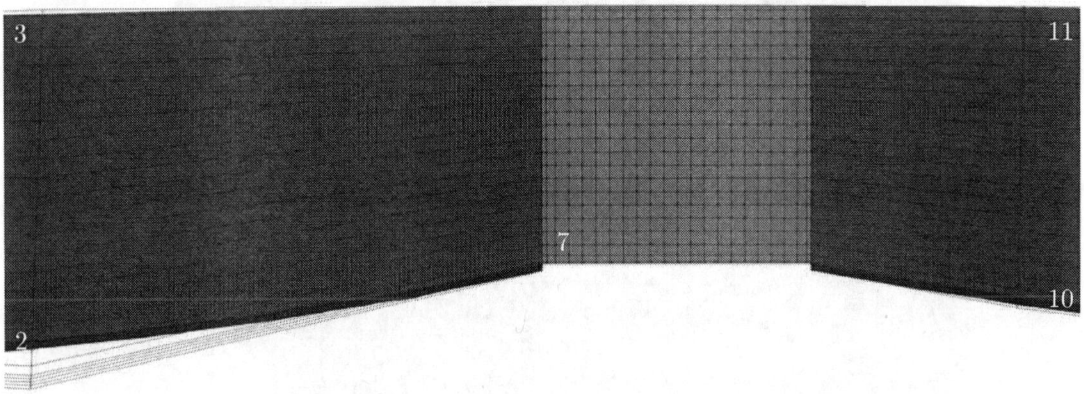

Figure 3. Close-up of the flow passage between casing and rotor tip. The B-Spline based geometry model is visualized by blue-colored patches. Black lines indicate the control net.

$\Gamma_k^S = \Gamma_l^N$ are parametrically compatible, that is, the parametric value u that describes the boundary segment is mapped to the same physical position by both mappings, e.g.

$$\mathbf{F}_k(u,0) = \mathbf{F}_l(u,1), \quad 0 \leq u \leq 1. \tag{5}$$

We will later relax this assumption to enable excessive mesh grading in the fluid domain along the boundaries and utilize the reparametrization (3) to restore parametric compatibility.

The ensemble of patches together with the list of patch neighbors is termed *multi-patch* topology. For the sake of better readability, let us drop subscript k and describe our isogeometric modeling and simulation approach for a single patch (unless explicitly stated otherwise) keeping in mind that all steps have to be applied to all patches Ω_k separately.

2.3. B-Spline based geometry modeling

In our isogeometric framework we resort to tensor-product B-Splines rather than non-uniform rational B-Splines (NURBS), which are most commonly used in Computer-Aided Design (CAD) applications but would render our approach more complicated. To begin with, let $U = [u_1, u_2, \ldots, u_{n+p+1}]$ be a sequence of non-decreasing values between zero and one with the additional constraint that the first and last $p+1$ entries are identical. This yields the so-called open knot vector U, which is used to construct univariate B-Spline basis functions $\{N_{i,p}(u)\}_{i=1}^n$ of degree p with the aid of the Cox-de-Boor recursion formula [10]:

$$p = 0: \quad N_{i,0}(u) = \begin{cases} 1 & \text{if } u_i \leq u < u_{i+1}, \\ 0 & \text{otherwise,} \end{cases} \tag{6}$$

$$p > 0: \quad N_{i,p}(u) = \frac{u - u_i}{u_{i+p} - u_i} N_{i,p-1}(u) + \frac{u_{i+p+1} - u}{u_{i+p+1} - u_{i+1}} N_{i+1,p-1}(u). \tag{7}$$

The above definition assumes the convention "$\frac{0}{0} = 0$" to prevent division by zero. B-Spline basis functions $\{N_{j,q}(v)\}_{j=1}^m$ of degree q are constructed from the knot vector $V = [v_1, v_2, \ldots, v_{m+q+1}]$.

Figure 4. Close-up of the flow passage between casing and rotor tip. The B-Spline based geometry model is visualized by blue-colored patches. Black lines indicate the control net.

Our overall aim is to represent the geometry mapping (4) as a bivariate B-Spline surface that is expressed in terms of the tensor product of univariate B-Spline basis functions. That is,

$$\mathbf{F}_h(u,v) = \sum_{i=1}^{n}\sum_{j=1}^{m} \mathbf{c}_{i,j} N_{i,p}(u) N_{j,q}(v), \quad 0 \leq u, v \leq 1. \tag{8}$$

It should be noted that the above mapping does not necessarily yield an exact representation of the true physical domain Ω but rather an approximation $\Omega_h \approx \Omega$, which is indicated by subscript 'h'. The quality of the approximation can often be enhanced by increasing the number of basis functions and distributing the entries of the generating knot vectors U and V in a way that accounts for geometric details; cf. Figures 3 and 4. However, an *exact* B-Spline representation of a domain that is described, say, by trigonometric functions (see Figure 1) is impossible.

In the above expression, $\mathbf{c}_{i,j} \in \mathbb{R}^2$ are the *control points* that 'steer' the deviation of the physical patch from the unit square as sketched schematically in Figure 5. It should be noted that only the four corner points are interpolated, e.g., $\mathbf{F}_h(0,0) = \mathbf{c}_{1,1}$, whereas all other control points can attain arbitrary values that may even lie outside of the physical patch, cf. Figures 3–5. Control points are connected in horizontal and vertical parametric direction by the so-called *control net*, which is indicated by black lines. Let us remark that these are not the grid lines of the computational mesh but only serve the purpose of giving control to the parametrization (8).

The univariate boundary parametrizations are obtained from the bivariate planar

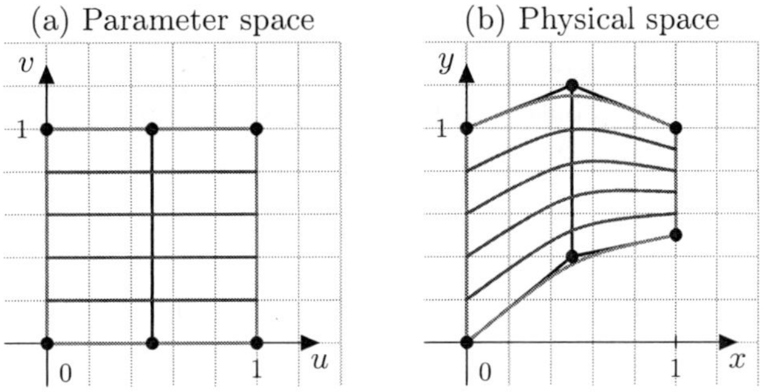

Figure 5. Mapping between (a) parameter and (b) physical domain. Black lines indicate the control net with control points $\mathbf{c}_{i,j}$ given by dots. Boundary segments are in red. Grid lines with parameter values varying in u-direction are sketched in blue for illustration purposes.

parametrization (8) by setting one of the two parameters to zero or one, e.g.

$$\mathbf{F}_h^S(u) := \mathbf{F}_h(u,0) = \sum_{i=1}^{n}\sum_{j=1}^{m} \mathbf{c}_{i,j} N_{i,p}(u) N_{j,q}(0) = \sum_{i=1}^{n} \mathbf{c}_{i,1} N_{i,p}(u), \quad 0 \le u \le 1. \tag{9}$$

Let us switch to vector notation, i.e. $\mathbf{x} := (x,y)$ and $\mathbf{u} := (u,v)$, to improve readability and also abbreviate the tensor-product of univariate B-Spline basis functions as $B_I(\mathbf{u}) := N_{i,p}(u)N_{j,q}(v)$, which allows us to express the geometry mapping (8) in a more compact way as follows

$$\mathbf{x} = \mathbf{F}_h(\mathbf{u}) = \sum_{I=1}^{n\times m} \mathbf{c}_I B_I(\mathbf{u}), \quad \mathbf{u} \in \hat{\Omega}. \tag{10}$$

Here, the global index $I := j \times (n-1) + i$ is computed from the local indices of univariate basis functions. Since the forward mapping $\mathbf{F}_h$ is assumed to be bijective, the inverse mapping

$$\mathbf{u} = \mathbf{F}_h^{-1}(\mathbf{x}), \quad \mathbf{x} \in \Omega_h. \tag{11}$$

from the physical domain Ω_h onto the parameter domain $\hat{\Omega}$ exists. Since it pulls any physical coordinate back into the parameter space $\hat{\Omega}$ it is termed the 'pull-back' mapping.

To further simplify the notation we drop subscript 'h' whenever it is clear from the context whether the exact object or its approximation is meant.

2.4. Geometry pre-processor

A geometry pre-processor has been developed in [11, 12, 13] that generates planar mappings of the form (8) from a point cloud $\{\mathbf{p}_a\}$ that represents the boundary profile $\partial\Omega$. In practice, this point cloud is generated by an in-house tool for designing screw machine rotor profiles provided by the Chair of Fluidics at TU Dortmund University but other point clouds are also accepted.

The geometry pre-processor uses elliptic grid generation techniques and extends them into an *isogeometric* variational framework, thereby especially exploiting the higher continuity, namely, C^{p-1} of B-Spline basis functions of degree p. The main algorithmic steps are as follows:

1. Generate individual parametrizations of the four boundary segments by fitting univariate B-Splines to the discrete points ensuring that the curves meet at the respective corners

$$\{\mathbf{p}_a^N\} \mapsto \mathbf{F}^N(u), \qquad \{\mathbf{p}_a^S\} \mapsto \mathbf{F}^S(u), \tag{12}$$

$$\{\mathbf{p}_a^E\} \mapsto \mathbf{F}^E(v), \qquad \{\mathbf{p}_a^W\} \mapsto \mathbf{F}^W(v). \tag{13}$$

IOP Conf. Series: Materials Science and Engineering **425** (2018) 012032 doi:10.1088/1757-899X/425/1/012032

2. Reparameterize opposite boundary segments (if necessary) to improve grid orthogonality.
3. Merge knot vectors of opposite boundary segments to obtained unified B-Spline bases.
4. Solve the vector-valued nonlinear problem [14] on the computational domain $\Omega_h \approx \Omega$

$$\mathbf{g}_{vv}\mathbf{x}_{uu} - 2\mathbf{g}_{uv}\mathbf{x}_{uv} + \mathbf{g}_{uu}\mathbf{x}_{vv} = 0, \quad \text{s.t.} \quad \mathbf{x}|_{\partial\Omega_h} = \partial\Omega \tag{14}$$

   with $\mathbf{g}_{uv} := \mathbf{F}_u \cdot \mathbf{F}_v$ using a variational approach to directly obtain the 'inner' control points of (8) as unknowns and imposing the boundary condition on the $\mathbf{c}_{i,j}$'s at the boundary.

It can be shown [15] that the exact solution to (14) yields indeed a bijective mapping. It is assumed that this property carries over to the numerical solution at least almost everywhere.

The efficient and robust implementation of the isogeometric mesh pre-processor involves many nifty details, which are described in [11, 12, 13] together with an extension of the underlying algorithm for generating volumetric parametrizations of twin-screw machine geometries from a sequence of planar ones sampled at different locations along the rotation axes. In practice, the input point cloud typically represents the entire geometry and may not even satisfy any ordering. It therefore needs to be segmented into point clouds that can be associated with the different segments of the different patches and possibly further pre-processed (e.g., removal of duplicate points) before the above algorithm can be applied to the individual patches/segments.

2.5. Isogeometric variational formulation

As stated in the previous section, the PDE problem (14) is solved by a variational approach, which is derived in a similar way as in *parametric* finite elements. In short, the residual of the strong form of the PDE problem is multiplied by a test function w from the test space W, integrated over the physical domain Ω and required to vanish for all admissible test functions $w \in W$. For illustration purposes, consider the generic linear PDE problem

$$\mathcal{L}\varphi = f \quad \text{in } \Omega, \tag{15}$$

which is cast into the variational formulation: Find a solution $\varphi \in V$ such that

$$\int_\Omega w\,[f - \mathcal{L}\varphi]\,\mathrm{d}\mathbf{x} = 0 \quad \text{for all } w \in W. \tag{16}$$

Here, V denotes the space of admissible solutions, which needs to account for any Dirichlet boundary conditions that are imposed on the differential equation (15). Neumann boundary conditions are easily implemented by performing integration by parts and overriding the corresponding expressions in the arising boundary integral terms. Next, trial and test spaces V and W are approximated by the finite dimensional polynomial space $\mathcal{S}^{p,q}(U,V)$, which is spanned by the tensor-product B-Spline basis functions $\{B_I\}_{I=1}^{n\times m}$ constructed with the Cox-de-Boor recursion formula (6)–(7) based on the knot vectors U and V, respectively.

The major difference between classical FEM and IGA is the use of the *same* basis for constructing the geometry mapping (8) and approximating the solution to problem (16), i.e.

$$\varphi_h(\mathbf{x}) = \sum_{I=1}^{n\times m} \varphi_I B_I \circ \mathbf{F}^{-1}(\mathbf{x}), \quad \mathbf{x} \in \Omega_h. \tag{17}$$

Since the B_I's are defined on $\hat{\Omega}$, the physical coordinates need to be pulled back. The solution coefficients φ_I are obtained by inserting (17) into (16), replacing w by the $n \times m$ basis functions and solving the resulting linear system of the form $A_h\varphi_h = f_h$ for φ_h. Due to the properties of B-Spline basis functions, the so-defined solution coefficients are *not* interpolatory, that is, $\varphi_h(u_i, v_j) \neq \varphi_{i,j}$, which is the case for classical FEM with nodal Lagrange basis functions.

In what follows we apply this generic approach to derive the isogeometric variational formulations for the equations of gas dynamics. The derivation of the variational formulation for the governing equation of the geometry pre-processor (14) is described in [11, 12, 13].

Compressible Euler equations. Consider the divergence form of the problem at hand

$$\frac{\partial U}{\partial t} + \nabla \cdot \mathcal{F}(U) = 0 \quad \text{in } \Omega^{\text{fluid}} \times [0, t_{\text{end}}], \tag{18}$$

where $U : \mathbb{R}^d \to \mathbb{R}^{d+2}$ denotes the state vector of conservative variables

$$U = \left[U_1, \ldots, U_{d+2}\right]^\top = \left[\rho, \rho\mathbf{v}, \rho E\right]^\top, \tag{19}$$

and $\mathcal{F} : \mathbb{R}^{d+2} \to \mathbb{R}^{(d+2)\times d}$ stands for the tensor of inviscid fluxes

$$\mathcal{F} = \begin{bmatrix} F_1^1 & \cdots & F_1^d \\ \vdots & \ddots & \vdots \\ F_{d+2}^1 & \cdots & F_{d+2}^d \end{bmatrix} = \begin{bmatrix} \rho\mathbf{v} \\ \rho\mathbf{v}\otimes\mathbf{v} + p\mathcal{I} \\ \rho E\mathbf{v} + p\mathbf{v} \end{bmatrix} \tag{20}$$

with mass density ρ, velocity $\mathbf{v}$, total energy E, and $\mathcal{I}$ denoting the d-dimensional identity tensor. For an ideal polytropic gas, the pressure p is given by the equation of state

$$p = (\gamma - 1)\rho e, \tag{21}$$

where $\gamma = c_p/c_v$ denotes the heat capacity ratio, which equals $\gamma = 1.4$ for dry air and

$$\rho e = \left(\rho E - 0.5\rho|\mathbf{v}|^2\right) \tag{22}$$

is the internal energy per unit volume, from which temperature T can be computed as follows

$$T = \frac{e}{c_v}. \tag{23}$$

The governing equations are equipped with initial conditions prescribed at time $t = 0$

$$U(\mathbf{x}, 0) = U_0(\mathbf{x}) \ \text{ in } \Omega^{\text{fluid}}, \tag{24}$$

and boundary conditions of Dirichlet and Neumann type, respectively

$$U = \mathcal{G}(U, U_\infty) \quad \text{on } \Gamma_D^{\text{fluid}}, \tag{25}$$

$$\mathbf{n} \cdot \mathcal{F} = \mathcal{F}_n(U, U_\infty) \quad \text{on } \Gamma_N^{\text{fluid}}. \tag{26}$$

Here, $\mathbf{n}$ is the outward unit normal vector and U_∞ denotes the vector of 'free stream' solution values, which are calculated and imposed as described in detail in [16].

After applying integration by parts to incorporate the Neumann-type boundary fluxes, the variational formulation of the problem at hand reads: Find $U(t) \in V$ such that

$$\int_{\Omega^{\text{fluid}}} w\frac{dU}{dt} - \nabla w \cdot \mathcal{F}(U)\,\mathrm{d}\mathbf{x} + \int_{\Gamma_N^{\text{fluid}}} w\mathcal{F}_n(U, U_\infty)\,\mathrm{d}s = 0 \tag{27}$$

for all test functions $w \in W$. Let us approximate the state vector and the inviscid fluxes as

$$U_h(\mathbf{x}, t) = \sum_{I=1}^{n\times m} U_I(t) B_I \circ \mathbf{F}^{-1}(\mathbf{x}), \tag{28}$$

$$\mathcal{F}_h(\mathbf{x}, t) = \sum_{I=1}^{n\times m} \mathcal{F}_I(t) B_I \circ \mathbf{F}^{-1}(\mathbf{x}), \tag{29}$$

where $\mathcal{F}_I(t) := \mathcal{F}(U_I(t))$ denotes the value of (20) evaluated at the I-th solution coefficient $U_I(t)$ at time t. This approach is known as Fletcher's group formulation [17] and it is frequently used in finite element methods for conservation laws to approximate the first-order flux function.

To further simplify the notation, let us define the consistent mass matrix $M_C := \{m_{IJ}\}$ and the discretized divergence operator $\mathbf{C} := \{\mathbf{c}_{IJ}\}$ whose entries are given by

$$m_{IJ} = \int_{\hat{\Omega}} B_I(\mathbf{u}) B_J(\mathbf{u}) \, |\det \mathbf{DF}(\mathbf{u})| \, \mathrm{d}\mathbf{u}, \tag{30}$$

$$\mathbf{c}_{IJ} = \int_{\hat{\Omega}} \left(\nabla B_I(\mathbf{u}) \mathbf{DF}(\mathbf{u})^{-1}\right) B_J(\mathbf{u}) \, |\det \mathbf{DF}(\mathbf{u})| \, \mathrm{d}\mathbf{u}. \tag{31}$$

The above expressions naturally arise from transforming all integrals over the physical domain Ω^{fluid} into integrals over the parametric domain $\hat{\Omega}$. Here, $\mathbf{DF}$ denotes the Jacobian matrix of the geometry mapping (8) and $\det \mathbf{DF}$ is its determinant, which is strictly non-zero as long as $\mathbf{F}$ is bijective. In contrast to parametric finite elements, where this technique is applied individually for each element ($\mathbf{F}$ is defined at the element level), IGA employs a patch-wise global mapping. In a practical implementation, the integral terms are evaluated by applying numerical quadrature, e.g., a sufficiently accurate Gauss quadrature rule, to each 'virtual' element $\omega_{i,j} = [u_i, u_{i+1}] \times [v_j, v_{j+1}]$, where u_i and v_j denote the knots introduced in Section 2.3.

At the end of the day, the semi-discrete form of the Euler equations reads [18]

$$M_C \frac{dU_h}{dt} - \sum_{b=1}^{d} C^b F^b(U_h) + S(U_h) = 0, \tag{32}$$

where superscript $b = 1, \ldots, d$ refers to the b-th spatial component of the discrete divergence operator $\mathbf{C}$ and the tensor of inviscid fluxes $\mathcal{F}$, respectively. Here, $S(U_h)$ accounts for the contribution of the boundary fluxes; see publication [16] for more details.

Equation (32) is discretized in time by an explicit strong stability preserving Runge-Kutta (SSP-RK) method of order three [19]. The resulting scheme is known to produce non-physical oscillations in the vicinity of shocks and steep gradients, which can lead to negative density and pressure values and, hence, to the complete failure of the simulation. This shortcoming is cured by applying algebraic flux correction (AFC), which has been originally developed for low-order FEM [16] and has been extended to isogeometric schemes recently [18, 20]. Preliminary results are given in Section 4.1 to demonstrate the general applicability of our isogeometric approach.

A word on parametric multi-physics coupling. Even though the numerical results presented in this publication are restricted to single-phase compressible flows, we give a brief conceptual outlook on how IGA simplifies the exchange of data between multiple disciplines. Let

$$q_h := (\nabla T_h(\mathbf{x}, t)) \cdot \mathbf{n}^{\text{fluid}} = \sum_{I=1}^{n \times m} T_I(t) \nabla \left(B_I \circ \mathbf{F}^{-1}(\mathbf{x})\right) \cdot \mathbf{n}^{\text{fluid}} \tag{33}$$

be the numerically computed temperature gradient in outward normal direction at the boundary segment Γ_N^{fluid} of the fluid domain Ω^{fluid} at time t and assume that its values should be prescribed as flux boundary condition $(\kappa \nabla T) \cdot \mathbf{n}^{\text{heat}} = -\kappa q$ at segment Γ_N^{heat} for the heat equation

$$\rho c_p \frac{dT}{dt} - \nabla \cdot (\kappa \nabla T) = \dot{q}_V \quad \text{in } \Omega^{\text{heat}} \times [0, t_{\text{end}}], \tag{34}$$

where $\dot{q}_V$ the heat energy generated per unit volume per unit time and κ the thermal conductivity of the material. The variational form reads: Find $T(t) \in V$ such that

$$\int_{\Omega^{\text{heat}}} w \left[\rho c_p \frac{dT}{dt} - \dot{q}_V\right] + \nabla w \cdot (\kappa \nabla T) \, \mathrm{d}\mathbf{x} + \int_{\Gamma_N^{\text{heat}}} w \kappa q \, \mathrm{d}s = 0 \tag{35}$$

for all test functions $w \in W$. After the discretization of the temperature variable similarly to (17) and substitution of (33) into the boundary integral we obtain the semi-discrete system

$$M_C \frac{dT_h}{dt} + KT_h + S(T_h) = 0 \tag{36}$$

where $K = \{k_{IJ}\}$ denotes the stiffness matrix whose entries are given by

$$k_{IJ} = \int_{\hat{\Omega}} \left(\nabla B_I(\mathbf{u})\mathbf{DF}(\mathbf{u})^{-1}\right) \cdot \left(\nabla B_J(\mathbf{u})\mathbf{DF}(\mathbf{u})^{-1}\right) |\det \mathbf{DF}(\mathbf{u})| \, \mathrm{du} \tag{37}$$

and $S(T_h)$ accounts for the boundary contribution. As before, the integral terms are evaluated by transforming them to the parametric space and applying numerical quadrature. Let $\mathbf{F}^{\text{fluid}} : [0,1] \to \Gamma_N^{\text{fluid}}$ and $\mathbf{F}^{\text{heat}} : [0,1] \to \Gamma_N^{\text{heat}}$ be two bijective parametrizations of the common interface between the fluid and heat domain. Since $\Gamma_N^{\text{heat}} \equiv \Gamma_N^{\text{fluid}}$ in the isogeometric multi-patch geometry model, the parametric values are related by the following expression

$$\mathbf{F}^{\text{fluid}}(t^{\text{fluid}}) = \mathbf{F}^{\text{heat}}(t^{\text{heat}}). \tag{38}$$

Hence, the temperature gradient stemming from the fluid solution can be directly evaluated in the boundary integral term of the heat equation without extra interpolation steps, i.e.

$$s_I = \int_0^1 B_I(t)\, \kappa\, q_h(\mathbf{F}^{\text{heat}}(t)) \, |\det \mathbf{Df}^{\text{heat}}| \, dt. \tag{39}$$

Equation (38) can also be used to apply a similar parametric coupling approach when imposing the wall temperature of the rotor as Dirichlet boundary condition to the fluid solver.

3. Isogeometric mesh generator for ANSYS CFX

The complete realization of the isogeometric modeling and analysis pipeline as described in the previous section requires that all computational building blocks, i.e., the geometry pre-processor and the flow solver, are implemented and operational at the maturity level of commercial tools in order to enable realistic simulations of twin-screw machines. To validate the quality of the isogeometric geometry pre-processor, we have realized an isogeometric mesh generator, which converts the parametric master geometry model into a block-structured mesh for ANSYS CFX^{TM} [3]. The block structure immediately follows from the multi-patch topology; cf. Figure 2.

The computational mesh for a single block, say, k is created by constructing a tensor-product mesh with equidistant mesh widths Δu and Δv in the parameter domain $\hat{\Omega}_k$ and transforming it with the corresponding forward mapping $\mathbf{F}_k(u,v)$ into the physical domain Ω_k. The resulting planar quadrilateral mesh is extruded in the z-direction to obtain the mesh shown in Figure 6.

Since the mapping is assumed to be bijective, an admissible tensor-product mesh in $\hat{\Omega}$ will yield an admissible mesh in Ω_k with the same connectivity pattern and curved mesh lines. The term admissible means that no twisted mesh cells occur but it does not account for mesh quality.

For use in ANSYS CFX^{TM}, curved grid lines are replaced by straight lines after which the admissibility of the mesh needs to be re-assessed by checking that the (oriented) element volume is strictly positive for each hexahedra (cf. Eq. (11.28) in [21]). In our experience, non-admissible elements rarely occurr for highly curved elements with extreme aspect ratios. A simple cure is to split the respective element into two in the parameter domain and map the newly introduced vectices to their positions in the physical domain by virtue of the master geometry mapping.

Despite all effort that is invested into the creation of the mapping (8) to ensure that vertical and horizontal grid lines intersect each other 'orthogonally' and are perpendicular to the boundary by applying boundary re-parametrization, no attention is paid to additional mesh

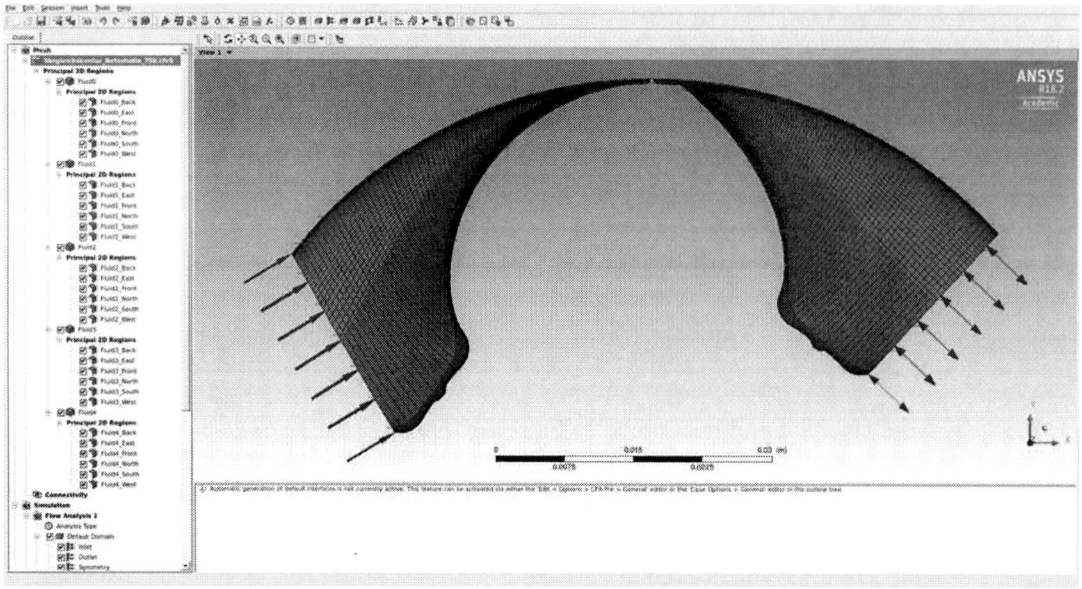

Figure 6. Block-structured hexahedral mesh for fluid domain (5-patch blue region) in CFX5-Pre. Adaptive mesh refinement towards boundary layers is imposed via pre-parametrization.

quality criteria like the accurate resolution of boundary layers. Figure 5 demonstrates that grid lines that run horizontally in the $u - v$ plane, and thus have constant distance to the domain boundary $\hat{\Gamma}$, can have varying distance to the boundary Γ in the $x - y$ plane.

Even though it is possible to augment the homogeneous problem (14) with a source term to modulate the mesh density locally, there are no clear guidelines on how to choose the source term in order to ensure a fixed distance between any two parallel grid lines or the first grid line and the wall. Moreover, this approach would require the numerical solution of the nonlinear PDE problem (14) for each computational grid, which is computationally expensive.

We therefore introduced an additional re-parametrization step, cf. Figure 7, which does not influence the shape of the geometry but only the distribution of grid lines. Let $\mathbf{G}(s,t) = (u,v)$

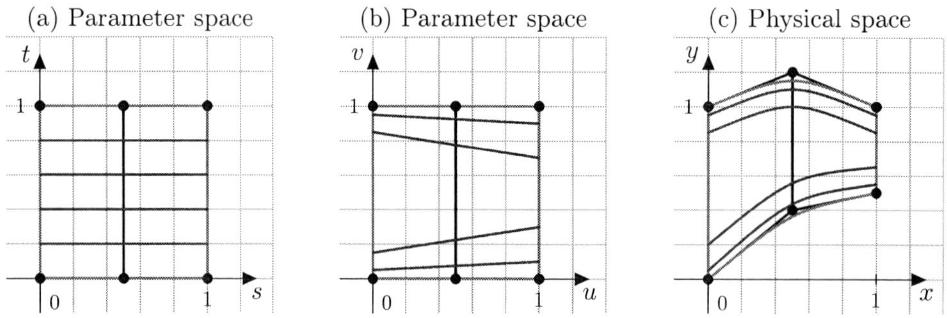

Figure 7. Mapping between (a)+(b) parameter and (c) physical domains. Black lines indicate the control net with control points $\mathbf{c}_{i,j}$ given by dots. Boundary segments are in red. Grid lines with parameter values varying in u-direction are sketched in blue for illustration purposes.

be a bijective mapping from $\tilde{\Omega} = [0,1]^2$ to $\hat{\Omega} = [0,1]^2$ that is defined as transfinite interpolation

$$\mathbf{G}(s,t) = \begin{pmatrix} (1-t)\mathbf{G}^W(s) + t\mathbf{G}^E(s) \\ (1-s)\mathbf{G}^S(t) + s\mathbf{G}^N(t) \end{pmatrix}, \quad 0 \le s,t \le 1, \tag{40}$$

where $\mathbf{G}^s$ are univariate mappings defined along the four boundary segments of the unit square.

The resulting mapping $(x,y) = \mathbf{F} \circ \mathbf{G}(s,t)$ is bijective as long as both $\mathbf{F}(u,v)$ and $\mathbf{G}(s,t)$ exhibit this property. As an example, the mesh grading function $(0 \le \gamma \le 1)$

$$\mathbf{G}_{LR}(\gamma) = 0.5 + \frac{\operatorname{atan}(\alpha(\gamma - 0.5))}{2\operatorname{atan}(\alpha/2)} \tag{41}$$

with $\alpha > 0$ being a user-definable parameter, clusters the grid lines towards the two ends of the unit interval, whereas mesh grading towards one end of the interval is achieved by

$$\mathbf{G}_L(\gamma) = 1 + \frac{\tanh((\gamma-1)\beta/2)}{\tanh(\beta/2)}, \tag{42}$$

$$\mathbf{G}_R(\gamma) = 1 - \frac{\tanh((0-\gamma)\beta/2)}{\tanh(\beta/2)}, \tag{43}$$

with $\beta > 0$. The computational mesh depicted in Figure 6 has been generated using the above functions and piece-wise combinations thereof, e.g., to achieve a moderate grading of the horizontal mesh width Δu_i over the entire length of the inlet section (patch numbers 2 and 3) with a rapid reduction close to the 'step' profile in the gap section; cf. Figure 9.

The quality of the resulting mesh has been assessed using ANSYS ICEM CFD^{TM}, which reports a mean quality value of 0.9561; cf. Figure 8. Numerical simulation results obtained on this mesh, which consists of 42,808 hexahedral elements are presented in Section 4.2.

4. Numerical Results

This section presents preliminary results from single-phase compressible flow simulations computed by our isogeometric solver that has been implemented in the open-source G+Smo (Geometry + Simulation Modules) library [6] and by the commercial tool ANSYS CFX^{TM} [3].

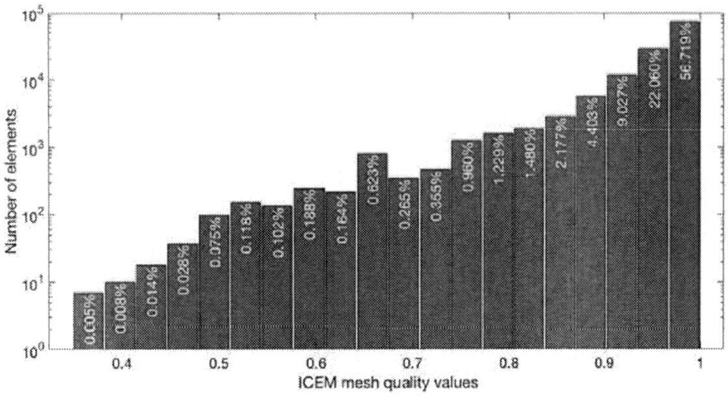

Figure 8. ANSYS ICEM^{TM} mesh quality criterion: Min = 0.3489, max = 1, mean = 0.9561.

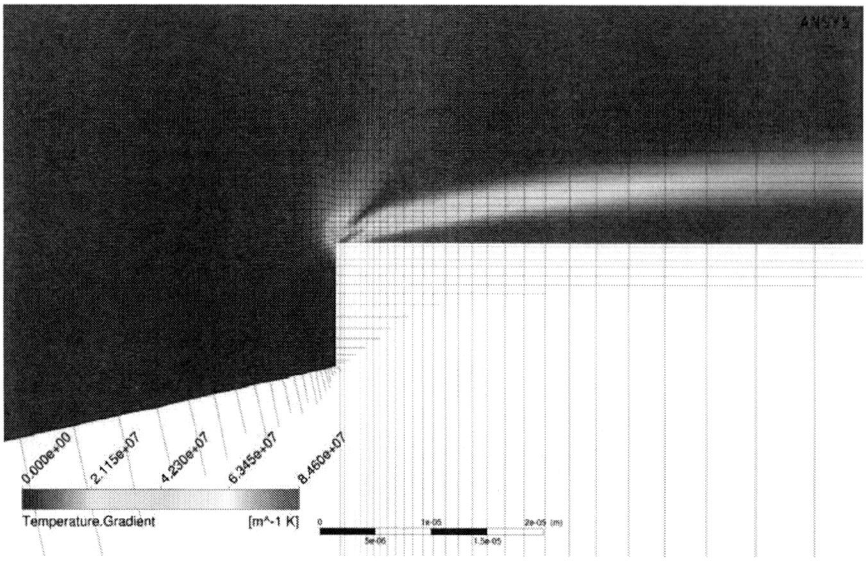

Figure 9. Temperature gradient in normal direction to the wall.

4.1. G+Smo results

The isogeometric flow solver has been tested for a variation of Sod's shock tube problem [7], which is a classical benchmark in computational gas dynamics. Consider a closed system (free-slip boundary conditions are imposed on the entire boundary Γ) with a fluid that is initially at rest and separated into two regions with $\rho_L = 1.0$, $p_L = 1.0$ and $\rho_R = 0.125$, $p_R = 0.1$ by a diaphragm located in the middle of the domain. Its abrupt removal causes the formation of a rarefaction wave, a contact discontinuity, and a shock wave. This setup has been considered on the VKI U-bend geometry [22], which has been modeled by 66×66 equidistantly distributed bi-quadratic tensor-product B-Splines. The density profile at time $t_{\text{end}} = 0.231$ that is shown in Figure 10 has been produced with our isogeometric approach, using the AFC technology for stabilization and a third-order explicit SSP-RK scheme with time-step size $\Delta t = 0.001$.

The black contour lines indicate the location of the three characteristic wave types, whereas 'parametric grid lines' (image of $\mathbf{F}(u, v)$ with one of the two parameters being kept constant) are given in white. The adopted AFC stabilization suppresses the generation of spurious oscillations reliably and furthermore ensures that the numerical solution values remain bounded by the physical upper and lower bounds, i.e. $0 < \rho \leq 1$, $0 < p$, and $0 < \rho E$.

It should be furthermore noted that the C^1 continuous parameterization of the curved boundary by a bi-quadratic B-Spline function yields a unique definition of the outward unit normal vector $\mathbf{n}(\mathbf{x}) = \mathbf{n}(\mathbf{F}(\mathbf{u}))$ in every point on the boundary Γ. Moreover, $\mathbf{n}$ is a continuous function of the boundary parameter value $\mathbf{u}$. As a consequence, the approximate solution is free of 'numerical artifacts', which often occur for polygonal boundary representations in P_1/Q_1 finite elements, where the C^0 'kinks' between two consecutive boundary segments acts as microscopic compression or expansion corners and, moreover, give rise to undetermined normal vectors in the vertices. In practice, this problem is often overcome by averaging the local normal vectors of the adjacent boundary segments but, still, the so-defined normal vector is not a continuous function of the boundary parameterization, which is the case for IGA.

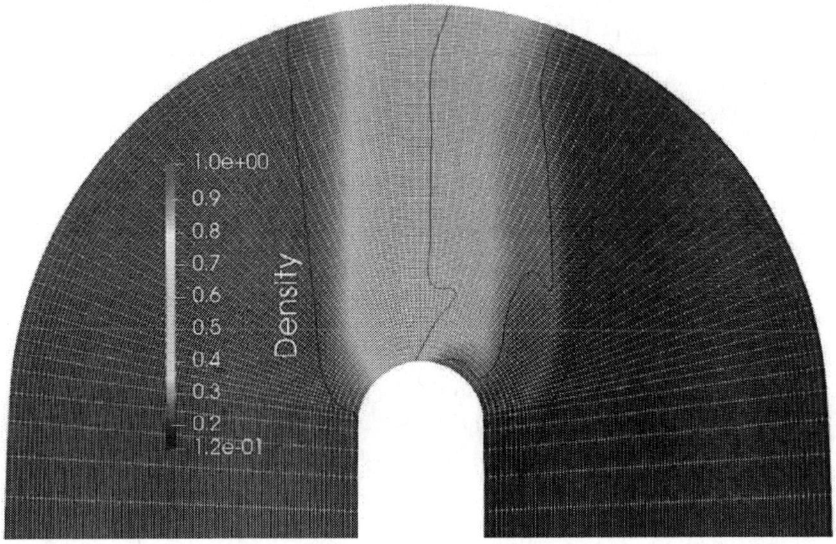

Figure 10. G+Smo solution to Sod's shock tube problem [7] computed on the VKI U-bend [22] geometry with 66 × 66 bi-quadratic tensor-product B-Splines and third-order SSP-RK scheme.

4.2. ANSYS CFX results
As a second test case, we have performed a fluid flow simulation for the test rig with asymmetric SRM rotor profile (cf. Figure 2) with ANSYS CFX^{TM} [3]. The computational grid (cf. Figure 6) consists of 42,808 block-structured elements and was generated from the parametric master geometry model with the aid of our isogeometric mesh generator. Further details of the simulation are given in Table 1. Contour plots (51 contour levels) for the pressure, Mach number, and temperature distribution in the gap region are presented in Figure 11 (a)-(c). A close-up view of the temperature gradient is depicted in Figure 9 showing a steep increase by orders of magnitudes at the corner singularity. This might be attributed to the numerically computed first-order approximation of the gradient with extremely small mesh widths in normal direction. Work on improved gradient reconstruction schemes based on high-order Splines is in progress.

The flow becomes supersonic ($Ma_{\max} = 1.775$) behind the smallest part of the flow passage, which goes a long with a significant drop in temperature by about 120 °C relative to the prescribed temperature of 25°C prescribed at the inlet and outlet and the solid walls.

5. Conclusions and Outlook
In this paper we have sketched the overall design principles of our isogeometric modeling and analysis pipeline for rotary twin-screw machines that is currently being developed by an international consortium consisting of research groups in Germany and the Netherlands. The central idea is to link all computational tools to the same parametric master geometry model that can be used in an isogeometric analysis framework or converted to a highly accurate computational mesh for ANSYS CFX^{TM} with the aid of our isogeometric mesh generator.

Preliminary results have demonstrated the applicability of the developed geometry pre-processor and the isogeometric mesh generator. Moreover, the isogeometric flow solver has been successfully applied to inviscid compressible flows, which have been studied in curved geometries. The extension to viscous turbulent flows and more complex multi-patch topologies is ongoing work. Current research focusses on the two-way coupling of the fluid flow solvers with an isogeometric thermo-dynamical simulation of the expansion of the screw machine rotors [23].

Basic Settings and Fluid Model	
Material	Air Ideal Gas
Reference Pressure	0 [bar]
Buoyancy Model	Non Buoyant
Heat Transfer	Total Energy (Incl. Viscous Work Term)
Turbulence	Shear Stress Transport (medium intensity 5%)
Wall Functions	Automatic
Static Pressure	1 [bar]
Static Temperature	25 [°C]
Inlet Boundary Condition	
Flow Regime	Subsonic
Mass flux rate	$0.04gs^{-1}$
Heat Transfer	Static Temperature 25 [°C]
Outlet Boundary Condition	
Flow Regime	Subsonic
Opening Pressure	1 [bar]
Opening Temperature	25 [°C]
Wall Boundary Condition	
Wall	No Slip, Smooth
Fixed Temperature	25 [°C]

Table 1. Configuration of the ANSYS CFX™ flow solver for the test rig simulations.

Acknowledgement

The authors would like to thank the anonymous reviewers for their valuable feedback. Support from Matthias Utri from TU Dortmund University by providing the test rig data and assisting in the ANSYS CFX™ simulations is gratefully acknowledged. We finally thank Andrzej Jaeschke from TU Lodz for performing the isogeometric flow simulations with G+Smo.

This project (MOTOR) has received funding from the European Unions Horizon 2020 research and innovation programme under grant agreement No 678727.

(a) Pressure

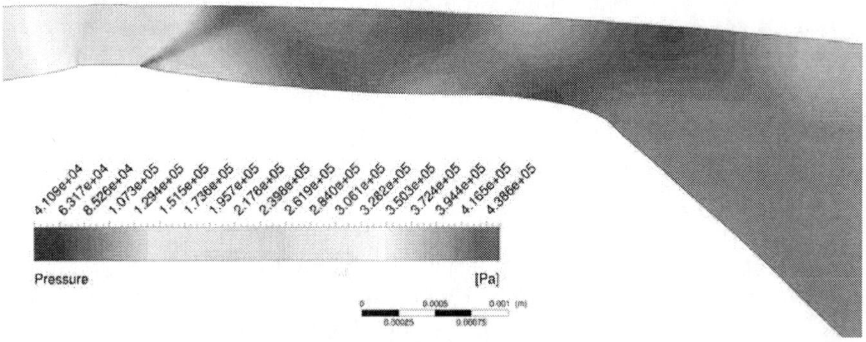

(b) Mach number

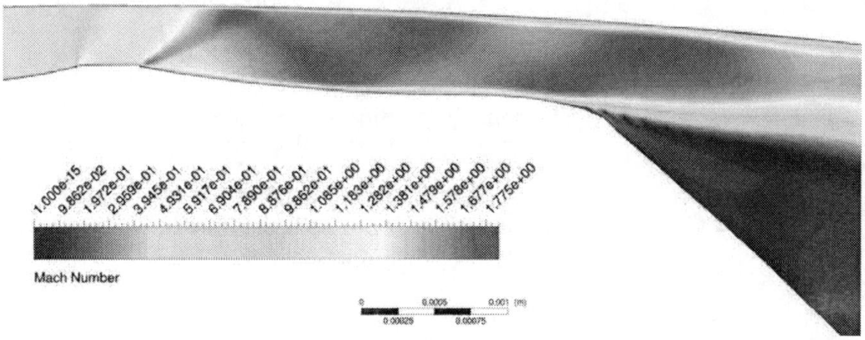

(c) Temperature

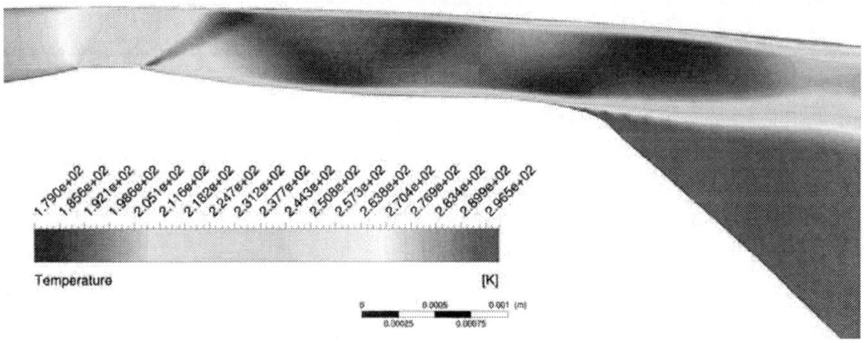

Figure 11. ANSYS CFX^{TM} solution computed on a block-structured hexahedral mesh: (a) pressure, (b) Mach number, and (c) temperature profile in the narrow gap of the test rig.

IOP Conf. Series: Materials Science and Engineering **425** (2018) 012032 doi:10.1088/1757-899X/425/1/012032

Notation

Ω	Exact physical domain (= screw machine geometry)
$\Gamma = \partial\Omega$	Boundary of the physical domain
Ω_h	Approximate computational physical domain
$\Gamma_h = \partial\Omega_h$	Boundary of the computational domain
$\hat{\Omega}$	Parameter domain of the geometry mapping
$\tilde{\Omega}$	Parameter domain of the re-parametrization mapping
$\mathbf{x} := (x, y)$	Coordinates in physical domains Ω and Ω_h
$\mathbf{u} := (u, v)$	Coordinates in parameter domain $\hat{\Omega}$
$\mathbf{s} := (s, t)$	Coordinates in parameter domain $\tilde{\Omega}$
U, V	Knot vector in horizontal and vertical direction, resp.
u_i, v_j	Knot entry of the open knot vector U and V, resp.
$N_{i,p}$	i-th B-Spline basis function of polynomial order p
B_I	I-th tensor-product B-Spline basis function
$\mathbf{F}$	Geometry mapping from $\hat{\Omega}$ to $\Omega_{(h)}$
$\mathbf{G}$	Reparameterization mapping from $\tilde{\Omega}$ to $\hat{\Omega}$
$\mathbf{c}_{i,j}$	Control points of the geometry mapping
$\mathbf{p}_a$	Discrete point in physical coordinates
Subscripts	
p, q	Polynomial order of B-Spline basis functions
i, j	Index of univariate B-Spline basis functions
I, J	Global index of basis functions
k, l	Patch number
u, v	First derivative in the parameter space
x, y	First derivative in the physical space
Superscripts	
$s \in (S, N, W, E)$	Face of a planar patch

IOP Conf. Series: Materials Science and Engineering **425** (2018) 012032 doi:10.1088/1757-899X/425/1/012032

Bibliography

[1] TwinMesh^{TM} `https://www.twinmesh.com`
[2] SCORG^{TM} `http://pdmanalysis.co.uk`
[3] ANSYS CFX `https://www.ansys.com/products/fluids/ansys-cfx`
[4] STAR-CCM+ `https://mdx.plm.automation.siemens.com/star-ccm-plus`
[5] Hughes T J, Cottrell J A and Bazilevs Y 2005 *Computer Methods in Applied Mechanics and Engineering* **194** 4135–4195
[6] Jüttler B, Langer U, Mantzaflaris A, Moore S E and Zulehner W 2014 *Proceedings in Applied Mathematics and Mechanics* **14** 961–962
[7] Sod G 1978 *Journal of Computational Physics* **27** 1 – 31
[8] Sachs R 2002 *Experimentelle Untersuchung von Gasströmungen in Schraubenmaschinen* Ph.D. thesis TU Dortmund University, Faculty of Mechanical Engineering
[9] Utri M, Höckenkamp S and Brümmer A 2018 *Proceedings of the 10th International Conference on Screw Machines (ICSM 2018)* (Dortmund, Germany)
[10] de Boor C 1971 Subroutine package for calculating with B-splines Tech. Rep. LA-4728-MS Los Alamos Scient. Lab.
[11] Hinz J, Möller M and Vuik C 2018 *Proceedings of the 10th International Conference on Screw Machines (ICSM 2018)* (Dortmund, Germany)
[12] Hinz J, Möller M and Vuik C 2018 *Computer Aided Geometric Design* **65** 48–75
[13] Hinz J, Möller M and Vuik C 2018 *Proceedings of the 7th European Conference on Computational Fluid Dynamics (ECFD 2018)* (Glasgow, United Kingdom)
[14] Thompson J F, Soni B K and Weatherill N P 1998 *Handbook of grid generation* (CRC press)
[15] Azarenok B N 2009 *Computational Mathematics and Mathematical Physics* **49** 797–809
[16] Kuzmin D, Möller M and Gurris M 2012 *Flux-Corrected Transport* (Springer) chap Algebraic flux correction II. Compressible flow problems
[17] Fletcher C 1983 *Computer Methods in Applied Mechanics and Engineering* **37** 225–243
[18] Möller M and Jaeschke A 2018 *Proceedings of the 19th International Conference on Finite Elements in Flow Problems (FEF 2017)* (Rome, Italy)
[19] Gottlieb S, Shu C and Tadmor E 2001 *SIAM Review* **43** 89–112
[20] Jaeschke A and Möller M 2018 *Proceedings of the 19th International Conference on Finite Elements in Flow Problems (FEF 2017)* (Rome, Italy)
[21] Wesseling P 2001 *Principles of Computational Fluid Dynamics* (*Springer Series in Computational Mathematics* vol 29) (Springer)
[22] Verstraete T 2016 The VKI U-bend optimization test case Tech. rep. Von Karman Institute for Fluid Dyanmics
[23] Shamanskiy A and Simeon B 2018 *Proceedings of the 10th International Conference on Screw Machines (ICSM 2018)* (Dortmund, Germany)

IOP Conf. Series: Materials Science and Engineering **425** (2018) 012033 doi:10.1088/1757-899X/425/1/012033

Fluid flow through male rotor housing clearances of dry running screw machines using dimensionless numbers

Matthias Utri, Simon Höckenkamp, Andreas Brümmer

Chair of Fluidics, TU Dortmund University
Leonhard-Euler-Str. 5, 44227 Dortmund, Germany

E-mail: matthias.utri@tu-dortmund.de, simon.hoeckenkamp@tu-dortmund.de, andreas.bruemmer@tu-dortmund.de

Abstract. The efficiency of positive displacement machines is strongly influenced by clearance flows. The prediction of the mass flow rate is a challenging task since it depends on thermodynamic and operating boundary conditions as well as the shape of the clearance path. When simulating positive displacement machines using chamber models, clearance flow is commonly estimated by using simplified approaches, such as equations for isentropic nozzle flow combined with a flow coefficient. Knowledge of these flow coefficients is important for an accurate simulation of the machine. For this reason one of the goals of the European Union funded project "MOTOR" (Multi-ObjecTive design Optimization of fluid eneRgy machines) is the determination of the flow coefficient for different clearance shapes. This paper deals with the simulation of the two-dimensional fluid flow in the housing clearance of twin-screw machines using the finite volume method for air as ideal gas. Dimensionless numbers are varied to show their individual influence on the mass flow. Results are compared with experimental data from reports in the literature and show good accordance. The primary factors affecting the results are Reynolds number and the moving boundary in the low Reynolds number region.

1. Introduction

The quality of simulations for positive displacement machines depends to a great extent on the correct prediction of clearance mass flow. Four different clearance types exist in screw machines, each with characteristic shapes and particular relative motion of the boundaries. Mass flow rate is often compared to the mass flow which would occur in the case of an isentropic and frictionless flow. For subcritical conditions, this isentropic nozzle flow can be calculated by

$$\dot{m}_{th} = \rho_c \cdot A_c \cdot c_c = p_{hp} \cdot A_c \cdot \sqrt{\frac{2 \cdot \kappa}{(\kappa - 1) \cdot R_s \cdot T_{hp}} \cdot \left((\Pi_c)^{\frac{2}{\kappa}} - (\Pi_c)^{\frac{\kappa+1}{\kappa}} \right)}, \quad \frac{p_{lp}}{p_{hp}} = \Pi_c > \Pi_{crit} \qquad (1)$$

for a flow from a high pressure volume to a low pressure volume [1]. For pressure ratios below the critical, the fluid reaches the speed of sound in the narrowest flow cross section and mass flow is independent of the low pressure:

$$\dot{m}_{th} = \rho_c \cdot A_c \cdot a_c = \rho_c \cdot A_c \cdot \sqrt{\kappa \cdot R_s \cdot T_c} = p_{hp} \cdot A_c \cdot \sqrt{\frac{2 \cdot \kappa}{(\kappa + 1) \cdot R_s \cdot T_{hp}} \cdot \left(\frac{2}{\kappa + 1} \right)^{\frac{1}{\kappa - 1}}}, \quad \Pi_c < \Pi_{crit} \qquad (2)$$

The equations assume the low pressure to be present in the minimum area of the flow cross section. The flow coefficient α relates the real mass flow, which can be determined by simulation or experiment, to this isentropic nozzle flow:

$$\alpha = \frac{\dot{m}_{real}}{\dot{m}_{th}} \tag{3}$$

A different definition of flow coefficient is used for mass flow measurement by means of orifices [2] and for the estimation of clearance mass flow in seals [3]. Several studies examine the clearance flow in positive displacement machines to present values of the flow coefficient and to give guidance in modelling these flows [4, 5, 6]. The flow coefficient for the plane-parallel front clearance has been examined by the authors [7]. Results reveal, that the mass flow depends mainly on clearance height and length, Reynolds number and circumferential speed of the rotors. The flow at the front clearance is highly influenced by wall friction due to the long flow path length and constant clearance gap, as well as inflow losses, especially for higher Reynolds numbers. The flow coefficient reaches values between 0.01 and 0.8 depending on the boundary conditions. In contrast, this paper investigates the flow through a male rotor housing clearance, where the smallest clearance height only occurs in a single cross section at the tip of the rotor lobe as shown in figure 1. Due to the characteristic shape of a converging and diverging flow path, flow coefficients are expected to vary considerable from those of the front clearance flow.

Fluid flow is simulated with Ansys CFX© software which uses the finite volume method. In order to validate the accuracy of the simulation, results are shown for a housing clearance contour which was examined in experiments [8]. Afterwards, characteristic numbers are introduced to generalise the dependencies of the flow coefficient. The goal of this study is to quantify the flow coefficient for a typical male rotor housing clearance and to identify the main influencing parameters by varying single characteristic numbers. The results will be used by the authors in the future to increase the accuracy of chamber model simulation.

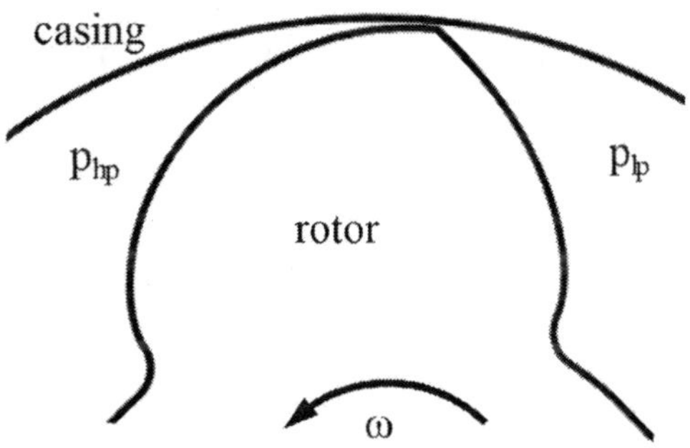

Figure 1. Representative male rotor housing clearance in 2D

2. Experimental validation of clearance simulation

Sachs and Kauder investigated different housing clearance contours in experiments [8, 9]. He used the Schlieren technique in an air test rig to visualise density gradients in the compressible flow, figure 2. Furthermore, mass flow rate was measured using a vortex flowmeter to determine flow coefficients. The clearance contour discussed in this section is without moving boundaries and has a minimum clearance gap of 0.4 mm. The tip of the male rotor shows a sealing border with sharp edges on each side, shown in figure 3. In this region the minimum clearance gap remains almost (due to the curved housing) constant over a length of 0.43 mm. The experiments were performed using air at 20°C at the clearance inlet. The low pressure conditions remained constant at 10^5 Pa and the high pressure conditions could be varied in order to vary the pressure ratio. Simulation results were reported by Kauder and Stratmann [10]. They tried to reproduce the Schlieren pictures by using the CFD-code FLUENT© and investigated different turbulence models, where the walls were treated as hydraulically smooth and adiabatic. They came to the conclusion that the Spalart Allmaras turbulence model delivers the best results concerning the flow picture, where the pressure ratios under investigation were supercritical.

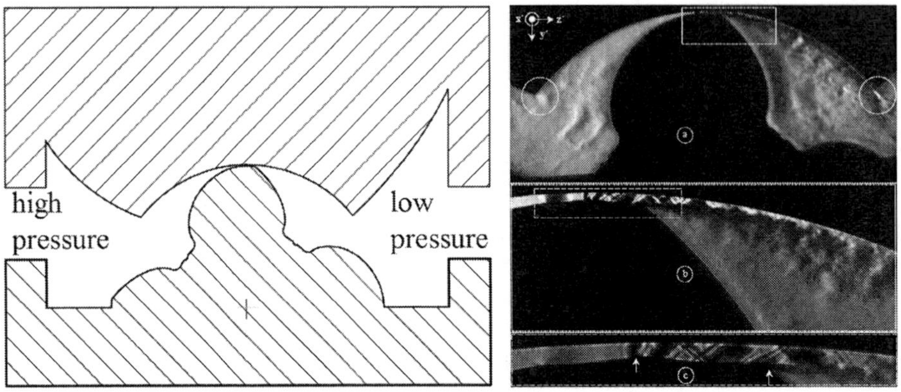

Figure 2. Clearance contour investigated by Sachs (left),
Example of a Schlieren picture for h = 0.4 mm and $\Pi_c = 0.25$ [8] (right)

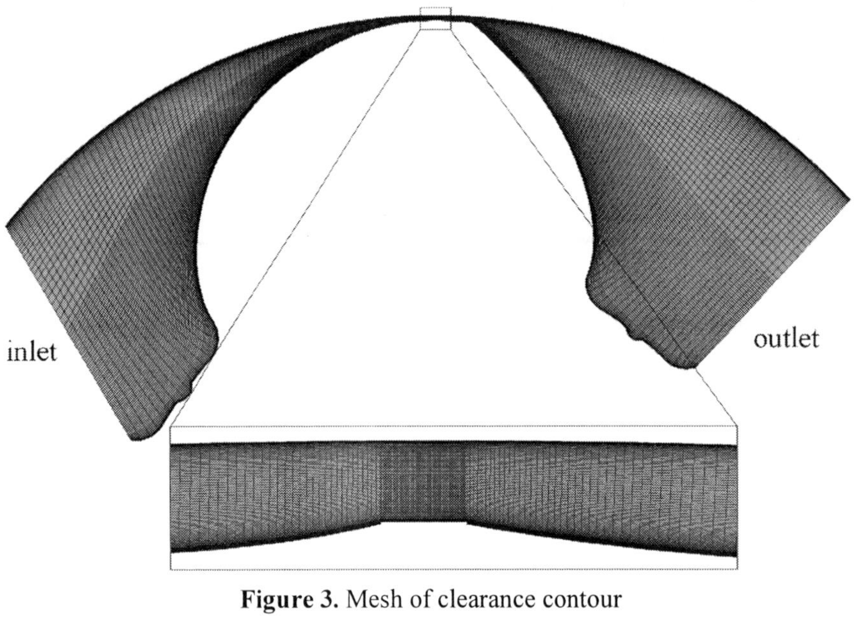

Figure 3. Mesh of clearance contour

Figure 3 shows the mesh which was used in this paper to validate the simulation. The mesh was generated using techniques described by Möller and Hinz [11]. The mesh consists of 43,000 cells, with a higher mesh density used in the narrowest clearance cross section and near the walls to capture details of the boundary layer and to avoid the use of wall function approximations. A grid study was performed to ensure that the solution is independent of the number of cells. Fluid was treated as an ideal gas and solid walls are considered adiabatic and hydraulically smooth.

Figure 4 shows a comparison of the flow pattern between the Schlieren photos by Sachs, the flow simulation by Kauder and Stratmann (Spalart Allmaras turbulence model), and flow simulation by the authors (Shear Stress Transport turbulence model). The pressure ratio under examination is $\Pi_c = p_{lp} / p_{hp} = 0.5$, for which Sachs determined an unsteady flow characteristic. In simulation, a low pressure of 10^5 Pa is set at the outlet and the model allows incoming and outgoing flows. A mass flow boundary is set at the inlet allowing incoming flows only, so that the high pressure conditions are a result of the iterative simulation. An angular compression shock is formed behind the sealing border and is reflected by the casing, inducing a flow separation at the casing. This phenomena is captured at the right position in the simulation as indicated in the views (b) and (c) of figure 4. When the reflection reaches the rotor contour, it induces an additional flow separation at the rotor. As in the study from Kauder and Stratmann, the simulation results show good agreement with the experiment concerning the flow pattern and justify the use of CFD to simulate clearance flow.

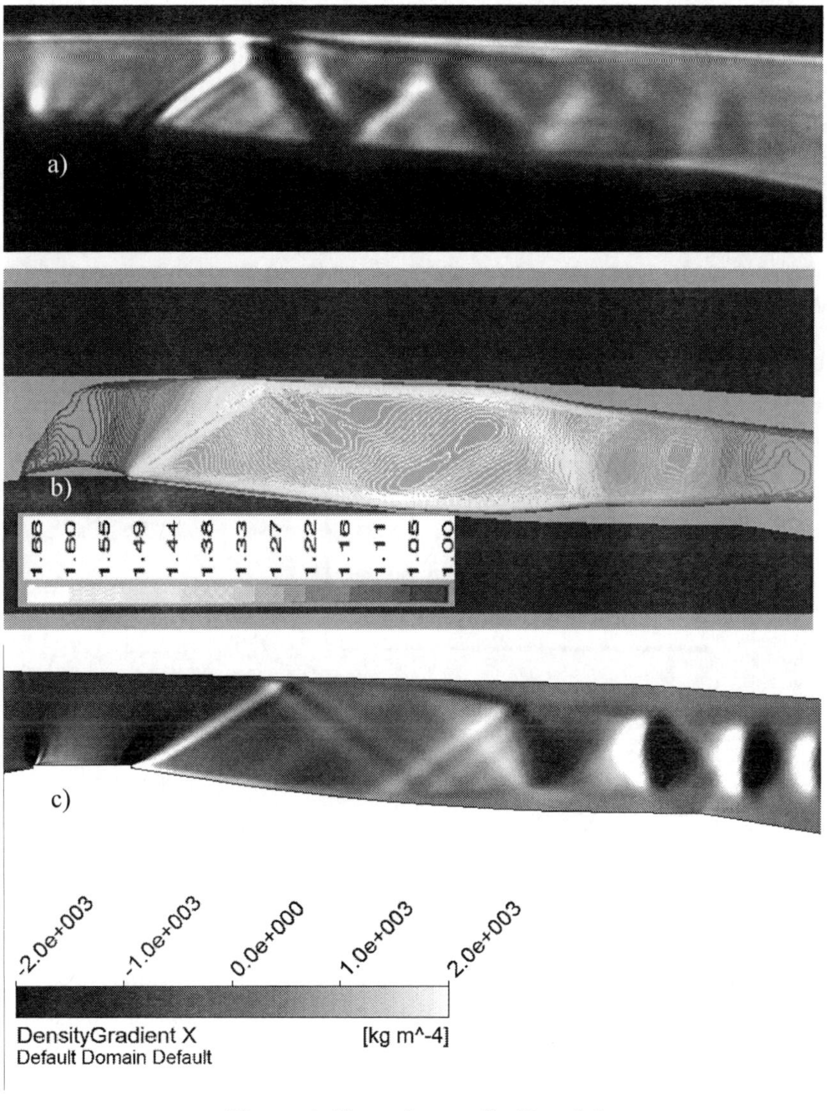

Figure 4. Flow pictures for $\Pi_c = 0.5$
(a) Schlieren picture (density gradient) by Sachs [8]
(b) Mach number distribution by Kauder and Stratmann with Spalart Allmaras turbulence model [10]
(c) Density gradient with SST turbulence model

In addition, the flow coefficient is evaluated for different pressure ratios to determine the turbulence model which best suits the experimental data. Figure 5 shows the flow coefficient as a function of pressure ratio and compares the experimental data with simulation performed by the authors. The Spalart Allmaras model, representing a one-equation model, and the SST turbulence model, representing a two-equation model, are compared. For pressure ratios below critical ($\Pi_{crit} = 0.528$ for dry air) there is little difference between simulation and experiment. The flow coefficient remains almost constant at $\alpha = 0.95$ for both turbulence models since the flow is choked. For these working points, the mass flow rate is independent of the low pressure conditions, meaning that a lowering of the low pressure would have no effect on mass flow rate. For higher pressure ratios, the flow coefficient increases up to values above 1.1. The determination of the theoretical mass flow (eq. (1)) assumes the low pressure conditions to be present in the narrowest flow cross section, where the model is based on the occurrence of a flow separation. In contrast, the real flow follows the contours of rotor and casing further downstream the narrowest flow section without flow separation. As a consequence, the pressure decreases under the conditions which are present at the low pressure side due to the increasing cross section as shown in the pressure distribution in figure 5. This leads to a lower pressure ratio than the isentropic equation supposed, resulting in a higher mass flow rate and flow coefficient. For pressure ratios above 0.7 there is some disagreement between the turbulence models and the difference between experiment and Spalart Allmaras turbulence model increases. A comparison of the flow pattern shows minor differences in the region of flow separation and in the pressure distribution, which can explain these differences. The goal of this work is to provide a satisfactory flow coefficient model over a wide range of parameters. Thus, the SST turbulence model is used for further investigations in this study.

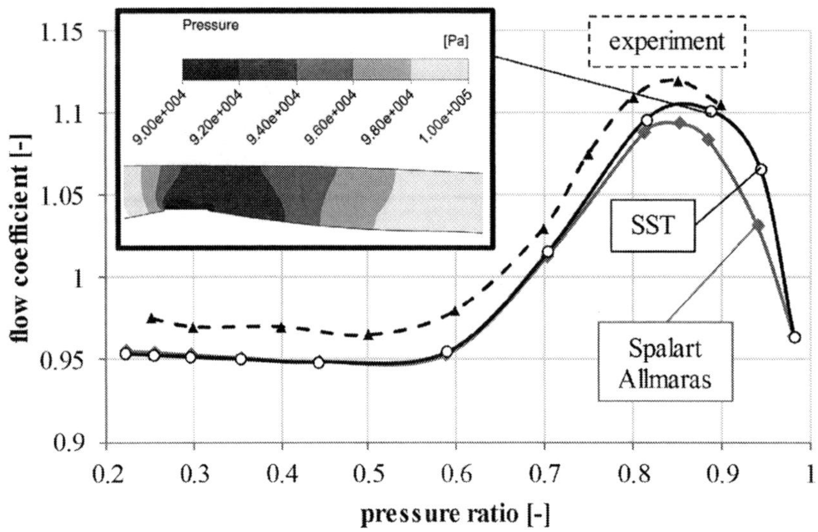

Figure 5. Flow coefficient as a function of pressure ratio (p_{lp} = const = 10^5 Pa) determined by experiment [8] and CFD simulation using the Spalart Allmaras and SST turbulence model

3. Clearance flow path

The clearance flow path examined in the following is the male rotor housing clearance of an asymmetric SRM-profile with a male rotor tip diameter of D = 72 mm. Analytical functions for defining this profile are described by Rinder [12]. The housing clearance is examined for a common rotor configuration with a rotor lead of l = 120 mm. The geometry of the rotor is shown in figure 6 (left). It is assumed that the housing clearance flow is orthogonal to the distorted rotor tip as implied in the figure. Figure 6 (right) shows the resulting clearance path in dashed lines. A change of the rotor lead, which is not discussed in this paper, would stretch or compress this clearance path. Since the clearance will be investigated by experiments in the future, the available clearance path area of the curved housing is translated to a clearance contour with a straight housing contour. This case is represented by the solid line and is the configuration examined in this study. The radius of curvature of the original housing is large, so the clearance paths do not differ much in the region of the narrowest flow cross section, where both paths show a sharp edge. The flow path can be described by two (bi)superellipse equations, one for the inflow until the narrowest flow cross section is reached, and one for the outflow, beginning at the narrowest flow cross section, eqs. (4) and (5). The point of origin is located on the tip of the rotor in the narrowest flow cross section.

$$\left|\frac{x}{13.5\,mm}\right|^{1.8} + \left(\frac{y + 10\,mm}{10\,mm}\right)^{1} = 1, \qquad -13.5\,mm \leq x \leq 0 \tag{4}$$

$$\left(\frac{x}{6.9\,mm}\right)^{1} + \left(\frac{y + 10\,mm}{10\,mm}\right)^{1} = 1, \qquad 0 < x \leq 6.9\,mm \tag{5}$$

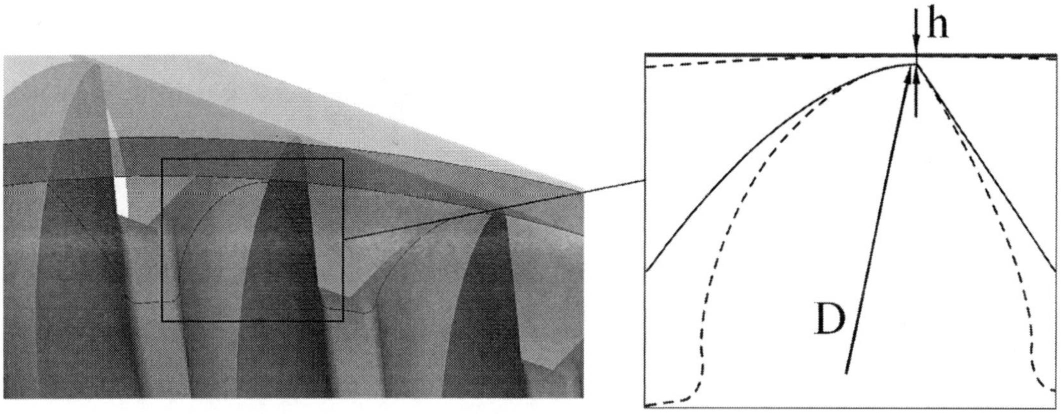

Figure 6. Exemplary screw machine under investigation (left), resulting original clearance path (dashed line) and via bisuperellipses approximated clearance path (solid line) (right)

With details of the flow path geometry determined, the next consideration is selection of details for the analytical study, in particular, the approach to specifying boundary conditions. Instead of varying the absolute boundary conditions, dimensionless numbers are used. These are defined in the next section.

4. Determination of characteristic numbers for housing clearance flow

The Pi-theorem [13] is used to determine independent parameters which need to be varied in order to examine the housing clearance geometry. Heat exchange between fluid and solid is not considered in this study. Furthermore, it is assumed that heat capacity and viscosity do not vary from the high to the low pressure side. This leads to the same dimensionless numbers Π_1 - Π_5 concerning thermodynamics and operation as presented for the front clearance flow in [7]. Since the geometry of the flow path differs from the front clearance, dimensionless numbers Π_6 - Π_8 concerning the geometry are different. The numbers are summarised in eq. (6):

$$\Pi_1 = \frac{\dot{m}_{real}}{\dot{m}_{th}} = \alpha,\ \Pi_2 = \frac{c_p}{c_v} = \kappa,\ \Pi_3 = \frac{p_{lp}}{p_{hp}} = \Pi_c,\ \Pi_4 = \frac{2\dot{m}_{real}}{b \cdot \eta_{hp}} = Re,\ \Pi_5 = \frac{u}{a_{hp}} = \mathrm{Ma}_u$$
$$\Pi_6 = \frac{h}{D},\ \Pi_7 = \frac{b}{D},\ \Pi_8 = \frac{l}{D} \tag{6}$$

The lead l of the rotor and the clearance width b are not varied in this study so Π_7 and Π_8 are constant. Air is the only fluid examined, so Π_2 also remains constant. The individual influence of the remaining dimensionless numbers Π_3 to Π_6 on the flow coefficient Π_1 is examined in this study by varying the boundary conditions of the CFD simulation.

5. Boundary conditions and variation parameters

The boundary conditions which remain constant for all simulation are summarised in table 1. As for the test rig contour presented in section 2, the fluid is dry air, treated as an ideal gas. Simulation is carried out assuming adiabatic and hydraulically smooth walls. Considering the results from figure 5, the SST turbulence model is chosen for the simulation.

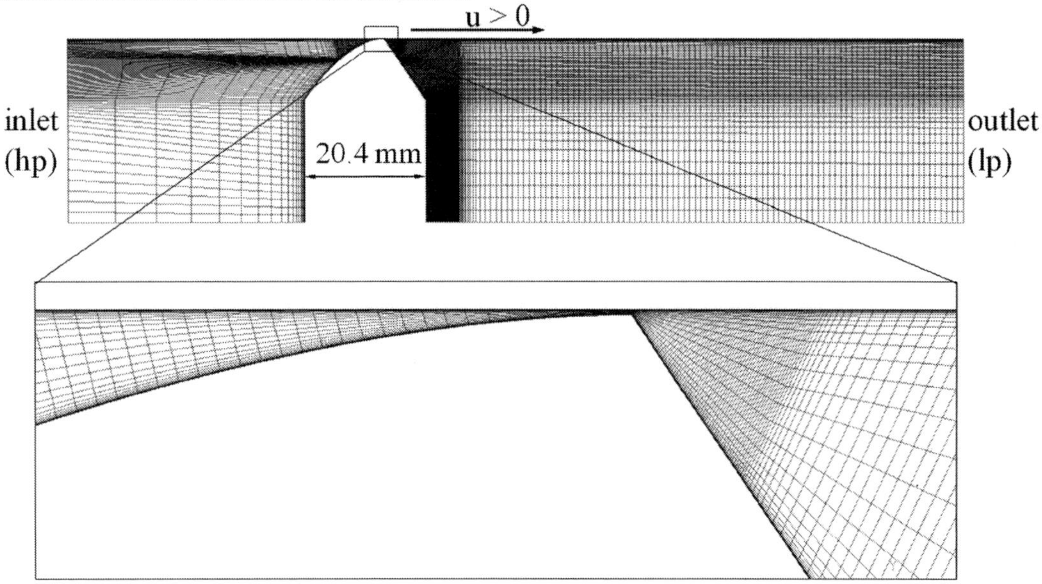

Figure 7. Representative mesh of the simulated flow path

Table 1. Constant boundary conditions

Parameter	Value
Geometry:	
- Profile	SRM (4+6)
- Rotor diameter D [mm]	72
- Rotor lead l [mm]	120
- Clearance width b [mm]	1 (one cell)
Fluid: dry air (ideal gas)	
- Specific gas constant [J/kg/K]	287
- Isentropic exponent [-]	1.4
- Inlet temperature [K]	298
- Outlet temperature [K]	298
- Dynamic viscosity [kg/m/s]	1.83e-5
Simulation:	
- Heat transfer fluid – solid	none (adiabatic)
- Turbulence model	SST
- Wall roughness	smooth walls
- Number of elements	31,000 - 41,000
- y^+ in clearance region	< 2

Figure 7 shows a mesh which was used in this study. The mesh consists of about 35,000 cells with the highest mesh density in the clearance region, where 79 cells in height direction are present. The clearance widths is one cell and without friction on the front cross section, so that an infinite clearance width is simulated. A grid study has been performed to ensure that the mesh has no influence on the results. The y^+ value is below two in the relevant clearance region for all simulated points.

Several parameters are varied in order to vary the dimensionless numbers from eq. (6). The rotor contour remains the same throughout the simulation while the clearance height h is varied between 0.02 and 0.2 mm in order to change the height-diameter-ratio (Π_6) and, due to a higher mass flow rate, the Reynolds number. Inlet and outlet pressures are varied to vary Reynolds number (Π_4) and pressure ratio (Π_3). Furthermore, circumferential speeds up to 120 m/s are investigated to vary the circumferential Mach number (Π_5). In practise, the fluid is accelerated to the circumferential speed of the rotors when flowing into the machine. Taking the relative system into account, it is meaningful to apply the moving boundary to the casing. A dependency between density and diameter of the clearance path, which is caused by centrifugal forces, is neglected in this study.

6. Simulation results with stationary boundaries

Results for non-moving boundaries are shown in figure 8. The Reynolds number is constant for the two larger values of height-diameter-ratio. This is equivalent to a constant clearance mass flow rate due to the constant viscosity. These results are achieved by varying the absolute pressures. The flow coefficient varies between 0.94 and 0.97, meaning that the assumption of an isentropic nozzle is almost justified for all pressure ratios. Furthermore, the influence of the height-diameter-ratio is minor. The flow coefficient remains almost constant for pressure ratios below 0.4 since the pressure ratio is supercritical and the flow is choked, as explained before for figure 5. An examination of the flow patterns (not shown) revealed that the flow separates slightly downstream the narrowest flow cross section. For $\Pi_c > 0.4$ the flow coefficient decreases slightly because the flow separates from the rotor contour directly at the sharp edge of the rotor.

Since mass flow rate and Reynolds number decrease with lower height-diameter-ratio, a lower Reynolds number is chosen for the low height-diameter-ratio; these results also shown in figure 8 (Re = 700). The flow coefficient is lower because the flow is more influenced by friction, leading to a reduction in flow velocity. Similar progressions for the flow coefficient were presented for circular convergent nozzles by Bragg [14], who validated his results by experiments. Altogether, the influence of pressure ratio and height-diameter-ratio is minor, especially for higher Reynolds numbers. These results are contrary to the findings for the front clearance flow [7], where a change in clearance height had a significant impact on the flow coefficient. This emphasizes that the male rotor housing clearance flow is not dominated by friction since the flow path is converging and the narrowest clearance is present in only one flow cross section.

Figure 9 shows the isolated influence of Reynolds number and height-diameter-ratio on the flow coefficient. The higher the Reynolds number, the lower is the relative height of the boundary layer of the flow as indicated in the vertical progression of the Mach number in the narrowest flow cross section in the inset on figure 9. For high Reynolds numbers, the flow is less affected by friction and shows a distinctive core flow, thereby increasing the flow coefficient. For low Reynolds numbers, the flow is strongly influenced by friction and shows the profile of a pure boundary layer flow similar to a laminar profile. Furthermore, no noticeable change from laminar to turbulent flow can be seen in the flow coefficient characteristics, although the turbulent kinetic energy (not shown) in the narrowest flow cross section increases with the Reynolds number.

When comparing at the same Reynolds number, no significant influence of the height-diameter-ratio on the flow coefficient can be observed. For the practical design of screw machines, lower clearance height still means lower absolute mass flow rate, resulting in reduced Reynolds number and flow coefficient.

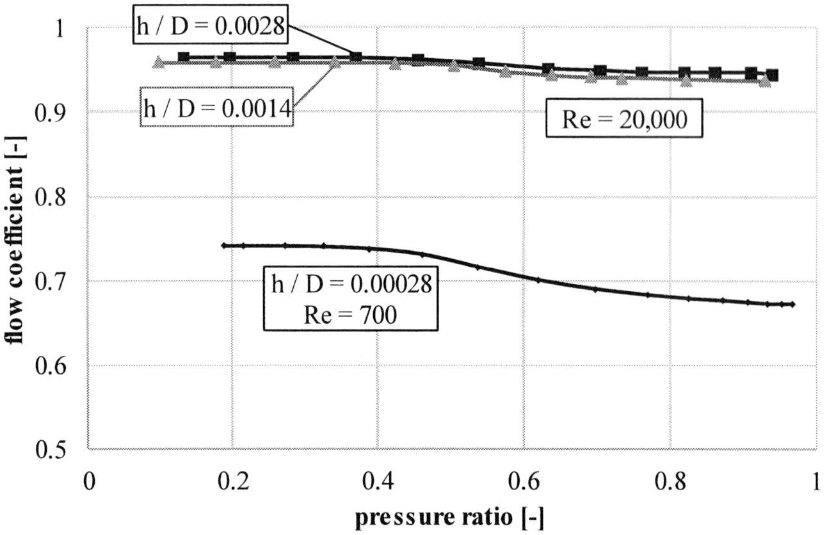

Figure 8. Flow coefficient as a function of pressure ratio for different height-diameter-ratios for Re = 20,000 and Re = 700

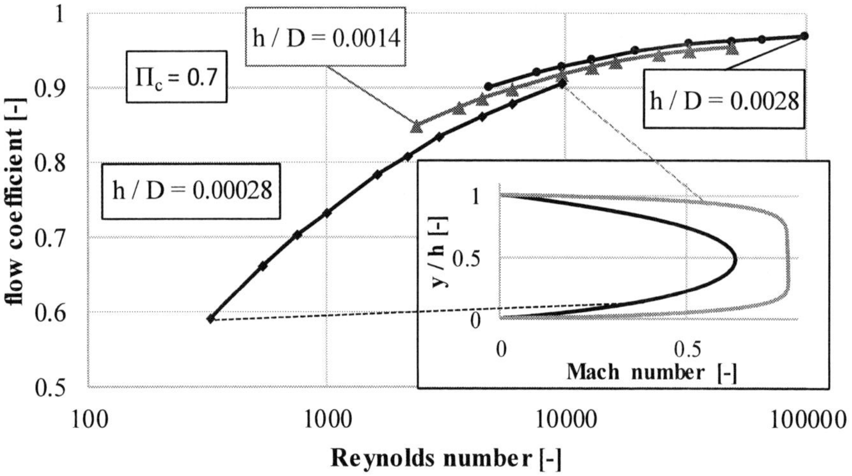

Figure 9. Flow coefficient as a function of Reynolds number for different height-diameter-ratios for $\Pi_c = 0.7$, Mach number distribution in the narrowest clearance section for Re = 9801 and Re = 327

7. Simulation results with moving boundary

The results presented in the previous section are valid for the clearance flow path defined by stationary boundaries. The movement of a boundary causes an additional Couette flow in the clearance. The results presented by Vimmr and Fryc showed no significant influence of the moving boundary on the clearance flow [6].

As explained before, the moving boundary is applied to the casing. Figure 10 shows results for a variation of the pressure ratio for constant Reynolds number and height-diameter-ratio. Theoretical mass flow rate is still calculated using eqs. (1) and (2), meaning that the theoretical mass flow rate is not modified by the moving boundary. For pressure ratios below 0.6 the moving boundary causes a constant shift to higher flow coefficients for a movement in flow direction and to lower flow coefficients for a movement against flow direction. The velocity profile of the flow is more affected in case the boundary moves against flow direction, since negative circumferential Mach numbers show a greater impact on the flow coefficient. For low pressure ratios, the influence of the moving boundary is small, changing only about one percent for a circumferential Mach number of 0.35 (u = 120 m/s).

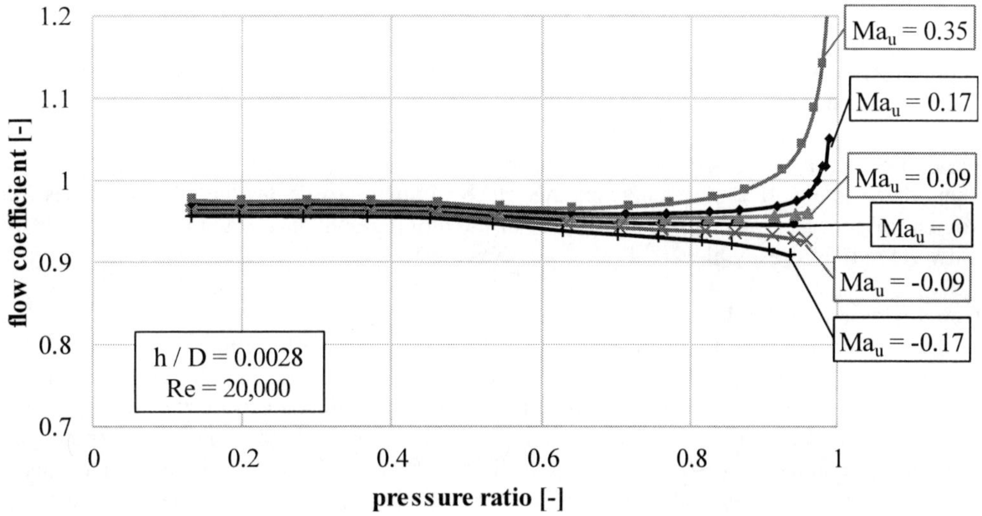

Figure 10. Flow coefficient as a function of pressure ratio for different circumferential Mach numbers for Re = 20,000 and h / D = 0.0028

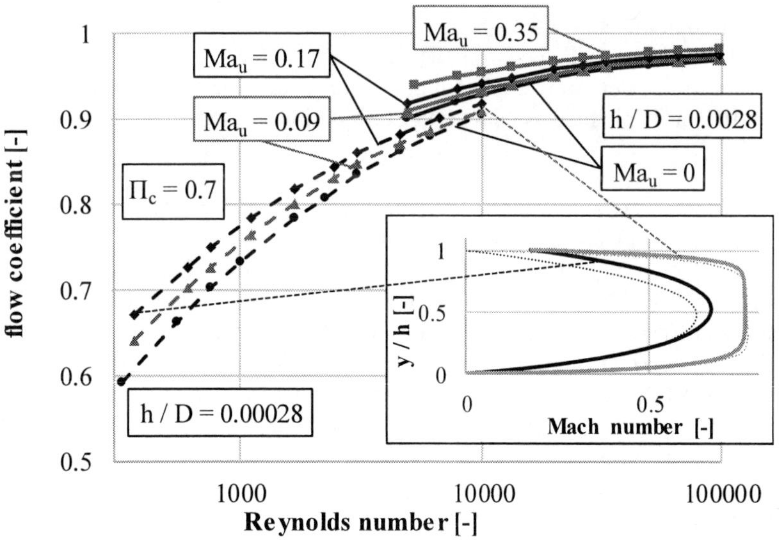

Figure 11. Flow coefficient as a function of Reynolds number for different height-diameter-ratios and circumferential Mach numbers for $\Pi_c = 0.7$, Mach number distribution in the narrowest clearance section for $Ma_u = 0.17$ for Re = 9801 and Re = 327

For higher pressure ratios the influence of the moving boundary increases because the ratio between velocity of the boundary and velocity of the flow increases. Furthermore, theoretical mass flow, which represents the denominator of the flow coefficient, tends to zero for a pressure ratio of one since the pressure induced mass flow disappears in the limiting case where high and low pressure conditions are equal, so that the flow coefficient tends to infinity. Altogether, compared to the results obtained for the front clearance [7], the influence of the moving boundary is minor because the velocity of the pressure induced flow is high due to the converging shape of the flow path.

Figure 11 shows the impact of the moving boundary for a variation of Reynolds number at a constant pressure ratio of 0.7. The shift to higher flow coefficients is greater for low Reynolds numbers which are represented by the lower height-diameter-ratios. The velocity profiles shown in figure 9 are shown in figure 11 as dotted lines and can be compared to the profiles obtained for the moving casing. The resulting Mach number distribution shows that the boundary layer is only little influenced for the higher Reynolds number since the gradient in the near wall region is already high. For the lower Reynolds number, there is a noticeable change in the Mach number profile leading to an increase in flow coefficient of about ten percent. The influence of the moving boundary is minor for $Re > 10^4$ because the flow velocity is already high.

8. Conclusion and outlook

This paper reports the use of CFD to investigate flow through male rotor housing clearances of screw machines. Flow pictures and flow coefficients from using a specialised test rig are reported in the literature and are taken as references to validate the simulation developed for the study reported here. The settings are adopted for the simulation of a typical male rotor housing clearance contour. Dimensionless numbers are varied systematically in order to show their individual influence. For stationary boundaries, the impact of the pressure ratio and the ratio of clearance height to rotor diameter on the flow coefficient are minor. Reynolds number is found to be an important factor because it determines the thickness of the boundary layer and therefore the profile of the flow. Investigations with a moving boundary revealed the strongest impact for high pressure ratios, due to small pressure induced mass flow rates, and, for lower Reynolds numbers, due to the friction affected boundary layer character of the flow. Although the results are only valid for a certain rotor profile, they can be adapted to many rotor profiles used in industry.

Further investigations will focus on the influence of the remaining parameters which were held constant in this study: isentropic exponent and rotor lead. Furthermore, an experimental validation of the flow coefficients is planned as well as an examination of the female rotor housing clearance.

References

[1] Saint Venant A J C and Wantzel L, 1839, *J. de l'Ecole Roy. Poly.* **16**, (series 1) 85
[2] DIN EN ISO 5167-2:2004: Measurement of fluid flow by means of pressure differential devices inserted in circular cross-section conduits running full - Part 2: Orifice plates
[3] Trutnovsky K and Komotori K 1981 *Berührungsfreie Dichtungen* VDI-Verlag Düsseldorf
[4] Prins J and Infante Ferreira C A 2003 Selected basic theory of gas leakage *Int. Conf. on Compr. And their System* IMechE Conference Transaction, London, pp. 521-530.
[5] Bell I H, Groll E A, Braun J E and Horton W T 2013 A computationally efficient hybrid leakage model for positive displacement compressors and expanders *Int. J. of Refrigeration* **36** pp 1965-1973
[6] Vimmr J and Fryc O 2006 Numerical Simulation of Leakage Flow Between Moving Rotor and Housing of Screw Compressor *Proceedings of the Conference Modelowanyie Inzynierskie* ISSN 1896-771X **32** pp. 461-468 Gliwice
[7] Utri M and Brümmer A 2018 Flow through Front Clearances of Dry Running Screw Machines Using Dimensionless Numbers *Int. Compressor Engineering Conf. at Purdue 2018* paper 1344
[8] Sachs R 2002 *Experimentelle Untersuchung von Gasströmungen in Schraubenmaschinen* Dissertation Dortmund University
[9] Kauder K and Sachs R 2002 Gas Flow through Gaps in Screw-type Machines *Conference on Screw Machines 2002, VDI Berichte 1715* VDI-Verlag Düsseldorf

[10] Kauder K and Stratmann D Theoretical Gas Flow through Gaps in Screw-type Machines *Conference on Screw Machines 2002, VDI Berichte 1715* VDI-Verlag Düsseldorf

[11] Möller M and Hinz J P 2018 Isogeometric Analysis Framework for the Numerical Simulation of Rotary Screw Machines. I. General Concept and Early Applications *Int. Conf. on Screw Machines 2018*

[12] Rinder L 1979 *Schraubenverdichter* (Wien, New York: Springer-Verlag)

[13] Buckingham E 1914 On physically similar systems; illustrations of the use of dimensional equations *Physical Review* **4** pp 345-376

[14] Bragg S L 1960 Effect of Compressibility and Convergent Nozzles *J. Mech. Sci* **2**, 35

Nomenclature

A	Area	(m^2)
a	Speed of sound	(m/s)
b	Clearance width	(m)
c	Flow velocity	(m/s)
c_p	Isobaric heat capacity	(J/kg/K)
c_v	Isochoric heat capacity	(J/kg/K)
D	Male rotor diameter	(m)
h	Height	(m)
l	Rotor lead	(m)
$\dot{m}$	Mass flow	(kg/s)
Ma	Mach number	(-)
Ma_u	Circumferential Mach number	(-)
p	Pressure	(Pa)
Re	Reynolds number	(-)
R_s	Specific gas constant	(J/kg/K)
T	Temperature	(K)
u	Circumferential speed	(m/s)
x, y	Coordinates	(m)
α	Flow coefficient	(-)
η	Dynamic viscosity	(Pa·s)
κ	Isentropic exponent	(-)
ρ	Density	(kg/m^3)
Π	Pressure ratio	(-)
Π	Characteristic number	(-)

Subscript

c	Clearance
crit	Critical
hp	High pressure
lp	Low pressure
real	Real flow
th	Theoretical

Acknowledgement

The authors would like to thank Matthias Möller and Jochen Hinz from TU Delft for providing the mesh of the test rig contour. The work leading to these results has received funding from the European Community's Horizon 2020 Programme (2014-2020) under grant agreement no 678727. The opinions expressed in the document are of the authors only and no way reflect the European Commission's opinions. The European Union is not liable for any use that may be made of the information.

AUTHOR INDEX